J.R. Pasek

LINEAR
SYSTEMS
ANALYSIS

McGRAW-HILL
BOOK COMPANY
New York
St. Louis
San Francisco
Düsseldorf
Johannesburg
Kuala Lumpur
London
Mexico
Montreal
New Delhi
Panama
Paris
São Paulo
Singapore
Sydney
Tokyo
Toronto

C. L. LIU

JANE W. S. LIU

Department of Computer Science
University of Illinois at Urbana-Champaign

Linear Systems Analysis

This book was set in Times New Roman.
The editors were Kenneth J. Bowman and Madelaine Eichberg;
the cover was designed by Pencils Portfolio, Inc.;
the production supervisor was Charles Hess.
The drawings were done by ANCO Technical Services.
Kingsport Press, Inc., was printer and binder.

Library of Congress Cataloging in Publication Data

Liu, Chung Laung.
Linear systems analysis.

Includes bibliographical references.
1. Electric networks. 2. Electric engineering—Mathematics. I. Liu, Jane W. S., joint author. II. Title.
TK454.2.L58 621.319'2 74-10555
ISBN 0-07-038120-8

LINEAR SYSTEMS ANALYSIS

1 2 3 4 5 6 7 8 9 0 KPKP 7 9 8 7 6 5

CONTENTS

PREFACE

This book was written as a textbook for an undergraduate course in linear systems theory. Such a course appears in most electrical engineering curricula at the junior-senior level, usually following a course in introductory circuit theory. We believe that this course can be one of the most important and exciting undergraduate courses in electrical engineering. It should provide a student with the necessary background for more advanced courses in control theory, communication theory, and circuit theory. Moreover, many of the concepts and techniques introduced are also useful in the study of probability theory, operations research, and computer science. This course probably is one of the first of several courses in which an engineering student has an opportunity to see how mathematics is applied in a nontrivial way to the solution of engineering problems. We hope that a student will not only learn how to apply the mathematical tools developed in the course but also appreciate the usefulness of these tools and understand the relationship between the physical phenomena and their mathematical descriptions. Another important purpose of this course is to show how a problem can be studied from different viewpoints and be solved by different methods. It is invariably the case that one gains more insight into a problem when one is able to look at and to tackle the problem from different angles and directions. We believe that study of linear systems theory will confirm this point.

Although most of the readers of this book probably have had a course in introductory circuit theory, such background is not essential in using this book. There are several examples drawn from linear circuit theory; however, they can be omitted without disrupting the presentation. As to the reader's mathematical background, a course in advanced calculus for engineers is quite adequate. On the other hand, we wish to emphasize the possibility of omitting a few sections for students who do not have such background. Indeed, by omitting the topics mentioned below, this book can be used by students who have taken only courses in calculus and differential equations. In Chap. 3, matrix algebra is used in the discussion of vector difference and differential equations. The omission of Secs. 3-3 and 3-8 to 3-11 will bypass the use of matrix notations. In fact, looking at a set of simultaneous scalar equations as a vector equation is a useful but not essential viewpoint in our discussion. In Chap. 9, some results from the theory of functions of a complex variable are used. These results will not be needed if Secs. 9-3 and 9-6 are omitted. The concept of region of absolute convergence presented in Chap. 6 is very important. However, an explicit evaluation of the inverse transformation formula using the method of residues is not crucial to the understanding of the material.

The book is intended for a one-semester course. Our experience is that all the material in this book can be covered at a reasonable pace. On the other hand, one can adjust the pace by omitting some of the topics. All sections marked with * can be omitted without disrupting the continuity. Furthermore, a less detailed discussion on some of the topics is possible as we shall comment on a chapter-by-chapter basis: In Chap. 1, the discussion on generalized functions in Sec. 1-6 can be omitted completely. In Chap. 2, the discussion on solution of differential equations can be brief since most students would have seen the material previously in a course in circuit theory or in a course in differential equations. In Chap. 3, the most important issues are the concept of states and to set up the state space description of systems. Discussion on the solution of state equations can be omitted. It would also be reasonable to discuss the solution of the state equations for discrete systems and omit that for continuous systems. Similarly, if Sec. 3-5 is discussed in detail, Sec. 3-12 can be omitted or be given only a brief treatment. In Chap. 4, a less detailed discussion on convolution algebra in Sec. 4-4 is possible. Many students may have previously studied the subject of Fourier series. In that case, the discussion in Chap. 6 can be brief. Moreover, Secs. 6-7 and 6-8 can be omitted completely, and so can Sec. 7-7. If one really is pressed for time, it is possible to omit Chap. 8 completely. However, our experience has been that students enjoyed very much seeing some interesting applications of the mathematical results they studied earlier. We would recommend a less quantitative discussion instead. By the time Chap. 9 is reached, the pace of the course can be quickened because students should have become quite familiar with the concepts of transformation by then. Section 9-3 and 9-6 can be omitted. Also, Sec. 10-4 can be omitted.

We want to thank Prof. William M. Siebert for introducing us to the subject, Prof. James N. Snyder, our department head, for his encouragement and support, and Profs. Steve W. Director and Burton J. Smith for their comments and suggestions. We also want to thank Mrs. Sharyn Cohen, Mrs. Connie Slovak, and Mrs. June Wingler for typing the manuscript.

C. L. LIU
JANE W. S. LIU

1

SIGNALS AND SYSTEMS

1-1 INTRODUCTION

To illustrate the coverage of this book, let us state some of the problems that we shall study: What is the principle of operation of the radio and television broadcast systems? How does one analyze the performance of an amplifier in a hi-fi set? At what rate do elementary particles interact in a nuclear reactor? How does the position of a space satellite change as a function of time? How does the operation of a company fluctuate according to its incomes, expenditures, and investment policy? How do the employment situation of a city and the migration of a labor force into the city affect each other? How does one compute the rate at which industrial waste is disposed by chemical treatment? How many descendants will a pair of rabbits produce in 10 years? Although these are problems from vastly different areas of studies, they all are concerned with how "systems" behave under some given conditions. A major goal of this book is to study the various mathematical techniques for analyzing the behavior of systems. Before discussing these techniques, we shall introduce in this chapter some basic concepts and terminologies.

1-2 SIGNALS AND SYSTEMS

A *signal* is a physical embodiment of a message. For example, a sequence of voltage pulses from a transistor amplifier in a digital computer is a signal which might represent a certain numerical quantity, a surge of electric current in the sensing circuit of a fire alarm system is a signal which might mean the detection of a dangerously high temperature, and a light spot on the surface of a cathode-ray display in a flight control radar system is a signal which might indicate the presence of an aircraft. Similarly, the sound wave from a loudspeaker, the light intensity distribution in a light diffraction pattern, and the air pressure reading from a pressure gauge are all examples of signals. A set of messages can be represented by different sets of signals. The problem of choosing a suitable set of signals for a given set of messages involves many considerations, such as economy and reliability. Here we shall not be concerned with this problem, which can be studied in a course in information theory. Rather, we are interested in the various ways of describing, characterizing, manipulating, and generating signals.

A *system* is a collection of physical devices that generates one or more signals, called the *output signals*, when stimulated by some other signals, called the *input signals*. The input signals are also called the *stimuli*, and the output signals are also called the *responses* of the system. In other words, a system can be viewed as a signal transducer that transforms input signals into corresponding output signals. For example, a transistor amplifier is a system which is a collection of transistors, resistors, and capacitors. When stimulated by an input voltage signal, the system will respond with an output voltage signal which is a scaled replica of the input signal. As other examples, a loudspeaker is a system that transforms an electric-current input into a sound-wave output, and a lens is a system that transforms a light signal into another light signal. We shall study in this book the properties of various systems and the relationship between their input and output signals.

1-3 CONTINUOUS AND DISCRETE SIGNALS

If we measure the output voltage of an amplifier in a radio set, we probably will find that the value of the signal fluctuates with time. We can describe this signal by expressing its value as a function of time. For example, a signal might vary sinusoidally with time:

$$v(t) = 10 \sin 120t \tag{1-1}$$

That is, at $t = 0$ the value of the signal is 0, at $t = \pi/240$ the value of the signal is 10, and so on. A signal is said to be a *time signal* if its value is a function of time. Thus

$$x(t) = e^{-t} \cos 2\pi t$$

$$y(t) = 6$$

are examples of time signals. Sometimes, the value of a signal varies with the spatial position. For example, in the case of the displacement of a vibrating string at a certain time instant, the value of the signal changes with spatial position, and we might have

$$d(x) = 10e^{-x} \cos 2\pi x$$

where x is the spatial distance from a certain origin. Also, the value of a signal might vary with both time and spatial position. For example, we might have

$$d(x, t) = 10e^{-x} \cos 2\pi x \sin 120t \tag{1-2}$$

Or the value of a signal might vary with the spatial position in three-dimensional space. For example, the light intensity distribution of a diffraction pattern at a given time instant can be described by

$$I(x, y, z) = \frac{1}{z^2}\left[\frac{\sin(x/z)}{x/z}\right]^2\left[\frac{\sin(y/z)}{y/z}\right]^2 \tag{1-3}$$

where x, y, z are the three coordinates in the three-dimensional space.

Mathematically, a signal is described by a function of one or more independent variables.† A signal is said to be a *one-dimensional* signal if its value is a function of a single independent variable, and it is said to be an *n-dimensional* signal if its value is a function of n independent variables. The signals in Eqs. (1-1) to (1-3) are, respectively, one-dimensional, two-dimensional, and three-dimensional. Most of our discussion will be limited to one-dimensional signals. We shall assume these one-dimensional signals to be time signals; that is, the independent variable will be the time variable t.

Being a physical quantity, a signal assumes real values at all time instants and consequently can be described by a real-valued function. However, for mathematical convenience, we generalize the notion of a real signal and suppose that a signal might assume complex values and thus can be described, in general, by a complex-valued function. For example, we might have

$$v(t) = (3 + j5)e^{(4+j\pi)t}$$

At $t = 0$, the value of the signal is $3 + j5$; at $t = \frac{1}{2}$, the value of the signal is

$$(3 + j5)e^{2+j(\pi/2)} = je^2(3 + j5) = e^2(-5 + j3)$$

and so on.

A signal is said to be (*time*) *continuous* within a certain time range if its value is specified for all instants in that time range.‡ As an example,

$$x(t) = \begin{cases} 0 & t < 0 \\ \cos \pi t & t > 0 \end{cases}$$

† Throughout our discussion all signals assume only scalar values.

‡ Except at the discontinuities.

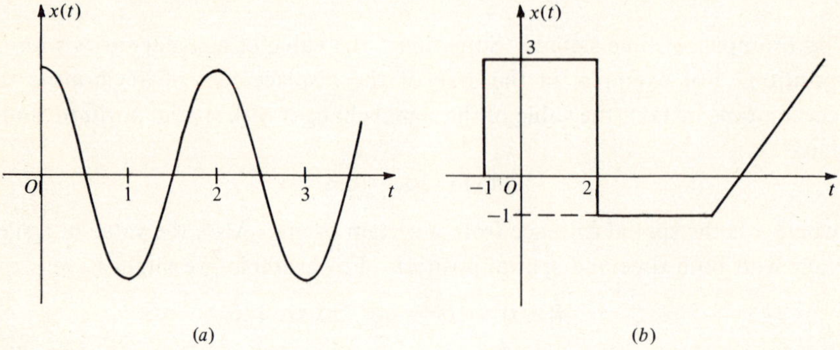

FIGURE 1-1

is a continuous signal within the range $-\infty \leq t \leq \infty$. This signal can also be represented graphically, as in Fig. 1-1*a*. From now on, unless otherwise specified by a continuous signal, we shall mean one that is continuous within the range $-\infty \leq t \leq \infty$. Note that a continuous signal may contain *discontinuities*. A continuous signal $x(t)$ is said to have a discontinuity at $t = t_0$ if for $\epsilon > 0$

$$\lim_{\epsilon \to 0} x(t_0 + \epsilon) \neq \lim_{\epsilon \to 0} x(t_0 - \epsilon)$$

We shall use $x(t_0+)$ to denote $\lim_{\epsilon \to 0} x(t_0 + \epsilon)$, and $x(t_0-)$ to denote $\lim_{\epsilon \to 0} x(t_0 - \epsilon)$. The value of the *discontinuity* at t_0 is defined to be

$$x(t_0+) - x(t_0-)$$

For example, the continuous signal in Fig. 1-1*a* contains a discontinuity at $t = 0$. Because $x(0-) = 0$ and $x(0+) = 1$, the value of this discontinuity is 1. As another example, the continuous signal in Fig. 1-1*b* contains two discontinuities, one at $t = -1$ and another at $t = 2$. The value of the discontinuity at -1 is 3, and the value of the discontinuity at 2 is -4.

A signal is said to be (*time*) *discrete* if its value is specified only at discrete time instants. Although, in general, these discrete time instants can be either equally or unequally separated, we shall assume that a discrete signal always has its values specified at $t = 0, \pm 1, \pm 2, \ldots$. (The values of t can either be taken as the actual time instants at which the values of the signal are specified or be taken as the indices of successive time instants that are not equally separated.) Graphically, a discrete signal can be represented as in Fig. 1-2, where the value of the signal at $t = -2$ is 0, at $t = -1$ is -1, at $t = 0$ is 1, at $t = 1$ is 3.5, at $t = 2$ is 6.25, at $t = 3$ is 9.125, and so on. Analytically, we shall write a discrete signal as $x(n)$, instead of $x(t)$, to emphasize

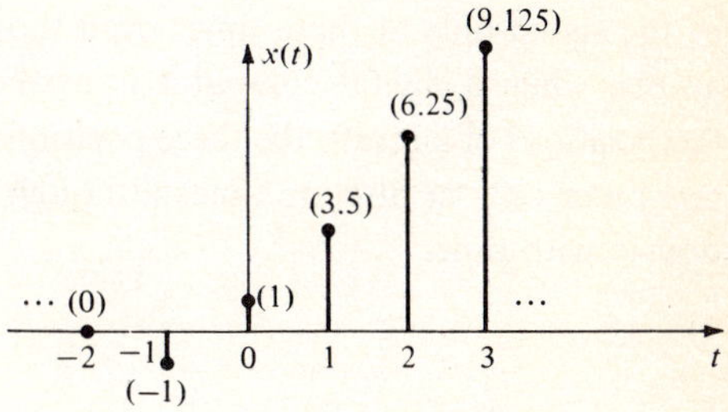

FIGURE 1-2

that it is a discrete signal whose value is specified only at integral values of n. For example, the signal in Fig. 1-2 can be written

$$x(n) = \begin{cases} 0 & n < -1 \\ 2^{-n} + 3n & n \geq -1 \end{cases}$$

We are interested in discrete signals because many signals are indeed time discrete. For example, if, in a process control system, a monitoring device measures the temperature inside a chemical reaction chamber once every 30 seconds, the output signal from the monitoring device is a time-discrete signal. As another example, the output of a sense amplifier in a magnetic-core memory unit is a sequence of 0s and 1s corresponding to the contents of the magnetic cores being read in successive memory cycles and is, therefore, a time-discrete signal. We also frequently encounter continuous signals whose values change only at discrete moments of time. Such signals can be described as discrete signals by their values at intervening discrete times. For example, we can describe the telegraph wave shown in Fig. 1-3a by its amplitudes at $t = 0, \pm 1, \pm 2, \ldots$ as shown in Fig. 1-3b. Sometimes, we are interested in the values of a continuous signal only at certain discrete times. In this case, we may choose to specify the value

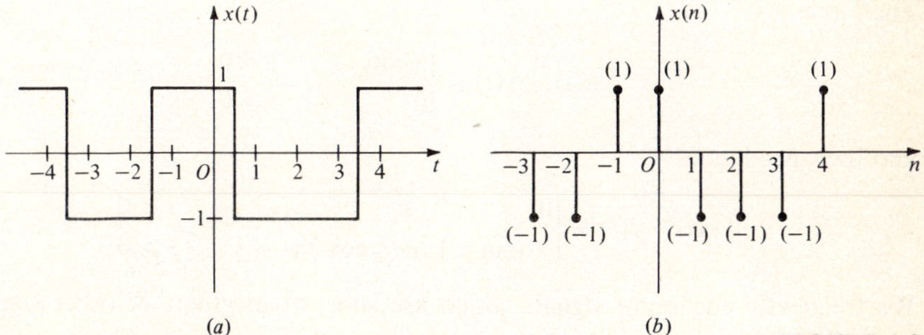

FIGURE 1-3

of the signal only at these times even though its value may vary in between. For example, when a digital computer is used in an air-traffic-control system to monitor the positions of aircraft, the three positional coordinates of an aircraft are read into the computer only at discrete times although the position of the aircraft changes continuously with time.

1-4 MANIPULATION OF SIGNALS

The *sum* of two signals is a signal whose value at any instant is equal to the sum of the values of the two signals at that instant. The *product* of two signals is a signal whose value at any instant is equal to the product of the values of the two signals at that instant. For example, the sum of the two signals

$$x(n) = \begin{cases} 0 & n < -1 \\ 2^{-n} + 5 & n \geq -1 \end{cases}$$

and

$$y(n) = \begin{cases} 3(2^n) & n < 0 \\ n + 2 & n \geq 0 \end{cases}$$

is

$$x(n) + y(n) = \begin{cases} 3(2^n) & n < -1 \\ \frac{17}{2} & n = -1 \\ 2^{-n} + n + 7 & n \geq 0 \end{cases}$$

and their product is

$$x(n) \cdot y(n) = \begin{cases} 0 & n < -1 \\ \frac{21}{2} & n = -1 \\ n2^{-n} + 2^{-n+1} + 5n + 10 & n \geq 0 \end{cases}$$

Similarly, let

$$x(t) = \begin{cases} 0 & t < 0 \\ \sin \pi t & t \geq 0 \end{cases}$$

and

$$y(t) = -\sin \pi t$$

Their sum is

$$x(t) + y(t) = \begin{cases} -\sin \pi t & t < 0 \\ 0 & t \geq 0 \end{cases}$$

and their product is

$$x(t) \cdot y(t) = \begin{cases} 0 & t < 0 \\ -(\sin \pi t)^2 = \frac{1}{2} \cos 2\pi t - \frac{1}{2} & t \geq 0 \end{cases}$$

We frequently encounter signals which are sums or products of other signals. For example, the input signal to a radio receiver is the sum of the signals broadcast by

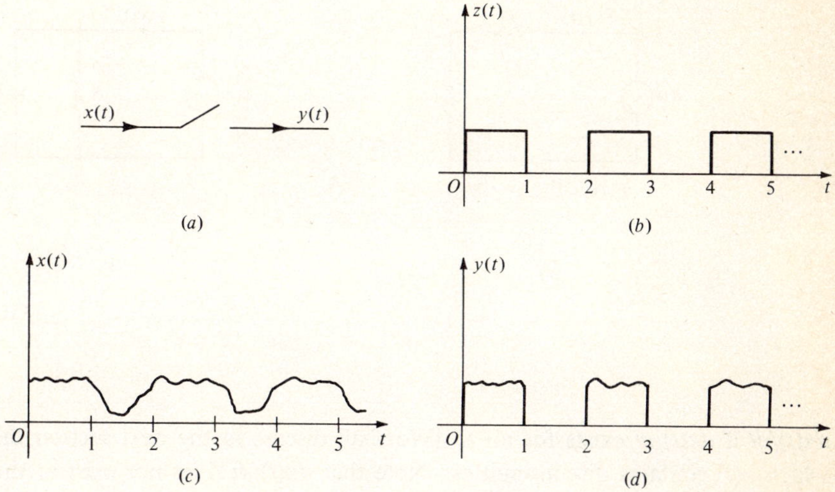

FIGURE 1-4

different radio stations. As another example, let us consider the simple circuit in Fig. 1-4*a*. If the switch is closed every other second starting at $t = 0$, the signal $y(t)$ is equal to the product of $x(t)$ and $z(t)$, where $z(t)$ is the signal shown in Fig. 1-4*b*. Thus, for the signal $x(t)$ in Fig. 1-4*c*, the corresponding signal $y(t)$ is shown in Fig. 1-4*d*. (In practice, the switch is not a mechanical but rather an electronic switch. Consequently, it can be closed and opened at a very fast rate.) A system whose output signal is the sum of its input signals is called an *adder*. A system whose output signal is the product of its input signals is called a *multiplier*.

The *time integral* of a continuous signal $x(t)$ is the signal $\int_{-\infty}^{t} x(\tau)\, d\tau$ whose value at any instant t is the accumulated area under the curve $x(\tau)$ from $-\infty$ to t. Figure 1-5 shows an example. The *derivative* of a continuous signal $x(t)$ is the time signal

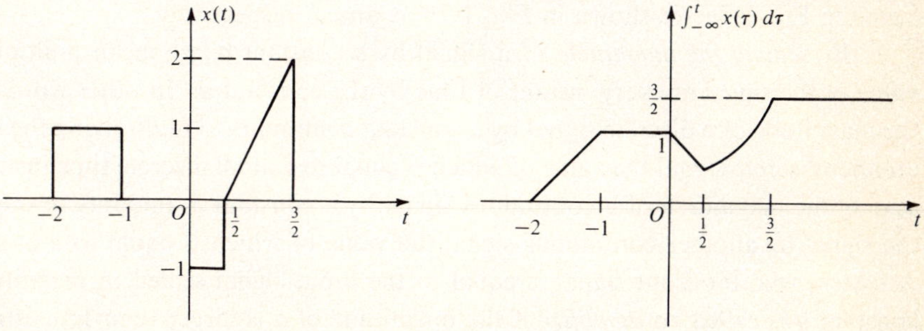

FIGURE 1-5

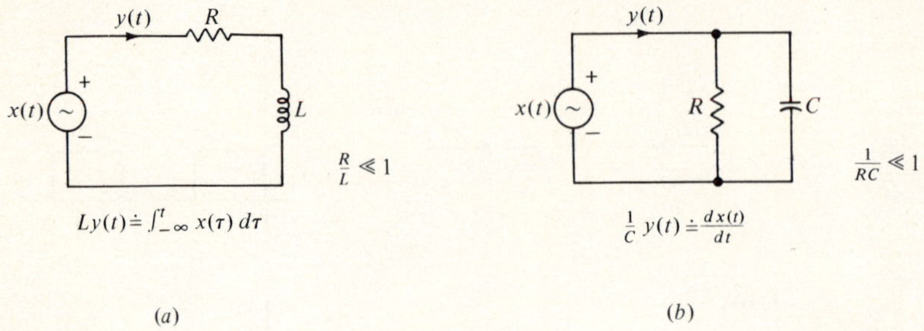

FIGURE 1-6

$dx(t)/dt$ if $dx(t)/dt$ exists for all t. [We shall discuss in the next section the case in which $x(t)$ contains discontinuities. Note that $dx(t)/dt$ does not exist at the discontinuities of $x(t)$.] The simple circuits in Fig. 1-6*a* and *b* can be used to generate the time integral and the time derivative of a signal and are called an *integrator* and a *differentiator*, respectively.

Corresponding to the notion of the time integral of a continuous signal, the *time summation* of a discrete signal $x(n)$ is a discrete signal the value of which at time instant n is equal to $\sum_{k=-\infty}^{n} x(k)$. Corresponding to the notion of the derivative of a continuous signal, the *forward difference* of a discrete signal $x(n)$ is a discrete signal, denoted $\Delta x(n)$, such that

$$\Delta x(n) = x(n+1) - x(n) \qquad \text{for all } n$$

and the *backward difference* of $x(n)$ is a discrete signal, denoted $\nabla x(n)$, such that

$$\nabla x(n) = x(n) - x(n-1) \qquad \text{for all } n$$

For example, the time summation, forward difference, and backward difference of the signal in Fig. 1-7*a* are shown in Fig. 1-7*b*, *c*, and *d*, respectively.

By *scaling the magnitude* of a signal by a constant a, we mean multiplying the value of the signal at every instant of time by the constant a. In other words, scaling the magnitude of a discrete signal by a constant a amounts to multiplying the signal by another discrete signal the value of which is equal to a at all discrete time instants, and scaling the magnitude of a continuous signal by a constant a amounts to multiplying the signal by another continuous signal the value of which is equal to a at all times. A system whose output signal is equal to the input signal scaled in magnitude by a constant a is called an *amplifier* if the magnitude of a is larger than 1; it is called an *attenuator* if the magnitude of a is less than 1.

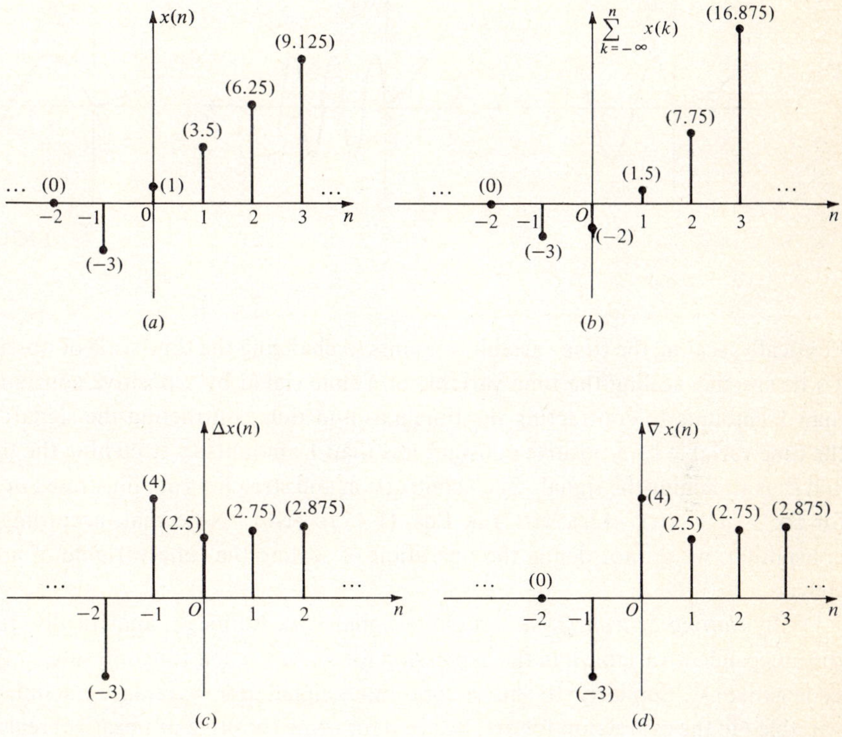

FIGURE 1-7

By *scaling the time variable* of a continuous signal by a positive constant a, we mean replacing the independent variable t in the expression for $x(t)$ by at. For example, given that

$$x(t) = \begin{cases} e^{2t} - 1 & t < 0 \\ \sin \pi t & t \geq 0 \end{cases} \tag{1-4}$$

we have

$$x(2t) = \begin{cases} e^{4t} - 1 & t < 0 \\ \sin 2\pi t & t \geq 0 \end{cases} \tag{1-5}$$

and

$$x\left(\frac{t}{2}\right) = \begin{cases} e^{t} - 1 & t < 0 \\ \sin \dfrac{\pi t}{2} & t \geq 0 \end{cases} \tag{1-6}$$

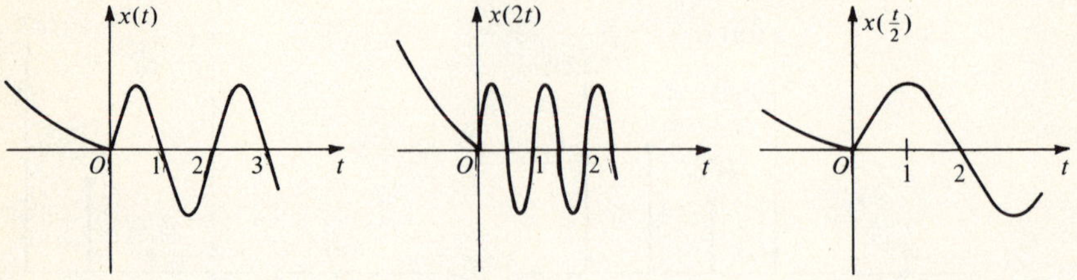

FIGURE 1-8

Physically, scaling the time variable amounts to changing the time scale of observation. To be specific, scaling the time variable of a time signal by a positive constant larger than 1 amounts to contracting the time axis and thus contracting the signal; scaling the time variable by a positive constant less than 1 amounts to stretching the time axis and thus stretching the signal. Such contraction and stretching are illustrated in Fig. 1-8 for the signals $x(t)$, $x(2t)$, $x(t/2)$ in Eqs. (1-4) to (1-6). Note that, according to our convention, we do not define the operation of scaling the time variable of a discrete signal.

By *shifting* or *translating* a discrete signal $x(n)$, we mean, analytically, replacing the independent variable n in the expression for $x(n)$ by $n + k$ for some integer (positive or negative) k. Similarly, to shift a continuous signal $x(t)$ we replace the independent variable t in the expression for $x(t)$ by $t + \delta$ for some (positive or negative) real number δ. Graphically, we see that replacing n by $n + k$ and t by $t + \delta$ amounts to *advancing* the signals for positive k and δ and to *delaying* the signals for negative n and δ. For example, let

$$x(n) = \begin{cases} 0 & n < -1 \\ 2^{-n} + 5 & n \geq -1 \end{cases} \tag{1-7}$$

We have

$$x(n+2) = \begin{cases} 0 & (n+2) < -1 \\ 2^{-(n+2)} + 5 & (n+2) \geq -1 \end{cases}$$

which simplifies to

$$x(n+2) = \begin{cases} 0 & n < -3 \\ \frac{1}{4}2^{-n} + 5 & n \geq -3 \end{cases}$$

and

$$x(n-1) = \begin{cases} 0 & (n-1) < -1 \\ 2^{-(n-1)} + 5 & (n-1) \geq -1 \end{cases}$$

which simplifies to

$$x(n-1) = \begin{cases} 0 & n < 0 \\ 2(2^{-n}) + 5 & n \geq 0 \end{cases}$$

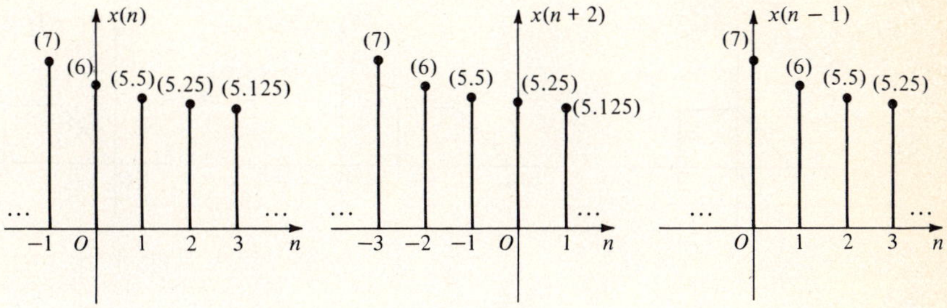

FIGURE 1-9

The signals $x(n)$, $x(n+2)$, $x(n-1)$ are shown in Fig. 1-9. Similarly, for the continuous signal

$$x(t) = \begin{cases} 0 & t < -1 \\ t+1 & -1 \le t < 0 \\ 1 & 0 \le t < 1 \\ 0 & 1 < t \end{cases} \tag{1-8}$$

we have

$$x(t+2) = \begin{cases} 0 & (t+2) < -1 \\ (t+2)+1 & -1 \le (t+2) < 0 \\ 1 & 0 \le (t+2) < 1 \\ 0 & 1 < (t+2) \end{cases}$$

which simplifies to

$$x(t+2) = \begin{cases} 0 & t < -3 \\ t+3 & -3 \le t < -2 \\ 1 & -2 \le t < -1 \\ 0 & -1 < t \end{cases}$$

Also, we have

$$x(t-1) = \begin{cases} 0 & t < 0 \\ t & 0 \le t < 1 \\ 1 & 1 \le t < 2 \\ 0 & 2 < t \end{cases}$$

The signals $x(t)$, $x(t+2)$, $x(t-1)$ are shown in Fig. 1-10. A system whose output signal is a delayed version of its input signal is called a *delay unit*. A system whose

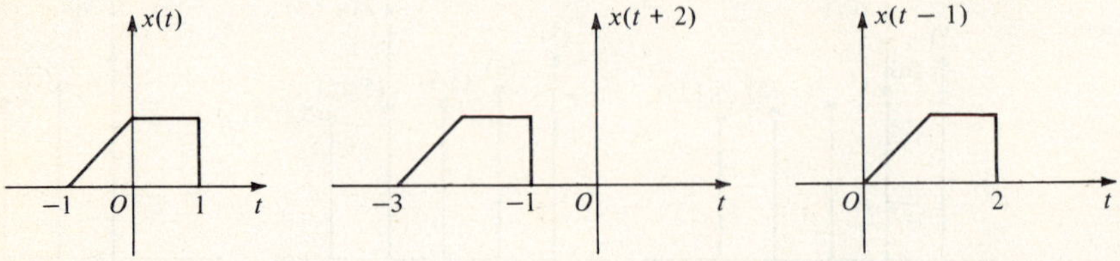

FIGURE 1-10

output signal is an advanced version of its input signal is known as a *predictor*. As will be discussed in Sec. 1-7, it is physically impossible to build a predictor.

To *transpose* a signal is to replace, in its analytic expression, the independent variable n by $-n$ for a discrete signal and to replace the independent variable t by $-t$ for a continuous signal. Intuitively, transposition means interchanging the "past" and the "future" of a time signal. It is intuitively clear that no physical system can perform such an operation. We introduce it here only for mathematical convenience. Graphically, to transpose a signal is to turn a time function around the axis $n = 0$ or $t = 0$. As an example, for the signal $x(n)$ in Eq. (1-7), we have

$$x(-n) = \begin{cases} 0 & n > 1 \\ 2^n + 5 & n \leq 1 \end{cases}$$

as illustrated in Fig. 1-11. Also, for the signal $x(t)$ in Eq. (1-8), we have

$$x(-t) = \begin{cases} 0 & t > 1 \\ -t + 1 & 1 \geq t > 0 \\ 1 & 0 \geq t > -1 \\ 0 & -1 > t \end{cases}$$

as illustrated in Fig. 1-12.

A discrete signal $x(n)$ is said to be *even* if $x(n) = x(-n)$ and is said to be *odd* if $x(n) = -x(-n)$. Similarly, a continuous signal $x(t)$ is said to be *even* if $x(t) = x(-t)$

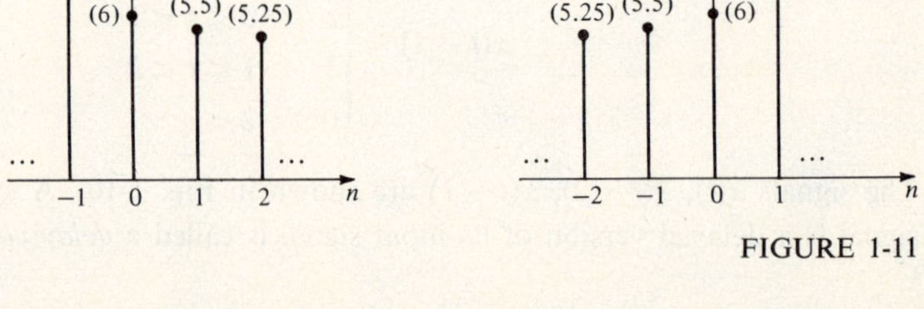

FIGURE 1-11

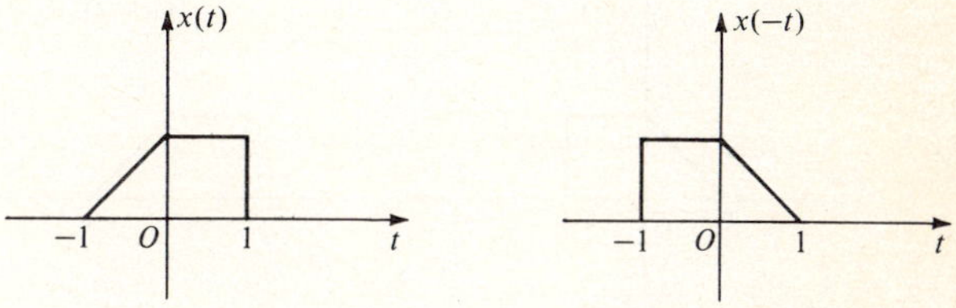

FIGURE 1-12

and to be *odd* if $x(t) = -x(-t)$. For example, $x(t) = \cos t$ is an even signal and $x(t) = \sin t$ is an odd signal. The signal $x(t)$ in Fig. 1-12 is neither even nor odd.

If we replace n by $-n + k$ in the analytic expression of a discrete signal, or replace t by $-t + \delta$ in the analytic expression of a continuous signal, the signal will be shifted *and* transposed. However, it should be noted that a positive k or δ causes the transposed signal to be *delayed*, and a negative k or δ causes the transposed signal to be *advanced*. As an example, let us consider the signal $x(t)$ in Eq. (1-8). Analytically, we have

$$x(-t+2) = \begin{cases} 0 & (-t+2) < -1 \\ (-t+2)+1 & -1 \le (-t+2) < 0 \\ 1 & 0 \le (-t+2) < 1 \\ 0 & 1 < (-t+2) \end{cases}$$

which simplifies to

$$x(-t+2) = \begin{cases} 0 & t > 3 \\ -t+3 & 3 \ge t > 2 \\ 1 & 2 \ge t > 1 \\ 0 & 1 > t \end{cases}$$

It is instructive to see how we determine the signal $x(-t + 2)$ graphically. We shall do this in two ways: For the signal $x(t)$ shown in Fig. 1-13*a*, we can first determine the advanced signal $x(t + 2)$ as in Fig. 1-13*b*, and then transpose the advanced signal to

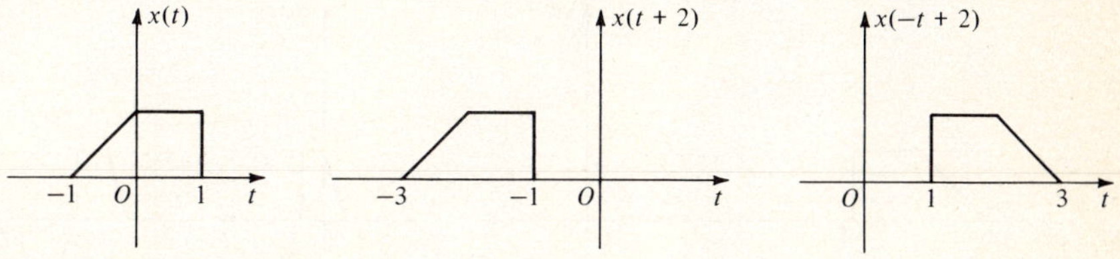

(*a*) (*b*) (*c*)

FIGURE 1-13

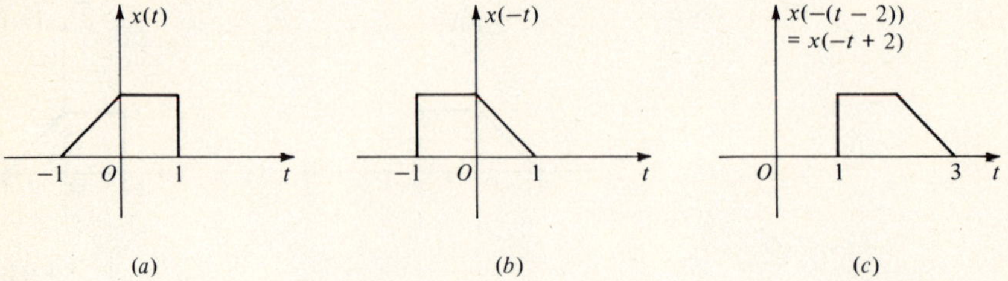

FIGURE 1-14

obtain $x(-t+2)$ as in Fig. 1-13*c*. On the other hand, for the same signal $x(t)$ in Fig. 1-14*a* we can first determine the transposed signal $x(-t)$ as in Fig. 1-14*b* and then delay the transposed signal to obtain $x(-t+2)$ as in Fig. 1-14*c*.

For a continuous signal $x(t)$, the signal $x(at+\delta)$ can be obtained either by scaling the time variable in the translated signal $x(t+\delta)$ by the constant a or by translating the time-scaled signal $x(at)$ by the amount δ/a. As an example, consider the continuous signal $x(t)$ in Eq. (1-8). The signal $x(3t+5)$ is

$$x(3t+5) = \begin{cases} 0 & t \le -2 \\ 3t+6 & -2 \le t \le -\frac{5}{3} \\ 1 & -\frac{5}{3} \le t \le -\frac{4}{3} \\ 0 & -\frac{4}{3} \le t \end{cases}$$

Graphically, we can determine $x(t+5)$ first, as shown in Fig. 1-15*a*, and then scale the time variable in $x(t+5)$ by a factor of 3; or we can determine $x(3t)$ first, as shown in Fig. 1-15*b*, and then translate the signal $x(3t)$ by an amount of $\frac{5}{3}$. In either case, we end up with the signal $x(3t+5)$ shown in Fig. 1-15*c*.

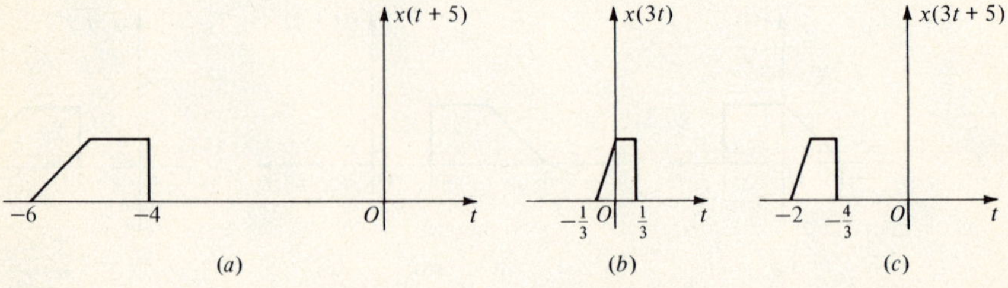

FIGURE 1-15

1-5 SINGULARITY FUNCTIONS

As was pointed out above, when there are discontinuities in a continuous signal $x(t)$, the derivative of $x(t)$ does not exist at these discontinuities. Consequently, $dx(t)/dt$ is not a well-defined time signal. There is nothing we can do, mathematically, to change the fact that $dx(t)/dt$ does not exist at the discontinuities of $x(t)$. However, as will be seen later, it is desirable, both conceptually and operationally, to include the derivative of continuous time signals containing discontinuities as signals in our consideration. In this section, we shall ignore mathematical rigor and define a sequence of "functions" which will enable us to describe the time derivatives of signals containing discontinuities. A mathematical justification of such definitions will be presented in the next section.

The *unit step*, denoted $u_{-1}(t)$, is a continuous time function the value of which is 0 for $t < 0$ and is 1 for $t > 0$, as shown in Fig. 1-16*a*. We define the *unit impulse* to be the derivative of the unit step. We shall denote the unit impulse by $u_0(t)$. Graphically, we use an arrow to denote the unit impulse, as shown in Fig. 1-16*b*. If we want to have some intuitive feeling about what the unit impulse is like, we can consider the unit step to be the limit of the time function shown in Fig. 1-17*a* when δ approaches 0. Thus, the unit impulse is the "limit" of the derivative of the time function in Fig. 1-17*a* when δ approaches 0, as illustrated in Fig. 1-17*b*. For any constant a, $au_0(t)$ is the derivative of the function $au_{-1}(t)$ the value of which is 0 for $t \leq 0$ and a for $t \geq 0$. We say that $au_0(t)$ is an impulse the *value* of which is a.

We now go on to define the *unit doublet*, denoted $u_1(t)$, to be the derivative of $u_0(t)$; the *unit triplet*, denoted $u_2(t)$, to be the derivative of $u_1(t)$; and, in general, the unit $(k + 1)$ tuplet, denoted $u_k(t)$, to be the derivative of $u_{k-1}(t)$ for $k \geq 0$. The functions $u_k(t)$ for $k \geq 0$ are known as *singularity functions*. We shall also use the graphical notation in Fig. 1-18 to represent singularity functions.

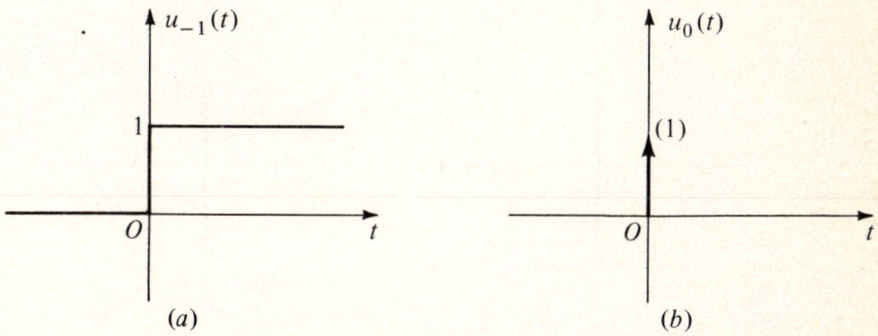

FIGURE 1-16

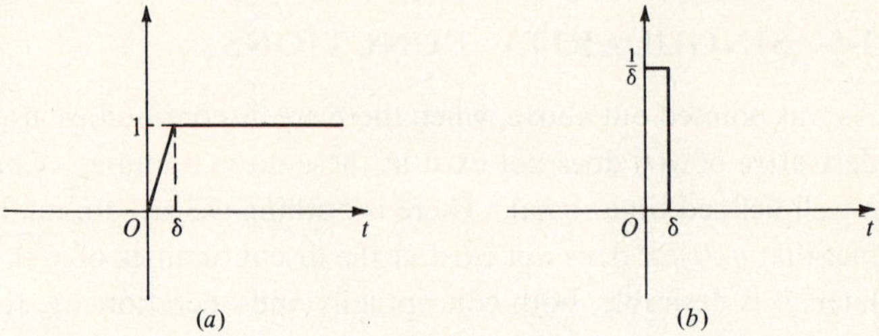

FIGURE 1-17

We can extend the definition of the singularity functions and define $u_k(t)$ to be the derivative of $u_{k-1}(t)$ for all k (positive as well as negative). Thus, we have

$$u_{-2}(t) = \begin{cases} t & t \geq 0 \\ 0 & t < 0 \end{cases}$$

$$u_{-3}(t) = \begin{cases} \frac{1}{2}t^2 & t \geq 0 \\ 0 & t < 0 \end{cases}$$

and, in general, for positive k

$$u_{-k}(t) = \begin{cases} \dfrac{1}{(k-1)!}\, t^{k-1} & t \geq 0 \\ 0 & t < 0 \end{cases}$$

Summing up, we have

$$\frac{du_k(t)}{dt} = u_{k+1}(t) \qquad \text{for all } k$$

$$\int_{-\infty}^{t} u_k(\tau)\, d\tau = u_{k-1}(t) \qquad \text{for all } k$$

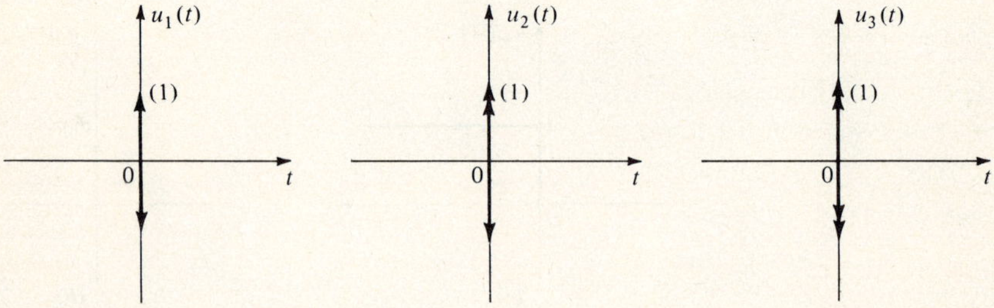

FIGURE 1-18

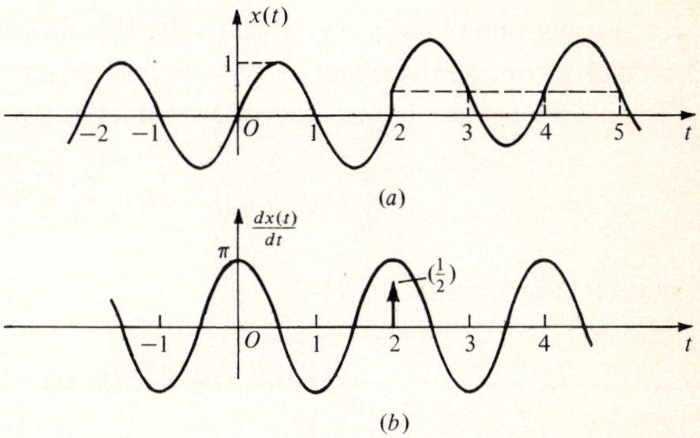

FIGURE 1-19

If we are to include singularity functions as signals in our consideration, we should also extend our definitions of the operations on time signals to include singularity functions. Of course, differentiation and integration of singularity functions are carried out according to the definitions above. The operations of adding two signals, scaling the magnitude of a signal, and shifting a signal are defined as they were before. Thus, the signal

$$2u_0(t) + 5u_0(t) - 4u_0(t)$$

is equal to $3u_0(t)$, and the signal

$$2u_0(t) + 5u_1(t)$$

cannot be further simplified. As another example, the signal in Fig. 1-19*a* can be written as

$$x(t) = \sin \pi t + \tfrac{1}{2}u_{-1}(t-2)$$

It follows that

$$\frac{dx(t)}{dt} = \pi \cos \pi t + \tfrac{1}{2}u_0(t-2)$$

as shown in Fig. 1-19*b*.

The product of two singularity functions is, in general, not defined. However, we define

$$x(t)u_0(t) = x(0)u_0(t) \tag{1-9}$$

if $x(t)$ is continuous at $t = 0$, and

$$x(t)u_0(t-\tau) = x(\tau)u_0(t-\tau)$$

if $x(t)$ is continuous at $t = \tau$. Intuitively, this means that an impulse "picks out" the value of a continuous signal $x(t)$ at the instant the impulse occurs.

To determine the product of a doublet and a continuous function, we note that

$$\frac{d}{dt}[x(t)u_0(t)] = x(t)\frac{du_0(t)}{dt} + \frac{dx(t)}{dt}u_0(t)$$

That is, according to Eq. (1-9),

$$\frac{d}{dt}[x(0)u_0(t)] = x(t)u_1(t) + x'(0)u_0(t)$$

or

$$x(0)u_1(t) = x(t)u_1(t) + x'(0)u_0(t)$$

which gives

$$x(t)u_1(t) = -x'(0)u_0(t) + x(0)u_1(t)$$

provided that $x(t)$ and $x'(t)$ are continuous at $t = 0$. Similarly, we have

$$x(t)u_1(t-\tau) = -x'(\tau)u_0(t-\tau) + x(\tau)u_1(t-\tau)$$

provided that $x(t)$ and $x'(t)$ are continuous at $t = \tau$. This is to say that the product of a doublet and a continuous function yields an impulse and a doublet, the values of which depend on the values of the continuous function and its derivative at the time the doublet occurs. For example, we have

$$u_0\left(t + \frac{1}{4}\right)\sin \pi t = -\frac{1}{\sqrt{2}}u_0\left(t + \frac{1}{4}\right)$$

$$u_1\left(t + \frac{1}{4}\right)\sin \pi t = -\frac{\pi}{\sqrt{2}}u_0\left(t + \frac{1}{4}\right) - \frac{1}{\sqrt{2}}u_1\left(t + \frac{1}{4}\right)$$

As another example, we have

$$\begin{aligned}\int_{-\infty}^{\infty} \sin(\omega t + \theta)u_0(t)\,dt &= \int_{-\infty}^{\infty} \sin\theta\, u_0(t)\,dt \\ &= \sin\theta \int_{-\infty}^{\infty} u_0(t)\,dt \\ &= \sin\theta\end{aligned}$$

Note that

$$\int_{-\infty}^{\infty} u_0(t)\,dt = \int_{-\infty}^{\tau} u_0(t)\,dt\bigg|_{\tau\to\infty} = u_{-1}(\tau)\bigg|_{\tau\to\infty} = 1$$

As a matter of fact, the result in this example can be generalized to

$$\int_{-\infty}^{\infty} x(t+\tau)u_0(t)\,dt = x(\tau) \tag{1-10}$$

if $x(t)$ is continuous at $t=\tau$, and for $k>0$,

$$\int_{-\infty}^{\infty} x(t+\tau)u_k(t)\,dt = (-1)^k \left.\frac{d^k x(t)}{dt^k}\right|_{t=\tau} \tag{1-11}$$

if $d^k x(t)/dt^k$ is continuous at $t=\tau$. We leave the verification of Eqs. (1-10) and (1-11) to the reader.

The products of other singularity functions with continuous functions can be determined in a similar manner. We shall leave the details to the reader. (See Prob. 1-13.)

As for scaling of the time variable, we follow the rule

$$u_k(at) = \frac{1}{a^{k+1}} u_k(t)$$

For example, we have

$$u_0(2t) = \tfrac{1}{2}u_0(t)$$

and

$$u_1(2t) = \tfrac{1}{4}u_1(t)$$

Transposition of singularity functions follows the rule

$$u_k(-t) = (-1)^k u_k(t) \qquad k \geq 0$$

In other words, $u_0(t), u_2(t), u_4(t), \ldots$ are even time functions, and $u_1(t), u_3(t), u_5(t), \ldots$ are odd time functions.†

*1-6 GENERALIZED FUNCTIONS

As was pointed out in the previous section, the inclusion of singularity functions as time signals results in a dilemma. On the one hand, we want to be able to manipulate (e.g., shift, scale, differentiate, integrate) singularity functions in the same way we manipulate other time signals. On the other hand, since singularity functions are not well-defined functions in the ordinary sense, mathematical operations defined for ordinary functions are not meaningful for singularity functions. Mathematicians have tried to resolve the difficulties in various ways. One of the most significant contribu-

† We wish to point out that such a definition is not an arbitrary one. It follows from the observation that the derivative of an even time function is odd, and the derivative of an odd time function is even. (See Prob. 1-7.)

tions is the theory of distributions developed by Laurent Schwartz in the early 1950s. We present in this section some of the basic ideas in the theory of distributions. It is not possible to give a complete treatment of the subject here. Rather, we want only to give the reader an opportunity to appreciate the flavor of a very elegant theory in mathematics.

We recall that a function of a real variable t is an assignment of values to all possible values of the real variable t. For example, for the function

$$x(t) = \sin \pi t + je^{-2t}$$

at $t = 0$, the value of the function is j; at $t = 0.5$, the value of the function is $1 + j0.368$; and so on. Clearly, specifying the values of a signal for all t is a complete way to describe a signal. From a physical point of view, one might say that we are not really interested in a signal by itself but, rather, we are interested in how a signal interacts with other signals. This brings up the possibility of describing a signal by specifying how it interacts with other signals. For this purpose, let us pick a set of functions, $\phi_1(t)$, $\phi_2(t)$, . . . , called *testing functions*. So that the testing functions will possess certain mathematical properties that are required in later developments, we choose the set of testing functions to be the set of all functions that satisfy the following conditions:

1 The functions are infinitely smooth, that is, they have continuous derivatives of all orders.

2 The value of each function is zero outside some finite interval. (It is *not* required, however, that each function be zero outside the *same* finite interval.)

An example of a testing function is

$$\phi(t) = \begin{cases} 0 & |t| \geq 1 \\ e^{1/(t^2 - 1)} & |t| < 1 \end{cases}$$

A *functional* on the set of testing functions is an assignment of values (complex, in the general case) to the testing functions. For example, Table 1-1 shows a functional

Table 1-1

Testing function	Value assigned by the functional
$\phi_1(t)$	$1 + j5.5$
$\phi_2(t)$	$1 + j5.5$
.....	....
$\phi_k(t)$	$1 + j5.5$
.....	....

that assigns the value $1 + j5.5$ to each of the testing functions. As another example, Table 1-2 shows a functional that assigns the value $\phi_k(2)$ to the testing function $\phi_k(t)$ for all k. Letter names, such as $f, g, \ldots$, can be used to denote functionals. However, anticipating the association of functionals with time signals, we shall denote functionals by $f(t), g(t), \ldots$, where the argument t is the argument of the testing functions $\phi_1(t)$, $\phi_2(t), \ldots$, although strictly speaking a functional does not have t as an argument. We shall use the notation $\langle g(t), \phi(t)\rangle$ to denote the value assigned to the testing function $\phi(t)$ by the functional $g(t)$.

A functional $g(t)$ is said to be *linear* if

$$\langle g(t), a\phi_i(t) + b\phi_j(t)\rangle = a\langle g(t), \phi_i(t)\rangle + b\langle g(t), \phi_j(t)\rangle$$

for all testing functions $\phi_i(t)$, $\phi_j(t)$ and all complex constants a, b. Note that if $\phi_i(t)$ and $\phi_j(t)$ are testing functions, then $a\phi_i(t) + b\phi_j(t)$ is also a testing function. A functional $g(t)$ is said to be *continuous* if for a sequence of testing functions $\phi_{i1}(t)$, $\phi_{i2}(t), \ldots$, which converges to the testing function $\phi_i(t)$, the sequence of numbers $\langle g(t), \phi_{i1}(t)\rangle$, $\langle g(t), \phi_{i2}(t)\rangle, \ldots$ converges to the number $\langle g(t), \phi_i(t)\rangle$. A linear continuous functional is called a *distribution* or a *generalized function*.

We propose to represent a signal as a distribution, with the intention of including both ordinary functions as well as singularity functions as distributions. To carry out such a proposition, we must first answer the question: How do we define a functional on the set of testing functions to represent a signal which is specified by an ordinary time function? For an ordinary function $g(t)$, let us define a corresponding functional $g(t)$, such that, for every testing function $\phi(t)$,

$$\langle g(t), \phi(t)\rangle = \int_{-\infty}^{\infty} g(t)\phi(t)\,dt$$

In other words, the functional $g(t)$ will assign to each testing function $\phi(t)$ the value of the integral $\int_{-\infty}^{\infty} g(t)\phi(t)\,dt$. Note that we purposely use the notation $g(t)$ for a time function and its corresponding functional to emphasize the fact that they are only different representations of the same signal. If $g(t)$ is integrable over every finite

Table 1-2

Testing function	Value assigned by the functional
$\phi_1(t)$	$\phi_1(2)$
$\phi_2(t)$	$\phi_2(2)$
.....	.....
$\phi_k(t)$	$\phi_k(2)$
.....	.....

interval, the integral $\int_{-\infty}^{\infty} g(t)\phi(t)\,dt$ exists for all $\phi(t)$. Moreover, it can be shown that the functional defined in this way is indeed linear and continuous and is thus a distribution.

Since we are to represent signals by distributions, we must now define how distributions are to be manipulated. Because a signal can be represented in two possible ways, namely, as an ordinary time function and as a distribution, the definitions of operations on distributions cannot be arbitrary and must be consistent with the corresponding definitions for ordinary time functions.

Addition of distributions: We define

$$\langle g_1(t) + g_2(t), \phi(t)\rangle \triangleq \langle g_1(t), \phi(t)\rangle + \langle g_2(t), \phi(t)\rangle$$

That is, the sum of two distributions $g_1(t)$ and $g_2(t)$, denoted $g_1(t) + g_2(t)$, assigns to $\phi(t)$ the sum of the values assigned to $\phi(t)$ by $g_1(t)$ and $g_2(t)$. To see that such a definition is consistent with that of addition of ordinary functions, we note that for two ordinary functions $g_1(t)$ and $g_2(t)$

$$\int_{-\infty}^{\infty} [g_1(t) + g_2(t)]\phi(t)\,dt = \int_{-\infty}^{\infty} g_1(t)\phi(t)\,dt + \int_{-\infty}^{\infty} g_2(t)\phi(t)\,dt$$

Scaling the magnitude of a distribution: We define

$$\langle ag(t), \phi(t)\rangle \triangleq \langle g(t), a\phi(t)\rangle$$

That is, the distribution $ag(t)$ assigns to the testing function $\phi(t)$ the value assigned by the distribution $g(t)$ to the testing function $a\phi(t)$. Again, such a definition is consistent with that of scaling the magnitude of an ordinary function since

$$\int_{-\infty}^{\infty} ag(t)\phi(t)\,dt = \int_{-\infty}^{\infty} g(t)a\phi(t)\,dt$$

Shifting a distribution: We define

$$\langle g(t-\tau), \phi(t)\rangle \triangleq \langle g(t), \phi(t+\tau)\rangle$$

That is, the distribution $g(t-\tau)$ assigns to the testing function $\phi(t)$ the value assigned by $g(t)$ to the testing function $\phi(t+\tau)$. To check the consistency of the definition, we note that for any ordinary function $g(t)$ we have

$$\int_{-\infty}^{\infty} g(t-\tau)\phi(t)\,dt = \int_{-\infty}^{\infty} g(\lambda)\phi(\lambda+\tau)\,d\lambda$$

by a change of variable $\lambda = t - \tau$.

Scaling the time variable: We define, for $a > 0$,

$$\langle g(at), \phi(t)\rangle \triangleq \left\langle g(t), \frac{1}{a}\phi\left(\frac{t}{a}\right)\right\rangle$$

That is, the distribution $g(at)$ assigns to the testing function $\phi(t)$ the value assigned by $g(t)$ to the function $(1/a)\phi(t/a)$. Note that the function $(1/a)\phi(t/a)$ is indeed a testing function if $\phi(t)$ is a testing function. Again, consistency of the definition is checked by observing that for $a > 0$

$$\int_{-\infty}^{\infty} g(at)\phi(t)\,dt = \int_{-\infty}^{\infty} g(\lambda)\,\frac{1}{a}\,\phi\!\left(\frac{\lambda}{a}\right) d\lambda$$

after a change of variable $\lambda = at$.

Transposition of a distribution: We define

$$\langle g(-t), \phi(t)\rangle \triangleq \langle g(t), \phi(-t)\rangle$$

That is, the distribution $g(-t)$ assigns to the testing function $\phi(t)$ the value assigned by $g(t)$ to the testing function $\phi(-t)$. For an ordinary function $g(t)$, we note that

$$\int_{-\infty}^{\infty} g(-t)\phi(t)\,dt = \int_{-\infty}^{\infty} g(\lambda)\phi(-\lambda)\,d\lambda$$

by a change of variable $\lambda = -t$.

Differentiation of a distribution: We define

$$\left\langle \frac{dg(t)}{dt}, \phi(t)\right\rangle \triangleq \left\langle g(t), -\frac{d\phi(t)}{dt}\right\rangle$$

That is, the distribution $dg(t)/dt$ assigns to the testing function $\phi(t)$ the value assigned by $g(t)$ to the function $-\,d\phi(t)/dt$. Because $\phi(t)$ is infinitely smooth, $d\phi(t)/dt$ is also a testing function. For an ordinary function $g(t)$ whose derivative $dg(t)/dt$ exists for all t, we have

$$\int_{-\infty}^{\infty} \frac{dg(t)}{dt}\,\phi(t) = g(t)\phi(t)\Big|_{-\infty}^{\infty} - \int_{-\infty}^{\infty} g(t)\,\frac{d\phi(t)}{dt}\,dt$$

$$= -\int_{-\infty}^{\infty} g(t)\,\frac{d\phi(t)}{dt}\,dt$$

Note that $g(t)\phi(t) = 0$ at $t = \infty$ and $t = -\infty$ because $\phi(t)$ is zero outside a finite interval. Thus, consistency of the definition is assured.

We are now ready to complete our presentation by showing the distributions corresponding to the singularity functions which we defined in a very intuitive manner in Sec. 1-5. Corresponding to the unit step $u_{-1}(t)$, we define the distribution $u_{-1}(t)$:

$$\langle u_{-1}(t), \phi(t)\rangle \triangleq \int_{0}^{\infty} \phi(t)\,dt$$

Let $u_0(t)$ denote the distribution which is the derivative of the distribution $u_{-1}(t)$. (Note that the derivative of a distribution always exists in contrast to the fact that the derivative of an ordinary function does not always exist. As a matter of fact, this is one of the major reasons for defining the singularity functions as distributions.) According to the rule of differentiating distributions, we have

$$\langle u_0(t), \phi(t)\rangle = \left\langle u_{-1}(t), -\frac{d\phi(t)}{dt}\right\rangle = -\int_0^\infty \frac{d\phi(t)}{dt}\,dt = \phi(0) - \phi(\infty) = \phi(0)$$

because $\phi(\infty) = 0$ for all $\phi(t)$. In other words, the tabular-form description of the distribution $u_0(t)$ is as given in Table 1-3.

Similarly, if we let $u_1(t)$ denote the distribution which is the derivative of the distribution $u_0(t)$, we have

$$\langle u_1(t), \phi(t)\rangle = \left\langle u_0(t), -\frac{d\phi(t)}{dt}\right\rangle = -\frac{d\phi(t)}{dt}\bigg|_{t=0}$$

and, in general,

$$\begin{aligned}\langle u_k(t), \phi(t)\rangle &= \left\langle u_{k-1}(t), -\frac{d\phi(t)}{dt}\right\rangle \\ &= \left\langle u_{k-2}(t), \frac{d^2\phi(t)}{dt^2}\right\rangle \\ &= \cdots \\ &= (-1)^k \frac{d^k\phi(t)}{dt^k}\bigg|_{t=0}\end{aligned}$$

We claim now that the distributions defined as successive derivatives of the distribution $u_{-1}(t)$ correspond to the singularity functions defined intuitively in Sec. 1-5. Such a claim can be asserted by showing that the distributions $u_0(t), u_1(t), \cdots$ indeed have those properties of singularity functions stated in Sec. 1-5. We shall leave the assertion to the interested reader.

Table 1-3

Testing function	Value assigned by the distribution
$\phi_1(t)$	$\phi_1(0)$
$\phi_2(t)$	$\phi_2(0)$
.....	.....
$\phi_k(t)$	$\phi_k(0)$
.....	.....

To conclude our very brief description of a deep mathematical subject, let us recapitulate the purpose of developing the theory of distributions. From an operational point of view, it is desirable to treat the successive derivatives of the step function $u_{-1}(t)$ as signals so that we can manipulate these derivatives in the same manner we manipulate ordinary time functions. However, mathematically, since these derivatives do not exist as time functions, it is meaningless to talk about how they are to be manipulated. The theory of distributions puts us on firm mathematical ground when we must acknowledge the existence of these derivatives and add, scale, and differentiate them. The reader should keep in mind that when we add, scale, and differentiate singularity functions from now on, we are actually adding, scaling, and differentiating the corresponding distributions which we might write $u_0(t) + 2u_1(t)$, $u_1(5t)$, $du_2(t)/dt$ as if they were ordinary time functions.

1-7 SYSTEMS

As was pointed out in Sec. 1-2, a system can be viewed as a signal transducer that transforms a set of input signals into a corresponding set of output signals. When we are interested only in the terminal behavior of a system, we shall represent it as a "black box." For example, Fig. 1-20*a* shows a system that accepts three input signals and generates two output signals, and Fig. 1-20*b* shows a single-input–single-output system. One of our major tasks in this book is to study the various ways of describing the input-output relationship of systems. Most of our discussion will be limited to systems with one input and one output, although many of the concepts developed later on can be extended immediately to systems with multiple inputs and outputs. For a given single-input–single-output system, we use the notation $H[x(t)] = y(t)$ to indicate that corresponding to the input signal $x(t)$ the output signal of the system is $y(t)$. In this section, we shall discuss some of the general properties of systems.

A system is said to be *continuous* if it accepts continuous time signals as inputs and generates continuous time signals as outputs. A system is said to be *discrete* if it

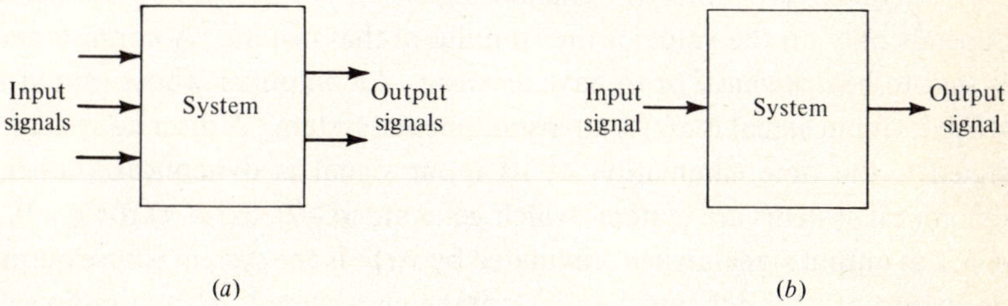

FIGURE 1-20

accepts discrete time signals as inputs and generates discrete time signals as outputs. A system is said to be *hybrid* if it accepts continuous (discrete) time signals as inputs and generates discrete (continuous) time signals as outputs. As examples, a loudspeaker which receives electrical signals as inputs and produces sound signals as outputs is a continuous system; a digital computer which reads sequences of digits and letters on punched cards as inputs and prints sequences of digits and letters on paper as outputs is a discrete system; and a television set is a hybrid system because the pictures it reproduces from the broadcast electromagnetic waves are actually discrete signals.

A system is said to be *causal*, or *realizable*, or *nonanticipatory* if it does not respond before being stimulated. That is, for a causal system, if an input signal is equal to zero for $t < t_0$ then the corresponding output signal is equal to zero for $t < t_0$. For example, a system in which

$$y(n) = x(n) + x(n-2)$$

is causal, and a system in which

$$y(n) = x(n) + x(n+2)$$

is noncausal. As another example, if

$$x(n) = \begin{cases} a & n = 0 \\ 0 & \text{otherwise} \end{cases}$$

and

$$y(n) = \begin{cases} 0 & n < -1 \\ a & n \geq -1 \end{cases}$$

are a pair of input and output signals of a system, the system is noncausal. Also, for an input signal $x(t)$, a system which generates $x(-t)$ as its output signal is noncausal, as is one which generates $x(t+5)$ as its output signal, whereas a system which generates $x(t-\tau)$ as its output signal for $\tau \geq 0$ is causal. It is obvious that all systems which can be built physically are causal systems. For example, a loudspeaker cannot produce music before the amplifier is turned on, and a digital computer cannot print out the result of computation before it is fed with the data.

A system is said to be *instantaneous* or *memoryless* if its response at any instant depends only on the value of the stimulus at that instant. A noninstantaneous system is said to be *dynamical* or to have *memory*. An amplifier whose output signal is $ax(t)$ when its input signal is $x(t)$ is an instantaneous system. A discrete system whose output signal is the time summation of its input signal is dynamical. Other examples of dynamical systems are systems which generate $x(-t)$, $x(t-\tau)$ for $\tau \neq 0$, and $x(at)$ for $a \neq 1$ as output signals when stimulated by $x(t)$. Is the system whose output signal is the derivative of its input signal an instantaneous system? Also, a radio set is an instantaneous system, and a vending machine whose inputs are the coins deposited by the customers and whose outputs are the merchandise is a dynamical system (because the vending machine can "remember" the amount of coins deposited).

A system is said to be *stable* (*in the bounded-input–bounded-output sense*) if the magnitude of the output signal is bounded at all time when the magnitude of the input signal is bounded at all time. Obviously, a system which generates the time integral of its input signal is not stable since the magnitude of the output signal can be unbounded even when the magnitude of the input signal is bounded. The reader can verify that a discrete system whose output signal is the forward or backward difference of its input signal is stable. Similarly, a system which shifts or transposes its input signal is stable. A system which scales the time variable of its input signal or scales the magnitude of its input signal by a finite constant is also stable.

A system is said to be *time-invariant* if a time-shifted input signal will yield a correspondingly time-shifted output signal. To be precise, for a time-invariant system $H[x(t)] = y(t)$ implies $H[x(t + \tau)] = y(t + \tau)$ for all τ. A system is said to be *time-varying* otherwise. Differentiators and amplifiers are time-invariant systems, but a system that transposes its input signal is a time-varying one. Because the characteristics of the transistors and the values of the resistors and capacitors change with time, a radio set is a time-varying system. However, if we assume that circuit components do not age, it can be approximated as a time-invariant system. A parking lot that charges different rates at different times of the day is another example of a time-varying system whose input is a discrete signal describing the arrival of cars and whose output is a discrete signal corresponding to the total receipt of the operation.

A system is said to be *homogeneous* if the response of the system to a signal $ax(t)$ is equal to a times the response of the system to the signal $x(t)$ for any constant a. That is,

$$H[ax(t)] = aH[x(t)]$$

A system is said to be *additive* if the response of the system to a signal $x_1(t) + x_2(t)$ is equal to the sum of the responses of the system to the signals $x_1(t)$ and $x_2(t)$. That is,

$$H[x_1(t) + x_2(t)] = H[x_1(t)] + H[x_2(t)]$$

A system is said to be *linear* if it is both homogeneous and additive. We leave it to the reader to show that a system is linear if for any input signals $x_1(t)$, $x_2(t)$ and any constants a_1, a_2

$$H[a_1 x_1(t) + a_2 x_2(t)] = a_1 H[x_1(t)] + a_2 H[x_2(t)]$$

We assume that the linearity condition can be extended to infinite sums and integrals. That is, if the input signal to a linear system, $x(t)$, can be expressed as an infinite sum

$$x(t) = \sum_{i=1}^{\infty} a_i x_i(t)$$

then
$$H[x(t)] = H\left[\sum_{i=1}^{\infty} a_i x_i(t)\right] = \sum_{i=1}^{\infty} a_i H[x_i(t)]$$

Similarly, if $x(t)$ can be expressed as an integral,

$$x(t) = \int_{-\infty}^{\infty} a(\lambda)\hat{x}(t, \lambda)\, d\lambda$$

where $\hat{x}(t, \lambda)$ is a family of time signals with λ being a parameter, then

$$H[x(t)] = H\left[\int_{-\infty}^{\infty} a(\lambda)\hat{x}(t, \lambda)\, d\lambda\right] = \int_{-\infty}^{\infty} a(\lambda)H[\hat{x}(t, \lambda)]\, d\lambda$$

We leave it to the reader to show that a system in which

$$y(t) = x(t) \sin 120\pi t$$

is linear, whereas a system in which

$$y(t) = x(t)x(t)$$

is not linear. The parking lot example mentioned above is a linear system, because the total receipt is proportional to the number of cars in the lot. However, if a discount rate is offered to customers who park more than one car in the lot, then it becomes a nonlinear system.

Throughout this book, our attention will be focused on the study of linear time-invariant systems, both because there is a very rich collection of mathematical techniques that can be applied to analyze the behavior of linear time-invariant systems, and because many systems encountered in practice are either linear time-invariant systems or can be approximated by linear time-invariant systems.

1-8 INTERCONNECTION OF SYSTEMS

"Large" systems are made up of "small" systems. (For the sake of clarity, these small systems are often called *subsystems* of the large one.) For example, a digital computer is made up of arithmetic units, control circuits, memory devices, and input-output terminals which can be viewed as subsystems of the overall digital computer system. A subsystem is an individual system in its own right and can, in turn, be made up of smaller systems. For example, an arithmetic unit in a digital computer is made up of adders, counters, indicators, and so on. As another example, a satellite tracking station is a system containing computers, antennas, communication equipment, display devices, and many more items as subsystems. From the point of view of designing and constructing large systems, it is a physical necessity to use small

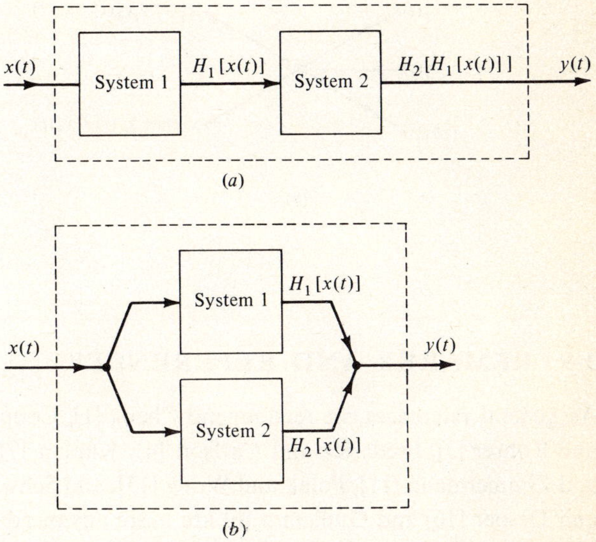

FIGURE 1-21

systems as building blocks. Also, from the point of view of analyzing and maintaining systems, it is desirable to be able to divide a complex system into simpler subsystems.

There are two basic ways in which systems can be interconnected: cascade connection and parallel connection. Figure 1-21*a* shows the general form of a cascade connection of two systems. Specifically, the output signal of system 1 is used as the input signal to system 2. Therefore, as indicated in Fig. 1-21*a*, $x(t)$ and $y(t)$ are, respectively, the input and output signals of the overall system. That is,

$$y(t) = H_2[H_1[x(t)]]$$

Figure 1-21*b* shows the general form of a parallel connection of two systems. Specifically, both system 1 and system 2 will receive the input signal of the overall system as their input signals, and the output signal of the overall system is the sum of the output signals of the two systems. That is, as indicated in Fig. 1-21*b*,

$$y(t) = H_1[x(t)] + H_2[x(t)]$$

Note that we shall consistently use the notation of a node to indicate a device that will add up all the signals coming into the node and distribute the sum as outgoing signals from the node, as demonstrated in Fig. 1-22*a* and *b*. We shall refer to such a device as a *distributing device*.

The problem of system connection will be explored in further detail in Chap. 10.

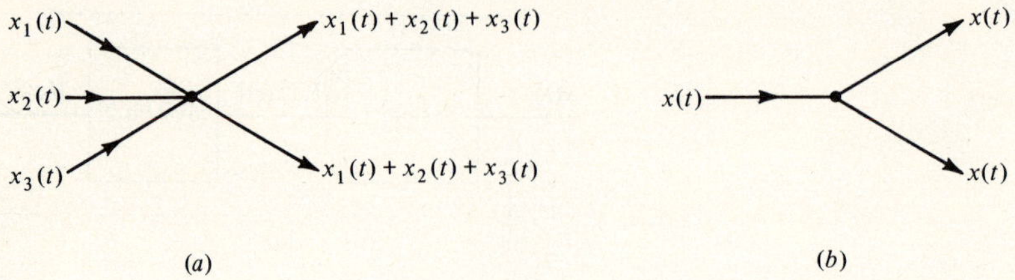

FIGURE 1-22

1-9 REMARKS AND REFERENCES

As general references, we recommend Cheng [1], Cooper and McGillem [2], Director and Rohrer [3], Frederick and Carlson [4], Kaplan [7], Kuo [8], Lathi [9], Mason and Zimmermann [11], Polak and Wong [13], and Schwarz and Friedland [15]. Zadeh and Desoer [16] and Guillemin [6] are more advanced references.

For a lucid discussion of the theory of distributions, see Zemanian [17]. Other important but more advanced references are Schwartz [14], Lighthill [10], and Gelfand and Shilov [5]. See also Appendix A of Zadeh and Desoer [16] and Appendix I of Papoulis [12] for a brief introduction to the subject.

[1] CHENG, D. K.: "Analysis of Linear Systems," Addison-Wesley Publishing Company, Inc., Reading, Mass., 1959.

[2] COOPER, G. R., and C. D. MCGILLEM: "Methods of Signal and System Analysis," Holt, Rinehart and Winston, Inc., New York, 1967.

[3] DIRECTOR, S. W., and R. A. ROHRER: "Introduction to System Theory," McGraw-Hill Book Company, New York, 1972.

[4] FREDERICK, D. K., and A. B. CARLSON: "Linear Systems in Communication and Control," John Wiley & Sons, Inc., New York, 1971.

[5] GELFAND, I. M., and G. E. SHILOV: "Generalized Functions," vols. 1–3, Academic Press, Inc., New York, 1964.

[6] GUILLEMIN, E. A.: "Theory of Linear Physical Systems," John Wiley & Sons, Inc., New York, 1963.

[7] KAPLAN, W.: "Operational Methods for Linear Systems," Addison-Wesley Publishing Company, Inc., Reading, Mass., 1962.

[8] KUO, B. C.: "Linear Networks and Systems," McGraw-Hill Book Company, New York, 1967.

[9] LATHI, B. P.: "Signals, Systems and Communication," John Wiley & Sons, Inc., New York, 1965.

[10] LIGHTHILL, M. J.: "Fourier Analysis and Generalized Functions," Cambridge University Press, New York, 1958.

[11] MASON, S. J., and H. J. ZIMMERMANN: "Electronics Circuits, Signals, and Systems," John Wiley & Sons, Inc., New York, 1960.

[12] PAPOULIS, A.: "The Fourier Integral and Its Applications," McGraw-Hill Book Company, New York, 1962.

[13] POLAK, E., and E. WONG: "Notes for a First Course on Linear Systems," Van Nostrand Reinhold Company, New York, 1970.

[14] SCHWARTZ, L.: "Théorie des distributions," vols. I and II, Actualités Scientifiques et Industrielles, Hermann and Cie, Paris, 1957, 1959.

[15] SCHWARZ, R. J., and B. FRIEDLAND: "Linear Systems," McGraw-Hill Book Company, New York, 1965.

[16] ZADEH, L. A., and C. A. DESOER: "Linear System Theory," McGraw-Hill Book Company, New York, 1963.

[17] ZEMANIAN, A. H.: "Distribution Theory and Transform Analysis," McGraw-Hill Book Company, New York, 1965.

PROBLEMS

1-1 Let $x(n)$ denote the total amount (in millions of dollars) in all the checking and savings accounts of a bank at the beginning of the nth month. Suppose that, within the nth month, n million dollars is deposited in the checking accounts and 2 million dollars is deposited in the savings accounts. Furthermore, suppose that the bank charges a 1-cent service charge for each dollar deposited in a checking account. Service charges are deducted from the accounts at the end of each month. A 2 percent monthly interest is paid to all savings account deposits at the end of each month. Find $x(2)$, $x(3)$, $x(4)$, and $x(5)$; suppose that $x(1)$ is equal to zero.

1-2 Let the discrete signal $x(n)$ denote the average number of descendants of a pair of rabbits in the nth month since their birth. Suppose that a rabbit matures and can reproduce when it is 1 month old and that it has an average lifetime of 4 months. In the 2d, 3d, and 4th month of their lives, each pair of rabbits will produce an average of two pairs of rabbits each month. Find $x(1)$, $x(2)$, $x(3)$, $x(4)$, $x(5)$.

1-3 A Ping-Pong ball is dropped to the floor from a height of 10 ft. Suppose that the ball always rebounds to reach half of the height from which it is dropped.

(*a*) Let $x(n)$ denote the height it reaches in the nth rebound. Sketch $x(n)$.

(*b*) Let $y(n)$ denote the loss in height during the nth rebound. Express $y(n)$ in terms of $x(n)$. Sketch $y(n)$.

(*c*) A second Ping-Pong ball identical to the first one is dropped from a height of 3 ft on the same floor at the same time as the first ball reaches the highest point of its third rebound. Let $z(n)$ denote the height the second ball reaches in its nth rebound. Express $z(n)$ in terms of $x(n)$.

1-4 In an air-traffic control system, the ground control station transmits a pulse of peak power P at the beginning of each interrogation period, as shown in Fig. 1P-1. The

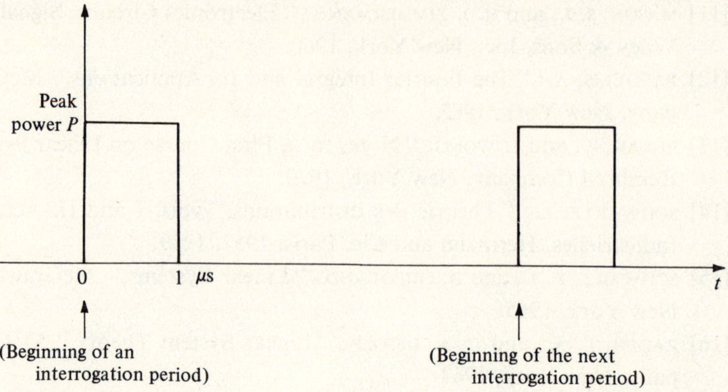

FIGURE 1P-1

transponder on board an airplane within the area covered by the ground control station picks up the pulse and responds by transmitting a sequence of pulses containing information on the identity of the plane, its flight number, its altitude, etc.

(*a*) Suppose that the power in the transmitted signal is uniform in all directions. Sketch the transmitted signal (in terms of peak power) at a distance r from the ground control station.

(*b*) Suppose that the airborne transponder is designed to sense an interrogation pulse of peak power equal to or larger than $P \times 10^{-4}$ per square mile. Find the maximum distance from the ground control station an airplane can respond to an interrogation.

(*c*) Suppose that a transponder transmits a sequence of responding pulses 5 μs after it senses an interrogation pulse. Suggest a way for the ground control station to determine the distance between the airplane and ground control station.

(*d*) Suppose that the durations of responding pulse sequences are always shorter than 10 μs. Find the shortest time interval between successive interrogation pulses such that there will be no confusion between the responses of airplanes to different interrogation pulses.

1-5 Let $x(n)$ be a discrete signal such that (1) $x(n) = 0$ for $n < 0$ and $n > 11$; (2) $x(n)$ is nondecreasing within the interval $0 \le n \le 11$. [That is, $x(0) \le x(1) \le x(2) \le \cdots \le x(11)$.]

Let

$$y(n) = x(n+8) + x(-n+8)$$

(*a*) Show that $y(n)$ is an even signal.

(*b*) Show that $y(n) = 0$ for $n < -8$ and $n > 8$.

(*c*) Given that

n	−8	0	1	2	3
$y(n)$	1	2	3	5	7

determine $x(n)$.

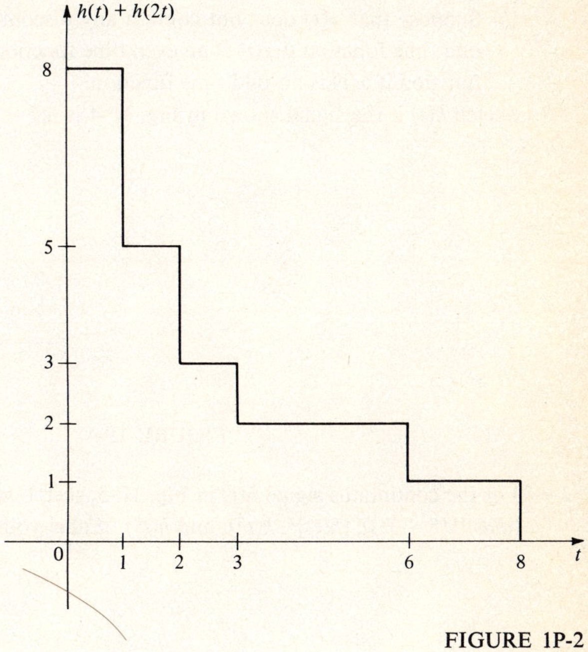

FIGURE 1P-2

1-6 Let $h(t)$ be a signal such that $h(t) = 0$ for $t < 0$ and $t > 8$, and $h(0+) = 4$. Given that $h(t) + h(2t)$ is as shown in Fig. 1P-2, determine $h(t)$.

1-7 (*a*) Sketch the derivatives of the time functions shown in Fig. 1P-3.

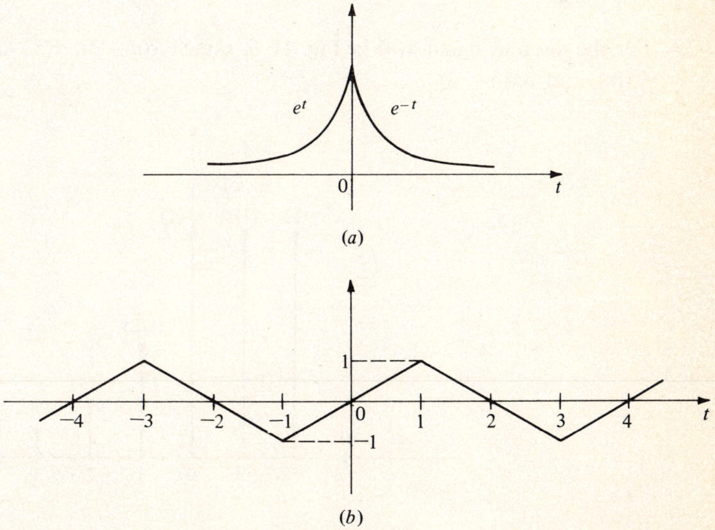

FIGURE 1P-3

(*b*) Suppose that $x(t)$ does not contain any discontinuities. Show that $dx(t)/dt$ is an odd time function if $x(t)$ is an even time function and that $dx(t)/dt$ is an even time function if $x(t)$ is an odd time function.

1-8 Sketch $h(t)$ if the signal shown in Fig. 1P-4 is $h(5-3t)$.

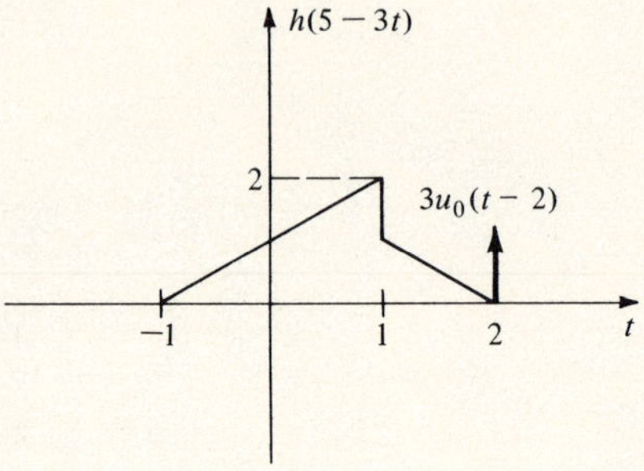

FIGURE 1P-4

1-9 For the continuous signal $h(t)$ in Fig. 1P-5, sketch $h(t-2)$, $\int_{-\infty}^{t} h(2-\tau)\,d\tau$, $h(3t+5)$, $(d/dt)[h(5-3t)]$, $[h(t)]^2$, $h(t^2)$, and $h(t^3)$. Label your sketches carefully.

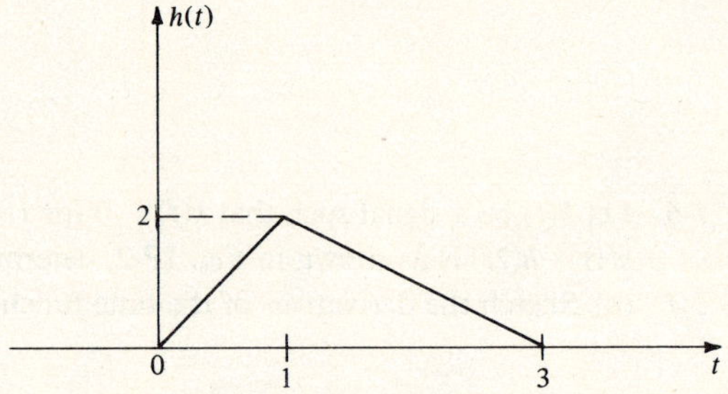

FIGURE 1P-5

1-10 For the discrete signal $x(n)$ in Fig. 1P-6, sketch $x(n-3)$, $x(5-n)$, $\sum_{k=-\infty}^{n} 4x(2-n)$, $\nabla x(n)$, and $\Delta x(n-3)$.

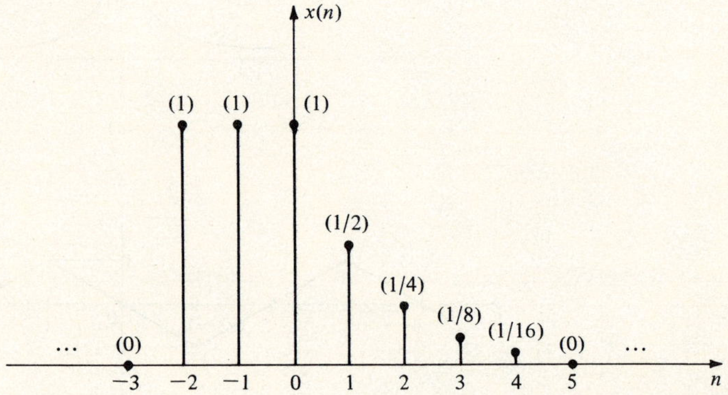

FIGURE 1P-6

1-11 Evaluate the following expressions:

(*a*) $\int_{-10}^{10} u_0(2t-3)(2t^2+t-5)\,dt$

(*b*) $\int_{-10}^{10} u_1(t+\frac{1}{4})(2t^2+t-5)\,dt$

1-12 Let $\phi(t)$ be a testing function.

(*a*) Express $\langle u_{-1}(t-2), \phi(t)\rangle$ as an integral.

(*b*) Evaluate $\langle u_{-1}(-t), \phi(t)\rangle$.

1-13 Consider a continuous signal $x(t)$ whose first k derivatives are continuous at $t=0$. Let $x^{(k)}(t)$ denote the kth derivative of $x(t)$.

(*a*) Show that

$$x(t)u_2(t) = x(0)u_2(t) - 2x^{(1)}(0)u_1(t) - x^{(2)}(0)u_0(t)$$

and

$$x(t)u_3(t) = x(0)u_3(t) - 3x^{(1)}(0)u_2(t) - 3x^{(2)}(0)u_1(t) - x^{(3)}(0)u_0(t)$$

(*b*) Show that for $k>0$

$$x(t)u_k(t) = x(0)u_k(t) - \binom{k}{1}x^{(1)}(0)u_{k-1}(t) - \binom{k}{2}x^{(2)}(0)u_{k-2}(t) - \cdots$$
$$- \binom{k}{k-1}x^{(k-1)}(0)u_1(t) - x^{(k)}(0)u_0(t)$$

1-14 In Sec. 1-6, the unit impulse $u_0(t)$ is defined as a distribution such that

$$\langle u_0(t), \phi(t)\rangle = \int_{-\infty}^{\infty} u_0(t)\phi(t)\,dt = \phi(0)$$

for all testing functions $\phi(t)$. The unit impulse can also be considered a limit of a sequence of ordinary functions $f_\delta(t)$ such that

$$\lim_{\delta\to 0}\int_{-\infty}^{\infty} f_\delta(t)\phi(t)\,dt = \phi(0)$$

for all testing functions $\phi(t)$.

(*a*) As was mentioned in Sec. 1-5, the unit impulse can be considered the limit of the rectangular function $g_\delta(t)$ shown in Fig. 1-17*b* as δ approaches 0. Show that

$$\lim_{\delta\to 0}\int_{-\infty}^{\infty} g_\delta(t)\phi(t)\,dt = \phi(0)$$

(*b*) Show that the unit impulse can also be considered a limit of $x_\delta(t)$, $y_\delta(t)$, or $z_\delta(t)$ shown in Fig. 1P-7 as $\delta \to 0$.

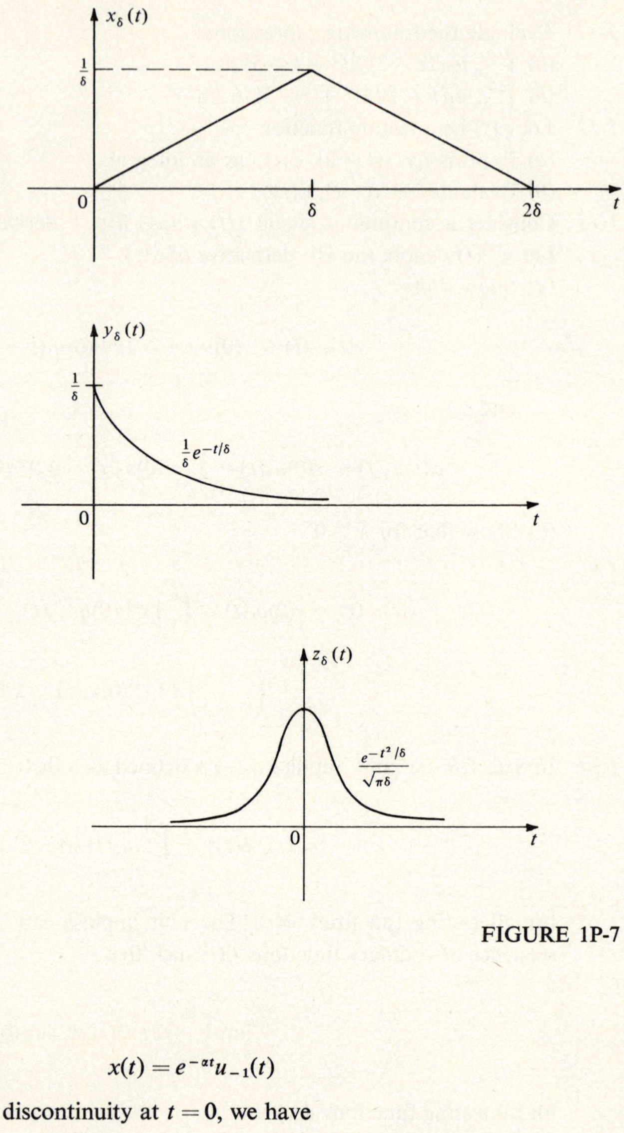

FIGURE 1P-7

1-15 Consider a time signal

$$x(t) = e^{-\alpha t}u_{-1}(t)$$

Because $x(t)$ contains a discontinuity at $t = 0$, we have

$$\frac{dx(t)}{dt} = u_0(t) - \alpha e^{-\alpha t}u_{-1}(t) \tag{1P-1}$$

In this problem, we shall verify this result in terms of the theory of distributions.

(*a*) Express $\langle x(t), \phi(t)\rangle$ as an integral.

(*b*) Show that

$$\left\langle \frac{dx(t)}{dt}, \phi(t) \right\rangle = \phi(0) - \alpha\langle x(t), \phi(t)\rangle$$

and thus verify the result in Eq. (1P-1).

1-16 Let $x(n)$ and $y(n)$ be the input and output signals of a discrete linear system. Given that

$$x(n) = \begin{cases} 1 & n = k \\ 0 & n \neq k \end{cases}$$

and

$$y(n) = \begin{cases} |k| & n = k \\ |k+1| & n = k+1 \\ 0 & \text{otherwise} \end{cases}$$

for all k:

(*a*) Find the output signal $y(n)$ corresponding to the input signal

$$x(n) = \begin{cases} -2 & n = -1 \\ 2 & n = 0 \\ -2 & n = 1 \\ 0 & \text{otherwise} \end{cases}$$

(*b*) Given that the output signal is

$$y(n) = \begin{cases} 3 & n = -1 \\ 0 & \text{otherwise} \end{cases}$$

and $x(n) = 0$ for $n < 0$, determine $x(0)$, $x(1)$, $x(2)$, $x(3)$, and $x(4)$.

1-17 Let

$$x_1(n) = \begin{cases} 1 & n = 0 \\ 2 & n = 1 \\ 0 & \text{otherwise} \end{cases}$$

be the input signal to a linear time-invariant system and

$$y_1(n) = \begin{cases} 1 & n = 0 \\ 0 & \text{otherwise} \end{cases}$$

be its corresponding output signal. Determine the output signal $y_2(n)$, given that $y_2(0) = \frac{1}{2}$, and the input signal is

$$x_2(n) = \begin{cases} 1 & n = 0 \\ 0 & \text{otherwise} \end{cases}$$

Note that it is not known whether the system is causal.

1-18 The response of a system to an input signal $x(t)$ is $x(-t)$. Is the system linear? Is it time-invariant? Is it causal? Explain your answers.

1-19 Consider a linear system whose response to an input signal $x(t)$ is $x(t + 5)$.

(*a*) Is the system time-invariant? Is it dynamical? Is it causal? Explain.

(*b*) Find the response of the system to $x(t^2)$ in terms of its response to $x(t)$.

1-20 The input to a causal linear time-invariant system, $x(t)$, and its corresponding output $y(t)$ are shown in Fig. 1P-8. Determine the step response of the system.

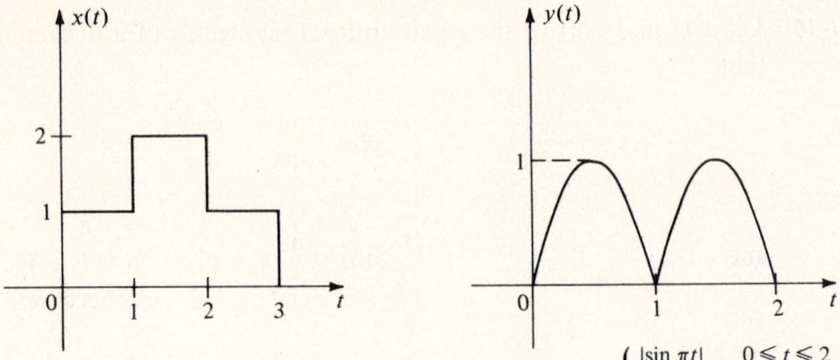

$$y(t) = \begin{cases} |\sin \pi t| & 0 \leqslant t \leqslant 2 \\ 0 & \text{otherwise} \end{cases}$$

FIGURE 1P-8

1-21 The response of a linear time-varying system at time t to an impulse applied at time τ is $e^{-(t+\tau)}u_{-1}(t-\tau)$. What is the response of this system when the input is $u_0(t) + 3u_0(t+1) + 3u_0(t+2)$?

1-22 Consider your instructor in this course as a discrete system whose input is a sequence of questions from the class and whose output is a sequence of answers. Is the system instantaneous? Causal? (Casual?) Time-invariant? Does this problem lead you to think about introducing additional ways to characterize systems? State some of them. (You may wish to submit the solution of this problem anonymously.)

1-23 Consider the internal combustion engine of an automobile as a continuous system whose input is the angular position of the gas pedal and whose output is the speed of the car. Discuss whether the system is instantaneous, causal, time-invariant, or linear. You may make some simple but realistic assumptions on the performance of a 1975 automobile.

1-24 The output signal $y(n)$ of a discrete system corresponding to an input signal $x(n)$ is

$$y(n) = \begin{cases} 0 & \text{if } \sum_{i=-\infty}^{n} x(i) < 100 \\ 1 & \text{if } \sum_{i=-\infty}^{n} x(i) \geq 100 \end{cases}$$

Is the system memoryless? Stable in the bounded-input–bounded-output sense? Time-invariant? Linear?

1-25 Consider a system whose response to a stimulus $x(t)$ is $\sqrt{[x(t)]^2 + [dx(t)/dt]^2}$. Is the system time-invariant? Homogeneous? Justify your answers.

1-26 In this problem, we shall show that an additive system is "almost" a homogeneous system. Let $H[x(t)]$ denote the responses of an additive system to a stimulus $x(t)$.

(*a*) Show that $H[ax(t)] = aH[x(t)]$ for any positive integer a.

(*b*) Show that $H[(a/b)x(t)] = (a/b)H[x(t)]$ for any positive integers a and b.

(*c*) Show that $H[0] = 0$.

(*d*) Show that $H[-x(t)] = -H[x(t)]$ and $H[-ax(t)] = -aH[x(t)]$ for any positive integer a.

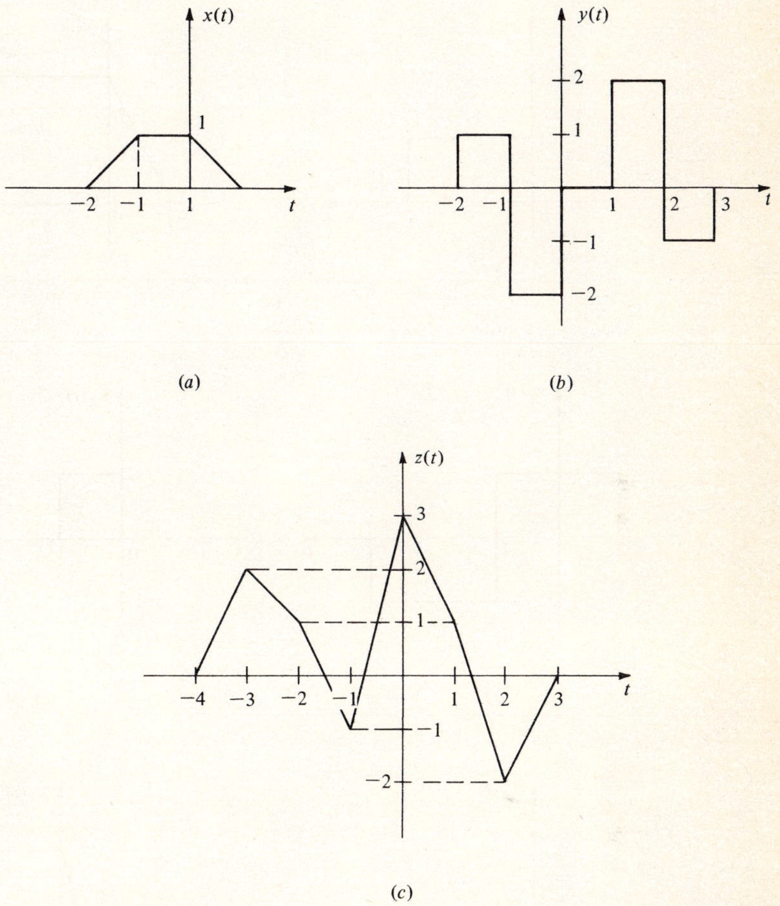

FIGURE 1P-9

1-27 Let $y(t)$ in Fig. 1P-9*b* be the output signal of a linear time-invariant system corresponding to the input signal $x(t)$ in Fig. 1P-9*a*.

(*a*) Sketch the output signal when the input signal is $2x(t+4)$.

(*b*) Sketch the output signal when the input signal is $z(t)$ in Fig. 1P-9*c*. Note that $z(t)$ can be expressed as

$$z(t) = \sum_{n=-\infty}^{\infty} z_n x(t-n)$$

for appropriately chosen constants z_n.

1-28 The signals $x(t)$ and $y(t)$ in Fig. 1P-10*a* are a pair of input and output signals of a linear time-invariant system.

(*a*) Sketch the output signal $y_1(t)$ corresponding to the input signal $x_1(t)$ in Fig. 1P-10*b*.

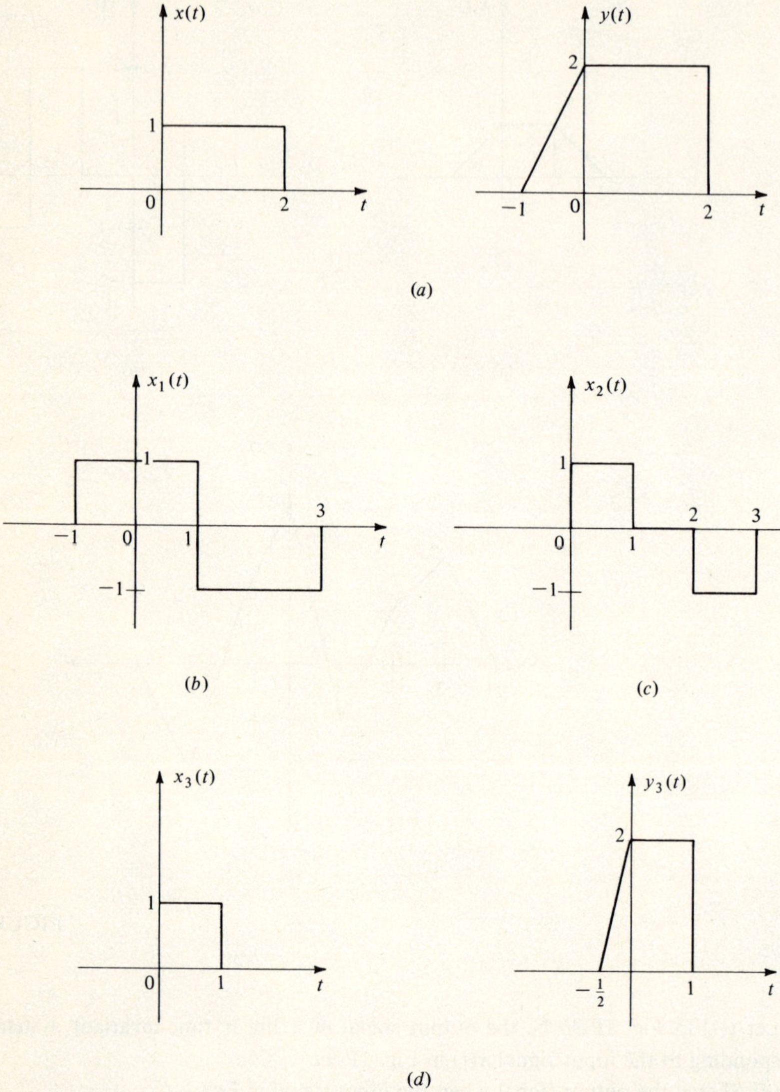

FIGURE 1P-10

(*b*) Sketch the output signal $y_2(t)$ corresponding to the input signal $x_2(t)$ in Fig. 1P-10*c*.

(*c*) A reader claims that for the input signal $x_3(t)$ shown in Fig. 1P-10*d* the corresponding output signal is $y_3(t)$, using the argument that if $x_3(t) = x(t/2)$ then $y_3(t) = y(t/2)$. Prove or disprove this claim.
[*Hint:* Use the result in part (*b*).]

1-29 Let A in Fig. 1P-11*a* denote a time-invariant linear system. Given that $x_1(t) = u_{-1}(t)$ and $y_1(t) = u_{-1}(t) - 2u_{-1}(t-1) + u_{-1}(t-2)$, determine $y_2(t)$ in the cascade system shown in Fig. 1P-11*b* when $x_2(t)$ is as shown in Fig. 1P-11*c*.

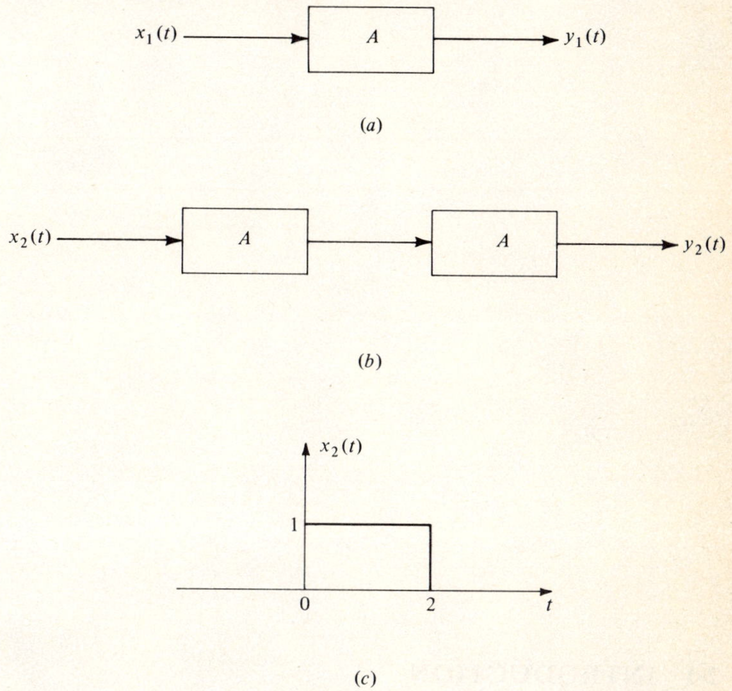

FIGURE 1P-11

2

DIFFERENTIAL AND DIFFERENCE EQUATIONS

2-1 INTRODUCTION

One of the major topics to be studied in this book is the input-output relationship of systems. In a most general way, the input-output relationship of a system can always be described by an exhaustive list consisting of all possible input signals to the system and their corresponding output signals. Such generality is clearly quite useless. An exhaustive listing is impossible to generate, and even if it could be generated, the problem of determining the output signal corresponding to a given input signal would amount to searching for a particular entry in an infinitely long list. Therefore, we shall limit our discussion to systems with certain regularities in their behavior so that their input-output relationship can be described in simple and compact ways.

If the input-output relationship of a system can be described by an equation (or a set of equations), then for a given input signal the equation (or the set of equations) can be solved to determine the corresponding output signal. For example, suppose that the input signal $x(n)$ and the output signal $y(n)$ of a discrete system are related by the equation

$$y(n) = x(n+1) - x(n) + x(n-1) + 5 \tag{2-1}$$

Then, given that

$$x(n) = \begin{cases} 0 & n < 2 \\ 2^{-n} & n \geq 2 \end{cases}$$

we obtain, from Eq. (2-1),

$$y(n) = \begin{cases} 5 & n < 1 \\ 5\frac{1}{4} & n = 1 \\ 4\frac{7}{8} & n = 2 \\ 3(2^{-n-1}) + 5 & n > 2 \end{cases}$$

As another example, if the input-output relationship of a continuous system is described by the equation

$$y(t) = x(t)\frac{dx(t)}{dt} + \frac{d^2x(t)}{dt^2}$$

then for any given input signal $x(t)$ the corresponding output signal $y(t)$ can be computed readily.

These two examples are rather simple in that the output signal of a system is expressed explicitly as a function of the input signal. In the more general case, the output signal might be related to the input signal by an equation such as that shown in the examples below:

$$\frac{d^2y(t)}{dt^2} + y(t)\frac{dy(t)}{dt} + y^2(t) = \frac{dx(t)}{dt} + 5x(t) \tag{2-2}$$

$$y^3(n) + 2y(n-1) + 5y(n-2) = x(n) + x(n-1) \tag{2-3}$$

In this chapter, we shall study systems whose input-output relationship can be described by *differential equations* (continuous systems) or *difference equations* (discrete systems). For a continuous system, a differential equation description relates the values of the input signal $x(t)$ and its derivatives $dx(t)/dt$, $d^2x(t)/dt^2$, ... to the values of the output signal $y(t)$ and its derivatives $dy(t)/dt$, $d^2y(t)/dt^2$, ..., for all t, as in Eq. (2-2). For a discrete system, a difference equation description relates the successive values of the input signal, $x(n)$, $x(n-1)$, $x(n-2)$, ..., to the successive values of the output signal, $y(n)$, $y(n-1)$, $y(n-2)$, ..., for all n, as in Eq. (2-3). Given the differential (difference) equation describing the input-output relationship of a system, we can determine the output signal corresponding to a given input signal by solving the differential (difference) equation. Unfortunately, there is no general procedure for solving differential (difference) equations. We shall limit our discussion to systems whose input-output relationship can be described by a class of differential (difference) equations known as *linear differential (difference) equations with constant coefficients.*

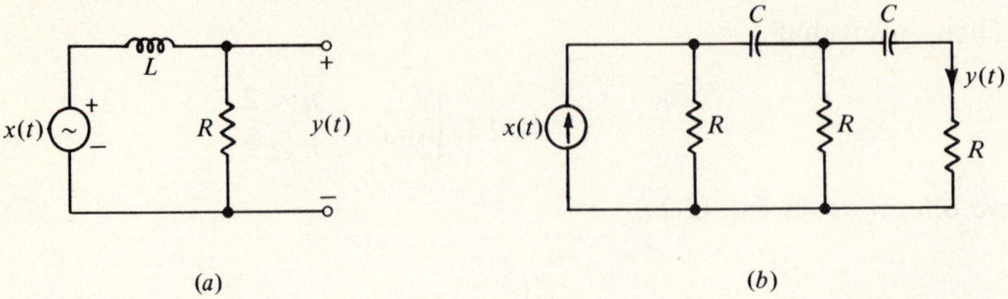

FIGURE 2-1

Not only is the solution method for this class of differential (difference) equations straightforward, but also they correspond to the class of systems in which we are most interested, namely, *linear time-invariant systems.*

Before discussing the method of solution for linear differential (difference) equations with constant coefficients, let us look at several examples of systems whose input-output relationship can be described by differential (difference) equations.

2-2 EXAMPLES OF SYSTEMS CHARACTERIZED BY DIFFERENTIAL AND DIFFERENCE EQUATIONS

A reader familiar with circuit theory may recall that the input-output relationship of an *RLC* network can be described by a differential equation. As an example, consider the circuit shown in Fig. 2-1*a*. The input signal [the voltage source $x(t)$] and the output signal [the voltage $y(t)$ across the resistor] are related by the differential equation

$$\frac{L}{R}\frac{dy(t)}{dt} + y(t) = x(t)$$

In the circuit in Fig. 2-1*b*, let the current source $x(t)$ be the input signal and the current $y(t)$ be the output signal. We have

$$3\frac{d^2y(t)}{dt^2} + \frac{4}{RC}\frac{dy(t)}{dt} + \frac{1}{R^2C^2}y(t) = \frac{d^2x(t)}{dt^2}$$

Just as in electric networks, the input-output relationship of many mechanical and electromechanical systems can also be described by differential equations. For example, for the system shown in Fig. 2-2 let the force applied to the mass, $x(t)$, be the input signal, and the displacement of the mass, $y(t)$, be the output signal. We have

$$M\frac{d^2y(t)}{dt^2} + \alpha\frac{dy(t)}{dt} + K[y(t) - y_0] = x(t)$$

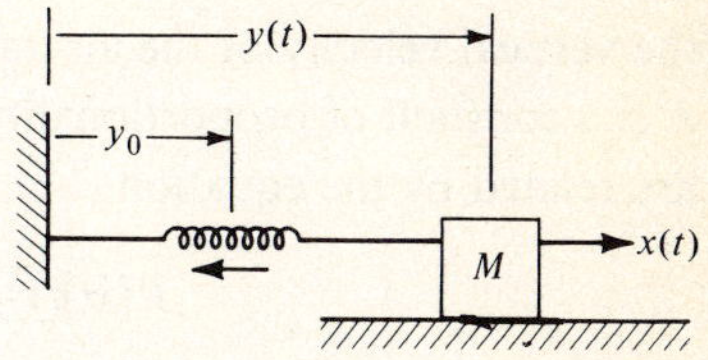

FIGURE 2-2 Frictional force = $\alpha \frac{dy(t)}{dt}$
Restoring force of the spring = $K[y(t) - y_0]$

where M is the weight of the mass, α is a proportional constant for the frictional force, K is a proportional constant for the restoring force of the spring, and y_0 is the initial displacement of the mass.

As another example, consider an automatic door control system which operates as follows: When a man steps on the doormat, a switch will be closed to allow a voltage $v(t)$ to excite the field of a dc servomotor. The field current $i(t)$ and the voltage $v(t)$ are related by the differential equation

$$v(t) = Ri(t) + L\frac{di(t)}{dt}$$

where R and L are the resistance and the inductance, respectively, of the field winding of the servomotor. The torque $\tau(t)$ generated by the servomotor is equal to $Ki(t)$, where K is a constant of proportionality. Let $\varphi(t)$ denote the angular position of the door. The torque $\tau(t)$ is balanced by the inertia of the door and the frictional force. The inertial force is proportional to $d^2\varphi(t)/dt^2$ (with constant of proportionality J), and the frictional force is proportional to $d\varphi(t)/dt$ (with constant of proportionality B). Thus, we have

$$\tau(t) = J\frac{d^2\varphi(t)}{dt^2} + B\frac{d\varphi(t)}{dt}$$

If we consider $v(t)$ and $\varphi(t)$ to be the input and output signals of the electromechanical system, respectively, the input-output relationship of the system is given by

$$LJ\frac{d^3\varphi(t)}{dt^3} + (RJ + LB)\frac{d^2\varphi(t)}{dt^2} + RB\frac{d\varphi(t)}{dt} = Kv(t)$$

We now present some examples of discrete systems, the input-output relationship of which can be described by difference equations. Consider an air-traffic control system in which the desired altitude of an aircraft, $x(n)$, is computed by a computer every second and is compared with the actual altitude of the aircraft, $y(n-1)$, determined by a tracking radar 1s previously. Depending on whether $x(n)$ is larger or smaller than $y(n-1)$, the altitude of the aircraft will be changed accordingly. Suppose that the vertical velocity of the aircraft at time n is equal to $K[x(n) - y(n-1)]$ ft/s, where K is a constant of proportionality. The input and output of the system, $x(n)$ and $y(n)$, are related by the equation

$$K[x(n) - y(n-1)] = y(n) - y(n-1)$$

That is,

$$y(n) + (K-1)y(n-1) = Kx(n)$$

As another example, let us model a savings bank by a discrete system with its input $x(n)$ being the total deposits in the nth month and its output $y(n)$ being the total assets at the end of the nth month. Suppose that the total assets of the bank are invested for a 105 percent monthly return. The input-output relationship of the system can be described by the difference equation

$$y(n) = x(n) + 1.05y(n-1)$$

We can also consider another model of the operation of the bank. Let us define the rate of growth of the bank in the nth month to be

$$\frac{y(n) - y(n-1)}{y(n-1)}$$

Suppose that the rate of growth in each month is equal to the sum of 5 percent of the increase in total deposits in that month and the rate of growth in the preceding month. We thus have

$$\frac{y(n) - y(n-1)}{y(n-1)} = 0.05[x(n) - x(n-1)] + \frac{y(n-1) - y(n-2)}{y(n-2)}$$

which simplifies to

$$y(n)y(n-2) - y^2(n-1) = 0.05y(n-1)y(n-2)x(n) - 0.05y(n-1)y(n-2)x(n-1)$$

Sometimes, the behavior of continuous systems can also be described conveniently by difference equations. For example, the input signal $v(t)$ and the output signal $i(t)$ of the circuit shown in Fig. 2-3a are related by the differential equation

$$\frac{di(t)}{dt} + i(t) = v(t) \tag{2-4}$$

Let $v(t)$ be the time signal shown in Fig. 2-3b. We solve Eq. (2-4) for each time interval $n \leq t \leq n+1$ for $n = 0, 1, 2, \ldots$ and obtain

$$i(t) = [i(n) - 2(-1)^n]e^{-(t-n)} + 2(-1)^n \tag{2-5}$$

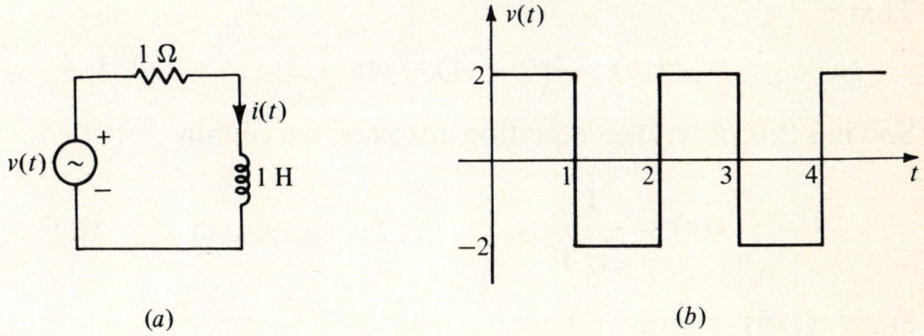

FIGURE 2-3

In order to determine an explicit expression for $i(t)$, we need to determine an explicit expression for $i(n)$. Setting $t = n + 1$ in Eq. (2-5), we obtain

$$i(n+1) - i(n)e^{-1} = 2(1 - e^{-1})(-1)^n \tag{2-6}$$

As will be seen later in this chapter, Eq. (2-6) can be solved for $i(n)$:

$$i(n) = \frac{2(1 - e^{-1})}{1 + e^{-1}}\,[e^{-n} - (-1)^n]$$

assuming that $i(0) = 0$. Thus, from Eq. (2-5), we obtain

$$i(t) = 2\left[\frac{1 - e^{-1}}{1 + e^{-1}}\,e^{-n} - \frac{2}{1 + e^{-1}}\,(-1)^n\right]e^{-(t-n)} + 2(-1)^n$$

As another example, suppose we want to determine the current $y(t)$ in the circuit in Fig. 2-4 in terms of the voltage source $x(t)$. We have

$$v(0) = y(t)$$
$$v(1) = 3y(t)$$

Moreover, at the $(n-1)$st node for $n = 2, 3, \ldots, 100$, we have the equation

$$\tfrac{1}{2}[v(n) - v(n-1)] = v(n-1) + \tfrac{1}{2}[v(n-1) - v(n-2)]$$

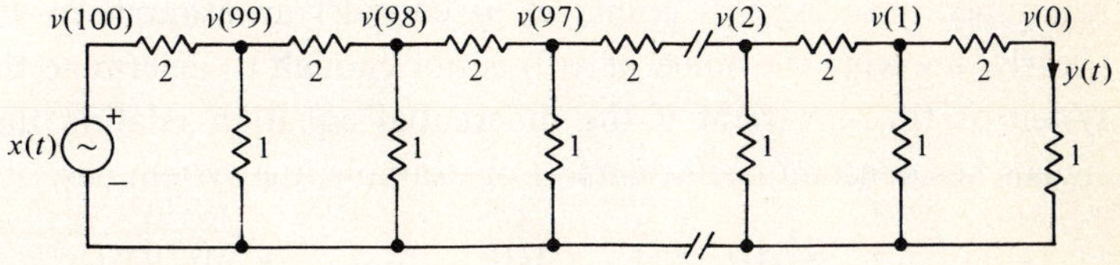

FIGURE 2-4

That is,

$$v(n) = 4v(n-1) - v(n-2) \qquad n = 2, 3, 4, \ldots, 100$$

Solving this difference equation for $v(n)$, we obtain

$$v(n) = \frac{1}{2\sqrt{3}}[(1+\sqrt{3})(2+\sqrt{3})^n - (1-\sqrt{3})(2-\sqrt{3})^n]y(t)$$

Since $v(100) = x(t)$, we have

$$x(t) = \frac{1}{2\sqrt{3}}[(1+\sqrt{3})(2+\sqrt{3})^{100} - (1-\sqrt{3})(2-\sqrt{3})^{100}]y(t)$$

or

$$y(t) = \frac{2\sqrt{3}}{[(1+\sqrt{3})(2+\sqrt{3})^{100} - (1-\sqrt{3})(2-\sqrt{3})^{100}]}x(t)$$

It should also be pointed out, however, that there are systems the input-output relationship of which cannot be described by a differential equation or a difference equation. Consider a discrete quantizer whose output $y(n)$ at any instant n is equal to the integral part of the input $x(n)$ if the fractional part of $x(n)$ is less than 0.5, and is equal to 1 plus the integral part of $x(n)$ if the fractional part of $x(n)$ is larger than or equal to 0.5. Clearly, the input-output relationship of such a system cannot be described by a difference equation or a differential equation.

2-3 DIFFERENTIAL EQUATION DESCRIPTION OF CONTINUOUS SYSTEMS

As was pointed out above, when the input-output relationship of a continuous system can be described by a differential equation, the output corresponding to a given input can be determined by solving such a differential equation. Furthermore, many of the properties of the system can be deduced directly from the differential equation. A system characterized by a differential equation is dynamical, because the value of the output $y(t)$ at any instant t depends not only on the value of $x(t)$ at that instant but also on the values of the derivatives of $y(t)$ and $x(t)$ at that instant which, in turn, are dependent on the values of $y(t)$ and $x(t)$ at other instants. [If the reader has difficulty seeing this point, let us consider an example in which $dy(t)/dt = x(t)$. Clearly, knowing the value of $x(0)$ is not enough to determine the value of $y(0)$.] A system is time-invariant if the differential equation relating the input and output signals has constant coefficients. For example, the system described by the equation

$$\frac{d^2y(t)}{dt^2} + 7y(t)\frac{dy(t)}{dt} + 5y^2(t) = 2\frac{d^2x(t)}{dt^2}\frac{dx(t)}{dt} - 5x(t)$$

is time-invariant, whereas the system described by the equation

$$5t^2 \frac{d^2y(t)}{dt^2} + 6 \frac{dy(t)}{dt} + ty(t) = \frac{dx(t)}{dt} + x(t) \tag{2-7}$$

is time-varying. Such an assertion is not difficult to see; we shall illustrate the point by a simple argument. Let $x(t)$ be an input signal to the system described by Eq. (2-7). Let $y(t)$ be the corresponding output signal. Suppose, at $t = 2$, $x(t) = 40$, $dx(t)/dt = 34$, $y(t) = 1$, $dy(t)/dt = 2$, $d^2y(t)/dt^2 = 3$. Clearly, these values satisfy Eq. (2-7) as can be verified immediately:

$$5(2)^2(3) + 6(2) + 2(1) = 40 + 34$$

Consider now $x(t - 5)$ as the input signal. It is clear that $y(t - 5)$ cannot possibly be the corresponding output signal because the values $x(t) = 40$, $dx(t)/dt = 34$, $y(t) = 1$, $dy(t)/dt = 2$, $d^2y(t)/dt^2 = 3$ no longer satisfy Eq. (2-7) for $t = 7$.

Our attention will be restricted to systems described by *linear differential equations with constant coefficients.* The general form of a linear differential equation with constant coefficients is

$$C_0 \frac{d^ry(t)}{dt^r} + C_1 \frac{d^{r-1}y(t)}{dt^{r-1}} + \cdots + C_{r-1} \frac{dy(t)}{dt} + C_r y(t)$$
$$= E_0 \frac{d^mx(t)}{dt^m} + E_1 \frac{d^{m-1}x(t)}{dt^{m-1}} + \cdots + E_{m-1} \frac{dx(t)}{dt} + E_m x(t) \tag{2-8}$$

where the C's and the E's are constants. The solution of this class of differential equations is quite straightforward. We assume that the reader has studied the subject and shall give only a brief treatment here.

The (total) solution of a linear differential equation with constant coefficients consists of two parts: the *homogeneous solution* and the *particular solution.* The homogeneous solution is a solution of Eq. (2-8) when $x(t)$ and all its derivatives are set to zero. That is, the homogeneous solution satisfies the equation

$$C_0 \frac{d^ry(t)}{dt^r} + C_1 \frac{d^{r-1}y(t)}{dt^{r-1}} + \cdots + C_{r-1} \frac{dy(t)}{dt} + C_r y(t) = 0 \tag{2-9}$$

The homogeneous solution consists of terms of the form $Ae^{\alpha t}$, where A is an arbitrary complex constant and α is a complex constant determined by the given differential equation. Substituting $y(t) = Ae^{\alpha t}$ into Eq. (2-9), we obtain

$$C_0 A\alpha^r e^{\alpha t} + C_1 A\alpha^{r-1} e^{\alpha t} + \cdots + C_{r-1} A\alpha e^{\alpha t} + C_r Ae^{\alpha t} = 0$$

which simplifies to

$$C_0 \alpha^r + C_1 \alpha^{r-1} + \cdots + C_{r-1}\alpha + C_r = 0 \tag{2-10}$$

Indeed, if α is a root of Eq. (2-10), $y(t) = Ae^{\alpha t}$ will satisfy Eq. (2-9). Equation (2-10) is called the *characteristic equation* of the differential equation (2-8). The roots of the characteristic equation, $\alpha_1, \alpha_2, \ldots, \alpha_r$, are called the *characteristic roots* of the differential equation.

When the characteristic roots of the differential equation are all distinct, the homogeneous solution will be

$$A_1 e^{\alpha_1 t} + A_2 e^{\alpha_2 t} + \cdots + A_r e^{\alpha_r t}$$

where $A_1, A_2, \ldots, A_r$ are constants to be determined by some known *boundary conditions*. We shall discuss how to determine these constants later. As an example, let us consider the differential equation

$$\frac{d^2y(t)}{dt^2} + 3\frac{dy(t)}{dt} + 2y(t) = x(t)$$

The characteristic equation is

$$\alpha^2 + 3\alpha + 2 = 0$$

The characteristic roots are

$$\alpha_1 = -1 \qquad \alpha_2 = -2$$

It follows that the homogeneous solution is

$$A_1 e^{-t} + A_2 e^{-2t}$$

When the characteristic roots are not all distinct the homogeneous solution will be of a slightly different form. To be specific, suppose that α_1 is a k-multiple root of the characteristic equation. Then, corresponding to α_1 there will be k terms in the homogeneous solution:

$$A_1 t^{k-1} e^{\alpha_1 t} + A_2 t^{k-2} e^{\alpha_1 t} + \cdots + A_{k-1} t e^{\alpha_1 t} + A_k e^{\alpha_1 t}$$

Clearly, the term $A_k e^{\alpha_1 t}$ satisfies Eq. (2-9). To show that the terms $A_{k-1} t e^{\alpha_1 t}, \ldots, A_2 t^{k-2} e^{\alpha_1 t}, A_1 t^{k-1} e^{\alpha_1 t}$ also satisfy Eq. (2-9), we note that because α_1 is a k-multiple root of the characteristic equation, it satisfies not only the characteristic equation but also its first, second, ..., and $(k-1)$st derivatives. That is, in addition to

$$C_0 \alpha_1^r + C_1 \alpha_1^{r-1} + \cdots + C_{r-1}\alpha_1 + C_r = 0 \tag{2-11}$$

we also have

$$C_0 r \alpha_1^{r-1} + C_1 (r-1)\alpha_1^{r-2} + \cdots + C_{r-1} = 0 \tag{2-12}$$

$$C_0 r(r-1)\alpha_1^{r-2} + C_1(r-1)(r-2)\alpha_1^{r-3} + \cdots + 2C_{r-2} = 0$$

$$\cdots\cdots\cdots\cdots\cdots\cdots\cdots\cdots\cdots\cdots$$

$$C_0 r(r-1)\cdots(r-k+2)\alpha_1^{r-k+1} + C_1(r-1)(r-2)\cdots(r-k+1)\alpha_1^{r-k} + \cdots + (k-1)!\,C_{r-k+1} = 0$$

Let us substitute $y(t) = A_{k-1}te^{\alpha_1 t}$ into the left-hand side of (2-9). Since

$$\frac{d}{dt}(A_{k-1}te^{\alpha_1 t}) = A_{k-1}(\alpha_1 te^{\alpha_1 t} + e^{\alpha_1 t})$$

$$\frac{d^2}{dt^2}(A_{k-1}te^{\alpha_1 t}) = A_{k-1}(\alpha_1{}^2 te^{\alpha_1 t} + 2\alpha_1 e^{\alpha_1 t})$$

$$\cdots\cdots\cdots\cdots\cdots\cdots\cdots\cdots\cdots\cdots\cdots\cdots$$

$$\frac{d^r}{dt^r}(A_{k-1}te^{\alpha_1 t}) = A_{k-1}(\alpha_1{}^r te^{\alpha_1 t} + r\alpha_1^{r-1}e^{\alpha_1 t})$$

we obtain

$$\begin{aligned} C_0\frac{d^r y(t)}{dt^r} &+ C_1\frac{d^{r-1}y(t)}{dt^{r-1}} + \cdots + C_{r-1}\frac{dy(t)}{dt} + C_r y(t) \\ &= A_{k-1}(C_0\alpha_1{}^r + C_1\alpha_1^{r-1} + \cdots + C_{r-1}\alpha_1 + C_r)te^{\alpha_1 t} \\ &\quad + A_{k-1}(C_0 r\alpha_1^{r-1} + C_1(r-1)\alpha_1^{r-2} + \cdots + C_{r-1})e^{\alpha_1 t} \end{aligned}$$

That this expression equals zero follows from Eqs. (2-11) and (2-12). Therefore, $A_{k-1}te^{\alpha_1 t}$ satisfies Eq. (2-9). That the other terms $A_1 t^{k-1}e^{\alpha_1 t}$, $A_2 t^{k-2}e^{\alpha_1 t}$, ..., $A_{k-2}t^2 e^{\alpha_1 t}$ also satisfy Eq. (2-9) can be shown in a similar manner. As an example, consider the differential equation

$$\frac{d^3y(t)}{dt^3} + 7\frac{d^2y(t)}{dt^2} + 16\frac{dy(t)}{dt} + 12y(t) = x(t)$$

The characteristic equation is

$$\alpha^3 + 7\alpha^2 + 16\alpha + 12 = 0$$

which can be written

$$(\alpha + 2)^2(\alpha + 3) = 0$$

Thus, the homogeneous solution is

$$A_1 te^{-2t} + A_2 e^{-2t} + A_3 e^{-3t}$$

We observe that the general form of the homogeneous solution depends only on the differential equation characterizing the system and does not depend on the input signal $x(t)$. For this reason, the homogeneous solution is often called the *natural response* (or *free response*) of the system. It should be pointed out, however, that the values of the constants A_1, A_2, ..., A_r do depend on the input $x(t)$, as will be seen later.

The particular solution depends explicitly on $x(t)$ and satisfies Eq. (2-8). Hence, the particular solution is called the *forced response* of the system. In simple cases, the particular solution can be obtained by inspection which we shall discuss briefly here.

In Chap. 11, a method for determining both the homogeneous solution and the particular solution at the same time will be presented. We illustrate the inspection procedure by an example. Suppose we are to determine the particular solution of the differential equation

$$\frac{d^2y(t)}{dt^2} + 2\frac{dy(t)}{dt} + 3y(t) = \frac{dx(t)}{dt} + x(t) \tag{2-13}$$

for $x(t) = t^2$. We guess that the particular solution is of the form

$$y(t) = P_1 t^2 + P_2 t + P_3$$

where P_1, P_2, P_3 are constants. Substituting the expressions of $x(t)$ and $y(t)$ into Eq. (2-13), we obtain

$$3P_1 t^2 + (4P_1 + 3P_2)t + (2P_1 + 2P_2 + 3P_3) = t^2 + 2t$$

Matching the coefficients of the powers of t on both sides, we obtain

$$3P_1 = 1$$
$$4P_1 + 3P_2 = 2$$
$$2P_1 + 2P_2 + 3P_3 = 0$$

These equations are solved to yield

$$P_1 = \tfrac{1}{3} \qquad P_2 = \tfrac{2}{9} \qquad P_3 = -\tfrac{10}{27}$$

For the same differential equation (2-13), suppose that $x(t) = e^t$. We guess that the particular solution is of the form

$$y(t) = Pe^t$$

Table 2-1

$x(t)$	Particular solution
t^n	$P_1 t^n + P_2 t^{n-1} + \cdots + P_n t + P_{n+1}$
e^{at}	Pe^{at} if a is not a characteristic root of the differential equation $P_1 te^{at} + P_2 e^{at}$ if a is a distinct characteristic root of the differential equation $P_1 t^{k-1} e^{at} + P_2 t^{k-2} e^{at} + P_3 t^{k-3} e^{at} + \cdots + P_k e^{at}$ if a is a $(k-1)$-multiple characteristic root of the differential equation
$\cos at$	$P_1 \cos at + P_2 \sin at$
$\sin at$	$P_1 \cos at + P_2 \sin at$

We leave it to the reader to show that $P = \frac{1}{3}$ by substituting the expressions of $x(t)$ and $y(t)$ into Eq. (2-13) and matching the coefficients of corresponding terms.

In summary, to determine the particular solution we guess first its general form and then determine the multiplying constants in the general form by matching coefficients. We list in Table 2-1 the general form of the particular solutions corresponding to some $x(t)$.

Finally, we should determine the constants $A_1, A_2, \ldots, A_r$ in the homogeneous solution. [Clearly, once these constants are determined, the total solution $y(t)$ is completely determined.] These constants can be determined by a set of boundary conditions. For an rth-order differential equation, the values of $y(t_0)$, $dy(t_0)/dt, \ldots,$ $d^{r-1}y(t_0)/dt^{r-1}$ for any t_0 are sufficient to determine the values of $A_1, A_2, \ldots, A_r$. Let us consider first the case in which the characteristic roots $\alpha_1, \alpha_2, \ldots, \alpha_r$ are all distinct. The total solution is of the form

$$A_1 e^{\alpha_1 t} + A_2 e^{\alpha_2 t} + \cdots + A_r e^{\alpha_r t} + p(t)$$

where $p(t)$ denotes the particular solution. The coefficients $A_1, A_2, \ldots, A_r$ can be determined by solving the set of simultaneous equations:

$$y(t_0) = A_1 e^{\alpha_1 t_0} + A_2 e^{\alpha_2 t_0} + \cdots + A_r e^{\alpha_r t_0} + p(t_0)$$

$$\frac{dy(t_0)}{dt} = A_1\alpha_1 e^{\alpha_1 t_0} + A_2\alpha_2 e^{\alpha_2 t_0} + \cdots + A_r\alpha_r e^{\alpha_r t_0} + \frac{dp(t_0)}{dt}$$

$$\cdots\cdots\cdots\cdots\cdots\cdots\cdots\cdots\cdots\cdots\cdots\cdots$$

$$\frac{d^{r-1}y(t_0)}{dt^{r-1}} = A_1\alpha_1^{r-1} e^{\alpha_1 t_0} + A_2\alpha_2^{r-1} e^{\alpha_2 t_0} + \cdots + A_r\alpha_r^{r-1} e^{\alpha_r t_0} + \frac{d^{r-1}p(t_0)}{dt^{r-1}}$$

According to the theory of simultaneous linear equations, the unknowns $A_1, A_2, \ldots,$ A_r can be solved uniquely if and only if the value of the determinant

$$\begin{vmatrix} e^{\alpha_1 t_0} & e^{\alpha_2 t_0} & \cdots & e^{\alpha_r t_0} \\ \alpha_1 e^{\alpha_1 t_0} & \alpha_2 e^{\alpha_2 t_0} & \cdots & \alpha_r e^{\alpha_r t_0} \\ \cdots & \cdots & \cdots & \cdots \\ \alpha_1^{r-1} e^{\alpha_1 t_0} & \alpha_2^{r-1} e^{\alpha_2 t_0} & \cdots & \alpha_r^{r-1} e^{\alpha_r t_0} \end{vmatrix} = e^{(\alpha_1+\alpha_2+\cdots+\alpha_r)t_0} \begin{vmatrix} 1 & 1 & \cdots & 1 \\ \alpha_1 & \alpha_2 & \cdots & \alpha_r \\ \alpha_1^2 & \alpha_2^2 & \cdots & \alpha_r^2 \\ \cdots & \cdots & \cdots & \cdots \\ \alpha_1^{r-1} & \alpha_2^{r-1} & \cdots & \alpha_r^{r-1} \end{vmatrix}$$

is nonzero. The determinant in the second expression is the well-known Vandermonde determinant whose value is equal to

$$(\alpha_1 - \alpha_2)(\alpha_1 - \alpha_3)\cdots(\alpha_1 - \alpha_r)(\alpha_2 - \alpha_3)(\alpha_2 - \alpha_4)\cdots(\alpha_2 - \alpha_r)\cdots(\alpha_{r-1} - \alpha_r)$$

$$= \prod_{\substack{i<j \\ 1\le i<r \\ 1<j\le r}} (\alpha_i - \alpha_j)$$

Since $\alpha_1, \alpha_2, \ldots, \alpha_r$ are distinct, the product of their differences is nonzero. Consequently, the unknowns $A_1, A_2, \ldots, A_r$ can be determined uniquely. A similar argument showing that the coefficients $A_1, A_2, \ldots, A_r$ can be determined uniquely from the boundary conditions can be made when some of the characteristic roots are multiple roots. We shall leave the details to the reader. (See Prob. 2-12.)

To complete our discussion, we show now that the total solution so determined is unique. That is, we want to show that the solution we found is not just one of several possible solutions to the differential equation. From a physical point of view, we certainly expect that a system responds to a given input signal with a unique output signal. (The physical significance of the boundary conditions is not obvious at this moment. It will be discussed in Chap. 3.) To show the uniqueness of the total solution, let us assume the contrary, that is, $y_1(t)$ and $y_2(t)$ are both total solutions of Eq. (2-8) which satisfy a given set of boundary conditions $y(t_0), dy(t_0)/dt, d^2y(t_0)/dt^2, \ldots, d^{r-1}y(t_0)/dt^{r-1}$. Let

$$z(t) = y_1(t) - y_2(t)$$

Since

$$C_0 \frac{d^r y_1(t)}{dt^r} + C_1 \frac{d^{r-1} y_1(t)}{dt^{r-1}} + \cdots + C_{r-1} \frac{dy_1(t)}{dt} + C_r y_1(t)$$
$$= E_0 \frac{d^m x(t)}{dt^m} + E_1 \frac{d^{m-1} x(t)}{dt^{m-1}} + \cdots + E_{m-1} \frac{dx(t)}{dt} + E_m x(t)$$

and

$$C_0 \frac{d^r y_2(t)}{dt^r} + C_1 \frac{d^{r-1} y_2(t)}{dt^{r-1}} + \cdots + C_{r-1} \frac{dy_2(t)}{dt} + C_r y_2(t)$$
$$= E_0 \frac{d^m x(t)}{dt^m} + E_1 \frac{d^{m-1} x(t)}{dt^{m-1}} + \cdots + E_{m-1} \frac{dx(t)}{dt} + E_m x(t)$$

we have

$$C_0 \frac{d^r z(t)}{dt^r} + C_1 \frac{d^{r-1} z(t)}{dt^{r-1}} + \cdots + C_{r-1} \frac{dz(t)}{dt} + C_r z(t) = 0 \tag{2-14}$$

Since

$$z(t_0) = y_1(t_0) - y_2(t_0) = 0$$
$$\frac{dz(t_0)}{dt} = \frac{dy_1(t_0)}{dt} - \frac{dy_2(t_0)}{dt} = 0$$
$$\cdots\cdots\cdots\cdots\cdots\cdots\cdots\cdots$$
$$\frac{d^{r-1} z(t_0)}{dt^{r-1}} = \frac{d^{r-1} y_1(t_0)}{dt^{r-1}} - \frac{d^{r-1} y_2(t_0)}{dt^{r-1}} = 0$$

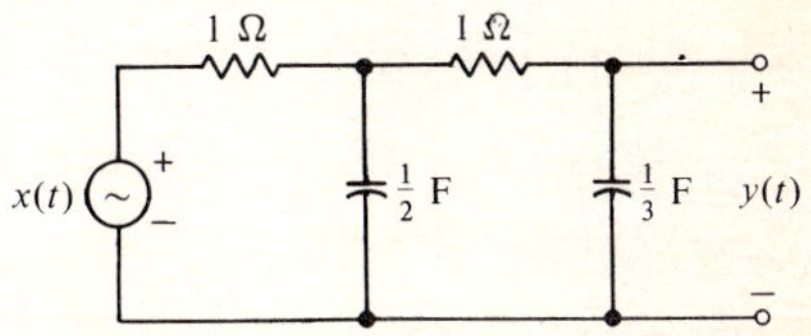

FIGURE 2-5

it follows from Eq. (2-14) that

$$\frac{d^r z(t_0)}{dt^r} = 0$$

Differentiating both sides of Eq. (2-14), we obtain

$$C_0 \frac{d^{r+1}z(t)}{dt^{r+1}} + C_1 \frac{d^r z(t)}{dt^r} + \cdots + C_{r-1} \frac{d^2 z(t)}{dt^2} + C_r \frac{dz(t)}{dt} = 0$$

Thus
$$\frac{d^{r+1}z(t_0)}{dt^{r+1}} = 0$$

Repeating the argument, we see that all derivatives of $z(t)$ at t_0 are equal to zero. Therefore, it is only possible that $z(t)$ is a constant function, and we can conclude that $z(t) = 0$ for all t. We thus have $y_1(t) = y_2(t)$, meaning that the total solution is indeed unique.

We conclude our discussion in this section with an example. Consider the electric circuit in Fig. 2-5 where the input $x(t)$, the voltage source, and the output $y(t)$, the voltage across the $\frac{1}{3}$-F capacitor, are related by the differential equation

$$\frac{d^2 y(t)}{dt^2} + 7 \frac{dy(t)}{dt} + 6y(t) = 6x(t)$$

Suppose that $x(t) = \sin 2t$. The characteristic equation of the differential equation is

$$\alpha^2 + 7\alpha + 6 = 0$$

with the characteristic roots

$$\alpha_1 = -1 \qquad \alpha_2 = -6$$

The homogeneous solution is

$$A_1 e^{-t} + A_2 e^{-6t}$$

According to Table 2-1, the particular solution is of the form

$$P_1 \sin 2t + P_2 \cos 2t$$

Substituting this expression into the differential equation, we obtain

$$-4P_1 \sin 2t - 4P_2 \cos 2t + 14P_1 \cos 2t - 14P_2 \sin 2t + 6P_1 \sin 2t + 6P_2 \cos 2t$$
$$= 6 \sin 2t$$

which simplifies to

$$(-4P_1 - 14P_2 + 6P_1 - 6) \sin 2t + (-4P_2 + 14P_1 + 6P_2) \cos 2t$$
$$= (2P_1 - 14P_2 - 6) \sin 2t + (14P_1 + 2P_2) \cos 2t = 0$$

The coefficients of the terms sin 2t and cos 2t must both equal 0. That is,

$$2P_1 - 14P_2 - 6 = 0$$
$$14P_1 + 2P_2 = 0$$

In other words,

$$P_1 = \tfrac{3}{50} \qquad P_2 = -\tfrac{21}{50}$$

Thus, the total solution is of the form

$$y(t) = A_1 e^{-t} + A_2 e^{-6t} + \tfrac{3}{50} \sin 2t - \tfrac{21}{50} \cos 2t$$

Suppose that the known boundary conditions are $y(0) = 0$ and $y'(0) = 0$. We have the equations

$$0 = A_1 + A_2 - \tfrac{21}{50}$$
$$0 = -A_1 - 6A_2 + \tfrac{6}{50}$$

which can be solved for A_1 and A_2:

$$A_1 = \tfrac{24}{50} \qquad A_2 = -\tfrac{3}{50}$$

The total solution of the differential equation, that is, the voltage across the $\frac{1}{3}$-F capacitor in Fig. 2-5, is then equal to

$$y(t) = \tfrac{12}{25} e^{-t} - \tfrac{3}{50} e^{-6t} + \tfrac{3}{50} \sin 2t - \tfrac{21}{50} \cos 2t$$

2-4 MATCHING THE BOUNDARY CONDITIONS

When the input signal $x(t)$ has different analytic expressions in different time regions, such as

$$x(t) = \begin{cases} 2e^{-2t} & t > 0 \\ e^{2t} & t < 0 \end{cases}$$

the output signal $y(t)$ can also be described by different analytic expressions in the corresponding time regions. Specifically, for each of the time regions we can solve the differential equation for the expression of $y(t)$ in that region by using the method of solution presented in Sec. 2-3. However, a little care should be exercised to match the boundary conditions in different time regions. Let us present first an illustrative example before we discuss the general procedure. Let

$$\frac{d^2y(t)}{dt^2} + 6\frac{dy(t)}{dt} + 5y(t) = \frac{d^2x(t)}{dt^2} - 2\frac{dx(t)}{dt} + x(t) \tag{2-15}$$

and

$$x(t) = \begin{cases} 2e^{-2t} & t > 0 \\ e^{2t} & t < 0 \end{cases} \tag{2-16}$$

For $t > 0$, the total solution of the differential equation is

$$y(t) = A_1 e^{-t} + A_2 e^{-5t} - 6e^{-2t} \tag{2-17}$$

For $t < 0$, the total solution of the differential equation is

$$y(t) = B_1 e^{-t} + B_2 e^{-5t} + \tfrac{1}{21}e^{2t} \tag{2-18}$$

Suppose that a set of boundary conditions is given as

$$y(1) = -0.435$$
$$y'(1) = 1.219$$

From Eq. (2-17), we obtain

$$-0.435 = A_1 e^{-1} + A_2 e^{-5} - 6e^{-2}$$
$$1.219 = A_1 e^{-1} - 5A_2 e^{-5} + 12e^{-2}$$

and the constants A_1 and A_2 are determined to be

$$A_1 = 1 \qquad A_2 = 1$$

Thus, for $t > 0$, the total solution of Eq. (2-15) is

$$y(t) = e^{-t} + e^{-5t} - 6e^{-2t} \tag{2-19}$$

To determine the constants B_1 and B_2 in Eq. (2-18), we need to know the value of $y(t_0)$ and $y'(t_0)$ for some $t_0 < 0$. As it turns out, we can compute the values of $y(0-)$ and $y'(0-)$† and from them determine the values of B_1 and B_2. We compute first $dx(t)/dt$ and $d^2x(t)/dt^2$ from Eq. (2-16):

$$\frac{dx(t)}{dt} = \begin{cases} -4e^{-2t} & t > 0 \\ u_0(t) & t = 0 \\ 2e^{2t} & t < 0 \end{cases}$$

$$\frac{d^2x(t)}{dt^2} = \begin{cases} 8e^{-2t} & t > 0 \\ u_1(t) - 6u_0(t) & t = 0 \\ 4e^{2t} & t < 0 \end{cases}$$

† Because $y(t)$ and $y'(t)$ might contain a discontinuity at $t = 0$, we should make a distinction between $y(0+)$ and $y(0-)$ and between $y'(0+)$ and $y'(0-)$.

It follows that the right-hand side of Eq. (2-15),

$$\frac{d^2x(t)}{dt^2} - 2\frac{dx(t)}{dt} + x(t)$$

contains the singularity functions $u_1(t) - 8u_0(t)$ at $t = 0$. Therefore, the left-hand side of Eq. (2-15)

$$\frac{d^2y(t)}{dt^2} + 6\frac{dy(t)}{dt} + 5y(t)$$

must also contain exactly these singularity functions at $t = 0$. In particular, the second derivative $d^2y(t)/dt^2$ must contain the singularity function $u_1(t)$, because if $dy(t)/dt$ or $y(t)$ contains the singularity function $u_1(t)$, $d^2y(t)/dt^2$ will contain a higher-order singularity function $u_2(t)$ or $u_3(t)$ which will not be matched by the singularity functions in the right-hand side of Eq. (2-15). That $d^2y(t)/dt^2$ contains $u_1(t)$ implies that $dy(t)/dt$ contains $u_0(t)$ and $y(t)$ contains a discontinuity equal to 1 at $t = 0$. However, since $dy(t)/dt$ contains the singularity function $u_0(t)$, the term $6dy(t)/dt$ in Eq. (2-15) will contain the singularity function $6u_0(t)$. So that the expression in the left-hand side of Eq. (2-15) will contain exactly the singularity functions $u_1(t) - 8u_0(t)$, $d^2y(t)/dt^2$ must also contain $-14u_0(t)$. [The reader can readily supply the argument on why the singularity function $-14u_0(t)$ should not be contained in $dy(t)/dt$ or $y(t)$.] It follows that $dy(t)/dt$ has a discontinuity equal to -14 at $t = 0$. Thus we have

$$y(0+) - y(0-) = 1$$
$$y'(0+) - y'(0-) = -14$$

From Eq. (2-19), we have

$$y(0+) = -4 \qquad y'(0+) = 6$$

It follows that

$$y(0-) = -5 \qquad y'(0-) = 20$$

Using these boundary conditions, we determine the two constants B_1 and B_2 in Eq. (2-18) as

$$B_1 = -\tfrac{4}{3} \qquad B_2 = -\tfrac{26}{7}$$

This example illustrates a general procedure for determining $y(t)$ when $x(t)$ is specified by different analytic expressions in different time regions. Corresponding to the expression for $x(t)$ in each of the time regions, a solution of the differential equation, with undetermined coefficients in the homogeneous solution, can be found. A set of boundary conditions in one of the time regions is needed so that the undetermined coefficients in the solution for that region can be determined. Boundary conditions for adjacent time regions can then be computed, and undetermined coefficients in the solution for these time regions can be determined. To be specific, suppose that $x(t)$ is given by different expressions in the three regions $t < t_1$, $t_1 < t < t_2$, and $t > t_2$. Also suppose that we are given a set of boundary conditions $y(t_0)$, $y'(t_0)$, $y''(t_0)$, ... for some t_0 within the range $t_1 < t < t_2$. The undetermined coefficients in $y(t)$ for the region $t_1 < t < t_2$ can be determined from the given boundary conditions. Knowing the values of $y(t_1+)$, $y'(t_1+)$, $y''(t_1+)$, ..., we can determine the values of $y(t_1-)$, $y'(t_1-)$, $y''(t_1-)$, ... from the discontinuities of $y(t)$, $y'(t)$, $y''(t)$, ... at t_1 which are computed by matching singularity functions on the two sides of the differential equation at t_1. The boundary conditions $y(t_1-)$, $y'(t_1-)$, $y''(t_1-)$, ... can then be used to determine the undetermined coefficients in the solution $y(t)$ for $t < t_1$. Similarly, knowing the values of $y(t_2-)$, $y'(t_2-)$, $y''(t_2-)$, ..., we can determine the values of $y(t_2+)$, $y'(t_2+)$, $y''(t_2+)$, ... from the discontinuities of $y(t)$, $y'(t)$, $y''(t)$, ... at t_2. The boundary conditions $y(t_2+)$, $y'(t_2+)$, $y''(t_2+)$, ... can then be used to determine the undetermined coefficients in $y(t)$ for $t > t_2$.

As another example, we consider a system described by the differential equation

$$\frac{dy(t)}{dt} + 3y(t) = x(t) \tag{2-20}$$

Let $x(t) = u_0(t)$ and, as boundary condition, $y(0-) = 0$. In the region $t < 0$, since $x(t) = 0$,

$$y(t) = A_1 e^{-3t}$$

From the boundary condition $y(0-) = 0$, we determine that $A_1 = 0$. Thus, $y(t) = 0$ for $t < 0$.

In the region $t > 0$, since $x(t) = 0$, we also have

$$y(t) = A_2 e^{-3t}$$

To determine the coefficient A_2, we note that at $t = 0$ the right-hand side of Eq. (2-20) contains the impulse $u_0(t)$. This impulse must be matched by an impulse in the left-hand side of Eq. (2-20). In particular, this impulse must be contained in $dy(t)/dt$. Therefore, $y(t)$ has a discontinuity equal to 1 at $t = 0$. Thus, we obtain the boundary condition $y(0+) = 1$. It follows that

$$y(t) = e^{-3t} \qquad t > 0$$

2-5 CAUSALITY AND LINEARITY

A system is said to be *initially at rest* if, for any stimulus that is equal to zero for $t < t_0$, the response of the system and all its derivatives are equal to zero at t_0-. We want to show that a system described by a linear differential equation with constant coefficients is causal if the system is initially at rest. Let $x(t)$ be an input signal that is equal to zero for $t < t_0$. We want to show that the corresponding output signal $y(t)$ is also equal to zero for $t < t_0$. Let us assume first that the characteristic roots of the differential equation, $\alpha_1, \alpha_2, \ldots, \alpha_r$, are all distinct. Thus for $t < t_0$,

$$y(t) = A_1 e^{\alpha_1 t} + A_2 e^{\alpha_2 t} + \cdots + A_r e^{\alpha_r t}$$

because the particular solution is equal to zero. To determine the constants $A_1, A_2, \ldots, A_r$, we note that

$$\begin{aligned} y(t_0-) &= 0 = A_1 e^{\alpha_1 t_0} + A_2 e^{\alpha_2 t_0} + \cdots + A_r e^{\alpha_r t_0} \\ y'(t_0-) &= 0 = A_1 \alpha_1 e^{\alpha_1 t_0} + A_2 \alpha_2 e^{\alpha_2 t_0} + \cdots + A_r \alpha_r e^{\alpha_r t_0} \\ &\cdots\cdots\cdots\cdots\cdots\cdots \\ y^{(r-1)}(t_0-) &= 0 = A_1 \alpha_1^{r-1} e^{\alpha_1 t_0} + A_2 \alpha_2^{r-1} e^{\alpha_2 t_0} + \cdots + A_r \alpha_r^{r-1} e^{\alpha_r t_0} \end{aligned} \tag{2-21}$$

Since the value of the determinant

$$\begin{vmatrix} e^{\alpha_1 t_0} & e^{\alpha_2 t_0} & \cdots & e^{\alpha_r t_0} \\ \alpha_1 e^{\alpha_1 t_0} & \alpha_2 e^{\alpha_2 t_0} & \cdots & \alpha_r e^{\alpha_r t_0} \\ \cdots & \cdots & \cdots & \cdots \\ \alpha_1^{r-1} e^{\alpha_1 t_0} & \alpha_2^{r-1} e^{\alpha_2 t_0} & \cdots & \alpha_r^{r-1} e^{\alpha_r t_0} \end{vmatrix}$$

is nonzero (see Sec. 2-3), according to the theory of simultaneous linear equations, the unknowns $A_1, A_2, \ldots, A_r$ in Eqs. (2-21) are all equal to zero. Therefore, we have

$$y(t) = 0 \qquad t < t_0$$

and can conclude that the system is causal. A similar argument can be made when the characteristic roots are not all distinct.

We want to investigate now whether a system described by a linear differential equation with constant coefficients is linear. It is not difficult to see that if for an input $x(t)$ the corresponding output is $y(t)$ then for the input $ax(t)$ the corresponding output is $ay(t)$, provided that the boundary conditions are also scaled by the multiplying constant a accordingly. To illustrate this point, consider a system described by

$$\frac{dy(t)}{dt} + 2y(t) = x(t) \tag{2-22}$$

Given that $x(t) = e^{-t}$ and $y(0) = 2$, the total solution is found to be

$$y(t) = e^{-2t} + e^{-t} \tag{2-23}$$

Given that $x(t) = 5e^{-t}$ and $y(0) = 10$, the total solution is then

$$y(t) = 5e^{-2t} + 5e^{-t}$$

which is the result in Eq. (2-23) scaled up by a factor of 5. Note, however, for $x(t) = 5e^{-t}$ and $y(0) = 2$, the total solution is

$$y(t) = -3e^{-2t} + 5e^{-t}$$

which is not related to the result in Eq. (2-23) by a multiplying constant.

Similarly, if for an input $x_1(t)$ the corresponding output is $y_1(t)$, and for an input $x_2(t)$ the corresponding output is $y_2(t)$, then for the input $x_1(t) + x_2(t)$ the corresponding output is $y_1(t) + y_2(t)$, provided that the boundary conditions are added accordingly. To illustrate this point, consider the system described by Eq. (2-22). Given that $x(t) = e^{-t}$ and $y(0) = 2$, the total solution is found to be

$$y(t) = e^{-2t} + e^{-t}$$

Given that $x(t) = \sin t$ and $y(0) = 1$, the total solution is found to be

$$y(t) = \tfrac{6}{5}e^{-2t} + \tfrac{2}{5}\sin t - \tfrac{1}{5}\cos t$$

Thus, for the input $x(t) = e^{-t} + \sin t$ and the boundary condition

$$y(0) = 2 + 1 = 3$$

the total solution is

$$y(t) = \tfrac{11}{5}e^{-2t} + e^{-t} + \tfrac{2}{5}\sin t - \tfrac{1}{5}\cos t$$

We want to show that a system described by a linear differential equation with constant coefficients is linear if it is initially at rest. Let $x(t)$ be an input signal that is equal to zero for $t < t_0$. Let $y(t)$ denote the response to $x(t)$, and let $z(t)$ denote the response to $ax(t)$ for some constant a. To show that $z(t)$ is indeed equal to $ay(t)$ we note first that the discontinuities of $z(t), z'(t), z''(t), \ldots$ at t_0 must be equal to a times the discontinuities of $y(t), y'(t), y''(t), \ldots$ at t_0. That is,

$$\begin{aligned}
z(t_0+) - z(t_0-) &= a[y(t_0+) - y(t_0-)] \\
z'(t_0+) - z'(t_0-) &= a[y'(t_0+) - y'(t_0-)] \\
z''(t_0-) - z''(t_0-) &= a[y''(t_0+) - y''(t_0+)] \\
&\cdots\cdots\cdots\cdots
\end{aligned}$$

Because $z(t_0-), z'(t_0-), z''(t_0-), \ldots, y(t_0-), y'(t_0-), y''(t_0-), \ldots$ are all zero, we have

$$\begin{aligned}
z(t_0+) &= ay(t_0+) \\
z'(t_0+) &= ay'(t_0+) \\
z''(t_0+) &= ay''(t_0+) \\
&\cdots\cdots
\end{aligned}$$

Since the boundary conditions for the input $ax(t)$ are equal to a times the boundary conditions for the input $x(t)$, according to our discussion above, the response to $ax(t)$, $z(t)$, is equal to a times the response to $x(t)$, $y(t)$.

Let $y_1(t)$ denote the response to an input $x_1(t)$, $y_2(t)$ denote the response to an input $x_2(t)$, and $z(t)$ denote the response to the input $x_1(t) + x_2(t)$. Suppose that $x_1(t) = 0$ for $t < t_0$ and $x_2(t) = 0$ for $t < t_1$. Without loss of generality, let us assume that $t_0 \le t_1$. Thus, $x_1(t) + x_2(t) = 0$ for $t < t_0$. To show that $z(t)$ is equal to $y_1(t) + y_2(t)$, we note that

$$z(t_0+) - z(t_0-) = [y_1(t_0+) + y_2(t_0+)] - [y_1(t_0-) + y_2(t_0-)]$$
$$z'(t_0+) - z'(t_0-) = [y_1'(t_0+) + y_2'(t_0+)] - [y_1'(t_0-) + y_2'(t_0-)]$$
$$z''(t_0+) - z''(t_0-) = [y_1''(t_0+) + y_2''(t_0+)] - [y_1''(t_0-) + y_2''(t_0-)]$$
$$\cdots\cdots\cdots\cdots\cdots\cdots\cdots\cdots\cdots\cdots\cdots\cdots$$

Because the system is initially at rest,

$$y_1(t_0-) = y_1'(t_0-) = y_1''(t_0-) = \cdots = 0$$
$$y_2(t_0-) = y_2'(t_0-) = y_2''(t_0-) = \cdots = 0$$
$$z(t_0-) = z'(t_0-) = z''(t_0-) = \cdots = 0$$

It follows that

$$z(t_0+) = y_1(t_0+) + y_2(t_0+)$$
$$z'(t_0+) = y_1'(t_0+) + y_2'(t_0+)$$
$$z''(t_0+) = y_1''(t_0+) + y_2''(t_0+)$$
$$\cdots\cdots\cdots\cdots\cdots\cdots$$

Since the boundary conditions for the input $x_1(t) + x_2(t)$ are equal to the sum of the corresponding boundary conditions for the inputs $x_1(t)$ and $x_2(t)$, $z(t)$ is equal to $y_1(t) + y_2(t)$. We have shown that the system is both homogeneous and additive; therefore, the system is linear.

At this point, a reader might feel that there is something mysterious about a system that is initially at rest. As a matter of fact, one might wonder about the physical significance of the boundary conditions which influence directly the solution of the differential equation. We shall have a thorough discussion of this point in Chap. 3.

2-6 DIFFERENCE EQUATION DESCRIPTION OF DISCRETE SYSTEMS

Just as in the case of a continuous system, many of the properties of a discrete system are exhibited in the difference equation relating the input and output signals of the system. Except in a degenerate case, a system described by a difference equation is dynamical because the value of $y(n)$ depends not only on the value of $x(n)$ but also on the values of $y(n-1), y(n-2), \ldots, x(n-1), x(n-2), \ldots$. A system is time-invariant if the coefficients of the difference equation are all constants. For example, the difference equation

$$y(n) + 5y(n-1)y(n-2) + 6y^2(n-2) = x(n) - 7x^2(n-1)$$

describes the input-output relationship of a time-invariant system, and the difference equation

$$y(n) + n^2y(n-1) + e^{-n}y(n-2) = nx(n)$$

describes the input-output relationship of a time-varying system. The notions of causality and linearity for systems described by difference equations are parallel to that for systems described by differential equations. We therefore shall not repeat the discussion in Sec. 2-5.

Given the difference equation relating the input and output signals of a system together with the values of $y(n_0), y(n_0+1), y(n_0+2), \ldots$ for some n_0, the values of $y(n)$ at successive time constants n can be computed in a step-by-step manner when the input signal $x(n)$ is known. Let us illustrate such a computation by an example. Suppose that the input-output relationship of a discrete system is described by

$$y(n) + y(n-1) = x(n) + 2x(n-2) \tag{2-24}$$

Given the input signal

$$x(n) = \begin{cases} n & n \geq 0 \\ 0 & n < 0 \end{cases}$$

and the boundary condition $y(0) = 2$, we obtain from Eq. (2-24), for $n = 1$,

$$y(1) + y(0) = x(1) + 2x(-1)$$

or

$$y(1) = -1$$

Similarly, for $n = 2$, Eq. (2-24) yields

$$y(2) + y(1) = x(2) + 2x(0)$$

or

$$y(2) = 2 + 1 = 3$$

and for $n = 3$

$$y(3) + y(2) = x(3) + 2x(1)$$

or

$$y(3) = 3 + 2 - 3 = 2$$

and so on for $n = 4, 5, 6, \ldots$. Also, for $n = 0$, Eq. (2-24) yields

$$y(0) + y(-1) = x(0) + 2x(-2)$$

or

$$y(-1) = -2$$

and for $n = -1$

$$y(-1) + y(-2) = x(0) + 2x(-3)$$

or

$$y(-2) = 2$$

and so on for $n = -2, -3, -4, \ldots$.

A step-by-step computation of the values of $y(n)$ as illustrated above is quite simple when a high-speed digital computer is available. However, it is still desirable in many cases to solve a difference equation for a closed-form expression for $y(n)$. The solution of difference equations is similar to the solution of differential equations. We shall limit our discussion to the solution of *linear difference equations with constant coefficients.* As will be seen, there is a close resemblance between the solution of linear difference equations with constant coefficients and the solution of linear differential equations with constant coefficients.

The general form of a linear difference equation with constant coefficients is

$$C_0 y(n) + C_1 y(n-1) + \cdots + C_{r-1} y(n-r+1) + C_r y(n-r)$$
$$= E_0 x(n) + E_1 x(n-1) + \cdots + E_{m-1} x(n-m+1) + E_m x(n-m) \quad (2\text{-}25)$$

where the C's and E's are constants. Such an equation is said to be of degree r [in the unknown $y(n)$], provided that C_0 and C_r do not equal zero. The total solution of Eq. (2-25) consists of two parts: the homogeneous solution and the particular solution. The homogeneous solution satisfies the equation

$$C_0 y(n) + C_1 y(n-1) + \cdots + C_{r-1} y(n-r+1) + C_r y(n-r) = 0 \qquad (2\text{-}26)$$

We claim that the homogeneous solution contains terms of the form $A\alpha^n$ for some appropriately chosen complex constants α. Substituting $y(n) = A\alpha^n$ into Eq. (2-26), we obtain

$$C_0 A\alpha^n + C_1 A\alpha^{n-1} + \cdots + C_{r-1} A\alpha^{n-r+1} + C_r A\alpha^{n-r} = 0$$

That is,

$$C_0 \alpha^r + C_1 \alpha^{r-1} + \cdots + C_{r-1}\alpha + C_r = 0 \qquad (2\text{-}27)$$

Therefore, if $\alpha_1, \alpha_2, \ldots, \alpha_r$ are the distinct roots of Eq. (2-27),

$$A_1 \alpha_1{}^n + A_2 \alpha_2{}^n + \cdots + A_r \alpha_r{}^n$$

will satisfy Eq. (2-26), where the A's are constants to be determined by a set of boundary conditions. Equation (2-27) is called the *characteristic equation* of the difference equation, and the roots of the characteristic equation are called the *characteristic roots.*

As an example, we consider a problem originated by Leonardo Fibonacci. Suppose that a pair of rabbits can produce a new pair of offspring every month, and rabbits become fertile after the age of 1 month. How many pairs of rabbits can be produced by a pair of newborn rabbits in a year's time? Let $y(n)$ denote the number of pairs of rabbits there are at the beginning of the nth month. Clearly, $y(1) = 1$. Also, $y(2) = 1$ because the pair of newborn rabbits will not become fertile until after the first month. It follows that $y(3) = 2$ because a pair of rabbits will be born in the 2d month, and $y(4) = 3$ because in the 3d month one of the two pairs of rabbits will produce a pair of offspring while the other pair is still newborn. To compute the value of $y(n)$, we ask the question: At the beginning of the $(n-1)$st month, how many pairs of rabbits among the $y(n-1)$ pairs are fertile? The answer is $y(n-2)$. It follows that at the beginning of the $(n-1)$st month there are $y(n-1) - y(n-2)$ pairs of newborn rabbits. We thus have

$$y(n) = 2y(n-2) + y(n-1) - y(n-2)$$

which simplifies to

$$y(n) - y(n-1) - y(n-2) = 0 \tag{2-28}$$

To solve the difference equation for an expression for $y(n)$, we note first that the characteristic equation is

$$\alpha^2 - \alpha - 1 = 0$$

It follows that the characteristic roots are

$$\alpha_1 = \frac{1+\sqrt{5}}{2} \qquad \alpha_2 = \frac{1-\sqrt{5}}{2}$$

and the homogeneous solution is

$$A_1\left(\frac{1+\sqrt{5}}{2}\right)^n + A_2\left(\frac{1-\sqrt{5}}{2}\right)^n \tag{2-29}$$

We shall discuss how to determine the constants A_1 and A_2 later on.

When the characteristic equation contains multiple roots, the homogeneous solution of a difference equation will be of slightly different form. Specifically, let α_1 be a k-multiple characteristic root; then its corresponding terms in the homogeneous solution are

$$A_1 n^{k-1}\alpha_1{}^n + A_2 n^{k-2}\alpha_1{}^n + \cdots + A_{k-1} n\alpha_1{}^n + A_k\alpha_1{}^n$$

Clearly the term $A_k\alpha_1{}^n$ satisfies Eq. (2-26). To show that the term $A_{k-1}n\alpha_1{}^n$ also satisfies Eq. (2-26), we substitute $y(n) = A_{k-1}n\alpha_1{}^n$ in the left-hand side of Eq. (2-26) and obtain

$$C_0 A_{k-1} n\alpha_1{}^n + C_1 A_{k-1}(n-1)\alpha_1^{n-1} + C_2 A_{k-1}(n-2)\alpha_1^{n-2} + \cdots + C_{r-1}A_{k-1}(n-r+1)\alpha_1^{n-r+1} + C_r A_{k-1}(n-r)\alpha_1^{n-r}$$

which is

$$A_{k-1}\alpha_1[C_0 n\alpha_1^{n-1} + C_1(n-1)\alpha_1^{n-2} + C_2(n-2)\alpha_1^{n-3} + \cdots$$
$$+ C_{r-1}(n-r+1)\alpha_1^{n-r} + C_r(n-r)\alpha_1^{n-r-1}] \quad (2\text{-}30)$$

Because α_1 is a k-multiple root of Eq. (2-27), it is also a k-multiple root of the equation

$$\alpha^{n-r}(C_0\alpha^r + C_1\alpha^{r-1} + \cdots + C_{r-1}\alpha + C_r) = 0 \quad (2\text{-}31)$$

Thus, α_1 not only satisfies Eq. (2-31) but also its derivative

$$C_0 n\alpha^{n-1} + C_1(n-1)\alpha^{n-2} + \cdots + C_{r-1}(n-r+1)\alpha^{n-r} + C_r(n-r)\alpha^{n-r-1} = 0$$

Therefore, the expression in brackets in (2-30) is equal to zero, and $A_{k-1}n\alpha_1{}^n$ indeed satisfies Eq. (2-26). That the other terms $A_1 n^{k-1}\alpha_1{}^n$, $A_2 n^{k-2}\alpha_1{}^n, \ldots$ also satisfy Eq. (2-26) can be shown in a similar manner.

As an example, consider the difference equation

$$y(n) + 6y(n-1) + 12y(n-2) + 8y(n-3) = x(n)$$

The characteristic equation is

$$\alpha^3 + 6\alpha^2 + 12\alpha + 8 = 0$$

The homogeneous solution is

$$y(n) = (A_1 n^2 + A_2 n + A_3)(-2)^n$$

because -2 is a triple characteristic root.

As to the particular solution, it can be determined by inspection in simple cases. As a matter of fact, the inspection procedure is quite similar to that for differential equations. Table 2-2 shows the general form of the particular solutions corresponding to some $x(n)$.

After determining the particular solution, we turn to the determination of the undetermined coefficients in the homogeneous solution. For an rth-order difference

Table 2-2

$x(n)$	**Particular solution**
n^k	$P_1 n^k + P_2 n^{k-1} + \cdots + P_k$
a^n	Pa^n if a is not a characteristic root of the difference equation
	$P_1 na^n + P_2 a^n$ if a is a distinct characteristic root of the difference equation
	$P_1 n^{k-1}a^n + P_2 n^{k-2}a^n + P_3 n^{k-3} + \cdots + P_k a^n$ if a is a $(k-1)$-multiple characteristic root of the difference equation

equation, the constants $A_1, A_2, \ldots, A_r$ in the homogeneous solution can be determined from the boundary conditions $y(n_0), y(n_0+1), \ldots, y(n_0+r-1)$ for any n_0. In other words, the values of $y(n)$ at r successive time instants are sufficient to determine the constants $A_1, A_2, \ldots, A_r$. The proof of this assertion is similar to that given for differential equations in Sec. 2-3, which we shall not include here. (It is discussed in detail in Ref. [10].)

As an example, let us complete our solution of the problem of rabbits discussed earlier in this section. After obtaining the homogeneous solution of the difference equation (2-28), we note that the particular solution is equal to zero because the right-hand side of Eq. (2-28) is zero. To determine the constants A_1 and A_2 in (2-29), we use the boundary conditions

$$y(1) = 1 \qquad y(2) = 1$$

Thus, we have

$$y(1) = 1 = A_1 \frac{1+\sqrt{5}}{2} + A_2 \frac{1-\sqrt{5}}{2}$$

$$y(2) = 1 = A_1 \left(\frac{1+\sqrt{5}}{2}\right)^2 + A_2 \left(\frac{1-\sqrt{5}}{2}\right)^2$$

Solving the A_1 and A_2, we obtain

$$A_1 = \frac{1}{\sqrt{5}} \qquad A_2 = -\frac{1}{\sqrt{5}}$$

Therefore, the total solution of Eq. (2-28) is

$$y(n) = \frac{1}{\sqrt{5}} \left(\frac{1+\sqrt{5}}{2}\right)^n - \frac{1}{\sqrt{5}} \left(\frac{1-\sqrt{5}}{2}\right)^n$$

Consequently, at the beginning of the 12th month, the number of pairs of rabbits is

$$y(12) = \frac{1}{\sqrt{5}} \left(\frac{1+\sqrt{5}}{2}\right)^{12} - \frac{1}{\sqrt{5}} \left(\frac{1-\sqrt{5}}{2}\right)^{12} = 144$$

As a final example, let us solve the difference equation

$$y(n) + 2y(n-1) = x(n) - x(n-1)$$

where $x(n) = n^2$ with the boundary condition $y(0) = 1$. Substituting the expression of $x(n)$ into the difference equation, we obtain

$$y(n) + 2y(n-1) = n^2 - (n-1)^2 = 2n - 1$$

The homogeneous solution is $A(-2)^n$. To determine the particular solution, we try a solution of the form $P_1 n + P_2$. Substituting this into the difference equation, we obtain

$$P_1 n + P_2 + 2[P_1(n-1) + P_2] = 2n - 1$$

which gives

$$3P_1n + 3P_2 - 2P_1 = 2n - 1 \tag{2-32}$$

Comparing the coefficients of n and the constant terms on the two sides of Eq. (2-32), we have

$$3P_1 = 2 \qquad \text{and} \qquad 3P_2 - 2P_1 = -1$$

That is,

$$P_1 = \tfrac{2}{3} \qquad P_2 = \tfrac{1}{9}$$

and the total solution is

$$y(n) = A(-2)^n + \tfrac{2}{3}n + \tfrac{1}{9}$$

From the given boundary condition, the constant A is determined to be $\frac{8}{9}$.

2-7 MATCHING THE BOUNDARY CONDITIONS

When the function $x(n)$ is specified by different analytic expressions in different time regions, we can determine the solution $y(n)$ in each region separately, as the following example illustrates. Let

$$y(n) + 2y(n-1) + y(n-2) = x(n) - x(n-1) \tag{2-33}$$

where

$$x(n) = \begin{cases} 2^n & n < 0 \\ 3^{-n} & n \geq 0 \end{cases}$$

For $n < 0$, the homogeneous solution is

$$(A_1n + A_2)(-1)^n$$

and the particular solution is

$$\tfrac{2}{9}(2^n)$$

Thus the total solution is

$$(A_1n + A_2)(-1)^n + \tfrac{2}{9}(2^n)$$

Suppose we are given the boundary conditions $y(-1) = \frac{1}{9}$ and $y(-2) = -\frac{17}{18}$. We solve the following equations for A_1 and A_2,

$$(-A_1 + A_2)(-1)^{-1} + \tfrac{2}{9}2^{-1} = \tfrac{1}{9}$$
$$(-2A_1 + A_2)(-1)^{-2} + \tfrac{2}{9}2^{-2} = -\tfrac{17}{18}$$

and obtain

$$A_1 = 1 \qquad A_2 = 1$$

For $n \geq 0$, the total solution is

$$(B_1 n + B_2)(-1)^n - \tfrac{1}{8}3^{-n}$$

To determine the constants B_1 and B_2, we compute first the values of $y(0)$ and $y(1)$. According to Eq. (2-33),

$$y(0) + 2y(-1) + y(-2) = x(0) - x(-1)$$

Thus

$$y(0) = 1 - 2^{-1} - 2\tfrac{1}{9} + \tfrac{17}{18} = \tfrac{11}{9}$$

Similarly

$$y(1) + 2y(0) + y(-1) = x(1) - x(0)$$

Thus

$$y(1) = 3^{-1} - 1 - 2\tfrac{11}{9} - \tfrac{1}{9} = -\tfrac{29}{9}$$

Using the boundary conditions $y(0) = \frac{11}{9}$ and $y(1) = -\frac{29}{9}$, we can determine the two constants B_1 and B_2 by solving the following equations:

$$B_2 - \tfrac{1}{8} = \tfrac{11}{9}$$

$$(B_1 + B_2)(-1)^1 - \tfrac{1}{8}3^{-1} = -\tfrac{29}{9}$$

It turns out that

$$B_1 = \tfrac{11}{6} \qquad B_2 = \tfrac{97}{72}$$

The general procedure for matching the boundary conditions should now become obvious. Let us assume that $x(n)$ is specified by two different analytic expressions in the two regions $n \geq n_0$ and $n < n_0$. After determining the total solution in one of the two regions, using a given set of boundary conditions in that region, a set of boundary conditions for the other region can be determined from the difference equation. (We observe that in the case of difference equations it is a very simple matter to determine a set of boundary conditions for a region when the total solution in an adjacent region has been determined. Yet we recall that in the case of differential equations the boundary conditions in a region are determined by the discontinuities of the solution and its derivatives at the boundary, where these discontinuities are determined by matching singularity functions in the two sides of the differential equation.) Once a set of boundary conditions is obtained, the total solution in the corresponding region can be determined.

2-8 REMARKS AND REFERENCES

To analyze the behavior of a system, we first set up a differential or difference equation describing the input-output relationship of the system from its physical specification, and we then solve the equation for the output signal corresponding to a given input signal. Although solving a differential or difference equation is purely a mathematical process, one should not lose sight of the physical meanings of the mathematical steps and the solution obtained. In particular, it probably is not obvious that for a system whose input-output relationship is described by a differential or difference equation, there is a part of the system's response whose general form is independent of the stimulus (the homogeneous solution). As will be discussed in Chap. 3, this part of the response is related to the "history" of the system which is retained in the system's "memory."

A large number of books on differential equations have been written. Some general references are Birkhoff and Rota [1], Coddington [3], and Kaplan [9]. For a discussion of differential equations and linear circuit theory, see Guillemin [6], Bose and Stevens [2], Huang and Parker [8], and Desoer and Kuh [4].

There are also many books on finite difference equations. See, for example, Levy and Lessman [10], Milne-Thomson [11], and Hildebrand [7, Chap. 3]. Applications of difference equations in probability theory can be found in Feller [5] and in combinatorial theory in Liu [12].

[1] BIRKHOFF, G., and G. C. ROTA: "Ordinary Differential Equations," 2d ed., Blaisdell Publishing Company, Waltham, Mass., 1969.

[2] BOSE, A. G., and K. N. STEVENS: "Introductory Network Theory," Harper & Row, New York, 1965.

[3] CODDINGTON, E. A.: "An Introduction to Ordinary Differential Equations," Prentice-Hall, Inc., Englewood Cliffs, N.J., 1961.

[4] DESOER, C. A., and E. S. KUH: "Basic Circuit Theory," McGraw-Hill Book Company, New York, 1969.

[5] FELLER, W.: "An Introduction to Probability Theory and Its Applications," 2d ed., vol. I, John Wiley & Sons, Inc., New York, 1957.

[6] GUILLEMIN, E. A.: "Introductory Circuit Theory," John Wiley & Sons, Inc., New York, 1953.

[7] HILDEBRAND, F. B.: "Methods of Applied Mathematics," Prentice-Hall, Inc., Englewood Cliffs, N.J., 1952.

[8] HUANG, T. S., and R. R. PARKER: "Network Theory, an Introduction," Addison-Wesley Publishing Company, Inc., Reading, Mass., 1971.

[9] KAPLAN, W.: "Ordinary Differential Equations," Addison-Wesley Publishing Company, Inc., Reading, Mass., 1958.

[10] LEVY, H., and F. LESSMAN: "Finite Difference Equations," The Macmillan Company, New York, 1961.

[11] MILNE-THOMSON, L. M.: "The Calculus of Finite Differences," The Macmillan Company, New York, 1933.

[12] LIU, C. L.: "Introduction to Combinatorial Mathematics," McGraw-Hill Book Company, New York, 1968.

PROBLEMS

2-1 Find the differential equation relating the input signal $x(t)$ and the output signal $y(t)$ in each of the circuits shown in Fig. 2P-1.

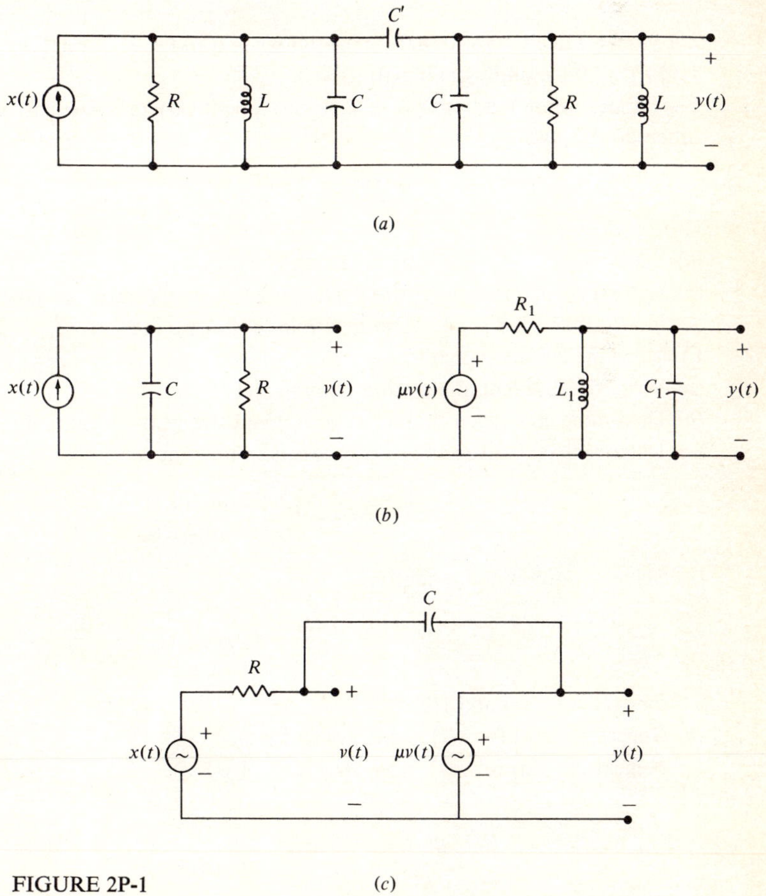

FIGURE 2P-1

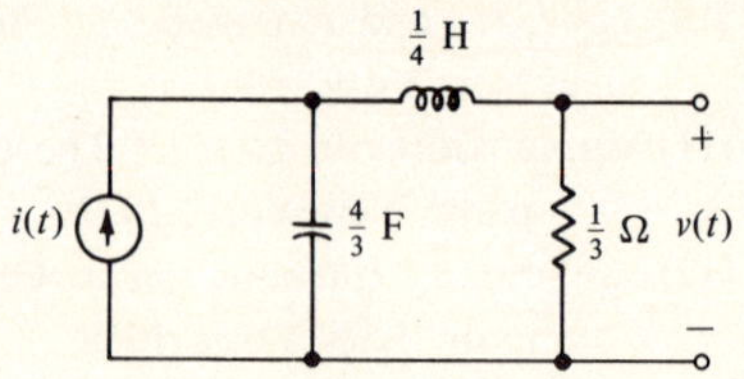

FIGURE 2P-2

2-2 Consider two systems whose input-output relationships are described by the differential equations

$$\frac{d^2y_1(t)}{dt^2} + 3\frac{dy_1(t)}{dt} + 2y_1(t) = x_1(t)$$

$$\frac{d^2y_2(t)}{dt^2} + 7\frac{dy_2(t)}{dt} + 10y_2(t) = x_2(t)$$

Let $x_1(t) = 12e^{-5t}$. Given that $x_2(t) = 0$ for $t \geq 0$ and $y_1(t) = y_2(t)$ for $t \geq 0$, determine $y_2(0+)$, $y_2'(0+)$, and $y_2(t)$ if $y_1(0+) = 0$.

2-3 The input-output relationship of a linear time-invariant system is described by the differential equation

$$\frac{dy(t)}{dt} + 2y(t) = \frac{dx(t)}{dt} + x(t)$$

(*a*) Let $x(t) = e^{-t}u_{-1}(t)$. Determine $y(t)$, given that the system is initially at rest.

(*b*) Let $x(t) = u_0(t) + Au_0(t-1)$. Determine the constant A, given that $y(t) = 0$ for $t < 0$ and for $t > 1$. Also determine $y(t)$ for $0 < t < 1$.

2-4 Consider the circuit shown in Fig. 2P-2.

(*a*) Write the differential equation relating $v(t)$ and $i(t)$.

(*b*) Determine $v(t)$, given that $i(t) = u_{-1}(t)$ and the system is initially at rest.

(*c*) Determine $v(t)$, given that the system is initially at rest and

$$i(t) = \begin{cases} 2 & 0 < t < 1 \\ 0 & \text{otherwise} \end{cases}$$

2-5 (*a*) Solve the differential equation

$$\frac{dy(t)}{dt} + 2y(t) = x(t)$$

for $x(t) = u_{-1}(t)$ and $y(0+) = 1$.

(*b*) Repeat part (*a*) for $x(t) = 5u_{-1}(t)$ and $y(0+) = 5$.

(*c*) Repeat part (*a*) for $x(t) = 5u_{-1}(t-5)$ and $y(5+) = 5$.

2-6 Let

$$\frac{d^2y(t)}{dt^2} + C_1\frac{dy(t)}{dt} + C_2\,y(t) = x(t)$$

Given that

$$x(t) = 10 \sin t u_{-1}(t)$$

and

$$y(t) = \sin t - 3 \cos t$$

for $t > 0$, determine C_1, C_2, and $y(t)$ for $t < 0$.

2-7 The input-output relationship of a certain system can be described by a second-order linear differential equation

$$\frac{d^2y(t)}{dt^2} + C_1 \frac{dy(t)}{dt} + C_2 y(t) = E_0 \frac{d^2x(t)}{dt^2} + E_1 \frac{dx(t)}{dt^2} + E_2 x(t)$$

Given that the output of the system is

$$y(t) = -e^{-t} + 3e^{-2t} \qquad \text{for } t > 0$$

when the input $x(t)$ is $u_0(t)$ and $y(0-) = 0$, $y'(0-) = 5$, determine the constants C_1, C_2, E_0, E_1, and E_2.

2-8 The input-output relationship of a continuous system is described by the differential equation

$$\frac{d^2y(t)}{dt^2} + C(t)\frac{dy(t)}{dt} + 6y(t) = x(t)$$

Let

$$C(t) = \begin{cases} 5 & t < 1 \\ 7 & t > 1 \end{cases}$$

$x(t) = 30e^{-4t}u_{-1}(t)$, and $y(0+) = 1$, $y'(0+) = -1$. Determine $y(t)$.

2-9 The input-output relationship of a continuous system is described by the differential equation

$$\frac{d^2y(t)}{dt^2} + C(t)\frac{dy(t)}{dt} + 6y(t) = x(t) + \frac{dx(t)}{dt}$$

Given that

$$C(t) = \begin{cases} 5 & t < 1 \\ 7 & t > 1 \end{cases}$$

$$x(t) = \begin{cases} 0 & t < 0 \\ 1 & 0 < t < 1 \\ 2 & t > 1 \end{cases}$$

and $y(0+) = y'(0+) = 0$, determine $y(t)$.

2-10 Let

$$\frac{dy(t)}{dt} + 2y(t) = x(t)$$

(*a*) Let $x_1(t) = e^{-t}$. Determine the corresponding $y_1(t)$, given that $y_1(0+) = 0$.
(*b*) Let $x_2(t) = dx_1(t)/dt$. Determine the corresponding $y_2(t)$, given that $y_2(0+) = 0$.

(*c*) Let $x_3(t) = e^{-t}u_{-1}(t)$. Determine the corresponding $y_3(t)$, given that the system is initially at rest.

(*d*) Let $x_4(t) = dx_3(t)/dt$. Determine the corresponding $y_4(t)$, given that the system is initially at rest.

2-11 A system is described by the differential equation

$$\frac{dy(t)}{dt} + ky(t) = x(t) + \epsilon(t)$$

where k is a positive constant, $x(t)$ is a known input signal, and $\epsilon(t)$ is unknown except for the fact that

$$|\epsilon(t)| \leq 1 \qquad \text{for } t \geq 0$$

Clearly, it is impossible to find $y(t)$ exactly for all t. Suppose that we approximate $y(t)$ for $t \geq 0$ by $\hat{y}(t)$, where

$$\frac{d\hat{y}(t)}{dt} + k\hat{y}(t) = x(t)$$

Let $\hat{y}(0-) = y(0-) = 0$. Show that the error $e(t) = |y(t) - \hat{y}(t)|$ is always less than $1/k$ for $t \geq 0$.

2-12 (*a*) As discussed in Sec. 2-3, the solution of the differential equation

$$\frac{d^2y(t)}{dt^2} + 2\alpha\frac{dy(t)}{dt} + \alpha^2 y(t) = x(t)$$

is

$$y(t) = A_1te^{\alpha t} + A_2e^{\alpha t} + p(t)$$

where $p(t)$ is the particular solution corresponding to the input signal $x(t)$. Show that the coefficients A_1 and A_2 can be determined uniquely when the boundary conditions $dy(t_0)/dt$ and $y(t_0)$ are given for any time instant t_0.

(*b*) The solution of the differential equation

$$\frac{d^4y(t)}{dt^4} + (4 + 2\alpha)\frac{d^3y(t)}{dt^3} + (3 + 8\alpha + \alpha^2)\frac{d^2y(t)}{dt^2} + (6\alpha + 4\alpha^2)\frac{dy(t)}{dt} + 3\alpha^2y(t) = 0$$

is

$$y(t) = A_1te^{\alpha t} + A_2e^{\alpha t} + A_3e^{-t} + A_4e^{-3t}$$

Show that the coefficients A_1, A_2, A_3, and A_4 can be determined uniquely when the boundary conditions $d^3y(t_0)/dt^3$, $d^2y(t_0)/dt^2$, $dy(t_0)/dt$, and $y(t_0)$ are given for any time instant t_0.

2-13 The input-output relationship of a system is described by the differential equation

$$\frac{d^2y(t)}{dt^2} + 3\frac{dy(t)}{dt} + 2y(t) = \frac{dx(t)}{dt} + 3x(t)$$

The output signal $y(t)$ for $t > 0$ corresponding to an input signal $x(t)$ is observed to be

$$y(t) = e^{-t} + 3e^{-2t} + 4e^{t} + 7e^{-7t}$$

(*a*) Was the system at rest prior to $t = 0$?

(*b*) Determine the input signal $x(t)$ for $t > 0$.

2-14 A sequence of binary digits is called a pattern, for example, 01 and 101. A pattern is said to occur at the nth digit of a sequence of binary digits if, in scanning the sequence from left to right, the pattern appears after the nth digit is scanned. After a pattern occurs, scanning starts all over again to search for the next occurrence of the pattern. For example, the pattern 101 occurs at the 4th and 8th digits in the sequence $1\underline{1}01\underline{01}0111$, but not at the 6th digit. Let $y(n)$ denote the number of n-digit binary sequences that have the pattern 111 occurring at the nth digit.

(*a*) Show that

$$y(n) + y(n-1) + y(n-2) = 2^{n-3}$$

for $n \geq 3$. Determine $y(n)$.

(*b*) Let $z(n)$ be the number of n-digit binary sequences that have the pattern 111 occurring for the first time at the nth digit. Show that

$$2^{n-3} = z(n) + z(n-1) + z(n-2) + 2^0 z(n-3) + 2^1 z(n-4) + \cdots + 2^{n-6} z(3)$$

2-15 Consider the multiplication of bacteria in a controlled environment. Let $y(n)$ denote the number of bacteria there are on the nth day. We define the rate of growth on the nth day to be $y(n) - 2y(n-1)$. If it is known that the rate of growth doubles every day, determine $y(n)$, given that $y(0) = 1$.

2-16 Consider the operation of a factory whose input is the new orders received each month, $x(n)$, and whose output is the monthly profit, $y(n)$. It is known that the average profit in every two successive months is equal to the average new order in that period.

(*a*) Given that

$$x(n) = \begin{cases} 0 & n < 0 \\ 2^n & n \geq 0 \end{cases}$$

and $y(0) = 0$, determine $y(n)$.

(*b*) Repeat part (*a*) for

$$x(n) = \begin{cases} 0 & n < 0 \\ 2^n & 0 \leq n \leq 10 \\ 2^{10} & n > 10 \end{cases}$$

2-17 A particle is moving in the horizontal direction. The distance it travels in each second is equal to two times the distance it traveled in the previous second. Let $x(n)$ denote the position of the particle at the nth second. Determine $x(n)$, given that $x(0) = 3$ and $x(3) = 10$.

2-18 Solve the following difference equations:

(*a*) $y(n) + 2y(n-1) = n - 2$, given that $y(0) = 1$.

(*b*) $y(n) + 2y(n-1) + y(n-2) = 3^n$, given that $y(-1) = 0$, $y(0) = 0$.

(*c*) $y(n) + 3y(n-1) + 2y(n-2) = x(n) + x(n-1)$ for

$$x(n) = \begin{cases} 0 & n < 0 \\ (-2)^n & n \geq 0 \end{cases}$$

given that $y(0) = 0$ and $y(1) = 0$.

2-19 Solve the difference equation

$$y(n) + 4y(n-1) + 4y(n-2) = x(n) + x(n-1)$$

given that

$$x(n) = \begin{cases} 0 & n < 0 \\ 1 & n \geq 0 \end{cases}$$

and $y(0) = 1$, $y(1) = 2$.

2-20 The input-output relationship of a discrete system is described by the difference equation

$$y(n) + 4y(n-1) + 4y(n-2) = x(n)$$

Given that

$$x(n) = \begin{cases} 0 & n < 0 \\ 1 & n \geq 0 \qquad n \neq 5 \\ 1 + \epsilon & n = 5 \end{cases}$$

determine the corresponding $y(n)$. It is known that the system is initially at rest.

2-21 The input-output relationship of a discrete system can be described by a second-order linear difference equation with constant coefficients. Corresponding to the input

$$x(n) = \begin{cases} 1 & n > 0 \\ 0 & n < 0 \end{cases}$$

the output is

$$y(n) = 2^n + 3(5^n) + 10 \qquad n \geq 0$$

(*a*) Given that the system is initially at rest, determine the difference equation.
(*b*) Determine the output corresponding to the input

$$x(n) = \begin{cases} 2 & 0 \leq n \leq 10 \\ 0 & \text{otherwise} \end{cases}$$

2-22 What is a suitable definition for a discrete system being initially at rest? Show that a discrete system whose input-output relationship can be described by a linear difference equation with constant coefficients is linear if the system is initially at rest.

3

STATE SPACE DESCRIPTION OF SYSTEMS

3-1 INTRODUCTION

We study in this chapter another way to describe the input-output relationship of systems. Let us begin our discussion with discrete systems. We recall that a memoryless system is one the output of which at any time instant n_0, $y(n_0)$, depends only on the input $x(n_0)$ at that instant. Therefore, the input-output relationship of a memoryless system can be described by an equation relating $x(n_0)$ and $y(n_0)$. On the other hand, the output of a dynamical system at the time instant n_0 does not depend on the input at n_0 alone. In order to determine the output of a dynamical system at n_0, we must know, besides the input to the system for $n \geq n_0$, that part of the *history* of the system prior to n_0 which will affect the behavior of the system at n_0. Although the history of a dynamical system prior to n_0 can be described, in a most complete way, by the initial condition of the system at $n = -\infty$ together with the input $x(n)$ within $-\infty \leq n < n_0$, as far as the behavior of the system at n_0 is concerned, such a detailed description of the past history is usually not necessary. As we have seen in Chap. 2, for a dynamical system whose input-output relationship is specified by a difference equation, the output signal for $n \geq n_0$ corresponding to a given input signal can be determined uniquely when the boundary conditions immediately prior to n_0 are known. In other words, the boundary conditions immediately prior to n_0 summarize that part of the past history of the system which will affect the behavior of the system for $n \geq n_0$. In this chapter, we shall discuss another way of describing the input-output relationship of dynamical systems, namely, the *state space* description. In such a description, the value of the output of a dynamical system at time n_0 is related algebraically to the values of the input at times n_0, $n_0 + 1$, $n_0 + 2$, ... as well as the values of a certain number of *system variables*. The values of these variables at time n_0 summarize the effect of the past history of the system on the output at time n_0.

Let us illustrate what we mean by an example: Consider a discrete system whose output at any time instant is equal to the difference between the largest value and the smallest value of the input the system has received so far. Clearly, the system is dynamical because its output at instant n does not depend on the value of the input at that instant, $x(n)$, alone. We ask now: How much information on the past input does the system retain in order to compute the output $y(n)$? If we let $\lambda_1(n)$ and $\lambda_2(n)$ denote the largest value and the smallest value among ..., $x(0)$, ..., $x(n-2)$, $x(n-1)$, respectively, then the value of the output at instant n is

$$y(n) = \max\,[\lambda_1(n), x(n)] - \min\,[\lambda_2(n), x(n)] \tag{3-1}$$

Indeed, we note in this example that to determine the value of the output at instant n it is not necessary to know all the past input values, ..., $x(0)$, $x(1)$, ..., $x(n-2)$, $x(n-1)$. The history of the system prior to instant n can be described compactly by the values of the two variables $\lambda_1(n)$ and $\lambda_2(n)$. Since $\lambda_1(n)$ and $\lambda_2(n)$ are the largest and the smallest value of the input up to the instant n, their values will change in successive time instants. In other words, the values of $\lambda_1(n)$ and $\lambda_2(n)$ must be updated at each time instant according to the equations

$$\lambda_1(n+1) = \max\,[\lambda_1(n), x(n)] \tag{3-2}$$

$$\lambda_2(n+1) = \min\,[\lambda_2(n), x(n)] \tag{3-3}$$

so that, at time instant $n + 1$, the value of $y(n + 1)$ can be computed from the values of $x(n+1)$, $\lambda_1(n+1)$, $\lambda_2(n+1)$, and so on. Equations (3-1) to (3-3) thus completely describe the behavior of the system.

Our discussion can be extended immediately to continuous systems. Consider the electric circuit in Fig. 3-1. We recall that the part of the history of the circuit prior to t_0 that will affect the behavior of the circuit for $t \geq t_0$ can be summarized by the value of the current in the inductor at t_0, $\lambda_1(t_0)$, and the value of the voltage across the capacitor at t_0, $\lambda_2(t_0)$. Indeed, we have

$$y(t_0) = \frac{x(t_0) - \lambda_2(t_0)}{R_1} - \lambda_1(t_0) = \frac{1}{R_1}\,x(t_0) - \frac{1}{R_1}\,\lambda_2(t_0) - \lambda_1(t_0)$$

or

$$y(t) = \frac{1}{R_1}\,x(t) - \frac{1}{R_1}\,\lambda_2(t) - \lambda_1(t)$$

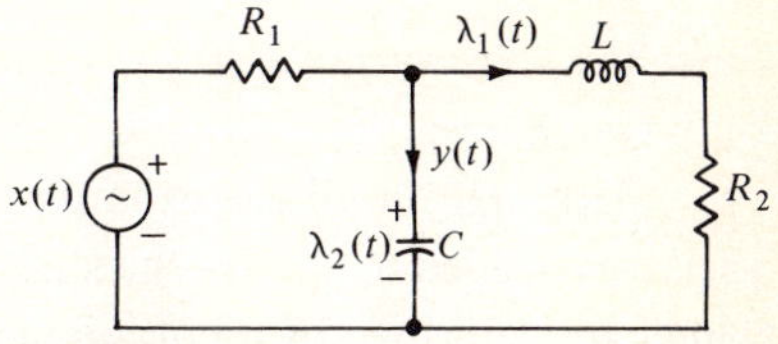

FIGURE 3-1

for any t. As time goes on, the summary of the history of the system is updated according to the equations

$$\frac{d\lambda_1(t)}{dt} = \frac{1}{L}\lambda_2(t) - \frac{R_2}{L}\lambda_1(t)$$

$$\frac{d\lambda_2(t)}{dt} = \frac{1}{CR_1}x(t) - \frac{1}{C}\lambda_1(t) - \frac{1}{CR_1}\lambda_2(t)$$

These two equations give the rates of change of the values of $\lambda_1(t)$ and $\lambda_2(t)$ from which new values of $\lambda_1(t)$ and $\lambda_2(t)$ can be computed. We shall come back to this point later.

We define the *state* of a system to be a summary of the part of its past history which will affect its future behavior. Thus, knowing the state of a system at any instant enables us to determine the behavior of the system from that instant on when the input signal is given. In our first example that part of the past history of the system is summarized as the largest and the smallest value of the input the system has so far received. In the second example it is the current in the inductor and the voltage across the capacitor. The state of a system can be represented by the values of a number of variables which are called *state variables*. That is, a combination of values of the state variables represents a state of the system. Because we can imagine each combination of values as a geometric point in a multidimensional space, a description of system behavior by its states is called a *state space description*.

A system might have an infinite number of states or only a finite number of states. In our first example, the state variables are $\lambda_1(n)$ and $\lambda_2(n)$. Since $\lambda_1(n)$ and $\lambda_2(n)$ can assume any real value, the system has an infinite number of states. Similarly, the system in the second example also has an infinite number of states specified by the values of the state variables $\lambda_1(t)$ and $\lambda_2(t)$. On the other hand, let us consider an *odd-even counter*, the output of which at time instant n is equal to 1 if there is an even number of nonnegative input values among $\ldots, x(0), x(1), \ldots, x(n-2), x(n-1), x(n)$ and is equal to zero otherwise. Let $\lambda(n)$ be a state variable which equals 1 if there is an even number of nonnegative input values among $\ldots, x(0), x(1), \ldots, x(n-2), x(n-1)$ and equals 0 otherwise. The output $y(n)$ can be expressed in terms of $\lambda(n)$ and $x(n)$ as

$$y(n) = \begin{cases} 1 & \text{if } \lambda(n) = 1 \text{ and } x(n) < 0, \text{ or if } \lambda(n) = 0 \text{ and } x(n) \geq 0 \\ 0 & \text{otherwise} \end{cases}$$

Since the state variable $\lambda(n)$ assumes only one of two possible values, the system has only two states.

Although in all the examples we have seen in this section the number of state variables needed to specify the state of a system is finite, there are systems the states of which are described by an infinite number of state variables. As an example, consider a system whose input-output relationship is given by the equation

$$y(n) = x(n - m)$$

where m is equal to the magnitude of the integral part of the value of $x(n)$. Thus, if $x(n) = 3.2$, then $y(n) = x(n - 3)$ and so on. Since at any instant n_0 the system must be able to recall the value of $x(n_0 - m)$ for an arbitrary value of m, an infinite number of state variables are needed to record the history of the system. We shall limit our discussion to systems the states of which can be specified by a finite number of state variables.

3-2 STATE SPACE DESCRIPTION OF DISCRETE SYSTEMS

In this section we study the state space description of the behavior of discrete systems. Let the state variables of a discrete system be denoted $\lambda_1(n), \lambda_2(n), \ldots, \lambda_r(n)$. At any instant n, the value of the output of the system, $y(n)$, can be computed from the values of the state variables at that instant together with the values of the inputs $x(n), x(n+1), x(n+2), \ldots$. That is, in functional notation,

$$y(n) = f(\lambda_1(n), \lambda_2(n), \ldots, \lambda_r(n), x(n), x(n+1), x(n+2), \ldots) \tag{3-4}$$

If the system is causal, $y(n)$ will not depend on the future input values $x(n+1), x(n+2), \ldots$. We then have

$$y(n) = f(\lambda_1(n), \lambda_2(n), \ldots, \lambda_r(n), x(n)) \tag{3-5}$$

An equation like Eq. (3-4) or Eq. (3-5) is called an *output equation* of the system.

As time goes on, the history of the system changes. In other words, the state of the system changes with time. Therefore, the values of the state variables must be updated accordingly. Since a summary of the history of the system at time instant $n + 1$ can be determined from a summary of the history of the system at time instant n together with the input value at time instant n,† the values of the state variables at time instant $n + 1$ are to be computed from the values of the state variables and the value of the input at time instant n. That is, in functional notation,

† Unless otherwise specified, we limit our discussion to causal systems.

$$\lambda_1(n+1) = g_1(\lambda_1(n), \lambda_2(n), \ldots, \lambda_r(n), x(n))$$
$$\lambda_2(n+1) = g_2(\lambda_1(n), \lambda_2(n), \ldots, \lambda_r(n), x(n))$$
$$\cdots\cdots\cdots\cdots\cdots\cdots\cdots\cdots\cdots\cdots\cdots\cdots$$
$$\lambda_r(n+1) = g_r(\lambda_1(n), \lambda_2(n), \ldots, \lambda_r(n), x(n))$$

Equations like these are called *state equations* of the system.

As was pointed out in the preceding section, a description of system behavior by an output equation and a set of state equations is called the state space description of the system. Given the state space description of a system, we can determine the output of the system, $y(n)$, for $n \geq n_0$ if we are given the input to the system, $x(n)$, for $n \geq n_0$, together with the state of the system at $n = n_0$. The state of the system at n_0 is called the *initial state* of the system.

As an example, consider a system with two state variables the output equation of which is

$$y(n) = 2\lambda_1(n) + 2\lambda_1(n)\lambda_2(n) - n^2\lambda_2(n) + x(n)$$

and the state equations are

$$\lambda_1(n+1) = \lambda_1(n) - \lambda_2(n)$$
$$\lambda_2(n+1) = \lambda_1(n)x(n)$$

Given that $\lambda_1(0) = 2$, $\lambda_2(0) = 1$, and $x(0) = 5$, we have

$$y(0) = 2(2) + 2(2)(1) - 0^2(1) + 5 = 13$$
$$\lambda_1(1) = 2 - 1 = 1$$
$$\lambda_2(1) = 2(5) = 10$$

Suppose that $x(1) = 10$. We have

$$y(1) = 2(1) + 2(1)(10) - 1^2(10) + 10 = 22$$
$$\lambda_1(2) = 1 - 10 = -9$$
$$\lambda_2(2) = 1(10) = 10$$

Given the subsequent input values, we can go on to compute $y(2)$, $\lambda_1(3)$, $\lambda_2(3)$, and so on.

The block diagram in Fig. 3-2 should help to further understand the significance of the state space description of discrete systems. At any instant n, the system computes the value of $y(n)$ from the values of $x(n)$, $\lambda_1(n)$, $\lambda_2(n), \ldots, \lambda_r(n)$. Also, the system updates the summary of its history by computing the values of $\lambda_1(n+1)$, $\lambda_2(n+1), \ldots, \lambda_r(n+1)$. The values of $\lambda_1(n+1)$, $\lambda_2(n+1), \ldots, \lambda_r(n+1)$ are delayed for one unit of time and will be used to compute the values of $y(n+1)$, $\lambda_1(n+2)$, $\lambda_2(n+2), \ldots, \lambda_r(n+2)$ at the instant $n+1$. As an example, let

$$y(t) = -\lambda_1(n) + 2x(n)$$

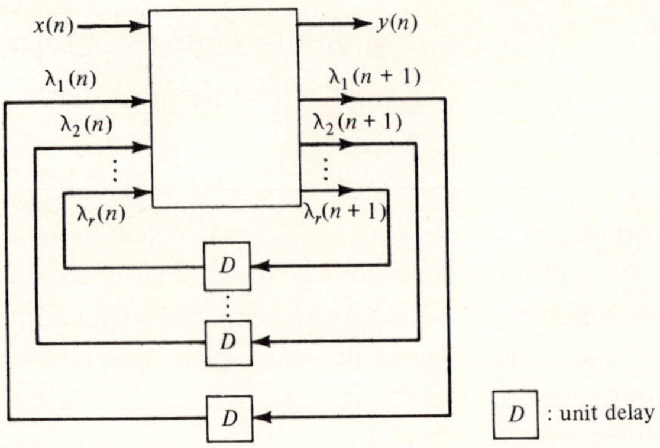

FIGURE 3-2

be the output equation and

$$\lambda_1(n+1) = \lambda_1(n)x(n)$$
$$\lambda_2(n+1) = \lambda_1(n) + 2\lambda_2(n)$$

be the state equations of a discrete system. The system can be described by the block diagram in Fig. 3-3, where ⊙ denotes a multiplier and ○ denotes a device that scales the magnitude of its input signal by the constant a.

Many of the properties of a system are exhibited in its output equation and state equations. A system is time-invariant if the coefficients in these equations are constants. For example, a system with an output equation

$$y(n) = \lambda(n) + 2x(n) - 7\lambda(n)x(n)$$

and a state equation

$$\lambda(n+1) = \lambda(n)[\lambda(n) + 2x(n)]$$

is time-invariant. On the other hand, a system with an output equation

$$y(n) = n\lambda(n) - x(n)$$

and a state equation

$$\lambda(n+1) = \lambda(n) + n^2$$

is a time-varying system. A system is linear if the output equation and the state equations are linear equations of the input signals and state variables and if the system is initially at rest.† For example, a system with an output equation

$$y(n) = n^2\lambda(n) + x(n)$$

† That is, the values of the state variables are all zero at the initial time instant.

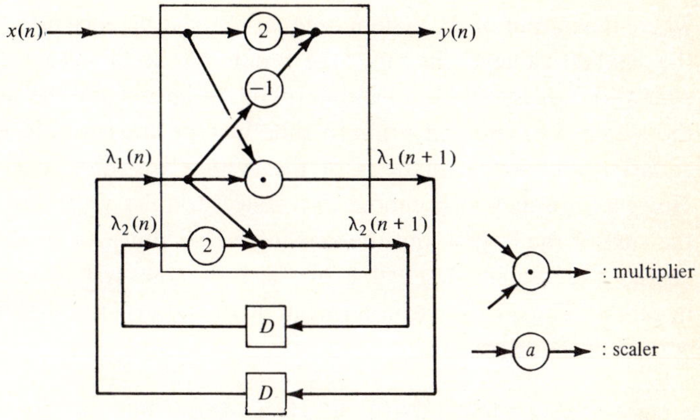

FIGURE 3-3

and a state equation

$$\lambda(n+1) = \lambda(n) + 2^{-n}x(n)$$

is linear if the system is initially at rest. Note, however, that a linear system might have a state space description that contains nonlinear output and/or state equations. (See Prob. 3-8.)

There are several points we should clarify: We note first that the output equation and the state equation formulation do not become more general if we let $y(n)$, $\lambda_1(n+1)$, $\lambda_2(n+1), \ldots, \lambda_r(n+1)$ be functions of some past values of the state variables as well. To see this point, suppose that we have a state equation

$$\lambda_1(n+1) = g_1(\lambda_1(n), \lambda_2(n-3), \ldots, \lambda_r(n), x(n)) \tag{3-6}$$

where the value of λ_1 at time $n+1$ depends on the value of λ_2 at time $n-3$. By introducing three new state variables, $\lambda_{r+1}(n)$, $\lambda_{r+2}(n)$, $\lambda_{r+3}(n)$, we have the additional state equations

$$\lambda_{r+1}(n+1) = \lambda_2(n)$$
$$\lambda_{r+2}(n+1) = \lambda_{r+1}(n)$$
$$\lambda_{r+3}(n+1) = \lambda_{r+2}(n)$$

Equation (3-6) can then be written

$$\lambda_1(n+1) = g_1(\lambda_1(n), \lambda_{r+3}(n), \ldots, \lambda_r(n), x(n))$$

because $\lambda_{r+3}(n) = \lambda_2(n-3)$.

Second, it should be pointed out that the state of a system can be specified by different sets of state variables. Consider the example of the discrete system in Sec. 3-1 where the output of the system is equal to the difference between the largest value and the smallest value of the input the system has so far received. An obvious choice of state variables is to let $\lambda_1(n)$ and $\lambda_2(n)$ be the largest and the smallest value of the input the system has received prior to time instant n, as we did in Sec. 3-1. The output equation and state equations of the system are then given by Eqs. (3-1) to (3-3). However, we may also choose two state variables $\gamma_1(n)$ and $\gamma_2(n)$ such that $\gamma_1(n)$ is the sum of the largest and the smallest value and $\gamma_2(n)$ is the difference between the largest and the smallest value of the input the system has received prior to time instant n. Thus, we have, in terms of $\gamma_1(n)$ and $\gamma_2(n)$, a slightly more complicated but correct output equation,

$$y(n) = \max\{\tfrac{1}{2}[\gamma_1(n) + \gamma_2(n)], x(n)\} - \min\{\tfrac{1}{2}[\gamma_1(n) - \gamma_2(n)], x(n)\}$$

and the corresponding state equations

$$\gamma_1(n+1) = \max\{\tfrac{1}{2}[\gamma_1(n) + \gamma_2(n)], x(n)\} + \min\{\tfrac{1}{2}[\gamma_1(n) - \gamma_2(n)], x(n)\}$$
$$\gamma_2(n+1) = \max\{\tfrac{1}{2}[\gamma_1(n) + \gamma_2(n)], x(n)\} - \min\{\tfrac{1}{2}[\gamma_1(n) - \gamma_2(n)], x(n)\}$$

It is immediately clear that the values of the state variables $\gamma_1(n)$, $\gamma_2(n)$ can be computed from the values of the state variables $\lambda_1(n)$, $\lambda_2(n)$:

$$\gamma_1(n) = \lambda_1(n) + \lambda_2(n)$$
$$\gamma_2(n) = \lambda_1(n) - \lambda_2(n)$$

and conversely:

$$\lambda_1(n) = \tfrac{1}{2}[\gamma_1(n) + \gamma_2(n)]$$
$$\lambda_2(n) = \tfrac{1}{2}[\gamma_1(n) - \gamma_2(n)]$$

Intuitively, since a set of state variables describes uniquely the state of a system, one expects to be able to transform one description into another. (We hasten to remark that we are not saying that the values of *any* set of state variables can be computed from those of any other set of state variables. This point will become clear after the discussion in the next paragraph.)

Third, although there is a minimum number of state variables that must be used to specify the state of a given system, it is perfectly all right to use more state variables than are necessary in a state space description of a system. In the preceding example, if we let $\delta_1(n)$ denote the largest value among $\ldots, x(0), \ldots, x(n-3), x(n-2)$; $\delta_2(n)$ denote the smallest value among $\ldots, x(0), \ldots, x(n-3), x(n-2)$; and $\delta_3(n)$ denote the input $x(n-1)$, we have as output equation

$$y(n) = \max[\delta_1(n), \delta_3(n), x(n)] - \min[\delta_2(n), \delta_3(n), x(n)]$$

and as state equations

$$\delta_1(n+1) = \max\,[\delta_1(n), \delta_3(n)]$$
$$\delta_2(n+1) = \min\,[\delta_2(n), \delta_3(n)]$$
$$\delta_3(n+1) = x(n)$$

Although such a formulation is redundant, it is nevertheless correct. We encourage the reader to convince himself (not just mathematically, but intuitively) that it is possible to compute the values of $\lambda_1(n)$ and $\lambda_2(n)$ [or $\gamma_1(n)$ and $\gamma_2(n)$] from the values of $\delta_1(n)$, $\delta_2(n)$, and $\delta_3(n)$, but not possible to compute the values of $\delta_1(n)$, $\delta_2(n)$, and $\delta_3(n)$ from the values of $\lambda_1(n)$ and $\lambda_2(n)$ [or $\gamma_1(n)$ and $\gamma_2(n)$].

*3-3 VECTOR DIFFERENCE EQUATIONS

As was pointed out in the preceding section, when the state equations and the output equation of a discrete system are given together with the initial state of the system, the response of the system to a given stimulus can be computed in a step-by-step manner. Specifically, we can compute $y(n_0)$, $\lambda_1(n_0+1)$, $\lambda_2(n_0+1)$, $\ldots$, $\lambda_r(n_0+1)$ from $\lambda_1(n_0)$, $\lambda_2(n_0)$, $\ldots$, $\lambda_r(n_0)$, $x(n_0)$. From $\lambda_1(n_0+1)$, $\lambda_2(n_0+1)$, $\ldots$, $\lambda_r(n_0+1)$, $x(n_0+1)$, we can then compute $y(n_0+1)$, $\lambda_1(n_0+2)$, $\lambda_2(n_0+2)$, $\ldots$, $\lambda_r(n_0+2)$, and so on. Such a step-by-step computation can be carried out compactly in matrix notation for time-invariant systems with linear state equations. Suppose that

$$\begin{aligned} \lambda_1(n+1) &= a_{11}\lambda_1(n) + a_{12}\lambda_2(n) + \cdots + a_{1r}\lambda_r(n) + b_1 x(n) \\ \lambda_2(n+1) &= a_{21}\lambda_1(n) + a_{22}\lambda_2(n) + \cdots + a_{2r}\lambda_r(n) + b_2 x(n) \\ &\cdots\cdots\cdots\cdots\cdots\cdots\cdots\cdots\cdots \\ \lambda_r(n+1) &= a_{r1}\lambda_1(n) + a_{r2}\lambda_2(n) + \cdots + a_{rr}\lambda_r(n) + b_r x(n) \end{aligned} \tag{3-7}$$

where the coefficients a_{ij} and b_i are constants. Let

$$\boldsymbol{\lambda}(n) = \begin{bmatrix} \lambda_1(n) \\ \lambda_2(n) \\ \vdots \\ \lambda_r(n) \end{bmatrix} \qquad \mathbf{B} = \begin{bmatrix} b_1 \\ b_2 \\ \vdots \\ b_r \end{bmatrix} \tag{3-8}$$

and

$$\mathbf{A} = \begin{bmatrix} a_{11} & a_{12} & \cdots & a_{1r} \\ a_{21} & a_{22} & \cdots & a_{2r} \\ \cdots & \cdots & \cdots & \cdots \\ a_{r1} & a_{r2} & \cdots & a_{rr} \end{bmatrix} \tag{3-9}$$

The r equations in (3-7) can now be written in matrix notation:

$$\boldsymbol{\lambda}(n+1) = \mathbf{A}\boldsymbol{\lambda}(n) + \mathbf{B}x(n) \tag{3-10}$$

Equation (3-10) can be viewed as a *single* difference equation in which the unknown is the vector $\boldsymbol{\lambda}(n)$. Consequently, such an equation is called a *vector difference equation* (as opposed to a *scalar difference equation*). From the given initial state of the system,

$$\boldsymbol{\lambda}(n_0) = \begin{bmatrix} \lambda_1(n_0) \\ \lambda_1(n_0) \\ \vdots \\ \lambda_r(n_0) \end{bmatrix}$$

we can compute $\boldsymbol{\lambda}(n)$ in a step-by-step manner:

$$\begin{aligned}
\boldsymbol{\lambda}(n_0+1) &= \mathbf{A}\boldsymbol{\lambda}(n_0) + \mathbf{B}x(n_0) \\
\boldsymbol{\lambda}(n_0+2) &= \mathbf{A}\boldsymbol{\lambda}(n_0+1) + \mathbf{B}x(n_0+1) \\
&= \mathbf{A}^2\boldsymbol{\lambda}(n_0) + \mathbf{A}\mathbf{B}x(n_0) + \mathbf{B}x(n_0+1) \\
\boldsymbol{\lambda}(n_0+3) &= \mathbf{A}\boldsymbol{\lambda}(n_0+2) + \mathbf{B}x(n_0+2) \\
&= \mathbf{A}^3\boldsymbol{\lambda}(n_0) + \mathbf{A}^2\mathbf{B}x(n_0) + \mathbf{A}\mathbf{B}x(n_0+1) + \mathbf{B}x(n_0+2)
\end{aligned} \tag{3-11}$$

$$\cdots\cdots\cdots\cdots\cdots\cdots\cdots\cdots\cdots\cdots\cdots\cdots$$

$$\begin{aligned}
\boldsymbol{\lambda}(n) &= \mathbf{A}\boldsymbol{\lambda}(n-1) + \mathbf{B}x(n-1) \\
&= \mathbf{A}^{n-n_0}\boldsymbol{\lambda}(n_0) + \mathbf{A}^{n-n_0-1}\mathbf{B}x(n_0) + \mathbf{A}^{n-n_0-2}\mathbf{B}x(n_0+1) + \cdots + \mathbf{B}x(n-1) \\
&= \mathbf{A}^{n-n_0}\boldsymbol{\lambda}(n_0) + \sum_{i=n_0}^{n-1} \mathbf{A}^{n-1-i}\mathbf{B}x(i)
\end{aligned}$$

As an example, let us consider the operation of a telephone company. In the nth year, the company raises $x(n)$ million dollars of new capital of which $0.75x(n)$ million dollars will be spent on installing new switching equipment for local service, and $0.25x(n)$ million dollars will be spent on installing new transmission cables to handle the increasing volume of long-distance telephone traffic. Each year, the company loses 20 cents on each dollar's worth of switching equipment. It earns a profit of 15 cents on each dollar's worth of transmission cables. However, this profit will be used to buy more switching equipment in the following year. We want to compute the net worth of the company in the nth year, $y(n)$, in millions of dollars. Clearly, the state of the system can be described by two state variables $\lambda_1(n)$ and $\lambda_2(n)$, where $\lambda_1(n)$ denotes the total worth of switching equipment and $\lambda_2(n)$ denotes the total worth of transmission cables the company owns in the nth year. We have the output equation

$$y(n) = \lambda_1(n) + \lambda_2(n)$$

and state equations

$$\lambda_1(n+1) = 0.80\lambda_1(n) + 0.15\lambda_2(n) + 0.75x(n)$$
$$\lambda_2(n+1) = \lambda_2(n) + 0.25x(n)$$

In matrix notation, the state equations can be written

$$\begin{bmatrix} \lambda_1(n+1) \\ \lambda_2(n+1) \end{bmatrix} = \begin{bmatrix} 0.8 & 0.15 \\ 0 & 1 \end{bmatrix} \begin{bmatrix} \lambda_1(n) \\ \lambda_2(n) \end{bmatrix} + \begin{bmatrix} 0.75 \\ 0.25 \end{bmatrix} x(n)$$

Suppose that we were given that $\lambda_1(0) = \lambda_2(0) = 0$ and

$$x(n) = 2^{-n+2} \qquad \text{for } n \geq 0$$

The successive values of the state variables are computed as

$$\begin{bmatrix} \lambda_1(1) \\ \lambda_2(1) \end{bmatrix} = \begin{bmatrix} 0.8 & 0.15 \\ 0 & 1 \end{bmatrix} \begin{bmatrix} 0 \\ 0 \end{bmatrix} + \begin{bmatrix} 0.75 \\ 0.25 \end{bmatrix} 4 = \begin{bmatrix} 3 \\ 1 \end{bmatrix}$$
$$\begin{bmatrix} \lambda_1(2) \\ \lambda_2(2) \end{bmatrix} = \begin{bmatrix} 0.8 & 0.15 \\ 0 & 1 \end{bmatrix} \begin{bmatrix} 3 \\ 1 \end{bmatrix} + \begin{bmatrix} 0.75 \\ 0.25 \end{bmatrix} 2 = \begin{bmatrix} 4.05 \\ 1.5 \end{bmatrix}$$
$$\begin{bmatrix} \lambda_1(3) \\ \lambda_2(3) \end{bmatrix} = \begin{bmatrix} 0.8 & 0.15 \\ 0 & 1 \end{bmatrix} \begin{bmatrix} 4.05 \\ 1.5 \end{bmatrix} + \begin{bmatrix} 0.75 \\ 0.25 \end{bmatrix} 1 = \begin{bmatrix} 4.215 \\ 1.75 \end{bmatrix}$$

The output at successive time instants is $y(0) = 0$, $y(1) = 4$, $y(2) = 5.55$, and $y(3) = 5.965$.

It is straightforward to carry out a step-by-step computation to determine the values of the output at successive time instants even when the output equation is a nonlinear equation in the state variables and the input. For instance, suppose we want to compute, in the preceding example, the amount in millions of dollars needed to acquire land to house the new facilities, which is given by the equation

$$y(n) = [(1 + 0.5n)\lambda_1(n)^2 + \lambda_2(n)]^{1/2} + x(n)$$

We have $y(0) = 4$, $y(1) = 5.808$, $y(2) = 6.857$, $y(3) = 7.294$, and so on.

The vector difference equation (3-10) can also be solved for a closed-form expression for the vector $\boldsymbol{\lambda}(n)$. Specifically, it can be shown that

$$\boldsymbol{\lambda}(n) = \mathbf{A}^n\mathbf{K} + \mathbf{P}$$

where $\mathbf{A}^n\mathbf{K}$ is the homogeneous solution, $\mathbf{P}$ is the particular solution, $\mathbf{K}$ is a column vector of constants to be determined by the initial state of the system, and $\mathbf{P}$ is a column vector determined by the input signal $x(n)$. We shall not discuss the derivation of this result because the steps are not particularly illuminating. Moreover, a parallel discussion for the solution of vector differential equations will be presented in Secs. 3-8 to 3-11.

By using matrix notation, the behavior of systems that have several inputs and outputs can also be conveniently described. Suppose that a system receives m input signals $x_1(n), x_2(n), \ldots, x_m(n)$ and generates k output signals $y_1(n), y_2(n), \ldots, y_k(n)$. Let the state equations and output equations of the system be

$$
\begin{aligned}
\lambda_1(n+1) &= a_{11}\lambda_1(n) + a_{12}\lambda_2(n) + \cdots + a_{1r}\lambda_r(n) + b_{11}x_1(n) \\
&\quad + b_{12}x_2(n) + \cdots + b_{1m}x_m(n) \\
\lambda_2(n+1) &= a_{21}\lambda_1(n) + a_{22}\lambda_2(n) + \cdots + a_{2r}\lambda_r(n) + b_{21}x_1(n) \\
&\quad + b_{22}x_2(n) + \cdots + b_{2m}x_m(n) \\
&\cdots\cdots\cdots\cdots\cdots\cdots\cdots\cdots\cdots\cdots \\
\lambda_r(n+1) &= a_{r1}\lambda_1(n) + a_{r2}\lambda_2(n) + \cdots + a_{rr}\lambda_r(n) + b_{r1}x_1(n) \\
&\quad + b_{r2}x_2(n) + \cdots + b_{rm}x_m(n)
\end{aligned}
\tag{3-12}
$$

and

$$
\begin{aligned}
y_1(n) &= c_{11}\lambda_1(n) + c_{12}\lambda_2(n) + \cdots + c_{1r}\lambda_r(n) + d_{11}x_1(n) \\
&\quad + d_{12}x_2(n) + \cdots + d_{1m}x_m(n) \\
y_2(n) &= c_{21}\lambda_1(n) + c_{22}\lambda_2(n) + \cdots + c_{2r}\lambda_r(n) + d_{21}x_1(n) \\
&\quad + d_{22}x_2(n) + \cdots + d_{2m}x_m(n) \\
&\cdots\cdots\cdots\cdots\cdots\cdots\cdots\cdots\cdots\cdots \\
y_k(n) &= c_{k1}\lambda_1(n) + c_{k2}\lambda_2(n) + \cdots + c_{kr}\lambda_r(n) + d_{k1}x_1(n) \\
&\quad + d_{k2}x_2(n) + \cdots + d_{km}x_m(n)
\end{aligned}
\tag{3-13}
$$

Let $\mathbf{B}$ denote the $r \times m$ matrix

$$
\begin{bmatrix}
b_{11} & b_{12} & \cdots & b_{1m} \\
b_{21} & b_{22} & \cdots & b_{2m} \\
\cdots & \cdots & \cdots & \cdots \\
b_{r1} & b_{r2} & \cdots & b_{rm}
\end{bmatrix}
$$

$\mathbf{C}$ denote the $k \times r$ matrix

$$
\begin{bmatrix}
c_{11} & c_{12} & \cdots & c_{1r} \\
c_{21} & c_{22} & \cdots & c_{2r} \\
\cdots & \cdots & \cdots & \cdots \\
c_{k1} & c_{k2} & \cdots & c_{kr}
\end{bmatrix}
$$

and **D** denote the $k \times m$ matrix

$$\begin{bmatrix} d_{11} & d_{12} & \cdots & d_{1m} \\ d_{21} & d_{22} & \cdots & d_{2m} \\ \cdots & \cdots & \cdots & \cdots \\ d_{k1} & d_{k2} & \cdots & d_{km} \end{bmatrix}$$

The r equations (3-12) can be written in matrix notation:

$$\boldsymbol{\lambda}(n+1) = \mathbf{A}\boldsymbol{\lambda}(n) + \mathbf{B}\mathbf{x}(n)$$

where $\boldsymbol{\lambda}(n)$ is the column vector in Eq. (3-8), **A** is the constant matrix in Eq. (3-9), and $\mathbf{x}(n)$ is the column vector

$$\begin{bmatrix} x_1(n) \\ x_2(n) \\ \vdots \\ x_m(n) \end{bmatrix}$$

Similarly, let

$$\mathbf{y}(n) = \begin{bmatrix} y_1(n) \\ y_2(n) \\ \vdots \\ y_k(n) \end{bmatrix}$$

Equations (3-13) can be written

$$\mathbf{y}(n) = \mathbf{C}\boldsymbol{\lambda}(n) + \mathbf{D}\mathbf{x}(n)$$

For a system with multiple inputs, corresponding to Eq. (3-11), we have

$$\boldsymbol{\lambda}(n) = \mathbf{A}^{n-n_0}\boldsymbol{\lambda}(n_0) + \sum_{i=n_0}^{n-1} \mathbf{A}^{n-1-i}\mathbf{B}\mathbf{x}(i)$$

As an example of systems with more than one input and one output, let us consider the operation of a bank. Let $x_1(n)$ denote the total amount of new deposits in the nth year, of which 50 percent are deposited in checking accounts, 30 percent in savings accounts, and 20 percent in trust accounts. Let $x_2(n)$ denote the total amount withdrawn from the bank in the nth year, of which 60 percent is from checking accounts, 30 percent is from savings accounts, and 10 percent is from trust accounts. The yearly interest rates for the savings accounts and trust accounts are 10 and 20 percent, respectively. At the end of a year, the bank automatically transfers for each customer half of the interest earned in his savings and trust accounts to his checking account, and the other half to his trust account. The bank charges a yearly service fee of 10 cents per dollar deposited in checking accounts. Such service fees are deducted from the checking accounts annually. Suppose that the bank spends 2 cents on bookkeeping expenses for each dollar deposited or withdrawn, and it makes a yearly profit of 50 cents per dollar by investing the money in all accounts. We want to compute $y_1(n)$, the average yearly profit of the bank, and $y_2(n)$, the yearly expense (interest paid plus expenses on bookkeeping less service charge). Let $\lambda_1(n)$, $\lambda_2(n)$, and $\lambda_3(n)$ denote the amount deposited in the checking, savings, and trust accounts, respectively. Using matrix notation, we have the vector state equation

$$\begin{bmatrix} \lambda_1(n+1) \\ \lambda_2(n+1) \\ \lambda_3(n+1) \end{bmatrix} = \begin{bmatrix} 0.9 & 0.05 & 0.1 \\ 0 & 1 & 0 \\ 0 & 0.05 & 1.10 \end{bmatrix} \begin{bmatrix} \lambda_1(n) \\ \lambda_2(n) \\ \lambda_3(n) \end{bmatrix} + \begin{bmatrix} 0.5 & -0.6 \\ 0.3 & -0.3 \\ 0.2 & -0.1 \end{bmatrix} \begin{bmatrix} x_1(n) \\ x_2(n) \end{bmatrix}$$

and the vector output equation

$$\begin{bmatrix} y_1(n) \\ y_2(n) \end{bmatrix} = \begin{bmatrix} 0.5 & 0.5 & 0.5 \\ -0.1 & 0.1 & 0.2 \end{bmatrix} \begin{bmatrix} \lambda_1(n) \\ \lambda_2(n) \\ \lambda_3(n) \end{bmatrix} + \begin{bmatrix} 0 & 0 \\ 0.02 & 0.02 \end{bmatrix} \begin{bmatrix} x_1(n) \\ x_2(n) \end{bmatrix}$$

3-4 SIMULTANEOUS DIFFERENCE EQUATIONS

For a given input signal $x(n)$, the set of state equations can be viewed as a set of simultaneous difference equations in the unknowns $\lambda_1(n), \lambda_2(n), \ldots, \lambda_r(n)$. Consequently, closed-form expressions for $\lambda_1(n), \lambda_2(n), \ldots, \lambda_r(n)$ can be obtained by solving the set of state equations. There is no general procedure for solving simultaneous difference equations. However, when the state equations are linear difference equations with constant coefficients, they can be solved for closed-form expressions for the state variables. These expressions can then be substituted into the output equation for an expression for $y(n)$.

Each of the solutions of the simultaneous difference equations (3-7) consists of a homogeneous solution and a particular solution. The homogeneous solutions satisfy Eqs. (3-7) when $x(n)$ is set to zero. The particular solutions depend on $x(n)$ and satisfy Eqs. (3-7). We now show that the homogeneous solutions are linear combinations of terms of the form $A\alpha^n$. Setting $x(n) = 0$ and substituting

$$\lambda_1(n) = A_1\alpha^n$$
$$\lambda_2(n) = A_2\alpha^n$$
$$\cdots\cdots\cdots\cdots$$
$$\lambda_r(n) = A_r\alpha^n$$

into Eqs. (3-7), we obtain

$$
\begin{aligned}
A_1\alpha^{n+1} &= a_{11}A_1\alpha^n + a_{12}A_2\alpha^n + \cdots + a_{1r}A_r\alpha^n \\
A_2\alpha^{n+1} &= a_{21}A_1\alpha^n + a_{22}A_2\alpha^n + \cdots + a_{2r}A_r\alpha^n \\
&\cdots\cdots\cdots\cdots\cdots\cdots\cdots\cdots\cdots\cdots \\
A_r\alpha^{n+1} &= a_{r1}A_1\alpha^n + a_{r2}A_2\alpha^n + \cdots + a_{rr}A_r\alpha^n
\end{aligned}
$$

that is,

$$
\begin{aligned}
(a_{11}-\alpha)A_1 + a_{12}A_2 + \cdots + a_{1r}A_r &= 0 \\
a_{21}A_1 + (a_{22}-\alpha)A_2 + \cdots + a_{2r}A_r &= 0 \\
\cdots\cdots\cdots\cdots\cdots\cdots\cdots\cdots\cdots \\
a_{r1}A_1 + a_{r2}A_2 + \cdots + (a_{rr}-\alpha)A_r &= 0
\end{aligned}
\tag{3-14}
$$

Thus, if the values of α, $A_1, A_2, \ldots, A_r$ are so chosen that Eqs. (3-14) are satisfied, then $A_1\alpha^n, A_2\alpha^n, \ldots, A_r\alpha^n$ will be homogeneous solutions of the state equations. We recall from our study of the theory of simultaneous equations that if the value of the determinant

$$
\begin{vmatrix}
a_{11}-\alpha & a_{12} & \cdots & a_{1r} \\
a_{21} & a_{22}-\alpha & \cdots & a_{2r} \\
\cdots & \cdots & \cdots & \cdots \\
a_{r1} & a_{r2} & \cdots & a_{rr}-\alpha
\end{vmatrix}
\tag{3-15}
$$

is nonzero, then the only solutions to Eqs. (3-14) are

$$A_1 = A_2 = \cdots = A_r = 0$$

Clearly, this is an uninteresting case. On the other hand, if the value of determinant (3-15) is zero, a nontrivial solution for $A_1, A_2, \ldots, A_r$ exists. Specifically, if the value of the determinant (3-15) is equal to zero for $\alpha = \alpha_1$ then, after setting the value of α to α_1 in Eqs. (3-14), we can solve these equations for $A_2, A_3 \ldots, A_r$ in terms of A_1.† We thus obtain

$$
\begin{aligned}
A_2 &= g_{21}A_1 \\
A_3 &= g_{31}A_1 \\
&\cdots\cdots \\
A_r &= g_{r1}A_1
\end{aligned}
$$

† The reader is reminded that only $r-1$ of the r equations (3-14) are independent.

where $g_{21}, g_{31}, \ldots, g_{r1}$ are constants. It follows that

$$\begin{aligned}
\lambda_1(n) &= A_1\alpha_1^{\,n} \\
\lambda_2(n) &= g_{21}A_1\alpha_1^{\,n} \\
\lambda_3(n) &= g_{31}A_1\alpha_1^{\,n} \\
&\cdots\cdots\cdots\cdots \\
\lambda_r(n) &= g_{r1}A_1\alpha_1^{\,n}
\end{aligned}$$

satisfy Eqs. (3-7) when $x(n)$ is set to zero.

To determine the general form of the homogeneous solutions, we note that the equation

$$\begin{vmatrix} a_{11}-\alpha & a_{12} & \cdots & a_{1r} \\ a_{21} & a_{22}-\alpha & \cdots & a_{2r} \\ \cdots & \cdots & \cdots & \cdots \\ a_{r1} & a_{r2} & \cdots & a_{rr}-\alpha \end{vmatrix} = 0 \tag{3-16}$$

has r roots. Let these roots be denoted $\alpha_1, \alpha_2, \ldots, \alpha_r$. If these roots are distinct, the homogeneous solutions will be

$$\begin{aligned}
\lambda_1(n) &= A_1\alpha_1^{\,n} + B_1\alpha_2^{\,n} + \cdots + R_1\alpha_r^{\,n} \\
\lambda_2(n) &= A_1 g_{21}\alpha_1^{\,n} + B_1 g_{22}\alpha_2^{\,n} + \cdots + R_1 g_{2r}\alpha_r^{\,n} \\
&\cdots\cdots\cdots\cdots\cdots\cdots\cdots\cdots \\
\lambda_r(n) &= A_1 g_{r1}\alpha_1^{\,n} + B_1 g_{r2}\alpha_2^{\,n} + \cdots + R_1 g_{rr}\alpha_r^{\,n}
\end{aligned}$$

where the coefficients $A_1, B_1, \ldots, R_1$ are to be determined by a set of boundary conditions such as the initial state of the system or the state of the system at some other time instant. The roots $\alpha_1, \alpha_2, \ldots, \alpha_r$ are called the characteristic roots of the set of Eqs. (3-7).†

As an example, consider the operation of the two storage warehouses of a chain of department stores. In each week, one-quarter of the merchandise in warehouse 1 is transferred to warehouse 2, and one-quarter of the merchandise in warehouse 2 is

† Those readers who have had a course in matrix theory will recall that $\alpha_1, \alpha_2, \ldots, \alpha_r$ are the characteristic values of the matrix

$$\begin{bmatrix} a_{11} & a_{12} & \cdots & a_{1r} \\ a_{21} & a_{22} & \cdots & a_{2r} \\ \cdots & \cdots & \cdots & \cdots \\ a_{r1} & a_{r2} & \cdots & a_{rr} \end{bmatrix}$$

and

$$A_1\begin{bmatrix} 1 \\ g_{21} \\ g_{31} \\ \vdots \\ g_{r1} \end{bmatrix} \qquad B_1\begin{bmatrix} 1 \\ g_{22} \\ g_{32} \\ \vdots \\ g_{r2} \end{bmatrix} \qquad \cdots \qquad R_1\begin{bmatrix} 1 \\ g_{2r} \\ g_{3r} \\ \vdots \\ g_{rr} \end{bmatrix}$$

are their corresponding characteristic vectors.

transferred to warehouse 1. Also, merchandise is delivered to warehouse 1 and merchandise is shipped from warehouse 2 on a weekly basis. Specifically, in the nth week $x(n)$ thousand cubic feet of merchandise is delivered to warehouse 1 and $2x(n)$ thousand cubic feet of merchandise is shipped from warehouse 2. Let $\lambda_1(n)$ and $\lambda_2(n)$ denote the total merchandise, in thousand cubic feet, stored in warehouses 1 and 2 in the nth week, respectively. Using $\lambda_1(n)$ and $\lambda_2(n)$ as state variables, we have as state equations

$$\begin{aligned} \lambda_1(n+1) &= \tfrac{3}{4}\lambda_1(n) + \tfrac{1}{4}\lambda_2(n) + x(n) \\ \lambda_2(n+1) &= \tfrac{1}{4}\lambda_1(n) + \tfrac{3}{4}\lambda_2(n) - 2x(n) \end{aligned} \tag{3-17}$$

To determine the homogeneous solutions, we determine first the roots of the equation

$$\begin{vmatrix} \frac{3}{4} - \alpha & \frac{1}{4} \\ \frac{1}{4} & \frac{3}{4} - \alpha \end{vmatrix} = 0$$

which simplifies to

$$\alpha^2 - \tfrac{3}{2}\alpha + \tfrac{1}{2} = 0 \tag{3-18}$$

The two roots of Eq. (3-18) are

$$\alpha_1 = 1 \qquad \alpha_2 = \tfrac{1}{2}$$

Thus, the homogeneous solutions are of the form

$$\begin{aligned} \lambda_1(n) &= A_1 1^n + B_1(\tfrac{1}{2})^n = A_1 + B_1(\tfrac{1}{2})^n \\ \lambda_2(n) &= A_2 1^n + B_2(\tfrac{1}{2})^n = A_2 + B_2(\tfrac{1}{2})^n \end{aligned}$$

However, according to Eqs. (3-14), A_1 and A_2 are related by

$$\begin{aligned} (\tfrac{3}{4} - 1)A_1 + \tfrac{1}{4}A_2 &= 0 \\ \tfrac{1}{4}A_1 + (\tfrac{3}{4} - 1)A_2 &= 0 \end{aligned}$$

In other words,

$$A_2 = A_1$$

Similarly, B_1 and B_2 are related by

$$\begin{aligned} (\tfrac{3}{4} - \tfrac{1}{2})B_1 + \tfrac{1}{4}B_2 &= 0 \\ \tfrac{1}{4}B_1 + (\tfrac{3}{4} - \tfrac{1}{2})B_2 &= 0 \end{aligned}$$

That is,

$$B_2 = -B_1$$

Therefore, the homogeneous solutions are

$$\begin{aligned} \lambda_1(n) &= A_1 + B_1(\tfrac{1}{2})^n \\ \lambda_2(n) &= A_1 - B_1(\tfrac{1}{2})^n \end{aligned}$$

The constants A_1 and B_1 are to be determined by a set of boundary conditions.

When the determinantal equation (3-16) has multiple roots, the homogeneous solutions will be in slightly different form. Instead of a general discussion involving complicated notation, we limit our presentation to an illustrative example. Let

$$\begin{aligned} \lambda_1(n+1) &= \lambda_1(n) - \lambda_2(n) \\ \lambda_2(n+1) &= \lambda_1(n) + 3\lambda_2(n) \end{aligned} \tag{3-19}$$

The equation

$$\begin{vmatrix} 1-\alpha & -1 \\ 1 & 3-\alpha \end{vmatrix} = 0$$

has a double root 2. Let

$$\begin{aligned} \lambda_1(n) &= A_1 2^n \\ \lambda_2(n) &= A_2 2^n \end{aligned}$$

We determine that $A_2 = -A_1$, as we did above. However, recognizing that the homogeneous solutions of $\lambda_1(n)$ and $\lambda_2(n)$ must contain still another term, we try the solutions

$$\begin{aligned} \lambda_1(n) &= B_1 n 2^n \\ \lambda_2(n) &= B_2 n 2^n \end{aligned}$$

Substituting these expressions into Eqs. (3-19), we obtain

$$\begin{aligned} B_1 n 2^{n+1} + B_1 2^{n+1} &= B_1 n 2^n - B_2 n 2^n \\ B_2 n 2^{n+1} + B_2 2^{n+1} &= B_1 n 2^n + 3B_2 n 2^n \end{aligned}$$

which simplify to

$$\begin{aligned} -(B_1 + B_2)n - 2B_1 &= 0 \\ (B_1 + B_2)n - 2B_2 &= 0 \end{aligned}$$

That these equations should be satisfied for all n leads to the trivial solutions

$$B_1 = 0 \qquad B_2 = 0$$

Let us next try the solutions

$$\begin{aligned} \lambda_1(n) &= B_1 n 2^n + C_1 2^n \\ \lambda_2(n) &= B_2 n 2^n + C_2 2^n \end{aligned}$$

Substituting these expressions into Eqs. (3-19), we obtain

$$\begin{aligned} B_1 n 2^{n+1} + B_1 2^{n+1} + C_1 2^{n+1} &= B_1 n 2^n + C_1 2^n - B_2 n 2^n - C_2 2^n \\ B_2 n 2^{n+1} + B_2 2^{n+1} + C_2 2^{n+1} &= B_1 n 2^n + C_1 2^n + 3B_2 n 2^n + 3C_2 2^n \end{aligned}$$

which simplify to

$$-(B_1 + B_2)n - (C_1 + C_2 + 2B_1) = 0$$
$$(B_1 + B_2)n + (C_1 + C_2 - 2B_2) = 0$$

So that these equations are satisfied for all n, we must have

$$B_1 + B_2 = 0$$
$$C_1 + C_2 + 2B_1 = 0$$
$$C_1 + C_2 - 2B_2 = 0$$

That is,

$$B_2 = -B_1$$
$$C_2 = -2B_1 - C_1$$

Therefore, the complete homogeneous solutions are

$$\lambda_1(n) = A_1 2^n + B_1 n2^n + C_1 2^n$$
$$\lambda_2(n) = -A_1 2^n - B_1 n2^n - 2B_1 2^n - C_1 2^n$$

Letting $D_1 = A_1 + C_1$, we have

$$\lambda_1(n) = D_1 2^n + B_1 n2^n$$
$$\lambda_2(n) = (-D_1 - 2B_1)2^n - B_1 n2^n$$

The constants B_1 and D_1 are to be determined by a set of boundary conditions.

A general method for determining the particular solutions will be discussed in Chap. 11. Here we mention only that, similar to the determination of the particular solution of a difference equation, it is possible to guess the general form of the particular solutions from the form of the input $x(n)$. To illustrate this point, let us consider the case in which the input $x(n)$ is of the form β^n. We guess that the particular solutions are of the form

$$\lambda_1(n) = P_1\beta^n$$
$$\lambda_2(n) = P_2\beta^n$$
$$\cdots\cdots\cdots\cdots$$
$$\lambda_r(n) = P_r\beta^n$$

where $P_1, P_2, \ldots, P_r$ are coefficients to be determined. Substituting these expressions into Eqs. (3-7), we obtain

$$P_1\beta^{n+1} = a_{11}P_1\beta^n + a_{12}P_2\beta^n + \cdots + a_{1r}P_r\beta^n + b_1\beta^n$$
$$P_2\beta^{n+1} = a_{21}P_1\beta^n + a_{22}P_2\beta^n + \cdots + a_{2r}P_r\beta^n + b_2\beta^n$$
$$\cdots\cdots\cdots\cdots\cdots\cdots\cdots\cdots\cdots\cdots\cdots\cdots$$
$$P_r\beta^{n+1} = a_{r1}P_1\beta^n + a_{r2}P_2\beta^n + \cdots + a_{rr}P_r\beta^n + b_r\beta^n$$

That is,

$$\begin{aligned}
(a_{11}-\beta)P_1 + a_{12}P_2 + \cdots + a_{1r}P_r &= -b_1 \\
a_{21}P_1 + (a_{22}-\beta)P_2 + \cdots + a_{2r}P_r &= -b_2 \\
\cdots\cdots\cdots\cdots\cdots\cdots\cdots\cdots\cdots\cdots \\
a_{r1}P_1 + a_{r2}P_2 + \cdots + (a_{rr}-\beta)P_r &= -b_r
\end{aligned} \tag{3-20}$$

From the set of equations, the constants $P_1, P_2, \ldots, P_r$ can be solved.†

For example, let $x(n)$ be 2^n in Eqs. (3-17). Substituting

$$\lambda_1(n) = P_1 2^n$$
$$\lambda_2(n) = P_2 2^n$$

into the state equations (3-17), we obtain

$$P_1 2^{n+1} = \tfrac{3}{4}P_1 2^n + \tfrac{1}{4}P_2 2^n + 2^n$$
$$P_2 2^{n+1} = \tfrac{1}{4}P_1 2^n + \tfrac{3}{4}P_2 2^n - 2(2^n)$$

which simplify to

$$\tfrac{5}{4}P_1 - \tfrac{1}{4}P_2 = 1$$
$$-\tfrac{1}{4}P_1 + \tfrac{5}{4}P_2 = -2$$

The solutions of this set of equations are

$$P_1 = \tfrac{1}{2} \qquad P_2 = -\tfrac{3}{2}$$

Therefore, the total solutions of Eqs. (3-17) are

$$\begin{aligned}
\lambda_1(n) &= A_1 + B_1(\tfrac{1}{2})^n + \tfrac{1}{2}(2^n) \\
\lambda_2(n) &= A_1 - B_1(\tfrac{1}{2})^n - \tfrac{3}{2}(2^n)
\end{aligned} \tag{3-21}$$

Finally, we note that the values of the state variables at any time instant can be used as boundary conditions to determine the coefficients in the homogeneous solutions. As an example, for the solutions in Eqs. (3-21) let us suppose that at $n = 0$ the system is in a state specified by $\lambda_1(0) = 2$ and $\lambda_2(0) = 3$. The coefficients in the homogeneous solutions are determined as

$$A_1 = 3 \qquad B_1 = -\tfrac{3}{2}$$

† Instead of going into the details, we leave to the reader the questions: Under what conditions can Eqs. (3-20) be solved for $P_1, P_2, \ldots, P_r$? What should the particular solutions be if Eqs. (3-20) cannot be solved for $P_1, P_2, \ldots, P_r$? [*Hint:* What happens if β is a characteristic root of Eqs. (3-7)?]

3-5 DIFFERENCE EQUATIONS AND STATE EQUATIONS

We saw in Chap. 2 that the input-output relationship of many discrete systems can be described by difference equations. We show now that a difference equation description of a discrete system can be transformed into a state space description. Let us consider first a simple case in which the difference equation is of the form

$$C_0 y(n) + C_1 y(n-1) + C_2 y(n-2) + \cdots + C_r y(n-r) = E_0 x(n)$$

where the right-hand side of the equation contains only the term $E_0 x(n)$. For a system described by such a difference equation, there is an obvious choice of state variables for the state space description. At time n, the part of its past history that affects the future behavior of the system can be summarized by the value of $y(n-1)$, $y(n-2)$, $\ldots$, $y(n-r)$, because the value of the output $y(n)$ at time n is determined uniquely by the values of these variables and the value of the input $x(n)$. Mathematically, we see that if we let

$$\begin{aligned} \lambda_1(n) &= y(n-r) \\ \lambda_2(n) &= y(n-r+1) \\ \lambda_3(n) &= y(n-r+2) \\ &\cdots\cdots\cdots \\ \lambda_{r-1}(n) &= y(n-2) \\ \lambda_r(n) &= y(n-1) \end{aligned}$$

we have the state equations

$$\begin{aligned} \lambda_1(n+1) &= \lambda_2(n) \\ \lambda_2(n+1) &= \lambda_3(n) \\ &\cdots\cdots\cdots \\ \lambda_{r-1}(n+1) &= \lambda_r(n) \\ \lambda_r(n+1) &= -\frac{1}{C_0}[C_r\lambda_1(n) + C_{r-1}\lambda_2(n) + \cdots + C_1\lambda_r(n) - E_0 x(n)] \end{aligned} \tag{3-22}$$

and the output equation

$$y(n) = -\frac{1}{C_0}[C_r\lambda_1(n) + C_{r-1}\lambda_2(n) + \cdots + C_1\lambda_r(n) - E_0 x(n)]$$

As an example, we note that a system described by the difference equation

$$3y(n) - 2y(n-1) + 2y(n-2) = 5x(n) \tag{3-23}$$

can also be described by the state equations

$$\lambda_1(n+1) = \lambda_2(n) \tag{3-24}$$

$$\lambda_2(n+1) = -\tfrac{2}{3}\lambda_1(n) + \tfrac{2}{3}\lambda_2(n) + \tfrac{5}{3}x(n) \tag{3-25}$$

and the output equation

$$y(n) = -\tfrac{2}{3}\lambda_1(n) + \tfrac{2}{3}\lambda_2(n) + \tfrac{5}{3}x(n)$$

There is another obvious way to choose the state variables. Writing the difference equation as

$$y(n) = -\frac{1}{C_0}[C_r y(n-r) + C_{r-1}y(n-r+1) + \cdots + C_1 y(n-1) - E_0 x(n)]$$

we see that if we choose as state variables

$$\gamma_1(n) = -\frac{1}{C_0}[C_r y(n-r) + C_{r-1}y(n-r+1) + \cdots + C_2 y(n-2) + C_1 y(n-1)]$$

$$\gamma_2(n) - -\frac{1}{C_0}[C_r y(n-r+1) + C_{r-1}y(n-r+2) + \cdots + C_2 y(n-1)]$$

$$\gamma_3(n) = -\frac{1}{C_0}[C_r y(n-r+2) + C_{r-1}y(n-r+3) + \cdots + C_3 y(n-1)] \tag{3-26}$$

....

$$\gamma_r(n) = -\frac{C_r}{C_0}y(n-1)$$

we have, as the output equation

$$y(n) = \gamma_1(n) + \frac{E_0}{C_0}x(n)$$

and, as state equations,

$$\gamma_1(n+1) = \gamma_2(n) - \frac{C_1}{C_0}y(n)$$

$$= \gamma_2(n) - \frac{C_1}{C_0}\gamma_1(n) - \frac{C_1}{C_0}\frac{E_0}{C_0}x(n)$$

$$\gamma_2(n+1) = \gamma_3(n) - \frac{C_2}{C_0}y(n)$$

$$= \gamma_3(n) - \frac{C_2}{C_0}\gamma_1(n) - \frac{C_2}{C_0}\frac{E_0}{C_0}x(n)$$

$$\gamma_3(n+1) = \gamma_4(n) - \frac{C_3}{C_0} y(n)$$

$$= \gamma_4(n) - \frac{C_3}{C_0}\gamma_1(n) - \frac{C_3}{C_0}\frac{E_0}{C_0} x(n)$$

$$\cdots\cdots\cdots\cdots\cdots\cdots\cdots\cdots\cdots\cdots$$

$$\gamma_r(n+1) = -\frac{C_r}{C_0} y(n)$$

$$= -\frac{C_r}{C_0}\gamma_1(n) - \frac{C_r}{C_0}\frac{E_0}{C_0} x(n)$$

As an example, the system described by Eq. (3-23) can also be described by the output equation

$$y(n) = \gamma_1(n) + \tfrac{5}{3}x(n) \tag{3-27}$$

and the state equations

$$\gamma_1(n+1) = \gamma_2(n) + \tfrac{2}{3}\gamma_1(n) + \tfrac{10}{9}x(n) \tag{3-28}$$

$$\gamma_2(n+1) = -\tfrac{2}{3}\gamma_1(n) - \tfrac{10}{9}x(n) \tag{3-29}$$

According to our discussion in Sec. 3-2, the state of the system described by Eq. (3-23) can either be specified by the values of $\lambda_1(n)$ and $\lambda_2(n)$ as in Eqs. (3-24) and (3-25) or be specified by the values of $\gamma_1(n)$ and $\gamma_2(n)$ as in Eqs. (3-28) and (3-29). It follows that we can compute the values of $\gamma_1(n)$ and $\gamma_2(n)$ from the values of $\lambda_1(n)$ and $\lambda_2(n)$, and vice versa. Indeed, we have the following relations:

$$\begin{aligned} \gamma_1(n) &= -\tfrac{2}{3}\lambda_1(n) + \tfrac{2}{3}\lambda_2(n) \\ \gamma_2(n) &= \qquad\qquad\quad -\tfrac{2}{3}\lambda_2(n) \end{aligned} \tag{3-30}$$

and

$$\begin{aligned} \lambda_1(n) &= -\tfrac{3}{2}\gamma_1(n) - \tfrac{3}{2}\gamma_2(n) \\ \lambda_2(n) &= \qquad\qquad\quad -\tfrac{3}{2}\gamma_2(n) \end{aligned}$$

In general, the two sets of state variables in Eqs. (3-22) and (3-26) are related according to the following set of equations:

$$\gamma_1(n) = -\frac{1}{C_0}[C_r\lambda_1(n) + C_{r-1}\lambda_2(n) + \cdots + C_1\lambda_r(n)]$$

$$\gamma_2(n) = -\frac{1}{C_0}[C_r\lambda_2(n) + C_{r-1}\lambda_3(n) + \cdots + C_2\lambda_r(n)]$$

$$\cdots\cdots\cdots\cdots\cdots\cdots\cdots\cdots\cdots\cdots$$

$$\gamma_r(n) = -\frac{1}{C_0}C_r\lambda_r(n)$$

We now consider the general case in which the input-output relationship of a discrete system is described by a difference equation of the form

$$C_0 y(n) + C_1 y(n-1) + C_2 y(n-2) + \cdots + C_r y(n-r)$$
$$= E_0 x(n) + E_1 x(n-1) + E_2 x(n-2) + \cdots + E_m x(n-m) \quad (3\text{-}31)$$

Extending directly the first procedure presented above, we define the state variables:

$$\lambda_1(n) = y(n-r)$$
$$\lambda_2(n) = y(n-r+1)$$
$$\cdots\cdots\cdots\cdots\cdots\cdots$$
$$\lambda_r(n) = y(n-1)$$
$$\lambda_{r+1}(n) = x(n-m)$$
$$\lambda_{r+2}(n) = x(n-m+1)$$
$$\cdots\cdots\cdots\cdots\cdots\cdots$$
$$\lambda_{r+m}(n) = x(n-1)$$

The output equation is

$$y(n) = -\frac{1}{C_0}[C_r\lambda_1(n) + C_{r-1}\lambda_2(n) + \cdots + C_1\lambda_r(n)$$
$$- E_m\lambda_{r+1}(n) - E_{m-1}\lambda_{r+2}(n) - \cdots - E_1\lambda_{r+m}(n) - E_0 x(n)]$$

The state equations are

$$\lambda_1(n+1) = \lambda_2(n)$$
$$\lambda_2(n+1) = \lambda_3(n)$$
$$\cdots\cdots\cdots\cdots\cdots\cdots\cdots\cdots\cdots\cdots\cdots\cdots$$
$$\lambda_{r-1}(n+1) = \lambda_r(n)$$
$$\lambda_r(n+1) = -\frac{1}{C_0}[C_r\lambda_1(n) + C_{r-1}\lambda_2(n) + \cdots + C_1\lambda_r(n)$$
$$- E_m\lambda_{r+1}(n) - E_{m-1}\lambda_{r+2}(n) - \cdots - E_1\lambda_{r+m}(n) - E_0 x(n)]$$
$$\lambda_{r+1}(n+1) = \lambda_{r+2}(n)$$
$$\lambda_{r+2}(n+1) = \lambda_{r+3}(n)$$
$$\cdots\cdots\cdots\cdots\cdots\cdots$$
$$\lambda_{r+m-1}(n+1) = \lambda_{r+m}(n)$$
$$\lambda_{r+m}(n+1) = x(n)$$

Although such a formulation is a correct one, as the reader probably has suspected, it uses an excessive number of state variables. Indeed, for $r \geq m$, a state space description using r state variables and, for $r < m$, a state space description using m state variables can be obtained. We present here the case $r \geq m$. (For the case $r \leq m$, see Prob. 3-9.)† Let us follow an extension of the second procedure discussed above in this section. To simplify the notation we write Eq. (3-31) as

$$C_0 y(n) + C_1 y(n-1) + C_2 y(n-2) + \cdots + C_r y(n-r)$$
$$= E_0 x(n) + E_1 x(n-1) + E_2 x(n-2) + \cdots + E_r x(n-r)$$

because some of the E's might be zero. Let

$$\gamma_1(n) = -\frac{1}{C_0}[C_r y(n-r) - E_r x(n-r) + C_{r-1} y(n-r+1)$$
$$- E_{r-1} x(n-r+1) + \cdots + C_1 y(n-1) - E_1 x(n-1)]$$
$$\gamma_2(n) = -\frac{1}{C_0}[C_r y(n-r+1) - E_r x(n-r+1) + C_{r-1} y(n-r+2)$$
$$- E_{r-1} x(n-r+2) + \cdots + C_2 y(n-1) - E_2 x(n-1)]$$

....

$$\gamma_r(n) = -\frac{1}{C_0}[C_r y(n-1) - E_r x(n-1)]$$

We have, as output equation,

$$y(n) = \gamma_1(n) + \frac{E_0}{C_0} x(n)$$

and, as the state equations,

$$\gamma_1(n+1) = \gamma_2(n) - \frac{C_1}{C_0} y(n) + \frac{E_1}{C_0} x(n)$$
$$= \gamma_2(n) - \frac{C_1}{C_0}\gamma_1(n) - \frac{C_1}{C_0}\frac{E_0}{C_0} x(n) + \frac{E_1}{C_0} x(n)$$
$$\gamma_2(n+1) = \gamma_3(n) - \frac{C_2}{C_0} y(n) + \frac{E_2}{C_0} x(n)$$
$$= \gamma_3(n) - \frac{C_2}{C_0}\gamma_1(n) - \frac{C_2}{C_0}\frac{E_0}{C_0} x(n) + \frac{E_2}{C_0} x(n)$$

....

$$\gamma_r(n+1) = -\frac{C_r}{C_0} y(n) + \frac{E_r}{C_0} x(n)$$
$$= -\frac{C_r}{C_0}\gamma_1(n) - \frac{C_r}{C_0}\frac{E_0}{C_0} x(n) + \frac{E_r}{C_0} x(n)$$

† We remind the reader that Eq. (3-31) is an rth-order difference equation and thus has exactly r characteristic roots. Therefore, the reader should satisfy himself that no matter whether $r + m$ or r or m state variables are used in the formulations, the same system response should be obtained for a given stimulus.

Thus, indeed, the state of the system can be specified by r state variables. An illustrative example will help us to see the significance of these formulations. Let

$$y(n) + y(n-1) = x(n) + 2x(n-1)$$

If we let

$$\lambda_1(n) = y(n-1)$$
$$\lambda_2(n) = x(n-1)$$

the output equation is

$$y(n) = -\lambda_1(n) + 2\lambda_2(n) + x(n)$$

and the state equations are

$$\lambda_1(n+1) = -\lambda_1(n) + 2\lambda_2(n) + x(n)$$
$$\lambda_2(n+1) = x(n)$$

But if we let

$$\gamma_1(n) = 2x(n-1) - y(n-1)$$

the output equation is

$$y(n) = x(n) + \gamma_1(n)$$

and the state equation is

$$\gamma_1(n+1) = x(n) - \gamma_1(n)$$

To see that it is redundant to use two state variables to specify the state of the system, we recognize that

$$\gamma_1(n) = -\lambda_1(n) + 2\lambda_2(n)$$

We can also obtain a difference equation that describes the input-output relationship of a linear time-invariant system from a state space description. Instead of a discussion in general terms, we restrict our presentation to an illustrative example. Let

$$y(n) = \lambda_1(n) - \lambda_2(n) + x(n) \tag{3-32}$$

be the output equation, and

$$\begin{aligned} \lambda_1(n+1) &= -\lambda_1(n) + 4\lambda_2(n) + 11x(n) \\ \lambda_2(n+1) &= -2\lambda_1(n) + 4\lambda_2(n) + 6x(n) \end{aligned} \tag{3-33}$$

be the state equations of a discrete system. Equation (3-32) can be written

$$y(n-1) = \lambda_1(n-1) - \lambda_2(n-1) + x(n-1) \tag{3-34}$$

or

$$y(n-2) = \lambda_1(n-2) - \lambda_2(n-2) + x(n-2) \tag{3-35}$$

What we want to do is to combine Eqs. (3-32) to (3-35) to obtain an equation that relates $x(n)$, $x(n-1)$, $x(n-2)$ and $y(n)$, $y(n-1)$, $y(n-2)$, using the state equations (3-33) to eliminate the state variables. Substituting Eqs. (3-33) into (3-32), we obtain

$$\begin{aligned} y(n) &= [-\lambda_1(n-1) + 4\lambda_2(n-1) + 11x(n-1)] - [-2\lambda_1(n-1) \\ &\quad + 4\lambda_2(n-1) + 6x(n-1)] + x(n) \\ &= \lambda_1(n-1) + 5x(n-1) + x(n) \end{aligned}$$

Another step of substitution yields

$$y(n) = -\lambda_1(n-2) + 4\lambda_2(n-2) + 11x(n-2) + 5x(n-1) + x(n) \tag{3-36}$$

Also, substituting Eqs. (3-33) into Eq. (3-34), we obtain

$$y(n-1) = \lambda_1(n-2) + 5x(n-2) + x(n-1) \tag{3-37}$$

From Eqs. (3-35) to (3-37), we have

$$\begin{aligned} &C_0 y(n) + C_1 y(n-1) + C_2 y(n-2) \\ &\quad = (-C_0 + C_1 + C_2)\lambda_1(n-2) + (4C_0 - C_2)\lambda_2(n-2) \\ &\quad\quad + C_0 x(n) + (5C_0 + C_1)x(n-1) + (11C_0 + 5C_1 + C_2)x(n-2) \end{aligned} \tag{3-38}$$

So that the terms $\lambda_1(n-2)$ and $\lambda_2(n-2)$ will be eliminated from Eq. (3-38), C_0, C_1, and C_2 must be so chosen that

$$\begin{aligned} -C_0 + C_1 + C_2 &= 0 \\ 4C_0 - C_2 &= 0 \end{aligned} \tag{3-39}$$

Thus, Eq. (3-38) becomes

$$C_0 y(n) - 3C_0 y(n-1) + 4C_0 y(n-2) = C_0 x(n) + 2C_0 x(n-1)$$

or

$$y(n) - 3y(n-1) + y(n-2) = x(n) + 2x(n-1)$$

which is a difference equation description of the input-output relationship of the given system.

This example should be abundantly clear in illustrating a general procedure for determining a difference equation description of a system from a state space description. There is, however, a point we should elaborate upon. One might ask: In the preceding example, how did we know at the beginning that the difference equation sought is a second-order difference equation, and, in general, for a given state space description of a system, how do we determine the order of the difference equation that describes its input-output relationship? Since the number of exponential terms in the homogeneous solution of $y(n)$ is equal to the order of the difference equation that describes the input-output relationship of the system, and is also equal to the number of state equations in the state space description, the order of the difference equation should equal the number of state equations.† Such a claim can also be verified directly from a purely mathematical consideration. Since the number of equations for determining the constants $C_0, C_1, C_2, \ldots$ such as those in Eqs. (3-39) is equal to the number of state variables used in the state space description, according to the theory of simultaneous equations, the unknowns $C_0, C_1, C_2, \ldots$ can be solved uniquely‡ if the number of unknowns is equal to the number of state variables.

3-6 STATE SPACE DESCRIPTION OF CONTINUOUS SYSTEMS

We turn now to the state space description of the behavior of continuous systems. At any time instant t, the state of a continuous system can be described by the values of its state variables $\lambda_1(t), \lambda_2(t), \ldots, \lambda_r(t)$. Since the state of a continuous system changes *continuously* with time, the values of the state variables also change continuously with timc. Thus, to specify how the values of the state variables change we should specify the *rates* of change of the values of the state variables. For a causal system, the rates of change of the values of state variables at any instant are functions of the state variables and the input at that instant. That is, in functional notation

$$
\begin{aligned}
\frac{d\lambda_1(t)}{dt} &= g_1(\lambda_1(t), \lambda_2(t), \ldots, \lambda_r(t), x(t)) \\
\frac{d\lambda_2(t)}{dt} &= g_2(\lambda_1(t), \lambda_2(t), \ldots, \lambda_r(t), x(t)) \\
&\cdots\cdots\cdots\cdots\cdots\cdots \\
\frac{d\lambda_r(t)}{dt} &= g_r(\lambda_1(t), \lambda_2(t), \ldots, \lambda_r(t), x(t))
\end{aligned}
\tag{3-40}
$$

The output $y(t)$ depends on the values of the state variables and the value of the input at that instant. That is, the output equation is of the form

$$y(t) = f(\lambda_1(t), \lambda_2(t), \ldots, \lambda_r(t), x(t)) \tag{3-41}$$

If the system is noncausal, $y(t)$ and $d\lambda_1/dt, d\lambda_2/dt, \ldots, d\lambda_r/dt$ may depend on the value of the input at time larger than t as well. Our attention will be restricted to causal systems.

† We assume that a minimum number of state variables is used in the state space description. For the case in which redundant state variables are used, see Prob. 3-21. We shall also discuss such a point further in Chaps. 4 and 9.

‡ Because the simultaneous equations (3-39) have zero constant terms, we can only solve $C_1, C_2, \ldots$ in terms of C_0.

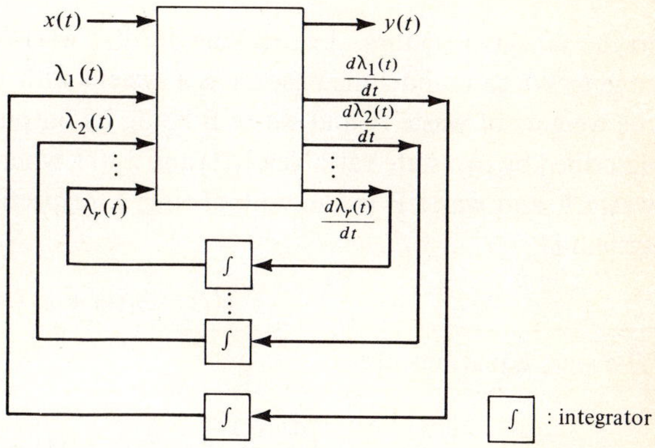

FIGURE 3-4

Just as in the case of discrete systems, many of the properties of a continuous system, such as time invariance and linearity, are exhibited in the state equations and the output equation. Since the analogy is quite obvious, we leave the details to the reader.

A block diagram of a continuous system is shown in Fig. 3-4. At any instant t, the system computes the values of $y(t)$, $d\lambda_1(t)/dt$, $d\lambda_2(t)/dt$, $\ldots$, $d\lambda_r(t)/dt$ from the values of $x(t)$, $\lambda_1(t)$, $\lambda_2(t)$, $\ldots$, $\lambda_r(t)$. The values of $d\lambda_1(t)/dt$, $d\lambda_2(t)/dt$, $\ldots$, $d\lambda_r(t)/dt$ are then used to update the values of $\lambda_1(t)$, $\lambda_2(t)$, $\ldots$, $\lambda_r(t)$. Specifically, the updated values of the state variables at $t + \Delta t$ are computed as

$$\lambda_1(t + \Delta t) = \lambda_1(t) + \int_t^{t+\Delta t} \frac{d\lambda_1(t)}{dt}\, dt$$

$$\lambda_2(t + \Delta t) = \lambda_2(t) + \int_t^{t+\Delta t} \frac{d\lambda_2(t)}{dt}\, dt$$

$$\cdots\cdots\cdots\cdots\cdots\cdots\cdots\cdots\cdots\cdots\cdots$$

$$\lambda_r(t + \Delta t) = \lambda_r(t) + \int_t^{t+\Delta t} \frac{d\lambda_2(t)}{dt}\, dt$$

Such updating computation is carried out by the integrators in Fig. 3-4.

Let us consider an example of disposing two kinds of chemical waste A and B in a tank with a supply of solvent. At any instant, waste A decreases at a rate equal to the difference between the total weight of the solvent and the total weight of waste A and waste B in the tank. Also, waste B decreases at a rate equal to the difference between two times the total weight of the solvent and the total weight of waste A plus two times the total weight of waste B in the tank. Suppose that the supply of solvent to the tank is $w(t)$ lb/s. Let $x(t)$ denote $\int_{-\infty}^{t} w(\tau)\,d\tau$, the total solvent in the tank at time t. We can model the process as a system with $x(t)$ being its input and the sum of the weights of waste A and waste B being its output. The state of the system can be described by two state variables $\lambda_1(t)$ and $\lambda_2(t)$, where $\lambda_1(t)$ and $\lambda_2(t)$ are the weights of waste A and waste B in the tank at time t, respectively. The output equation of the system is

$$y(t) = \lambda_1(t) + \lambda_2(t)$$

The state equations are

$$\frac{d\lambda_1(t)}{dt} = \lambda_1(t) + \lambda_2(t) - x(t)$$

$$\frac{d\lambda_2(t)}{dt} = \lambda_1(t) + 2\lambda_2(t) - 2x(t)$$

A block diagram description of the system is shown in Fig. 3-5.

Let us also select some examples from circuit theory, a subject the reader has already studied extensively. For the circuit shown in Fig. 3-6, the voltage across the capacitor can be used as a state variable in a state space description. To write the state equation, we note that the current in R_1 is $C[d\lambda(t)/dt]$, the current in R_2 is

$$\frac{1}{R_2}\left[CR_1\frac{d\lambda(t)}{dt} + \lambda(t)\right]$$

and the current in R_3 is

$$\frac{C(R_1 + R_2)}{R_2}\frac{d\lambda(t)}{dt} + \frac{1}{R_2}\lambda(t)$$

Therefore, we have

$$x(t) - \frac{C(R_1 + R_2)R_3}{R_2}\frac{d\lambda(t)}{dt} - \frac{R_3}{R_2}\lambda(t) = CR_1\frac{d\lambda(t)}{dt} + \lambda(t)$$

which simplifies to

$$\frac{d\lambda(t)}{dt} = \frac{R_2}{C(R_1R_3 + R_2R_3 + R_1R_2)}\left[-\frac{R_2 + R_3}{R_2}\lambda(t) + x(t)\right] \qquad (3\text{-}42)$$

Notice that the capacitor C indeed corresponds to an integrator in the block diagram in Fig. 3-4. The output of the integrator is the voltage across the capacitor, $\lambda(t)$. From $\lambda(t)$ and $x(t)$, the system computes, according to Eq. (3-42), the input to the integrator, $d\lambda(t)/dt$, which is the current in the capacitor.

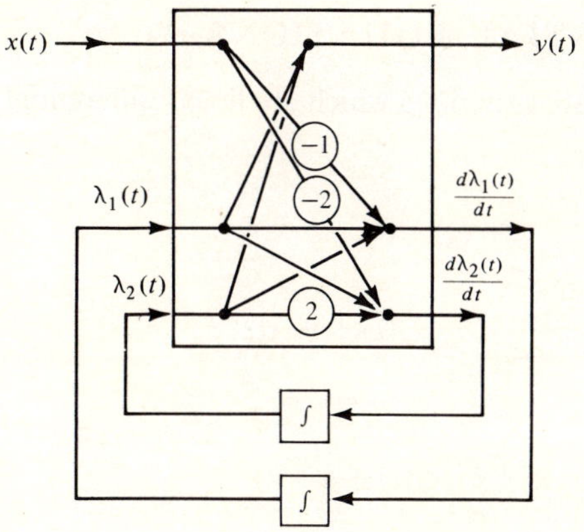

FIGURE 3-5

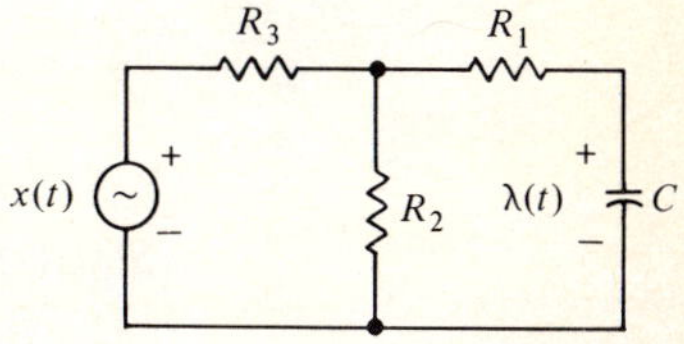

FIGURE 3-6

As another example, consider the circuit shown in Fig. 3-7. Let $\lambda_1(t)$ denote the voltage across the capacitor and $\lambda_2(t)$ denote the current in the inductor. Using $\lambda_1(t)$ and $\lambda_2(t)$ as state variables, the reader can readily check that the two state equations are

$$\frac{d\lambda_1(t)}{dt} = -\frac{1}{20}\lambda_1(t) + \frac{3}{20}\lambda_2(t)$$

$$\frac{d\lambda_2(t)}{dt} = -\frac{3}{25}\lambda_1(t) - \frac{6}{25}\lambda_2(t) + \frac{1}{5}x(t)$$

The two corresponding integrators are the capacitor and the inductor.

For this particular circuit, the currents in the two resistors, $\gamma_1(t)$ and $\gamma_2(t)$, can also be used as state variables. Such a point becomes clear when we note that

$$\lambda_2(t) = \gamma_1(t) + \gamma_2(t)$$

$$\lambda_1(t) = 3\gamma_2(t) - 2\gamma_1(t)$$

However, for such a choice of state variables, it is no longer possible to identify explicitly the integrators in the system.

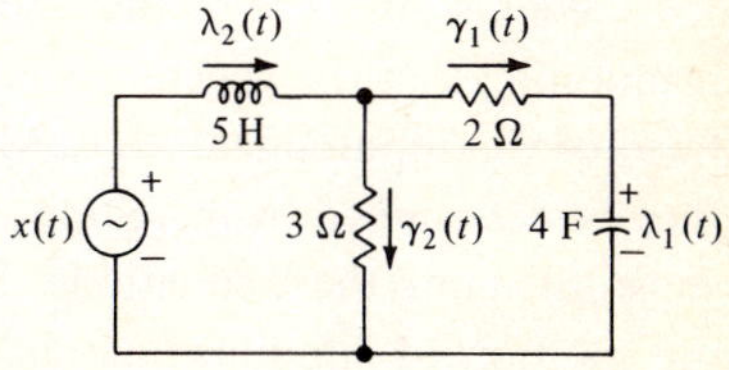

FIGURE 3-7

3-7 SIMULTANEOUS DIFFERENTIAL EQUATIONS

We consider in this section the solution of state equations which are linear differential equations with constant coefficients:

$$\begin{aligned}
\frac{d\lambda_1(t)}{dt} &= a_{11}\lambda_1(t) + a_{12}\lambda_2(t) + \cdots + a_{1r}\lambda_r(t) + b_1 x(t) \\
\frac{d\lambda_2(t)}{dt} &= a_{21}\lambda_1(t) + a_{22}\lambda_2(t) + \cdots + a_{2r}\lambda_r(t) + b_2 x(t) \\
&\cdots\cdots\cdots\cdots\cdots\cdots\cdots\cdots\cdots\cdots \\
\frac{d\lambda_r(t)}{dt} &= a_{r1}\lambda_1(t) + a_{r2}\lambda_2(t) + \cdots + a_{rr}\lambda_r(t) + b_r x(t)
\end{aligned} \tag{3-43}$$

The procedure for solving a set of simultaneous linear first-order differential equations is quite similar to that for solving a set of linear first-order difference equations, as discussed in Sec. 3-4. We shall give only a brief presentation here. Again, each of the solutions of the simultaneous equations consists of a homogeneous solution as well as a particular solution. The homogeneous solutions are of the form $e^{\alpha t}$ for some appropriately chosen constants α. It can be shown by direct substitution that when the roots $\alpha_1, \alpha_2, \ldots, \alpha_r$ of the determinantal equation

$$\begin{vmatrix}
a_{11}-\alpha & a_{12} & \cdots & a_{1r} \\
a_{21} & a_{22}-\alpha & \cdots & a_{2r} \\
\cdots & \cdots & \cdots & \cdots \\
a_{r1} & a_{r2} & \cdots & a_{rr}-\alpha
\end{vmatrix} = 0 \tag{3-44}$$

are all distinct, the homogeneous solutions are

$$\begin{aligned}
\lambda_1(t) &= A_1 e^{\alpha_1 t} + B_1 e^{\alpha_2 t} + \cdots + R_1 e^{\alpha_r t} \\
\lambda_2(t) &= A_1 g_{21} e^{\alpha_1 t} + B_1 g_{22} e^{\alpha_2 t} + \cdots + R_1 g_{2r} e^{\alpha_r t} \\
&\cdots\cdots\cdots\cdots\cdots\cdots\cdots\cdots\cdots\cdots \\
\lambda_r(t) &= A_1 g_{r1} e^{\alpha_1 t} + B_1 g_{r2} e^{\alpha_2 t} + \cdots + R_1 g_{rr} e^{\alpha_r t}
\end{aligned}$$

where the constants $g_{21}, \ldots, g_{2r}, g_{31}, \ldots, g_{3r}, \ldots, g_{r1}, \ldots, g_{rr}$ are determined in the same manner as that in Sec. 3-4, and the constants $A_1, B_1, \ldots, R_1$ are to be determined by a set of boundary conditions.

We also dispose of the discussion on finding the particular solution by merely mentioning that the general form of the particular solutions can be guessed from the form of the input signal $x(t)$. Multiplying constants in the particular solutions can then be determined by matching terms in Eqs. (3-43) when the particular solutions are substituted into these equations. Since such steps are, again, also identical to those for the solution of simultaneous difference equations, we shall leave the details to the reader.

3-8 SOLUTION OF VECTOR DIFFERENTIAL EQUATIONS: THE HOMOGENEOUS SOLUTION

Using matrix notation, we can write Eqs. (3-43) as a single equation. Let

$$\boldsymbol{\lambda}(t) = \begin{bmatrix} \lambda_1(t) \\ \lambda_2(t) \\ \vdots \\ \lambda_r(t) \end{bmatrix}$$

$$\mathbf{A} = \begin{bmatrix} a_{11} & a_{12} & \cdots & a_{1r} \\ a_{21} & a_{22} & \cdots & a_{2r} \\ \cdots & \cdots & \cdots & \cdots \\ a_{r1} & a_{r2} & \cdots & a_{rr} \end{bmatrix}$$

and

$$\mathbf{B} = \begin{bmatrix} b_1 \\ b_2 \\ \vdots \\ b_r \end{bmatrix}$$

Equations (3-43) can then be written compactly as

$$\frac{d\boldsymbol{\lambda}(t)}{dt} = \mathbf{A}\boldsymbol{\lambda}(t) + \mathbf{B}x(t) \tag{3-45}$$

Because the unknown $\boldsymbol{\lambda}(t)$ in Eq. (3-45) is a vector, such an equation is known as a *vector differential equation.* Moreover, it is a linear vector differential equation with constant coefficients, following the same definition we had for a linear *scalar* differential equation with constant coefficients. The solution of a linear vector differential equation with constant coefficients consists of two parts: a homogeneous solution and a particular solution. The homogeneous solution satisfies Eq. (3-45) when $x(t)$ is set to zero. We show now that the homogeneous solution is

$$\boldsymbol{\lambda}(t) = \left(\mathbf{I} + \mathbf{A}t + \mathbf{A}^2\frac{t^2}{2!} + \mathbf{A}^3\frac{t^3}{3!} + \cdots + \mathbf{A}^i\frac{t^i}{i!} + \cdots\right)\mathbf{K}\dagger \tag{3-46}$$

† It can be shown that the series $\mathbf{I} + \mathbf{A}t + \mathbf{A}^2(t^2/2!) + \mathbf{A}^3(t^3/3!) + \cdots$ converges absolutely for any matrix $\mathbf{A}$ and any finite t.

where **I** is the identity matrix and

$$\mathbf{K} = \begin{bmatrix} K_1 \\ K_2 \\ \vdots \\ K_r \end{bmatrix}$$

is a column vector of undetermined coefficients to be determined by given boundary conditions. Differentiating both sides of Eq. (3-46) with respect to t, we obtain

$$\frac{d\boldsymbol{\lambda}(t)}{dt} = \left(\mathbf{A} + \mathbf{A}^2 \frac{t}{1!} + \mathbf{A}^3 \frac{t^2}{2!} + \cdots + \mathbf{A}^i \frac{t^{i-1}}{(i-1)!} + \cdots\right)\mathbf{K} = \mathbf{A}\boldsymbol{\lambda}(t)$$

Thus, the vector $\boldsymbol{\lambda}(t)$ in Eq. (3-46) satisfies Eq. (3-45) when $x(t)$ is set to 0.

As an example, let

$$\frac{d\lambda_1(t)}{dt} = 2\lambda_1(t) - \lambda_2(t)$$

$$\frac{d\lambda_2(t)}{dt} = -4\lambda_1(t) + 5\lambda_2(t)$$

These two equations can be written as a vector differential equation:

$$\frac{d}{dt}\begin{bmatrix} \lambda_1(t) \\ \lambda_2(t) \end{bmatrix} = \begin{bmatrix} 2 & -1 \\ -4 & 5 \end{bmatrix}\begin{bmatrix} \lambda_1(t) \\ \lambda_2(t) \end{bmatrix}$$

Thus

$$\begin{aligned}
\boldsymbol{\lambda}(t) = \begin{bmatrix} \lambda_1(t) \\ \lambda_2(t) \end{bmatrix} &= \left(\begin{bmatrix} 1 & 0 \\ 0 & 1 \end{bmatrix} + \begin{bmatrix} 2 & -1 \\ -4 & 5 \end{bmatrix} t + \begin{bmatrix} 2 & -1 \\ -4 & 5 \end{bmatrix}^2 \frac{t^2}{2} \right.\\
&\qquad \left. + \begin{bmatrix} 2 & -1 \\ -4 & 5 \end{bmatrix}^3 \frac{t^3}{3!} + \cdots\right)\begin{bmatrix} K_1 \\ K_2 \end{bmatrix} \\
&= \left(\begin{bmatrix} 1 & 0 \\ 0 & 1 \end{bmatrix} + \begin{bmatrix} 2 & -1 \\ -4 & 5 \end{bmatrix} t + \begin{bmatrix} 8 & -7 \\ -28 & 29 \end{bmatrix} \frac{t^2}{2} \right.\\
&\qquad \left. + \begin{bmatrix} 44 & -43 \\ -172 & 173 \end{bmatrix} \frac{t^3}{6} + \cdots\right)\begin{bmatrix} K_1 \\ K_2 \end{bmatrix} \\
&= \begin{bmatrix} 1 + 2t + 4t^2 + \frac{22}{3}t^3 + \cdots & -t - \frac{7}{2}t^2 - \frac{43}{6}t^3 + \cdots \\ -4t - 14t^2 - \frac{86}{3}t^3 + \cdots & 1 + 5t + \frac{29}{2}t^2 + \frac{173}{6}t^3 + \cdots \end{bmatrix}\begin{bmatrix} K_1 \\ K_2 \end{bmatrix}
\end{aligned}$$

In other words,

$$\lambda_1(t) = K_1(1 + 2t + 4t^2 + \tfrac{22}{3}t^3 + \cdots) + K_2(-t - \tfrac{7}{2}t^2 - \tfrac{43}{6}t^3 + \cdots)$$
$$\lambda_2(t) = K_1(-4t - 14t^2 - \tfrac{86}{3}t^3 + \cdots) + K_2(1 + 5t + \tfrac{29}{2}t^2 + \tfrac{173}{6}t^3 + \cdots)$$

Although we have obtained a homogeneous solution of the vector differential equation (3-45), that it is an infinite series neither makes the manipulation easy nor gives much insight into the behavior of the system. We, therefore, would like to put the infinite series into closed-form expressions. Since we know from our discussion in Sec. 3-7 that $\lambda_1(t)$ and $\lambda_2(t)$ consist of exponential terms, we write

$$\lambda_1(t) = \left(\frac{4}{5}K_1 + \frac{1}{5}K_2\right)\left(1 + t + \frac{t^2}{2!} + \frac{t^3}{3!} + \cdots\right)$$
$$+ \left(\frac{1}{5}K_1 - \frac{1}{5}K_2\right)\left(1 + 6t + \frac{(6t)^2}{2!} + \frac{(6t)^3}{3!} + \cdots\right)$$
$$= A_1 e^t + B_1 e^{6t}$$

where $\quad A_1 = \frac{4}{5}K_1 + \frac{1}{5}K_2 \qquad B_1 = \frac{1}{5}K_1 - \frac{1}{5}K_2$

Similarly, we have

$$\lambda_2(t) = \left(\frac{4}{5}K_1 + \frac{1}{5}K_2\right)\left(1 + t + \frac{t^2}{2!} + \frac{t^3}{3!} + \cdots\right)$$
$$+ \left(-\frac{4}{5}K_1 + \frac{4}{5}K_2\right)\left[1 + 6t + \frac{(6t)^2}{2!} + \frac{(6t)^3}{3!} + \cdots\right]$$
$$= A_1 e^t - 4B_1 e^{6t}$$

Such manipulation of the infinite series representations of $\lambda_1(t)$ and $\lambda_2(t)$ into closed-form expressions certainly is not obvious, especially when the unknown column vector $\boldsymbol{\lambda}(t)$ has a large number of components. For a more systematic approach, let us define the exponential matrix function $e^{\mathbf{A}t}$ to be

$$e^{\mathbf{A}t} = \mathbf{I} + \mathbf{A}t + \mathbf{A}^2\frac{t^2}{2!} + \mathbf{A}^3\frac{t^3}{3!} + \cdots + \mathbf{A}^i\frac{t^i}{i!} + \cdots \tag{3-47}$$

Note that, at this moment, $e^{\mathbf{A}t}$ is nothing more than shorthand for the infinite series. It follows that the homogeneous solution of the vector differential equation (3-45) is

$$\boldsymbol{\lambda}(t) = e^{\mathbf{A}t}\mathbf{K}$$

So that we can obtain closed-form expressions for the homogeneous solution $\lambda_1(t)$, $\lambda_2(t), \ldots, \lambda_r(t)$, we want to compute the matrix function $e^{\mathbf{A}t}$ in closed form. We shall present two methods of computation in the next two sections. Another method will be presented in Chap. 11, after the discussion of Laplace transformation.

*3-9 COMPUTING THE MATRIX FUNCTION $e^{\mathbf{A}t}$ USING THE CAYLEY-HAMILTON THEOREM

The first method of computing the matrix function $e^{\mathbf{A}t}$ is based on the *Cayley-Hamilton theorem* in matrix algebra. Let $\mathbf{A}$ be an $r \times r$ matrix the elements of which are denoted a_{ij}. The determinantal equation (3-44) is called the *characteristic equation* of $\mathbf{A}$. It is easy to see that this equation is an rth-degree equation in the unknown α and can be written

$$\alpha^r + h_{r-1}\alpha^{r-1} + h_{r-2}\alpha^{r-2} + \cdots + h_1\alpha + h_0 = 0 \tag{3-48}$$

for some $h_{r-1}, h_{r-2}, \ldots, h_1, h_0$. The roots of this equation are called the *characteristic values* of the matrix **A**. The Cayley-Hamilton theorem states that *any matrix satisfies its characteristic equation.* In other words, since Eq. (3-48) is the characteristic equation of **A**, we can replace the unknown α by the matrix **A** and obtain the matrix equation

$$\mathbf{A}^r + h_{r-1}\mathbf{A}^{r-1} + h_{r-2}\mathbf{A}^{r-2} + \cdots + h_1\mathbf{A} + h_0\mathbf{I} = \mathbf{0} \tag{3-49}$$

Note that the constant h_0 in Eq. (3-48) is replaced by the matrix $h_0\mathbf{I}$ and the constant 0 is replaced by the matrix **0**. Thus, we obtain from Eq. (3-49)

$$\mathbf{A}^r = -(h_{r-1}\mathbf{A}^{r-1} + h_{r-2}\mathbf{A}^{r-2} + \cdots + h_1\mathbf{A} + h_0\mathbf{I}) \tag{3-50}$$

That is, the matrix $\mathbf{A}^r$ can be expressed as a polynomial in **A** whose degree is $r - 1$ or less. It follows that $\mathbf{A}^{r+1}$ can also be expressed as a polynomial in **A** whose degree does not exceed $r - 1$, since

$$\begin{aligned}
\mathbf{A}^{r+1} = \mathbf{A}\mathbf{A}^r &= -\mathbf{A}(h_{r-1}\mathbf{A}^{r-1} + h_{r-2}\mathbf{A}^{r-2} + \cdots + h_1\mathbf{A} + h_0\mathbf{I}) \\
&= -h_{r-1}\mathbf{A}^r - h_{r-2}\mathbf{A}^{r-1} - \cdots - h_1\mathbf{A}^2 - h_0\mathbf{A} \\
&= h_{r-1}(h_{r-1}\mathbf{A}^{r-1} + h_{r-2}\mathbf{A}^{r-2} + \cdots + h_1\mathbf{A} + h_0\mathbf{I}) \\
&\quad - h_{r-2}\mathbf{A}^{r-1} - \cdots - h_1\mathbf{A}^2 - h_0\mathbf{A} \\
&= (h_{r-1}^2 - h_{r-2})\mathbf{A}^{r-1} + (h_{r-1}h_{r-2} - h_{r-3})\mathbf{A}^{r-2} + \cdots \\
&\quad + (h_{r-1}h_2 - h_1)\mathbf{A}^2 + (h_{r-1}h_1 - h_0)\mathbf{A} + h_{r-1}h_0\mathbf{I}
\end{aligned}$$

In a similar manner, any power of **A** can be expressed as a polynomial of degree not exceeding $r - 1$ by repeatedly replacing $\mathbf{A}^r$ by the expression in Eq. (3-50).

As an example, consider the state equations

$$\begin{aligned}
\frac{d\lambda_1(t)}{dt} &= \lambda_1(t) + 4\lambda_2(t) \\
\frac{d\lambda_2(t)}{dt} &= 2\lambda_1(t) + 3\lambda_2(t)
\end{aligned} \tag{3-51}$$

The matrix **A** is

$$\mathbf{A} = \begin{bmatrix} 1 & 4 \\ 2 & 3 \end{bmatrix}$$

The characteristic equation of **A** is

$$\begin{vmatrix} 1-\alpha & 4 \\ 2 & 3-\alpha \end{vmatrix} = \alpha^2 - 4\alpha - 5 = 0$$

According to the Cayley-Hamilton theorem, we have

$$\mathbf{A}^2 - 4\mathbf{A} - 5\mathbf{I} = \mathbf{0} \tag{3-52}$$

Equation (3-52) can be checked readily:

$$\begin{bmatrix} 1 & 4 \\ 2 & 3 \end{bmatrix}^2 - 4\begin{bmatrix} 1 & 4 \\ 2 & 3 \end{bmatrix} - 5\begin{bmatrix} 1 & 0 \\ 0 & 1 \end{bmatrix}$$
$$= \begin{bmatrix} 9 & 16 \\ 8 & 17 \end{bmatrix} - \begin{bmatrix} 4 & 16 \\ 8 & 12 \end{bmatrix} - \begin{bmatrix} 5 & 0 \\ 0 & 5 \end{bmatrix}$$
$$= \begin{bmatrix} 0 & 0 \\ 0 & 0 \end{bmatrix}$$

It follows that all powers of **A** can be expressed as polynomials of **A** of degree not exceeding 1. Thus

$$\begin{aligned} \mathbf{A}^2 &= 4\mathbf{A} + 5\mathbf{I} \\ \mathbf{A}^3 &= \mathbf{A}\mathbf{A}^2 = \mathbf{A}(4\mathbf{A} + 5\mathbf{I}) = 4\mathbf{A}^2 + 5\mathbf{A} = 4(4\mathbf{A} + 5\mathbf{I}) + 5\mathbf{A} \\ &= 21\mathbf{A} + 20\mathbf{I} \\ \mathbf{A}^4 &= \mathbf{A}\mathbf{A}^3 = \mathbf{A}(21\mathbf{A} + 20\mathbf{I}) = 21\mathbf{A}^2 + 20\mathbf{A} = 21(4\mathbf{A} + 5\mathbf{I}) + 20\mathbf{A} \\ &= 104\mathbf{A} + 105\mathbf{I} \end{aligned}$$

and so on.

We now go on with the problem of computing the matrix function $e^{\mathbf{A}t}$. According to our discussion, the infinite sum of powers of **A** in Eq. (3-47) can be expressed as a polynomial in **A** in degree not exceeding $r - 1$:

$$e^{\mathbf{A}t} = c_0\mathbf{I} + c_1\mathbf{A} + c_2\mathbf{A}^2 + \cdots + c_{r-1}\mathbf{A}^{r-1}$$

where $c_0, c_1, c_2, \ldots, c_{r-1}$ are functions of t. (Note that $e^{\mathbf{A}t}$ is viewed as a function of **A** with t being a constant.) To determine $c_0, c_1, c_2, \ldots, c_{r-1}$, we make use of the following result: Suppose that matrix **A** satisfies the matrix equation

$$f(\mathbf{A}) = c_0\mathbf{I} + c_1\mathbf{A} + \cdots + c_{r-1}\mathbf{A}^{r-1}$$

for some matrix function $f(\mathbf{A})$. If α is a characteristic value of **A**, then

$$f(\alpha) = c_0 + c_1\alpha + c_2\alpha^2 + \cdots + c_{r-1}\alpha^{r-1}$$

Furthermore, if α is a k-multiple characteristic value of **A**, then

$$f(\alpha) = c_0 + c_1\alpha + c_2\alpha^2 + \cdots + c_{r-1}\alpha^{r-1}$$
$$\left.\frac{df(z)}{dz}\right|_{z=\alpha} = c_1 + 2c_2\alpha + 3c_3\alpha^2 + \cdots + (r-1)c_{r-1}\alpha^{r-2}$$
$$\left.\frac{d^2f(z)}{dz^2}\right|_{z=\alpha} = 2c_2 + 6c_3\alpha + \cdots + (r-1)(r-2)c_{r-1}\alpha^{r-3}$$
$$\cdots\cdots\cdots\cdots\cdots\cdots\cdots\cdots\cdots\cdots$$
$$\left.\frac{d^{k-1}f(z)}{dz^{k-1}}\right|_{z=\alpha} = (k-1)!\,c_{k-1} + k!\,c_k\alpha + \frac{(k+1)!}{2}c_{k+1}\alpha^2 + \cdots + \frac{(r-1)!}{(r-k)!}c_{r-1}\alpha^{r-k}$$

Therefore, if $\alpha_1, \alpha_2, \ldots, \alpha_r$ are distinct characteristic values of **A**, we can determine $c_0, c_1, c_2, \ldots, c_r$, which are functions of t, from the r equations:

$$\begin{aligned} e^{\alpha_1 t} &= c_0 + c_1\alpha_1 + c_2\alpha_1{}^2 + \cdots + c_{r-1}\alpha_1^{r-1} \\ e^{\alpha_2 t} &= c_0 + c_1\alpha_2 + c_2\alpha_2{}^2 + \cdots + c_{r-1}\alpha_2^{r-1} \\ &\cdots\cdots\cdots\cdots\cdots\cdots \\ e^{\alpha_r t} &= c_0 + c_1\alpha_r + c_2\alpha_r{}^2 + \cdots + c_{r-1}\alpha_r^{r-1} \end{aligned} \tag{3-53}$$

If α_1 is a k-multiple characteristic value of **A**, then the k equations

$$\begin{aligned} e^{\alpha_1 t} &= c_0 + c_1\alpha_1 + c_2\alpha_1{}^2 + \cdots + c_{r-1}\alpha_1^{r-1} \\ te^{\alpha_1 t} &= c_1 + 2c_2\alpha_1 + 3c_3\alpha_1{}^2 + \cdots + (r-1)c_{r-1}\alpha_1^{r-2} \\ &\cdots\cdots\cdots\cdots\cdots\cdots \\ t^{k-1}e^{\alpha_1 t} &= (k-1)!\,c_{k-1} + \frac{k!}{1!}c_k\alpha_1 + \cdots + \frac{(r-1)!}{(r-k)!}c_{r-1}\alpha_1^{r-k} \end{aligned}$$

together with the last $r-k$ of Eqs. (3-53) will enable us to determine $c_0, c_1, c_2, \ldots, c_{r-1}$. In a similar way we can generalize the procedure to the case when **A** has more than one multiple characteristic value.

For the state equations (3-51), since the two characteristics values of **A** are -1 and 5, we have

$$\begin{aligned} e^{-t} &= c_0 + c_1(-1) \\ e^{5t} &= c_0 + c_1(5) \end{aligned}$$

Therefore,

$$\begin{aligned} c_0 &= \tfrac{5}{6}e^{-t} + \tfrac{1}{6}e^{5t} \\ c_1 &= -\tfrac{1}{6}e^{-t} + \tfrac{1}{6}e^{5t} \end{aligned}$$

Thus, the homogeneous solution of the vector differential equation is

$$\begin{aligned} \boldsymbol{\lambda}(t) &= [(\tfrac{5}{6}e^{-t} + \tfrac{1}{6}e^{5t})\mathbf{I} + (-\tfrac{1}{6}e^{-t} + \tfrac{1}{6}e^{5t})\mathbf{A}]\mathbf{K} \\ &= \begin{bmatrix} \tfrac{2}{3}e^{-t} + \tfrac{1}{3}e^{5t} & -\tfrac{2}{3}e^{-t} + \tfrac{2}{3}e^{5t} \\ -\tfrac{1}{3}e^{-t} + \tfrac{1}{3}e^{5t} & \tfrac{1}{3}e^{-t} + \tfrac{2}{3}e^{5t} \end{bmatrix} \begin{bmatrix} K_1 \\ K_2 \end{bmatrix} \\ &= \begin{bmatrix} (\tfrac{2}{3}K_1 - \tfrac{2}{3}K_2)e^{-t} + (\tfrac{1}{3}K_1 + \tfrac{2}{3}K_2)e^{5t} \\ (-\tfrac{1}{3}K_1 + \tfrac{1}{3}K_2)e^{-t} + (\tfrac{1}{3}K_1 + \tfrac{2}{3}K_2)e^{5t} \end{bmatrix} \end{aligned}$$

If we solve the state equations by the method of solving simultaneous differential equations discussed in Sec. 3-7, we obtain

$$\begin{aligned} \lambda_1(t) &= A_1e^{-t} + B_1e^{5t} \\ \lambda_2(t) &= -\tfrac{1}{2}A_1e^{-t} + B_1e^{5t} \end{aligned}$$

Clearly, these expressions check out our result obtained by computing the matrix function $e^{\mathbf{A}t}$.

As another example, let

$$A = \begin{bmatrix} 1 & -1 \\ 1 & 3 \end{bmatrix}$$

To compute $e^{\mathbf{A}t}$, we shall determine c_0 and c_1 in

$$e^{\mathbf{A}t} = c_0 \mathbf{I} + c_1 \mathbf{A}$$

Since **A** has 2 as a double characteristic value, we have the equations

$$e^{2t} = c_0 + 2c_1$$
$$te^{2t} = \qquad c_1$$

Solving for c_0 and c_1, we obtain

$$c_0 = e^{2t} - 2te^{2t}$$
$$c_1 = te^{2t}$$

and

$$e^{\mathbf{A}t} = (e^{2t} - 2te^{2t})\mathbf{I} + te^{2t}\begin{bmatrix} 1 & -1 \\ 1 & 3 \end{bmatrix}$$
$$= \begin{bmatrix} e^{2t} - te^{2t} & -te^{2t} \\ te^{2t} & e^{2t} + te^{2t} \end{bmatrix}$$

*3-10 COMPUTING THE MATRIX FUNCTION $e^{\mathbf{A}t}$ BY DIAGONALIZATION OF THE MATRIX A

A second method to compute $e^{\mathbf{A}t}$ is based on the diagonalization of the matrix **A**. Let **D** be an $r \times r$ diagonal matrix

$$\mathbf{D} = \begin{bmatrix} \alpha_1 & 0 & 0 & \cdots & 0 \\ 0 & \alpha_2 & 0 & \cdots & 0 \\ \cdots & \cdots & \cdots & \cdots & \cdots \\ 0 & 0 & 0 & \cdots & \alpha_r \end{bmatrix}$$

Since

$$\mathbf{D}^i = \begin{bmatrix} \alpha_1{}^i & 0 & 0 & \cdots & 0 \\ 0 & \alpha_2{}^i & 0 & \cdots & 0 \\ \cdots & \cdots & \cdots & \cdots & \cdots \\ 0 & 0 & 0 & \cdots & \alpha_r{}^i \end{bmatrix}$$

we have

$$e^{\mathbf{D}t} = \mathbf{I} + \mathbf{D}t + \mathbf{D}^2 \frac{t^2}{2!} + \cdots + \mathbf{D}^i \frac{t^i}{i!} + \cdots$$

$$= \begin{bmatrix} \sum_{i=0}^{\infty} \frac{(\alpha_1 t)^i}{i!} & 0 & 0 & \cdots & 0 \\ 0 & \sum_{i=0}^{\infty} \frac{(\alpha_2 t)^i}{i!} & 0 & \cdots & 0 \\ \cdots & \cdots & \cdots & \cdots & \cdots \\ 0 & 0 & 0 & \cdots & \sum_{i=0}^{\infty} \frac{(\alpha_r t)^i}{i!} \end{bmatrix}$$

$$= \begin{bmatrix} e^{\alpha_1 t} & 0 & 0 & \cdots & 0 \\ 0 & e^{\alpha_2 t} & 0 & \cdots & 0 \\ \cdots & \cdots & \cdots & \cdots & \cdots \\ 0 & 0 & 0 & \cdots & e^{\alpha_r t} \end{bmatrix}$$

Let **A** be an $r \times r$ matrix with *distinct* characteristic values $\alpha_1, \alpha_2, \ldots, \alpha_r$. It is a well-known result in matrix theory that a nonsingular matrix **T** can be found such that $\mathbf{T}^{-1}\mathbf{AT}$ is a diagonal matrix **D** which has $\alpha_1, \alpha_2, \ldots, \alpha_r$ as the diagonal elements. That is,

$$\mathbf{T}^{-1}\mathbf{AT} = \mathbf{D} = \begin{bmatrix} \alpha_1 & 0 & 0 & \cdots & 0 \\ 0 & \alpha_2 & 0 & \cdots & 0 \\ \cdots & \cdots & \cdots & \cdots & \cdots \\ 0 & 0 & 0 & \cdots & \alpha_r \end{bmatrix} \tag{3-54}$$

Thus

$$e^{\mathbf{A}t} = e^{\mathbf{TDT}^{-1}t}$$

$$= \left[I + (\mathbf{TDT}^{-1})t + (\mathbf{TDT}^{-1})^2 \frac{t^2}{2!} + \cdots + (\mathbf{TDT}^{-1})^i \frac{t^i}{i!} + \cdots \right]$$

Since

$$(\mathbf{TDT}^{-1})^i = (\mathbf{TDT}^{-1})(\mathbf{TDT}^{-1}) \cdots (\mathbf{TDT}^{-1}) = \mathbf{TD}^i\mathbf{T}^{-1}$$

it follows that

$$e^{\mathbf{A}t} = T\left[\mathbf{I} + \mathbf{D}t + \mathbf{D}^2 \frac{t^2}{2!} + \cdots + \mathbf{D}^i \frac{t^i}{i!} + \cdots \right] \mathbf{T}^{-1}$$

$$= Te^{\mathbf{D}t}\mathbf{T}^{-1}$$

$$= \mathbf{T}\begin{bmatrix} e^{\alpha_1 t} & 0 & 0 & \cdots & 0 \\ 0 & e^{\alpha_2 t} & 0 & \cdots & 0 \\ \cdots & \cdots & \cdots & \cdots & \cdots \\ 0 & 0 & 0 & \cdots & e^{\alpha_r t} \end{bmatrix}\mathbf{T}^{-1}$$

To illustrate the computational procedure, let

$$\mathbf{A} = \begin{bmatrix} 1 & 4 \\ 2 & 3 \end{bmatrix}$$

For

$$\mathbf{T} = \begin{bmatrix} 2 & 1 \\ -1 & 1 \end{bmatrix} \qquad \mathbf{T}^{-1} = \begin{bmatrix} \frac{1}{3} & -\frac{1}{3} \\ \frac{1}{3} & \frac{2}{3} \end{bmatrix}$$

we have

$$\mathbf{T}^{-1}\mathbf{A}\mathbf{T} = \begin{bmatrix} -1 & 0 \\ 0 & 5 \end{bmatrix}$$

Therefore,

$$e^{\mathbf{A}t} = \begin{bmatrix} 2 & 1 \\ -1 & 1 \end{bmatrix}\begin{bmatrix} e^{-t} & 0 \\ 0 & e^{5t} \end{bmatrix}\begin{bmatrix} \frac{1}{3} & -\frac{1}{3} \\ \frac{1}{3} & \frac{2}{3} \end{bmatrix}$$

$$\begin{bmatrix} \frac{2}{3}e^{-t} + \frac{1}{3}e^{5t} & -\frac{2}{3}e^{-t} + \frac{2}{3}e^{5t} \\ -\frac{1}{3}e^{-t} + \frac{1}{3}e^{5t} & -\frac{1}{3}e^{-t} + \frac{2}{3}e^{5t} \end{bmatrix}$$

Such a way of computing $e^{\mathbf{A}t}$ also suggests the introduction of a new set of state variables. Let $\gamma_1(t), \gamma_2(t), \ldots, \gamma_r(t)$ be r variables such that

$$\begin{bmatrix} \gamma_1(t) \\ \gamma_2(t) \\ \vdots \\ \gamma_r(t) \end{bmatrix} = \mathbf{T}^{-1}\begin{bmatrix} \lambda_1(t) \\ \lambda_2(t) \\ \vdots \\ \lambda_r(t) \end{bmatrix}$$

That is,

$$\boldsymbol{\gamma}(t) = T^{-1}\boldsymbol{\lambda}(t) \tag{3-55}$$

where

$$\boldsymbol{\gamma}(t) = \begin{bmatrix} \gamma_1(t) \\ \gamma_2(t) \\ \vdots \\ \gamma_r(t) \end{bmatrix}$$

Since $$\boldsymbol{\lambda}(t) = T\boldsymbol{\gamma}(t) \tag{3-56}$$

we can compute $\boldsymbol{\gamma}(t)$ from $\boldsymbol{\lambda}(t)$ according to Eq. (3-55) and compute $\boldsymbol{\lambda}(t)$ from $\boldsymbol{\gamma}(t)$ according to Eq. (3-56). Consequently, we conclude that $\boldsymbol{\gamma}(t)$ is also a vector of state variables. From

$$\frac{d\boldsymbol{\lambda}(t)}{dt} = \mathbf{A}\boldsymbol{\lambda}(t) + \mathbf{B}x(t)$$

we obtain

$$\begin{aligned}\frac{d\boldsymbol{\gamma}(t)}{dt} = \frac{d}{dt}T^{-1}\boldsymbol{\lambda}(t) &= T^{-1}[\mathbf{A}\boldsymbol{\lambda}(t) + \mathbf{B}x(t)] \\ &= \mathbf{T}^{-1}\mathbf{AT}\boldsymbol{\gamma}(t) + \mathbf{T}^{-1}\mathbf{B}x(t) \\ &= \mathbf{D}\boldsymbol{\gamma}(t) + \mathbf{B}'x(t)\end{aligned}$$

where $$\mathbf{B}' = \mathbf{T}^{-1}\mathbf{B} = \begin{bmatrix} b'_1 \\ b'_2 \\ \vdots \\ b'_r \end{bmatrix}$$

Since $\mathbf{D}$ is a diagonal matrix, the state equations in terms of the state variables $\gamma_1(t), \gamma_2(t), \ldots, \gamma_r(t)$ are r *independent* (as against *simultaneous*) first-order differential equations:

$$\frac{d\gamma_1(t)}{dt} = \alpha_1\gamma_1(t) + b'_1x(t)$$

$$\frac{d\gamma_2(t)}{dt} = \alpha_2\gamma_2(t) + b'_2x(t)$$

$$\cdots\cdots\cdots\cdots\cdots\cdots$$

$$\frac{d\gamma_r(t)}{dt} = \alpha_r\gamma_r(t) + b'_rx(t)$$

We thus conclude that, for a given set of state variables, if the characteristic values of the matrix $\mathbf{A}$ in the corresponding state equations are all distinct, we can find another set of state variables so that the state equations in terms of this new set of state variables are independent first-order differential equations. As an example, let us consider the system described by the state equations (3-51). In this case,

$$\mathbf{A} = \begin{bmatrix} 1 & 4 \\ 2 & 3 \end{bmatrix}$$

$$\mathbf{T}^{-1} = \begin{bmatrix} \frac{1}{3} & -\frac{1}{3} \\ \frac{1}{3} & \frac{2}{3} \end{bmatrix}$$

If we let

$$\begin{bmatrix}\gamma_1(t)\\ \gamma_2(t)\end{bmatrix} = \begin{bmatrix}\frac{1}{3} & -\frac{1}{3}\\ \frac{1}{3} & \frac{2}{3}\end{bmatrix}\begin{bmatrix}\lambda_1(t)\\ \lambda_2(t)\end{bmatrix} = \frac{1}{3}\begin{bmatrix}\lambda_1(t) - \lambda_2(t)\\ \lambda_1(t) + 2\lambda_2(t)\end{bmatrix}$$

we obtain the state equations:

$$\frac{d\gamma_1(t)}{dt} = -\gamma_1(t)$$

$$\frac{d\gamma_2(t)}{dt} = 5\gamma_2(t)$$

When the characteristic values of **A** are not all distinct, the matrix **A** might not be diagonalizable. However, it can be transformed into its Jordan canonical form and the matrix function $e^{\mathbf{A}t}$ can be computed accordingly. We leave the details of the computation to Prob. 3-23.

3-11 SOLUTION OF VECTOR DIFFERENTIAL EQUATIONS: THE PARTICULAR SOLUTION

The particular solution of the vector differential equation

$$\frac{d\boldsymbol{\lambda}(t)}{dt} = \mathbf{A}\boldsymbol{\lambda}(t) + \mathbf{B}x(t) \tag{3-57}$$

depends on the input $x(t)$. In simple cases we can guess the form of the particular solution $\boldsymbol{\lambda}(t)$ from the form of $x(t)$. We illustrate the procedure by examining the case in which $x(t)$ is of the form $e^{\beta t}$, where β is not a characteristic value of **A**. In this case the particular solution is of the form

$$\boldsymbol{\lambda}(t) = \mathbf{Z}e^{\beta t}$$

where

$$\mathbf{Z} = \begin{bmatrix} Z_1 \\ Z_2 \\ \vdots \\ Z_r \end{bmatrix}$$

is a column vector of multiplying constants. Substituting $\boldsymbol{\lambda}(t) = \mathbf{Z}e^{\beta t}$ into Eq. (3-57), we obtain

$$\beta\mathbf{Z}e^{\beta t} = \mathbf{A}\mathbf{Z}e^{\beta t} + \mathbf{B}e^{\beta t}$$

or

$$\mathbf{Z} = -\mathbf{B}[\mathbf{A} - \beta\mathbf{I}]^{-1}$$

Notice that $\mathbf{A} - \beta\mathbf{I}$ has an inverse because β is not a characteristic value of **A**.

We show now, in the general case, the particular solution is

$$e^{\mathbf{A}t} \int_{-\infty}^{t} e^{-\mathbf{A}\tau} \mathbf{B} x(\tau)\, d\tau$$

Multiplying both sides of Eq. (3-57) by $e^{-\mathbf{A}t}$, we obtain

$$e^{-\mathbf{A}t} \frac{d\boldsymbol{\lambda}(t)}{dt} = e^{-\mathbf{A}t}\mathbf{A}\boldsymbol{\lambda}(t) + e^{-\mathbf{A}t}\mathbf{B}x(t)$$

or

$$e^{-\mathbf{A}t} \frac{d\boldsymbol{\lambda}(t)}{dt} - e^{-\mathbf{A}t}\mathbf{A}\boldsymbol{\lambda}(t) = e^{-\mathbf{A}t}\mathbf{B}x(t)$$

that is,

$$\frac{d}{dt}[e^{-\mathbf{A}t}\boldsymbol{\lambda}(t)] = e^{-\mathbf{A}t}\mathbf{B}x(t) \tag{3-58}$$

Integrating both sides of Eq. (3-58), we obtain

$$\boldsymbol{\lambda}(t) = e^{\mathbf{A}t} \int_{-\infty}^{t} e^{-\mathbf{A}\tau}\mathbf{B}x(\tau)\, d\tau$$

Note that $e^{-\mathbf{A}\tau}\mathbf{B}x(\tau)$ is a column vector of r components, and to integrate such a column vector is simply to integrate its entries one by one.

The total solution of Eq. (3-57) is, therefore,

$$\boldsymbol{\lambda}(t) = e^{\mathbf{A}t}\mathbf{K} + e^{\mathbf{A}t} \int_{-\infty}^{t} e^{-\mathbf{A}t}\mathbf{B}x(\tau)\, d\tau \tag{3-59}$$

To determine the vector **K**, we need to know the initial state of the system, that is, the value of $\boldsymbol{\lambda}(t)$ at t_0, $\boldsymbol{\lambda}(t_0)$. Since, according to Eq. (3-59),

$$\boldsymbol{\lambda}(t_0) = e^{\mathbf{A}t_0}\mathbf{K} + e^{\mathbf{A}t_0} \int_{-\infty}^{t_0} e^{-\mathbf{A}\tau}\mathbf{B}x(\tau)\, d\tau$$

$$\mathbf{K} = e^{-\mathbf{A}t_0}\boldsymbol{\lambda}(t_0) - \int_{-\infty}^{t_0} e^{-\mathbf{A}\tau}\mathbf{B}x(\tau)\, d\tau$$

the total solution of Eq. (3-57) is then

$$\boldsymbol{\lambda}(t) = e^{\mathbf{A}(t-t_0)}\boldsymbol{\lambda}(t_0) - e^{\mathbf{A}t} \int_{-\infty}^{t_0} e^{-\mathbf{A}\tau}\mathbf{B}x(\tau)\, d\tau + e^{\mathbf{A}t} \int_{-\infty}^{t} e^{-\mathbf{A}\tau}\mathbf{B}x(\tau)\, d\tau$$

$$= e^{\mathbf{A}(t-t_0)}\boldsymbol{\lambda}(t_0) + e^{\mathbf{A}t} \int_{t_0}^{t} e^{-\mathbf{A}\tau}\mathbf{B}x(\tau)\, d\tau \tag{3-60}$$

3-12 DIFFERENTIAL EQUATIONS AND STATE EQUATIONS

The input-output relationship of a continuous system described by a differential equation can also be described by a set of state equations and an output equation. The procedure of transforming a differential equation description into a state space description is quite similar to that of transforming a difference equation description into a state space description as was discussed in Sec. 3-5. Therefore, our presentation will be brief here. Consider a differential equation of the form

$$C_0 \frac{d^r y(t)}{dt^r} + C_1 \frac{d^{r-1} y(t)}{dt^{r-1}} + \cdots + C_{r-1} \frac{dy(t)}{dt} + C_r y(t) = E_0 x(t)$$

If we define the state variables to be

$$\begin{aligned}
\lambda_1(t) &= y(t) \\
\lambda_2(t) &= \frac{dy(t)}{dt} \\
\lambda_3(t) &= \frac{d^2 y(t)}{dt^2} \\
&\cdots\cdots\cdots \\
\lambda_{r-1}(t) &= \frac{d^{r-2} y(t)}{dt^{r-2}} \\
\lambda_r(t) &= \frac{d^{r-1} y(t)}{dt^{r-1}}
\end{aligned}$$

then the state equations are

$$\begin{aligned}
\frac{d\lambda_1(t)}{dt} &= \lambda_2(t) \\
\frac{d\lambda_2(t)}{dt} &= \lambda_3(t) \\
\frac{d\lambda_3(t)}{dt} &= \lambda_4(t) \\
&\cdots\cdots\cdots \\
\frac{d\lambda_{r-1}(t)}{dt} &= \lambda_r(t) \\
\frac{d\lambda_r(t)}{dt} &= -\frac{1}{C_0}[C_r \lambda_1(t) + C_{r-1}\lambda_2(t) + \cdots + C_1 \lambda_{r-1}(t) - E_0 x(t)]
\end{aligned}$$

and the output equation is

$$y(t) = \lambda_1(t)$$

As an example, let

$$\frac{d^2y(t)}{dt^2} + 2\frac{dy(t)}{dt} + y(t) = 3x(t)$$

be a differential equation describing the input-output relationship of a continuous system. Let the state variables be

$$\lambda_1(t) = y(t)$$

$$\lambda_2(t) = \frac{dy(t)}{dt}$$

In the state space description, the state equations are

$$\frac{d\lambda_1(t)}{dt} = \lambda_2(t)$$

$$\frac{d\lambda_2(t)}{dt} = -\lambda_1(t) - 2\lambda_2(t) + 3x(t)$$

and the output equation is

$$y(t) = \lambda_1(t)$$

In the general case, the input-output relation of a continuous system can be described by a differential equation of the form

$$C_0\frac{d^ry(t)}{dt^r} + C_1\frac{d^{r-1}y(t)}{dt^{r-1}} + \cdots + C_{r-1}\frac{dy(t)}{dt} + C_ry(t)$$

$$= E_0\frac{d^mx(t)}{dt^m} + E_1\frac{d^{m-1}x(t)}{dt^{m-1}} + \cdots + E_{m-1}\frac{dx(t)}{dt} + E_mx(t)$$

A procedure for transferring the differential equation description into the state space description is quite similar to that for transferring a difference equation into the state space description discussed in Sec. 3-5. We shall limit ourselves to some illustrative examples. Consider the differential equation

$$\frac{dy(t)}{dt} + 2y(t) = 3\frac{dx(t)}{dt} + 4x(t)$$

Integrating both sides of the equation, we can rewrite the equation as

$$y(t) = -2\int_{-\infty}^{t} y(\tau)\,d\tau + 3x(t) + 4\int_{-\infty}^{t} x(\tau)\,d\tau$$

Letting

$$\lambda(t) = -2\int_{-\infty}^{t} y(\tau)\,d\tau + 4\int_{-\infty}^{t} x(\tau)\,d\tau$$

be a state variable, we have as our output equation

$$y(t) = \lambda(t) + 3x(t)$$

and as state equation

$$\frac{d\lambda(t)}{dt} = -2y(t) + 4x(t) = -2\lambda(t) - 2x(t)$$

Also, consider the example

$$\frac{d^2y(t)}{dt^2} + 2\frac{dy(t)}{dt} + 3y(t) = 4\frac{d^2x(t)}{dt^2} + 5\frac{dx(t)}{dt} + 6x(t) \tag{3-61}$$

Integrating both sides of the equation twice, we obtain

$$y(t) = -2\int_{-\infty}^{t} y(\tau)\,d\tau - 3\int_{-\infty}^{t} d\tau \int_{-\infty}^{\tau} y(\delta)\,d\delta + 4x(t)$$
$$+ 5\int_{-\infty}^{t} x(\tau)\,d\tau + 6\int_{-\infty}^{t} d\tau \int_{-\infty}^{\tau} x(\delta)\,d\delta$$

If we let

$$\lambda_1(t) = -2\int_{-\infty}^{t} y(\tau)\,d\tau - 3\int_{-\infty}^{t} d\tau \int_{-\infty}^{\tau} y(\delta)\,d\delta + 5\int_{-\infty}^{t} x(\tau)\,d\tau + 6\int_{-\infty}^{t} d\tau \int_{-\infty}^{\tau} x(\delta)\,d\delta$$

we have

$$\frac{d\lambda_1(t)}{dt} = -2y(t) - 3\int_{-\infty}^{t} y(\tau)\,d\tau + 5x(t) + 6\int_{-\infty}^{t} x(\tau)\,d\tau$$

If we let

$$\lambda_2(t) = -3\int_{-\infty}^{t} y(\tau)\,d\tau + 6\int_{-\infty}^{t} x(\tau)\,d\tau$$

we have

$$\frac{d\lambda_2(t)}{dt} = -3y(t) + 6x(t)$$

Thus, in the state space description, the output equation is

$$y(t) = \lambda_1(t) + 4x(t) \tag{3-62}$$

and the state equations are

$$\frac{d\lambda_1(t)}{dt} = -2y(t) + 5x(t) + \lambda_2(t) = -2\lambda_1(t) + \lambda_2(t) - 3x(t)$$

$$\frac{d\lambda_2(t)}{dt} = -3y(t) + 6x(t) = -3\lambda_1(t) - 6x(t) \tag{3-63}$$

We can also obtain from a given state space description a differential equation description of continuous systems. We ask the reader to give a procedure for doing so and to check the procedure by recovering the differential equation (3-61) from the output equation (3-62) and the state equations (3-63).

3-13 REMARKS AND REFERENCES

There are several reasons for studying the state space description of system behavior after we have studied the differential equation description in Chap. 2. The differential equation is strictly a description of the terminal behavior of a system. On the other hand, some of the internal activities of a system are exhibited through the values of the state variables in the state space description. Such a description not only gives further insight into the behavior of a system but also provides information that is useful in many cases. For example, although the position of a spacecraft is the output in which we are interested, the speed and acceleration of the spacecraft are also useful information in many considerations. In many cases, it is conceptually simpler to set up a state space description of a given physical system than to set up a differential equation description, because a state space description can be viewed as a division of a differential equation description into steps. With the matrix notation, the state space formulation is also most suitable for describing the behavior of multiple-input–multiple-output systems.

From a mathematical point of view, when we use a state space description instead of a difference or differential equation description, we are trading the problem of solving an rth-order difference or differential equation for the problem of solving r simultaneous first-order difference or differential equations. We urge the reader to compare the methods of solution presented in Chaps. 2 and 3 and to convince herself that, for a given physical system, the state space description and the difference or differential equation description will always yield the same response to a given stimulus.

As general references, see DeRusso, Roy, and Close [4] or Ogata [8]. More advanced references are Brockett [3] and Athans and Falb [1]. The topic of simultaneous differential equation systems can be found in most books on differential equations; see, for example, Boyce and DiPrima [2]. The concept of state is also used in the synthesis and analysis of sequential circuits (see Hill and Peterson [7]); in the modeling of logical machines (see Hennie [6]); and in the modeling of a class of random processes, known as Markov processes (see Feller [5]).

[1] ATHANS, M., and P. L. FALB: "Optimal Control," McGraw-Hill Book Company, New York, 1966.

[2] BOYCE, W. E., and R. C. DIPRIMA: "Elementary Differential Equations and Boundary Value Problems," 2d ed., John Wiley & Sons, Inc., New York, 1969.

[3] BROCKETT, R. W.: "Linear Dynamical Systems," John Wiley & Sons, Inc., New York, 1970.

[4] DERUSSO, P. M., R. J. ROY, and C. M. CLOSE: "State Variables for Engineers," John Wiley & Sons, Inc., New York, 1965.

[5] FELLER, W.: "An Introduction to Probability Theory and Its Applications," 2d ed., vol. 1, John Wiley & Sons, Inc., New York, 1957.

[6] HENNIE, F. C.: "Finite-state Models for Logical Machines," John Wiley & Sons, Inc., New York, 1968.

[7] HILL, F. J., and G. R. PETERSON: "Introduction to Switching Theory and Logical Design," John Wiley & Sons, Inc., New York, 1968.

[8] OGATA, K.: "State Space Analysis of Control Systems," Prentice-Hall, Inc., Englewood Cliffs, N.J., 1967.

PROBLEMS

3-1 A vending machine sells soft drinks at the price of 15 cents a can. The machine accepts nickels, dimes, and quarters. When a coin is deposited, the machine will put it into one of three bins, depending on whether it is a nickel, a dime, or a quarter. If a quarter or two dimes is deposited and there are not enough nickels or dimes in the bins for the change, the machine will return the coins and flash the "use correct change" sign. Consider the vending machine a discrete system whose input signals are the numbers of nickels, dimes, and quarters deposited during each sale and whose output signal is equal to 1 if the machine can make change and is equal to 0 otherwise. Write the state equations and the output equation describing this system.

3-2 To study the population size of deer in a small conservation area, 100 deer are caught, marked with red paint, and then released on January 1 every year. Let $x(n)$ denote the number of unmarked deer among those caught at the beginning of the nth year. [Thus, $x(1) = 100$.] Suppose that the average number of offspring born to a deer in the nth year is equal to $4[1 - x(n-3)/100]$. Suppose that, on the average, one-fourth of the deer alive on January 1 of each year die of natural causes and $x(n)$ deer are allowed to be killed in the hunting season in October. Let $x(n)$ be the input signal of the system.

(*a*) Write the state equations and the ouput equation describing the system if we want to calculate the average size of the population of the deer at the end of each year.

(*b*) Write the state equations describing the system if we want to estimate the probability of catching 95 unmarked deer on January 1 of the nth year.

3-3 A space vehicle is descending vertically in the terminal descent phase of a lunar hovering mission. The input signal to the system is the thrust $x(t)$ applied to brake the vehicle so that its velocity becomes zero at an altitude h from the lunar surface. Then, the vehicle will orbit the moon at an altitude h. Let us describe the state of the space vehicle by the values of three state variables: $\lambda_1(t)$, the altitude; $\lambda_2(t)$, the velocity; and $\lambda_3(t)$, the mass of the spacecraft (the mass of the vehicle decreases with time because of the consumption of fuel). Suppose that the burning of α lb of fuel per second will generate 1 lb of thrust. Write the state equations describing this system.

3-4 There are two kinds of particles inside a nuclear reactor. In every second, an α particle will split into three β particles, and a β particle will split into an α particle and two β particles. If there is a single α particle in the reactor at $t = 0$, how many particles are there altogether at $t = 100$?

3-5 (*a*) The input-output relationship of a discrete system is described by the difference equation

$$y(n) + 3y(n-1) + 2y(n-2) = x(n) - x(n-1)$$

Determine the unit response of the system, assuming that the system is initially at rest.

(*b*) Determine a state space description of the system in part (*a*). Solve the state equations and the output equation for the state variables and the output $y(n)$ when the input is the unit function, assuming that the system is initially at rest, i.e., the values of all state variables as well as the value of the output are all equal to zero initially.

3-6 Consider the operation of a factory which can be modeled as a discrete system. For convenience, let us assume that all quantities are measured in thousands of dollars. In each month, the input is the total amount of new orders, and the output is the total production of that month. The backlog in each month is equal to the backlog plus the new orders minus the production, all of the preceding month. The production in each month is equal to the sum of two times the production and the value of machines available (also measured in thousands of dollars) of the preceding month. In each month, new machines whose total value is equal to two times the backlog of the preceding month will be purchased, and all old machines are kept. (If there is a negative backlog, it means old machines will be sold.)

(*a*) Using the backlog $\lambda_1(n)$, the production $\lambda_2(n)$, and the total value of machines in the factory, $\lambda_3(n)$, as state variables, write the state equations and output equation.

(*b*) Assume that, at the beginning of the operation, there is a backlog of 10 (thousands of dollars), no new production, and two thousand dollars worth of machines in the factory. The new orders in three successive months are 5, 7, 2 (in thousands of dollars). Use matrix notation to compute the successive values of the state variables and the output.

(*c*) Determine the general form of the homogeneous solutions of $\lambda_1(n)$, $\lambda_2(n)$, and $\lambda_3(n)$.

3-7 The output equation of a discrete system is

$$y(n) = \lambda_1(n) + \lambda_2(n)$$

and the state equations are

$$\lambda_1(n+1) = \lambda_1(n) - 2\lambda_2(n) + x(n)$$
$$\lambda_2(n+1) = A\lambda_1(n) + B\lambda_2(n)$$

It is given that $x(n) = 0$ and $y(n) = 8(-1)^n - 5(-2)^n$ for $n \geq 0$.

(*a*) Determine the constants A and B.

(*b*) Determine $\lambda_1(n)$ and $\lambda_2(n)$ in closed form.

3-8 Consider a system described by the following equations:

$$y(n) = \lambda_1(n)\lambda_2(n) + x(n)$$
$$\lambda_1(n+1) = \lambda_1(n) - \lambda_2(n)$$
$$\lambda_2(n+1) = -\lambda_1(n) - \lambda_2(n)$$

(*a*) Determine the homogeneous solutions, given that $\lambda_1(0) = 2$ and $\lambda_2(0) = 2$.

(*b*) Determine $y(n)$ for $n \geq 0$, given that $x(n) = 3^n$ for $n \geq 0$.

(*c*) Find a difference equation relating $x(n)$ and $y(n)$. Determine $y(n)$ directly and check your answer with that in part (*b*).

(*d*) Repeat parts (*b*) and (*c*) if $x(n) = 2^n$ for $n \geq 0$. [What is the particular solution for the difference equation in part (*c*)?]

3-9 Consider a discrete system whose input-output relationship is described by a difference equation of the form

$$C_0 y(n) + C_1 y(n-1) + C_2 y(n-2) + \cdots + C_r y(n-r)$$
$$= E_0 x(n) + E_1 x(n-1) + E_2 x(n-2) + \cdots + E_m x(n-m)$$

When $r \leq m$, this equation can be written

$$E_0 x(n) - C_0 y(n) + E_1 x(n-1) - C_1 y(n-1) + E_2 x(n-2) - C_2 y(n-2)$$
$$+ \cdots + E_m x(n-m) - C_m y(n-m) = 0$$

Find the state space description of this system, using m state variables.

3-10 In the circuit shown in Fig. 3P-1 let $L = 1$ H, $C = 1$ F, $E = 2$ V. Let the initial voltage across the capacitor be 9 V, and the initial current in the inductor be zero. Investigate the behavior of the circuit by plotting the voltage across the capacitor against the current in the inductor.

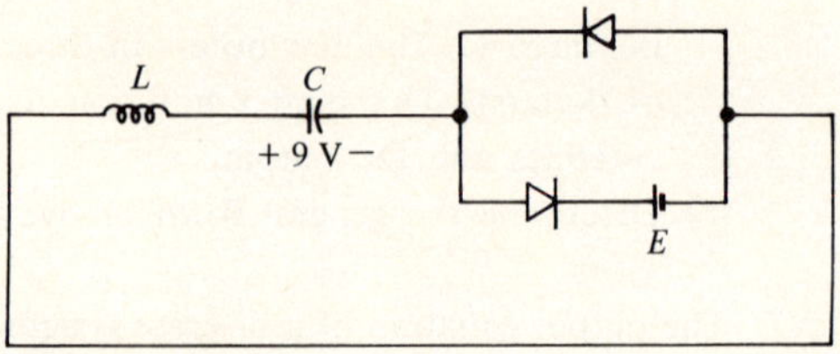

FIGURE 3P-1

3-11 Consider the state equations

$$\frac{d\lambda_1(t)}{dt} = \lambda_1(t) + 2\lambda_2(t)$$

$$\frac{d\lambda_2(t)}{dt} = -\lambda_1(t) + 4\lambda_2(t)$$

Given that $\lambda_1(0) = 3$ and $\lambda_2(0) = 2$, solve the state equations by the following methods:

(*a*) Determine the α's, A's, B's in $\lambda_1(t) = A_1 e^{\alpha_1 t} + B_1 e^{\alpha_2 t}$, and $\lambda_2(t) = A_2 e^{\alpha_1 t} + B_2 e^{\alpha_2 t}$.

(*b*) Evaluate $e^{\mathbf{A}t}$ by determining c_0, c_1 in $e^{\mathbf{A}t} = c_0\mathbf{I} + c_1\mathbf{A}$.

(*c*) Use the change of state variables:

$$\lambda_1(t) = 2\gamma_1(t) + \gamma_2(t)$$

$$\lambda_2(t) = \gamma_1(t) + \gamma_2(t)$$

3-12 The equivalent circuit of the amplifier in Fig. 3P-2*a* is shown in Fig. 3P-2*b*. Let $R = R_g = 2 \times 10^3\ \Omega$, $R_i = 10^4\ \Omega$, $\mu = 2 \times 10^3$, $C = 10^{-5}$ F.

(*a*) Using the voltage across the capacitor as a state variable, write the output equation and the state equation. Note that $\mu v_1(t)$ is a dependent voltage source.

(*b*) Determine $v_0(t)$ for $t \geq 0$ if $v_i(t) = u_{-1}(t)$ and the circuit is initially at rest.

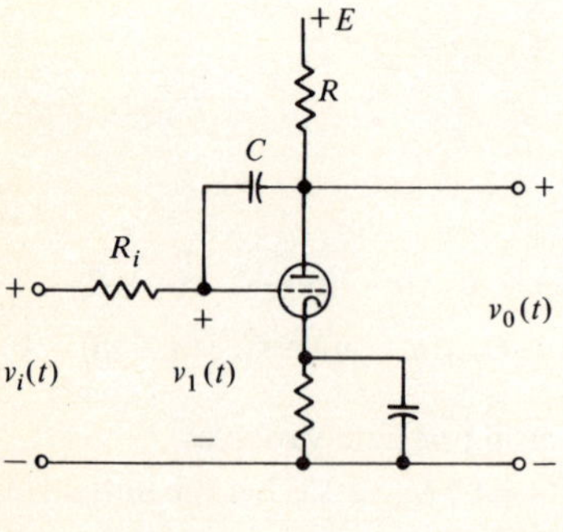

(*a*)

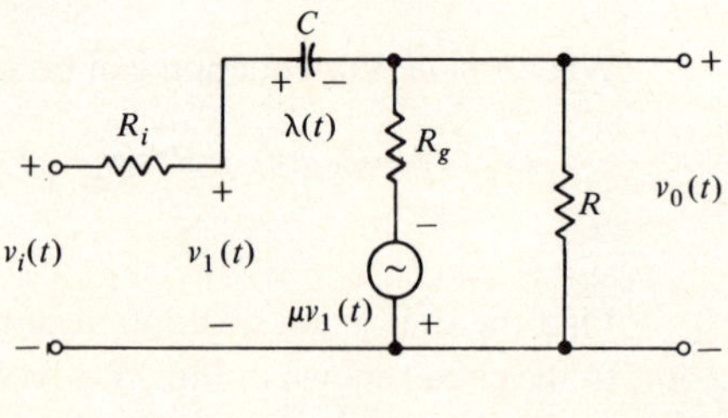

(*b*)

FIGURE 3P-2

3-13 Consider two linear time-invariant systems whose vector state equations and output equations are

$$\frac{d}{dt}\boldsymbol{\lambda}(t) = \begin{bmatrix} 5 & 1 & 0 \\ 6 & 0 & 1 \\ 1 & 0 & 0 \end{bmatrix} \boldsymbol{\lambda}(t) + \begin{bmatrix} 2 \\ 1 \\ 3 \end{bmatrix} x(t)$$

$$y(t) = [1 \quad 0 \quad 0]\boldsymbol{\lambda}(t)$$

and

$$\frac{d}{dt}\boldsymbol{\gamma}(t) = \begin{bmatrix} 5 & 1 & 0 \\ 6 & 0 & 1 \\ 1 & 0 & 0 \end{bmatrix} \boldsymbol{\gamma}(t) + \begin{bmatrix} 1 \\ 0 \\ 0 \end{bmatrix} x(t)$$

$$z(t) = [2 \quad 1 \quad 3]\boldsymbol{\gamma}(t)$$

Show that $y(t) = z(t)$ for $t \geq 0$ if $\boldsymbol{\lambda}(0) = \begin{bmatrix} 2 \\ 1 \\ 3 \end{bmatrix}$ and $\boldsymbol{\gamma}(0) = \begin{bmatrix} 1 \\ 0 \\ 0 \end{bmatrix}$.

3-14 Consider the circuit shown in Fig. 3P-3. The box labeled A denotes a new kind of circuit element whose terminal current is equal to the second derivative of its terminal voltage.

(*a*) Write the state equations for an appropriately chosen set of state variables.

(*b*) Determine the natural frequencies of the network from the state equations. (For your information, one of them is equal to -2.)

(*c*) Let $x(t) = 0$. Suppose that the initial conditions at $t = 0$ are such that for large t the voltage across the capacitor is approximately $12e^{-2t}$. What is, approximately, the voltage across the device A for large t? (If you obtain your solution from the state equations, check your answer by direct substitution, and vice versa.)

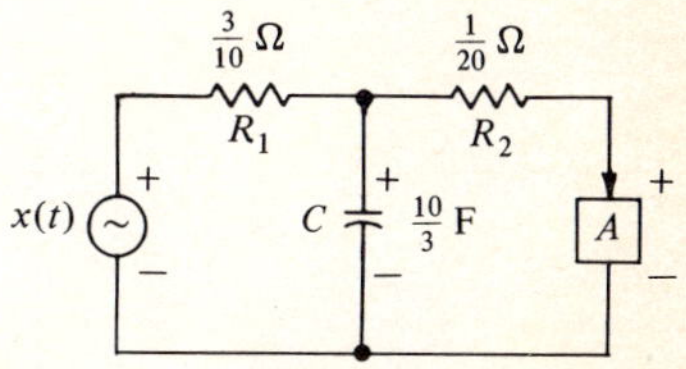

FIGURE 3P-3

3-15 Consider the movement of a particle in a two-dimensional plane. Let $\lambda_1(t)$ and $\lambda_2(t)$ denote the horizontal and vertical position of the particle at time t, respectively.

(*a*) Suppose that the motion of the particle is governed by the equations

$$\frac{d\lambda_1(t)}{dt} = 4\lambda_2(t)$$

$$\frac{d\lambda_2(t)}{dt} = -\lambda_1(t)$$

Determine $\lambda_1(t)$ and $\lambda_2(t)$, and then plot the trajectory of the particle in the plane, given that the particle is at the point (4, 0) when $t = 0$.

(*b*) Suppose that the motion of the particle is governed by the equations

$$\frac{d\lambda_1(t)}{dt} = g_1(\lambda_1(t), \lambda_2(t)) - x(t)$$

$$\frac{d\lambda_2(t)}{dt} = g_2(\lambda_1(t), \lambda_2(t))$$

where g_1 and g_2 are two nonlinear functions:

$$g_1(\lambda_1(t), \lambda_2(t)) = \begin{cases} -1 & \lambda_1(t) \geq 0, \lambda_2(t) \geq 0 \\ -1 & \lambda_1(t) < 0, \lambda_2(t) \geq 0 \\ 2 & \lambda_1(t) < 0, \lambda_2(t) < 0 \\ 1 & \lambda_1(t) \geq 0, \lambda_2(t) < 0 \end{cases}$$

$$g_2(\lambda_1(t), \lambda_2(t)) = \begin{cases} 4 & \lambda_1(t) \geq 0, \lambda_2(t) \geq 0 \\ -2 & \lambda_1(t) < 0, \lambda_2(t) \geq 0 \\ -2 & \lambda_1(t) < 0, \lambda_2(t) < 0 \\ 2 & \lambda_1(t) > 0, \lambda_2(t) < 0 \end{cases}$$

Let
$$x(t) = \begin{cases} 1 & 0 < t < 2.5 \\ 0 & \text{otherwise} \end{cases}$$

Plot the trajectory of the particle in the plane, given that the particle is at the point (3, 0) when $t = 0$.

[*Remark:* Observe that by adjusting the duration of the signal $x(t)$ the particle can be "kicked" into a chosen orbit.]

3-16 In the circuit shown in Fig. 3P-4, determine $y(t)$ and $z(t)$ for $t \geq 0$, given that $y(0+) = 3$ and $z(0+) = 3$.

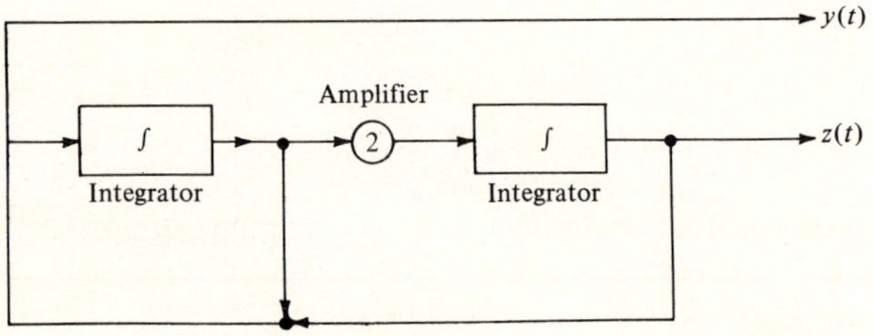

FIGURE 3P-4

3-17 Let

$$\frac{d\lambda_1(t)}{dt} = \lambda_1(t) + \lambda_2(t) - 3x(t)$$

$$\frac{d\lambda_2(t)}{dt} = 2\lambda_1(t) - \lambda_2(t)$$

be the state equations of a continuous system. For $x(t) = u_{-1}(t)$, it is known that the values of both $\lambda_1(t)$ and $\lambda_2(t)$ remain unchanged from $t = 0$ to $t = \infty$. Determine the boundary conditions $\lambda_1(0+)$ and $\lambda_2(0+)$.

3-18 The incremental equivalent circuit of an *RC*-coupled amplifier is shown in Fig. 3P-5. Let $R = 10^6\ \Omega$, $C = 10^{-6}$ F. Note that Ki_b is a dependent current source.

(*a*) Write the state equations for the circuit, using the voltages across the capacitors as state variables.

(*b*) For what values of K will i_b be a decaying sinusoid? (If you wonder why this circuit never oscillates, you may recall that for sustained oscillation at least one more section of the *RC* circuit is needed.)

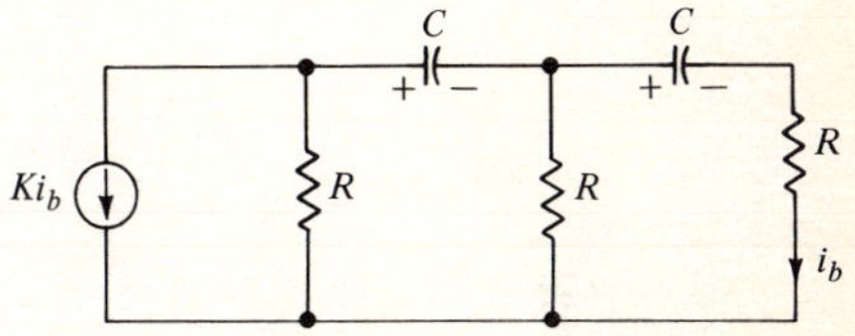

FIGURE 3P-5

3-19 Determine the state space representation of the following systems:

(*a*) $y(n) + 4y(n-1) + 3y(n-2) = x(n) + x(n-1)$

(*b*) $\dfrac{d^2y(t)}{dt^2} + 4\dfrac{dy(t)}{dt} + 3y(t) = \dfrac{dx(t)}{dt} + x(t)$

3-20 Determine a state space description of a system whose input-output relationship is described by the differential equation

$$\frac{d^2y(t)}{dt^2} + 2y(t)\frac{dy(t)}{dt} + 3[y(t)]^2 = 4x(t)$$

3-21 (*a*) Find the state space description of the *RLC* circuit shown in Fig. 3P-6, using the voltages across the capacitors and the current in the inductor as state variables. Determine the homogeneous solutions of the state equations.

(*b*) Find the differential equation description of the circuit from its state space description.

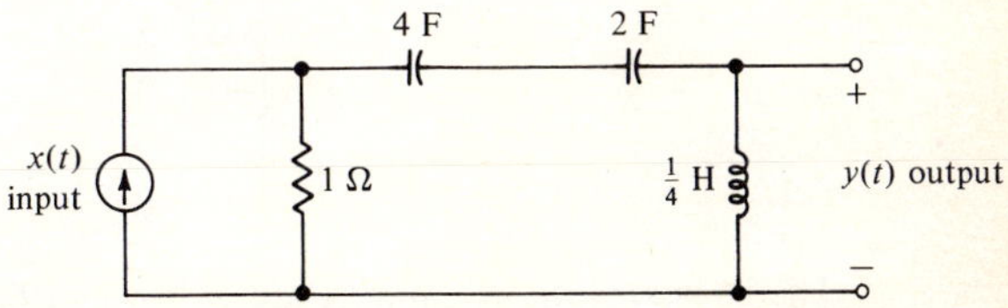

FIGURE 3P-6

3-22 (a) For the system shown in Fig. 3P-7, write the output equation and the state equations, using the outputs of the integrators, $\lambda_1(t)$ and $\lambda_2(t)$, as state variables.

(b) Let $\lambda_1(0+) = \lambda_2(0+) = 1$. Let $x(t) = 2e^{-3t}$ for $t \geq 0$. Determine $\lambda_1(t)$ and then $\lambda_2(t)$ for $t \geq 0$ by solving the state equations.

(c) Write a single differential equation relating $x(t)$ and $y(t)$. Determine $y(t)$ for $t \geq 0$ by solving the differential equation directly.

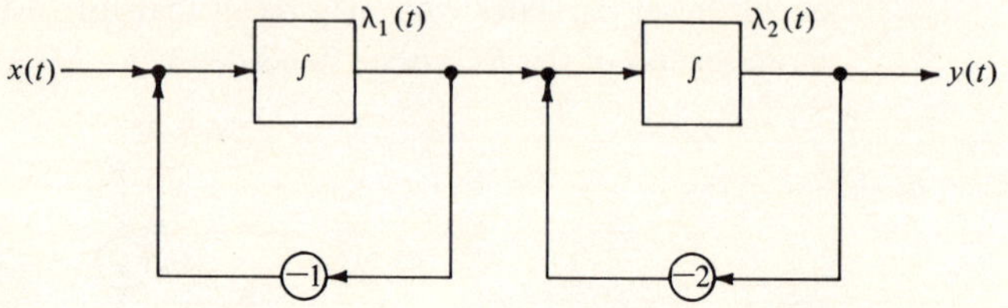

FIGURE 3P-7

3-23 (a) For the matrix

$$\mathbf{A} = \begin{bmatrix} 2 & 1 \\ 0 & 2 \end{bmatrix}$$

show that

$$\mathbf{A}^n = \begin{bmatrix} 2^n & n2^{n-1} \\ 0 & 2^n \end{bmatrix}$$

for $n = 1, 2, 3, \ldots,$ and

$$e^{\mathbf{A}t} = \begin{bmatrix} e^{2t} & te^{2t} \\ 0 & e^{2t} \end{bmatrix}$$

(b) Let $\mathbf{A}$ be a 5×5 matrix whose characteristic values are such that $\alpha_1 = \alpha_2 = \alpha_3$, $\alpha_4 = \alpha_5$, and $\alpha_4 \neq \alpha_1$. There exists a nonsingular matrix $\mathbf{T}$ such that

$$\mathbf{A} = \mathbf{TJT}^{-1}$$

where

$$\mathbf{J} = \begin{bmatrix} \alpha_1 & c_1 & 0 & 0 & 0 \\ 0 & \alpha_1 & c_2 & 0 & 0 \\ 0 & 0 & \alpha_1 & 0 & 0 \\ 0 & 0 & 0 & \alpha_4 & c_3 \\ 0 & 0 & 0 & 0 & \alpha_4 \end{bmatrix}$$

The constants c_1, c_2, and c_3 are either 1 or 0. $\mathbf{J}$ is called the Jordan canonical form of $\mathbf{A}$. For the case $c_1 = c_2 = c_3 = 1$, show that

$$e^{\mathbf{A}t} = e^{\mathbf{TJT}^{-1}t} = \mathbf{T}e^{\mathbf{J}t}\mathbf{T}^{-1}$$

$$= \mathbf{T}\begin{bmatrix} e^{\alpha_1 t} & te^{\alpha_1 t} & t^2e^{\alpha_1 t} & 0 & 0 \\ 0 & e^{\alpha_1 t} & te^{\alpha_1 t} & 0 & 0 \\ 0 & 0 & e^{\alpha_1 t} & 0 & 0 \\ 0 & 0 & 0 & e^{\alpha_4 t} & te^{\alpha_4 t} \\ 0 & 0 & 0 & 0 & e^{\alpha_4 t} \end{bmatrix}\mathbf{T}^{-1}$$

3-24 We introduce in this problem the notion of *controllability* of a system. A linear time-invariant system is said to be state controllable if for any initial state at time $t = 0$ there is an input signal that will drive the system into any chosen state in finite time. Clearly, this is an important property of systems when we consider, for example, the problem of controlling the position of a spacecraft.

Consider the state equations of a discrete system:

$$\lambda_1(n+1) = \lambda_2(n) + x(n)$$

$$\lambda_2(n+1) = -\lambda_1(n) + 3x(n)$$

(*a*) Let $\lambda_1(0) = 1$ and $\lambda_2(0) = 1$. Can the system be driven into the state $\lambda_1 = 0$, $\lambda_2 = 0$ in one unit of time? If so, how should the value of $x(0)$ be chosen?

(*b*) Repeat part (*a*) for $\lambda_1(0) = -6$ and $\lambda_2(0) = 2$.

(*c*) State a general condition on the values of $\lambda_1(0)$ and $\lambda_2(0)$ so that the system can be driven into the state $\lambda_1(1) = 0$, $\lambda_2(1) = 0$.

(*d*) Let $\lambda_1(0) = 4$ and $\lambda_2(0) = 10$. Determine $x(0)$ and $x(1)$ so that $\lambda_1(2) = 0$, $\lambda_2(2) = 0$.

(*e*) Consider the discrete system described by the vector state equation

$$\begin{bmatrix} \lambda_1(n+1) \\ \lambda_2(n+1) \end{bmatrix} = \begin{bmatrix} 3 & -2 \\ -1 & 1 \end{bmatrix}\begin{bmatrix} \lambda_1(n) \\ \lambda_2(n) \end{bmatrix} + \begin{bmatrix} 2 \\ 1 \end{bmatrix}x(n)$$

Find the input signal that will drive this system from an initial state $\lambda_1(0)$ and $\lambda_2(0)$ at $n = 0$ to $\lambda_1(2) = 0$ and $\lambda_2(2) = 0$ at $n = 2$. Express $x(0)$ and $x(1)$ in terms of $\lambda_1(0)$ and $\lambda_2(0)$.

(*f*) Consider the system described by the state equation

$$\begin{bmatrix} \lambda_1(n+1) \\ \lambda_2(n+1) \end{bmatrix} = \begin{bmatrix} 1 & 1 \\ 0 & 1 \end{bmatrix}\begin{bmatrix} \lambda_1(n) \\ \lambda_2(n) \end{bmatrix} + \begin{bmatrix} 0 \\ 1 \end{bmatrix}x(n)$$

Suppose that the initial state of the system is $\lambda_1(0) = 1$ and $\lambda_2(0) = 0$. Show that it is not possible to find an input signal that will drive the system to the state $\lambda_1(n) = 0$ and $\lambda_2(n) = 0$ for any n.

(*g*) Let

$$\boldsymbol{\lambda}(n+1) = \mathbf{A}\boldsymbol{\lambda}(n) + \mathbf{B}x(n)$$

be the vector state equation of a discrete system. Let $\mathbf{A}$ be an $r \times r$ *nonsingular* matrix. Show that the system can be driven into the state $\lambda_1 = \lambda_2 = \cdots = \lambda_r = 0$

in r time units if and only if the column vectors $\mathbf{A}^{-1}\mathbf{B}, (\mathbf{A}^{-1})^2\mathbf{B}, (\mathbf{A}^{-1})^3\mathbf{B}, \ldots,$ $(\mathbf{A}^{-1})^r\mathbf{B}$ are linearly independent. That is, the value of the determinant

$$|\mathbf{A}^{-1}\mathbf{B} \;\vdots\; (\mathbf{A}^{-1})^2\mathbf{B} \;\vdots\; (\mathbf{A}^{-1})^3\mathbf{B} \;\vdots\; \cdots \;\vdots\; (\mathbf{A}^{-1})^r\mathbf{B}|$$

is nonzero.

(*h*) We say that a state $\boldsymbol{\lambda}_i$ is state controllable at t_0 if there exists an input $x(t)$ that drives the system from the state $\boldsymbol{\lambda}_i$ at t_0 to any final state $\boldsymbol{\lambda}_f$ in a finite time interval. Show that, for a system described by linear state equations, a state $\boldsymbol{\lambda}_i$ is state controllable at t_0 if and only if there is an input signal $x(t)$ that drives the system from $\boldsymbol{\lambda}_i$ to the final state $\boldsymbol{\lambda}_f = \mathbf{0}$ in a finite time interval.

(*i*) A system is said to be completely state controllable at t_0 if every state is state controllable at t_0. Show that a time-invariant system is completely state controllable at any time t_0 if and only if it is completely state controllable at $t = 0$.

4

CONVOLUTION

4-1 INTRODUCTION

We recall our discussion in Sec. 2-1 that a most general way to describe the input-output relationship of a system is to list exhaustively all the input-output pairs of signals of the system. As was shown in Chaps. 2 and 3, such an exhaustive listing is not necessary when the input and output signals of a system are related by a difference (differential) equation or by an output equation together with a set of state equations. In this chapter, we present another compact way to describe the input-output relationship of linear time-invariant systems.

For the moment, let us limit our discussion to discrete systems. Suppose that it is possible to choose a set of discrete signals, $\ldots, \xi_{-1}(n), \xi_0(n), \xi_1(n), \xi_2(n), \ldots, \xi_k(n), \ldots$, such that any discrete signal $x(n)$ can be expressed as

$$x(n) = \cdots + a_{-1}\xi_{-1}(n) + a_0\,\xi_0(n) + a_1\xi_1(n) + a_2\,\xi_2(n) + \cdots + a_k\,\xi_k(n) + \cdots$$

where $\ldots, a_{-1}, a_0, a_1, \ldots, a_k, \ldots$ are multiplying constants. Clearly, when the responses of a *linear* system to the signals $\ldots, \xi_{-1}(n), \xi_0(n), \xi_1(n), \xi_2(n), \ldots, \xi_k(n), \ldots$ are known, the response of the system to the signal $x(n)$ can be determined. To be explicit, if $\ldots, h_{-1}(n), h_0(n), h_1(n), h_2(n), \ldots, h_k(n), \ldots$ are the responses of the system to the signals $\ldots, \xi_{-1}(n), \xi_0(n), \xi_1(n), \xi_2(n), \ldots, \xi_k(n), \ldots$, then the response of the system to the signal $x(n)$ is

$$y(n) = \cdots + a_{-1}h_{-1}(n) + a_0 h_0(n) + a_1 h_1(n) + a_2 h_2(n) + \cdots + a_k h_k(n) + \cdots$$

At this point, it seems that very little has been accomplished as far as obtaining a compact description of the input-output relationship of linear systems is concerned, since in order to specify the behavior of a system we still have to specify the responses $\ldots, h_{-1}(n), h_0(n), h_1(n), h_2(n), \ldots, h_k(n), \ldots$. However, for linear systems which are also *time-invariant*, two very attractive possibilities to simplify the description arise. Suppose that the signals $\xi_i(n)$ are so chosen that the response of the system to each of these signals is equal to a constant times the signal. That is,

$$\begin{gathered}
\cdots\cdots\cdots\cdots\cdots \\
h_{-1}(n) = \lambda_{-1}\xi_{-1}(n) \\
h_0(n) = \lambda_0 \xi_0(n) \\
h_1(n) = \lambda_1 \xi_1(n) \\
h_2(n) = \lambda_2 \xi_2(n) \\
\cdots\cdots\cdots\cdots\cdots \\
h_k(n) = \lambda_k \xi_k(n) \\
\cdots\cdots\cdots\cdots\cdots
\end{gathered}$$

where $\ldots, \lambda_{-1}, \lambda_0, \lambda_1, \lambda_2, \ldots, \lambda_k, \ldots$ are multiplying constants. In this case, the input-output relationship of the system can be specified by the list of constants $\ldots, \lambda_{-1}, \lambda_0, \lambda_1, \lambda_2, \ldots, \lambda_k, \ldots$. Such a possibility will be discussed in Chap. 5.

In this chapter, we consider another choice of the signals $\xi_i(n)$ which also leads to a compact description of the input-output relationship of linear time-invariant systems. Let the signals $\xi_i(n)$ be delayed or advanced versions of the signal $\xi_0(n)$ for all $i \neq 0$. To be specific, let

$$\begin{gathered}
\cdots\cdots\cdots\cdots\cdots \\
\xi_{-1}(n) = \xi_0(n+1) \\
\xi_1(n) = \xi_0(n-1) \\
\xi_2(n) = \xi_0(n-2) \\
\cdots\cdots\cdots\cdots\cdots \\
\xi_k(n) = \xi_0(n-k) \\
\cdots\cdots\cdots\cdots\cdots
\end{gathered}$$

Then, for a time-invariant system, the responses of the system to the signals $\xi_i(n)$ are also delayed or advanced versions of the response of the system to the signal $\xi_0(n)$. That is,

$$
\begin{aligned}
&\cdots\cdots\cdots\cdots\cdots\cdots \\
h_{-1}(n) &= h_0(n+1) \\
h_1(n) &= h_0(n-1) \\
h_2(n) &= h_0(n-2) \\
&\cdots\cdots\cdots\cdots\cdots\cdots \\
h_k(n) &= h_0(n-k) \\
&\cdots\cdots\cdots\cdots\cdots\cdots
\end{aligned}
$$

where $h_0(n)$ is the system's response to the input $\xi_0(n)$. It follows that for an input signal $x(n)$ such that

$$
\begin{aligned}
x(n) = \cdots + a_{-1}\xi_0(n+1) + a_0\,\xi_0(n) + a_1\xi_0(n-1) \\
+ a_2\,\xi_0(n-2) + \cdots + a_k\,\xi_0(n-k) + \cdots
\end{aligned}
\tag{4-1}
$$

the corresponding response will be

$$
\begin{aligned}
y(n) = \cdots + a_{-1}h_0(n+1) + a_0\,h_0(n) + a_1h_0(n-1) \\
+ a_2\,h_0(n-2) + \cdots + a_k\,h_0(n-k) + \cdots
\end{aligned}
$$

The critical question is then: Is it possible to choose $\xi_0(n)$ such that all discrete signals can be expressed as in Eq. (4-1)? The answer to this question is an affirmative one. As a matter of fact, it is easy to see that

$$
u_0(n) = \begin{cases} 1 & n = 0 \\ 0 & n \neq 0 \end{cases}
$$

is one such choice for $\xi_0(n)$.† The discrete signal $u_0(n)$ is called the *unit function*. The response of a system to the unit function is called the *unit response* of the system, which will be denoted $h(n)$ hereafter. Since any discrete signal $x(n)$ can be expressed as

$$
x(n) = \sum_{i=-\infty}^{\infty} x(i)u_0(n-i)
$$

we have, for a linear time-invariant system,

$$
y(n) = \sum_{i=-\infty}^{\infty} x(i)h(n-i) \tag{4-2}
$$

† The reader can readily show that

$$
u_{-1}(n) = \begin{cases} 0 & n < 0 \\ 1 & n \geq 0 \end{cases}
$$

is another possible choice.

As an example, let the unit response of a discrete linear time-invariant system be

$$h(n) = \begin{cases} 1 & n = -1 \\ 2 & n = 0 \\ 2 & n = 1 \\ 0 & \text{otherwise} \end{cases}$$

Then, corresponding to the input

$$x(n) = \begin{cases} 2 & n = 0 \\ 0 & n = 1 \\ -3 & n = 2 \\ 4 & n = 3 \\ 0 & \text{otherwise} \end{cases}$$

the output is

$$y(n) = 2h(n) - 3h(n-2) + 4h(n-3)$$

The computation of $y(n)$ is shown graphically in Fig. 4-1.

The summation in Eq. (4-2) is known as the *convolution summation* of the signals $x(n)$ and $h(n)$, which will be denoted $x(n) * h(n)$. The meaning of the convolution summation is quite clear: Each of the values $\ldots, x(-1), x(0), x(1), x(2), \ldots$ contributes to the value of the output signal at any time instant. Specifically, the value of the output signal at instant n is a sum of $x(-1)h(n+1)$, the contribution of $x(-1)$, and $x(0)h(n)$, the contribution of $x(0)$, and $x(1)h(n-1)$, the contribution of $x(1)$, $\ldots$, and so on.

As another example, let us consider the problem of depositing money in a savings account at an interest rate of 0.5 percent per month, with interest compounded monthly. Suppose that we deposit \$50 in the savings account each month for a period of 5 years. We want to know the total amount in the account 4 years after the first deposit and 20 years after the first deposit. Consider the account as a discrete system the input of which, $x(n)$, is the monthly deposit and the output of which, $y(n)$, is the total amount in the account. Thus, we have

$$x(n) = \begin{cases} 50 & n = 0, 1, 2, \ldots, 59 \\ 0 & \text{otherwise} \end{cases}$$

Since the unit response of the system is

$$h(n) = \begin{cases} (1.005)^n & n \geq 0 \\ 0 & n < 0 \end{cases}$$

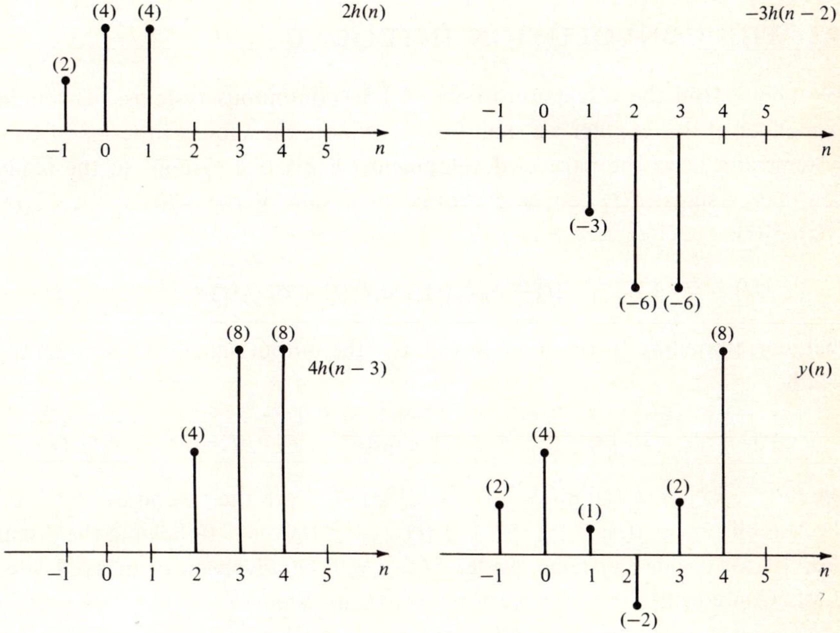

FIGURE 4-1

we have

$$
\begin{aligned}
y(n) &= x(n) * h(n) \\
&= \sum_{i=-\infty}^{\infty} x(i)h(n-i) \\
&= \begin{cases} 0 & n < 0 \\ \sum_{i=0}^{n} 50(1.005)^i \dagger & 0 \le n \le 59 \\ \sum_{i=n-59}^{n} 50(1.005)^i & 60 \le n \end{cases}
\end{aligned}
$$

Therefore, at $n = 47$

$$y(47) = \sum_{i=0}^{47} 50(1.005)^i$$

and at $n = 239$

$$y(239) = \sum_{i=180}^{239} 50(1.005)^i$$

† It is not totally obvious how we obtain this expression from the expressions of $x(n)$ and $h(n)$. This point will be clarified in Sec. 4-2.

4-2 THE CONVOLUTION INTEGRAL

We now extend the discussion in Sec. 4-1 to continuous systems. Throughout the remainder of this chapter we shall further explore the subject in terms of continuous systems and leave the parallel development for discrete systems to the reader. If a continuous signal $x(t)$ can be expressed as a sum of the signals $\ldots, \xi_{-1}(t), \xi_0(t), \xi_1(t), \xi_2(t), \ldots, \xi_k(t), \ldots,$

$$x(t) = \cdots + a_{-1}\xi_{-1}(t) + a_0\,\xi_0(t) + a_1\xi_1(t) + a_2\,\xi_2(t) + \cdots + a_k\,\xi_k(t) + \cdots$$

then corresponding to the input signal $x(t)$ the output signal $y(t)$ of a linear system will be

$$y(t) = \cdots + a_{-1}h_{-1}(t) + a_0\,h_0(t) + a_1h_1(t) + a_2\,h_2(t) + \cdots + a_k\,h_k(t) + \cdots$$

where $\ldots, h_{-1}(t), h_0(t), h_1(t), h_2(t), \ldots, h_k(t), \ldots$ are the responses of the system to the stimuli $\ldots, \xi_{-1}(t), \xi_0(t), \xi_1(t), \xi_2(t), \ldots, \xi_k(t), \ldots$. Following the discussion in Sec. 4-1 for discrete systems, we let $\xi_i(t)$, $i \neq 0$, be advanced or delayed versions of $\xi_0(t)$. Consequently, for a linear time-invariant system, $h_i(t)$, $i \neq 0$, are also advanced or delayed versions of $h_0(t)$. We want to choose $\xi_0(t)$ such that a large number of continuous signals $x(t)$ can be expressed as a weighted sum of the $\xi_i(t)$'s. Suppose that $\xi_0(t)$ is a narrow pulse, as shown in Fig. 4-2, which is denoted $u(t)$. Let

$$\begin{aligned}
&\cdots\cdots\cdots\cdots\cdots\cdots \\
\xi_{-1}(t) &= u(t + \delta) \\
\xi_1(t) &= u(t - \delta) \\
\xi_2(t) &= u(t - 2\delta) \\
&\cdots\cdots\cdots\cdots\cdots\cdots \\
\xi_k(t) &= u(t - k\delta) \\
&\cdots\cdots\cdots\cdots\cdots\cdots
\end{aligned}$$

Clearly, any staircaselike time signal $x(t)$ such as the one shown in Fig. 4-3 can be expressed as a sum of the functions $\xi_i(t)$:

$$x(t) = \sum_{i=-\infty}^{\infty} \frac{x(i\delta)}{1/\delta} u(t - i\delta) = \sum_{i=-\infty}^{\infty} x(i\delta)u(t - i\delta)\delta \tag{4-3}$$

If the response of a linear time-invariant system to $u(t)$ is $h_\delta(t)$, then the output of the system corresponding to the input $x(t)$ is

$$y(t) = \sum_{i=-\infty}^{\infty} x(i\delta)h_\delta(t - i\delta)\delta \tag{4-4}$$

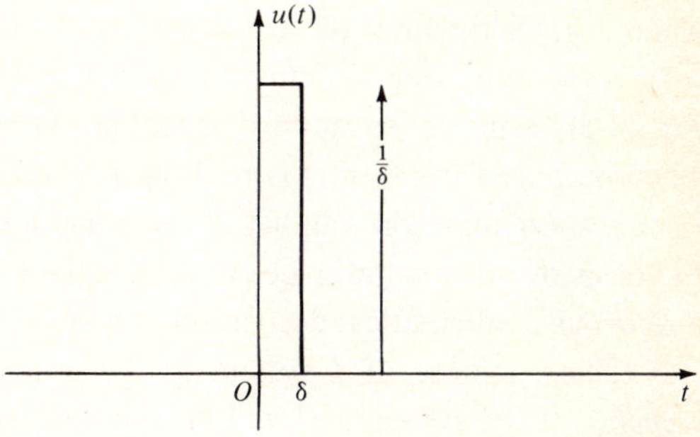

FIGURE 4-2

It is obvious that not every continuous signal can be expressed as a weighted sum as in Eq. (4-3). However, if a continuous signal is sufficiently smooth,† it can be approximated by such a weighted sum. Moreover, the approximation in Eq. (4-3) becomes better and better as δ becomes smaller and smaller. When δ approaches zero, $x(t)$ can be written as an integral which is the limit of the sum in Eq. (4-3). To be explicit, when $i\delta$ becomes τ and δ becomes $d\tau$, the sum in Eq. (4-3) becomes

$$x(t) = \int_{-\infty}^{\infty} x(\tau)u_0(t - \tau)\, d\tau \tag{4-5}$$

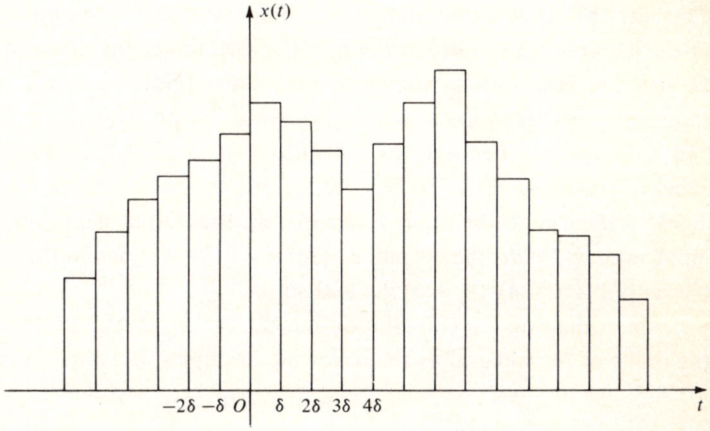

FIGURE 4-3

† A continuous signal is said to be *sufficiently smooth*, or to be *of bounded variation*, if it has a finite number of maxima and minima within any finite interval. For example, the signal $x(t) = \sin t$ is sufficiently smooth. However, the signal $x(t) = \sin(1/t)$ is not, because the number of maxima and minima within an interval that includes $t = 0$ is not finite.

where $u_0(t)$ is the limit of the narrow pulse $u(t)$ when δ approaches 0. We recall that $u_0(t)$ is the unit impulse defined in Sec. 1-5. [Intuitively, the argument used to derive Eq. (4-5) seems to be quite reasonable. However, as was pointed out in Sec. 1-5, when δ approaches 0, $u(t)$ is no longer a well-defined function. Consequently, there is a question about the validity of the integral in Eq. (4-5) and of the results derived on the basis of this integral. Let us assure the reader that, although we did not present a rigorous mathematical derivation here, our discussion is valid and can be justified.†]

The response of a system to the unit impulse $u_0(t)$ is known as the *impulse response* of the system and will be denoted $h(t)$. According to Eq. (4-5), for a linear time-invariant system whose impulse response is $h(t)$ the output signal corresponding to an input signal $x(t)$ is

$$\begin{aligned} y(t) = H[x(t)] &= H\left[\int_{-\infty}^{\infty} x(\tau)u_0(t-\tau)\,d\tau\right] \\ &= \int_{-\infty}^{\infty} x(\tau)H[u_0(t-\tau)]\,d\tau \\ &= \int_{-\infty}^{\infty} x(\tau)h(t-\tau)\,d\tau \end{aligned} \tag{4-6}$$

[This integral is the limit of the sum in Eq. (4-4) as δ approaches zero.] The integral in Eq. (4-6) is called the *convolution integral* of the two signals $x(t)$ and $h(t)$, which will also be denoted $x(t) * h(t)$.

The significance of the convolution integral can be demonstrated graphically. Consider the input signal $x(t)$ and the impulse response $h(t)$ in Fig. 4-4*a*. The value of the integral in Eq. (4-6) at $t = t_0$ is the area under the curve which is the product of $x(\tau)$ and $h(t_0 - \tau)$, as shown in Fig. 4-4*b*. [Note that $h(t_0 - \tau)$ is obtained by translating $h(-\tau)$ by an amount t_0, forward for positive t_0 and backward for negative t_0.] Indeed, the value of the output at t_0, $y(t_0)$, is due to the accumulated effect of the input $x(t)$. To be specific, due to the input value $x(0)\,d\tau$, the output is $x(0)h(t_0)\,d\tau$; due to the input value $x(\delta)\,d\tau$, the output is $x(\delta)h(t_0 - \delta)\,d\tau$; due to the input value $x(2\delta)\,d\tau$, the output is $x(2\delta)h(t_0 - 2\delta)\,d\tau$; due to the input value $x(-\delta)\,d\tau$, the output is $x(-\delta)h(t_0 + \delta)\,d\tau$; and so on.

In evaluating the convolution integral in Eq. (4-6), one should be careful about the limits of the integral as the following examples illustrate: Let

$$x(t) = e^{-t}u_{-1}(t)$$

and

$$h(t) = \sin t u_{-1}(t)$$

as shown in Fig. 4-5*a*. To determine $x(t) * h(t)$, we shall evaluate the integral

$$\int_{-\infty}^{\infty} e^{-\tau}u_{-1}(\tau)\sin(t-\tau)u_{-1}(t-\tau)\,d\tau \tag{4-7}$$

† In terms of the theory of distributions as was mentioned briefly in Sec. 1-6.

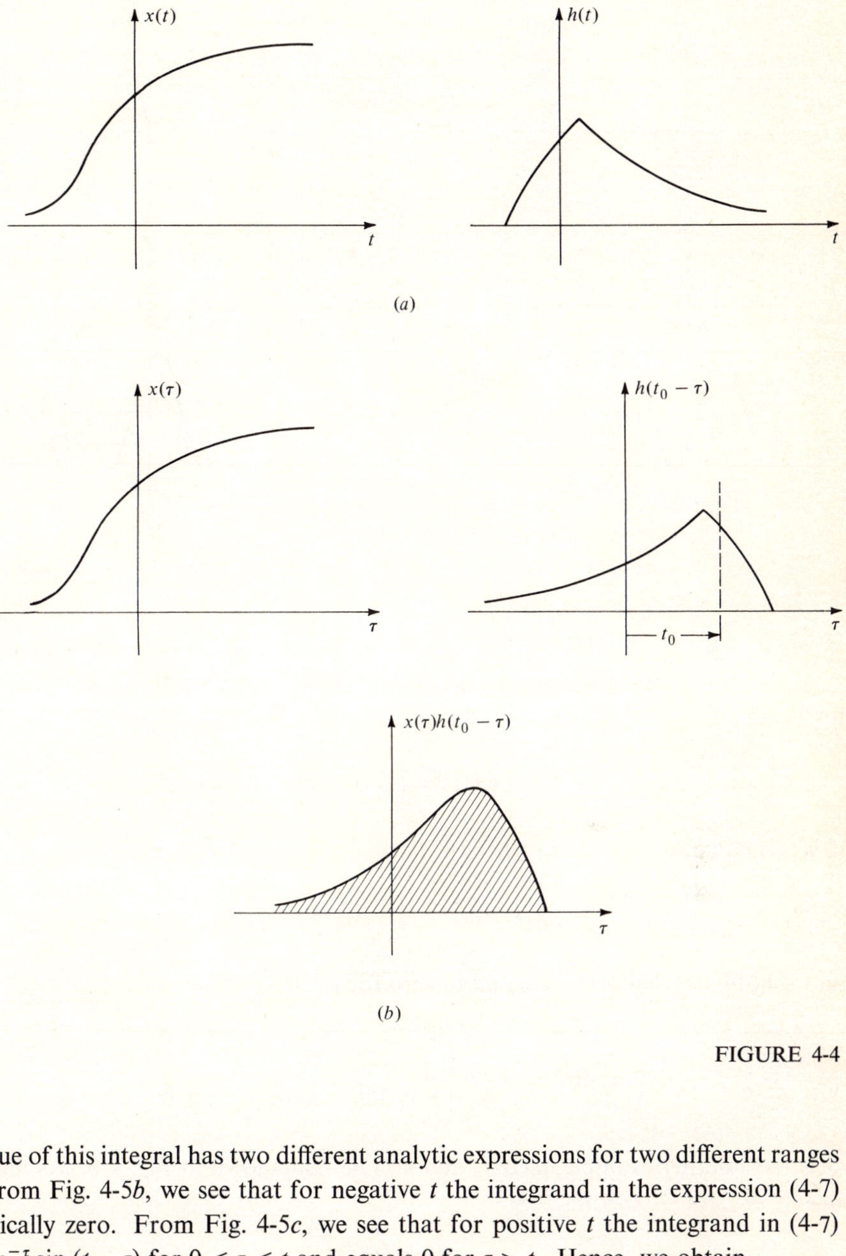

FIGURE 4-4

The value of this integral has two different analytic expressions for two different ranges of t. From Fig. 4-5b, we see that for negative t the integrand in the expression (4-7) is identically zero. From Fig. 4-5c, we see that for positive t the integrand in (4-7) equals $e^{-\tau} \sin(t-\tau)$ for $0 \le \tau \le t$ and equals 0 for $\tau > t$. Hence, we obtain

$$x(t) * h(t) = \begin{cases} 0 & t < 0 \\ \displaystyle\int_0^t e^{-\tau} \sin(t-\tau)\, d\tau = \tfrac{1}{2}(e^{-t} + \sin t - \cos t) & t \ge 0 \end{cases}$$

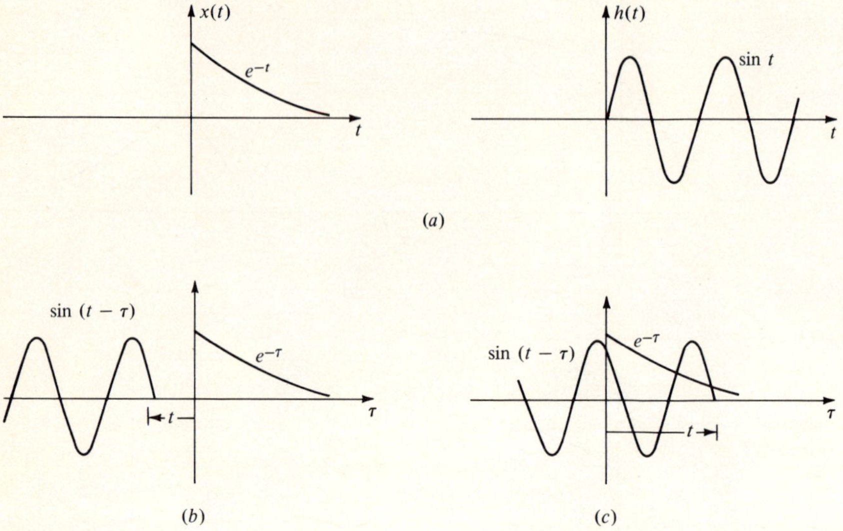

FIGURE 4-5

From this example, we can deduce the general results: If $x(t)$ is equal to zero for $t < 0$,

$$x(t) * h(t) = \int_0^{\infty} x(\tau)h(t - \tau)\, d\tau$$

if $h(t)$ is equal to zero for $t < 0$,

$$x(t) * h(t) = \int_{-\infty}^{t} x(\tau)h(t - \tau)\, d\tau \tag{4-8}$$

and if both $x(t)$ and $h(t)$ are equal to zero for $t < 0$

$$x(t) * h(t) = \begin{cases} 0 & t < 0 \\ \displaystyle\int_0^t x(\tau)h(t - \tau)\, d\tau & t \geq 0 \end{cases}$$

As another example, let $x(t)$ and $h(t)$ be that shown in Fig. 4-6. As illustrated in Fig. 4-7, for $-\infty < t \leq -\frac{1}{2}$

$$x(t) * h(t) = 0$$

for $-\frac{1}{2} \leq t \leq 1$

$$x(t) * h(t) = \int_{-\frac{1}{2}}^{t} 1\left(\frac{1}{2}\right)(t - \tau)\, d\tau = \frac{t^2}{4} + \frac{t}{4} + \frac{1}{16}$$

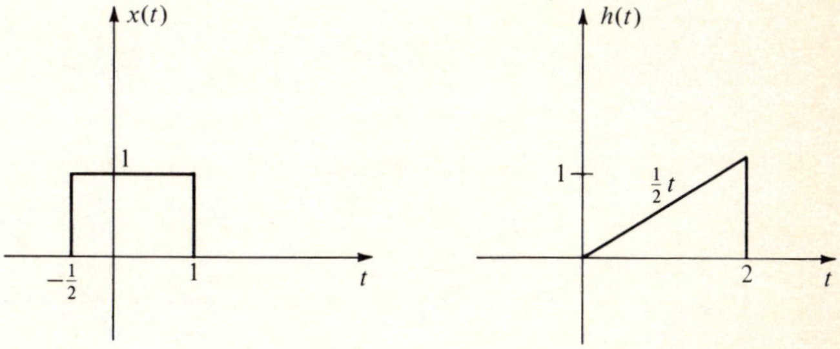

FIGURE 4-6

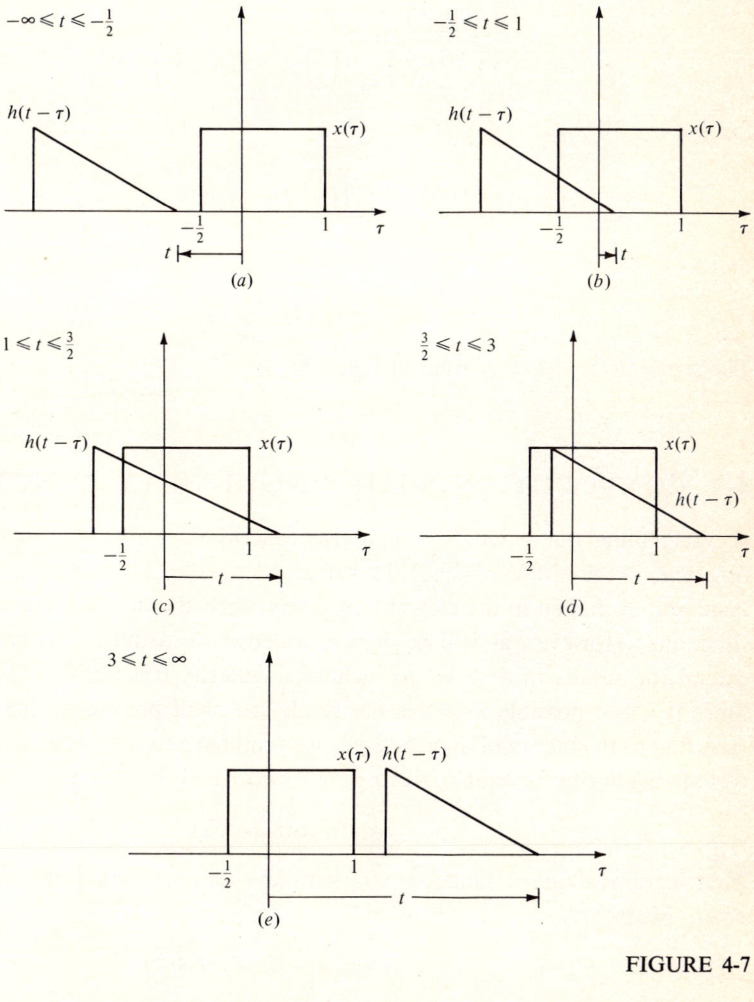

FIGURE 4-7

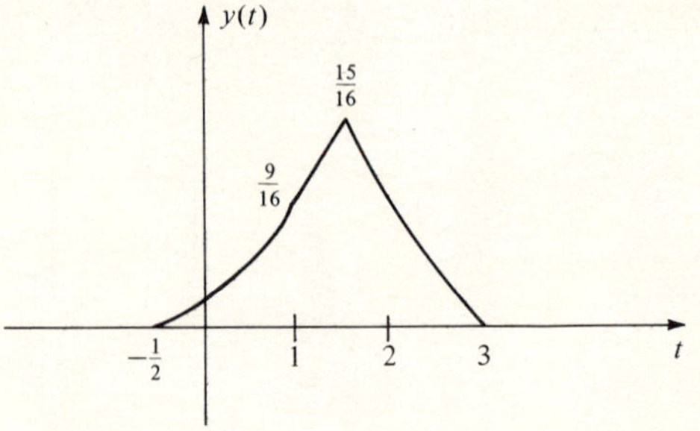

FIGURE 4-8

for $1 \le t \le \frac{3}{2}$

$$x(t) * h(t) = \int_{-\frac{1}{2}}^{1} 1\left(\frac{1}{2}\right)(t - \tau)\, d\tau = \frac{3}{4}t - \frac{3}{16}$$

for $\frac{3}{2} \le t \le 3$

$$x(t) * h(t) = \int_{t-2}^{1} 1\left(\frac{1}{2}\right)(t - \tau)\, d\tau = -\frac{t^2}{4} + \frac{t}{2} + \frac{3}{4}$$

and for $3 \le t \le \infty$

$$x(t) * h(t) = 0$$

The signal $x(t) * h(t)$ is plotted in Fig. 4-8.

4-3 CONVOLUTION WITH SINGULARITY FUNCTIONS

As was pointed out in Sec. 1-5, singularity functions are not well-defined mathematical functions in the ordinary sense. Therefore, a convolution integral cannot be manipulated and evaluated in the conventional way when the integrand contains singularity functions. However, as will be demonstrated in subsequent sections, it is useful to extend the results in Sec. 4-2 to include singularity functions in our consideration. Since it is not possible to prove the results we shall present in this section without resorting to the theory of distributions, we shall have to accept these results as properties of singularity functions. According to Eq. (1-9), Eq. (4-5) can be written

$$x(t) * u_0(t) = x(t)$$

That is, convolving a function $x(t)$ with the unit impulse yields the function $x(t)$ itself. Moreover,

$$x(t) * u_0(t + \delta) = x(t + \delta)$$

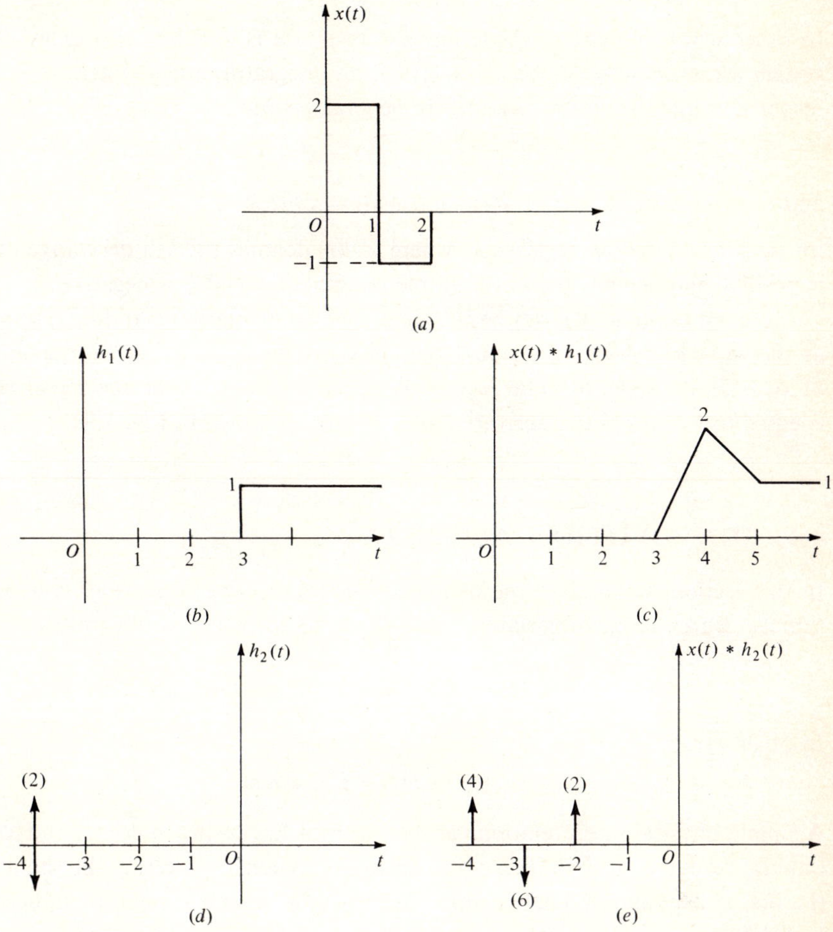

FIGURE 4-9

That is, convolving a signal with $u_0(t+\delta)$ results in translating the signal backward for time δ. Similarly, we have

$$x(t) * u_{-1}(t) = \int_{-\infty}^{t} x(\tau)\, d\tau$$

and

$$x(t) * u_{-1}(t+\delta) = \int_{-\infty}^{t} x(\tau+\delta)\, d\tau$$

Also, we have

$$x(t) * u_1(t) = x'(t)$$

and

$$x(t) * u_1(t+\delta) = x'(t+\delta)$$

In other words, a system whose impulse response is $u_0(t+\delta)$ is a delay element, a system whose impulse response is $u_{-1}(t)$ is an integrator, and a system whose impulse response is $u_1(t)$ is a differentiator. In general, we have

$$x(t) * u_k(t) = x^{(k)}(t) \tag{4-9}$$

and

$$x(t) * u_k(t+\delta) = x^{(k)}(t+\delta)$$

for positive as well as negative k, where $x^{(k)}(t)$ denotes the kth derivative of $x(t)$ if k is positive and denotes the kth-multiple integral of $x(t)$ if k is negative.

As an example, let $x(t)$ be the signal shown in Fig. 4-9*a*, and $h_1(t)$ be $u_{-1}(t-3)$ as shown in Fig. 4-9*b*. The signal $x(t) * h_1(t)$, shown in Fig. 4-9*c*, is the time integral of $x(t-3)$. If we let $h_2(t)$ be $2u_1(t+4)$, as shown in Fig. 4-9*d*, the signal $x(t) * h_2(t)$ is equal to two times the derivative of $x(t+4)$, as shown in Fig. 4-9*e*.

4-4 CONVOLUTION ALGEBRA

In this section, we study some of the properties of the operation of convolving two signals. By a change of variable $\lambda = t - \tau$ in the convolution integral, we obtain

$$\int_{-\infty}^{\infty} x(\tau)h(t-\tau)\,d\tau = \int_{-\infty}^{\infty} h(\lambda)x(t-\lambda)\,d\lambda$$

In other words,

$$x(t) * h(t) = h(t) * x(t) \tag{4-10}$$

A simple physical interpretation can be given to Eq. (4-10). Let $h(t)$ and $x(t)$ be the impulse responses of two linear time-invariant systems. If $x(t)$ is the input signal to the first system and $h(t)$ is the input signal to the second system, then, according to Eq. (4-10), the output signals of the two systems will be the same. The block diagrams in Fig. 4-10 illustrate the situation.

Suppose that the impulse response of a system $h(t)$ can be written

$$h(t) = h_1(t) + h_2(t)$$

Then, corresponding to the input $x(t)$ the output of the system is

$$\begin{aligned} x(t) * h(t) &= \int_{-\infty}^{\infty} x(\tau)h(t-\tau)\,d\tau \\ &= \int_{-\infty}^{\infty} x(\tau)[h_1(t-\tau) + h_2(t-\tau)]\,d\tau \\ &= \int_{-\infty}^{\infty} x(\tau)h_1(t-\tau)\,d\tau + \int_{-\infty}^{\infty} x(\tau)h_2(t-\tau)\,d\tau \\ &= x(t) * h_1(t) + x(t) * h_2(t) \end{aligned} \tag{4-11}$$

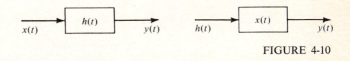

FIGURE 4-10

In other words, the output of the system can be viewed as the sum of the outputs of two systems whose impulse responses are $h_1(t)$ and $h_2(t)$, respectively, when the inputs to these two systems are both equal to $x(t)$. It follows from our discussion in Sec. 1-8 that the system whose impulse response is $h_1(t) + h_2(t)$ is a parallel connection of two subsystems whose impulse responses are $h_1(t)$ and $h_2(t)$, as illustrated in the block diagram in Fig. 4-11. Conversely, we note that the impulse response of a system which is a parallel connection of two subsystems is equal to the sum of the impulse responses of the subsystems.

Interchanging the order of integration in $[x(t) * h_1(t)] * h_2(t)$, we obtain†

$$\begin{aligned}[x(t) * h_1(t)] * h_2(t) &= \int_{-\infty}^{\infty} \left[\int_{-\infty}^{\infty} x(\lambda)h_1(\tau - \lambda)\, d\lambda\right] h_2(t - \tau)\, d\tau \\ &= \int_{-\infty}^{\infty} x(\lambda)\left[\int_{-\infty}^{\infty} h_1(\tau - \lambda)h_2(t - \tau)\, d\tau\right] d\lambda \\ &= \int_{-\infty}^{\infty} x(\lambda)\left[\int_{-\infty}^{\infty} h_1(\tau)h_2(t - \lambda - \tau)\, d\tau\right] d\lambda \\ &= x(t) * [h_1(t) * h_2(t)] \end{aligned} \tag{4-12}$$

† Notice that changing the order of integration is a critical step in the derivation of Eq. (4-12). This step is not always valid. For example, let

$$x(t) = e^{-t} \qquad h_1(t) = e^{-t}u_{-1}(t) \qquad \text{and} \qquad h_2(t) = (2e^{-3t} - e^{-2t})u_{-1}(t)$$

Since

$$x(t) * h_1(t) = \int_{-\infty}^{t} e^{-\tau}e^{-(t-\tau)}\, d\tau = \infty$$

for all t, we have

$$[x(t) * h_1(t)] * h_2(t) = \infty$$

But, for $0 \leqq t$

$$\begin{aligned} h_1(t) * h_2(t) &= \int_0^t e^{-\tau}(2e^{-3(t-\tau)} - e^{-2(t-\tau)}\, d\tau \\ &= e^{-2t} - e^{-3t} \end{aligned}$$

and for $t \leqq 0$

$$h_1(t) * h_2(t) = 0$$

Therefore

$$\begin{aligned} x(t) * [h_1(t) * h_2(t)] &= \int_{-\infty}^{t} e^{-\tau}(e^{-2(t-\tau)} - e^{-3(t-\tau)})\, d\tau \\ &= \tfrac{1}{2}e^{-t} \end{aligned}$$

The condition under which the order of integration can be changed may not be meaningful to those readers who have not had a course in advanced calculus. Let us simply assure the reader that this condition is satisfied in all cases we are to encounter.

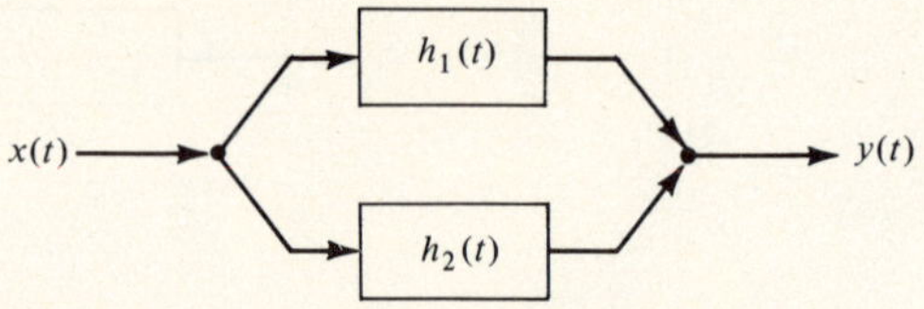

FIGURE 4-11

Let $h_1(t)$ and $h_2(t)$ be the impulse responses of two systems that are connected in cascade as shown in Fig. 4-12*a*. For an input signal $x(t)$, the output signal of the overall system is $[x(t) * h_1(t)] * h_2(t)$, which, according to Eq. (4-12), is equal to $x(t) * [h_1(t) * h_2(t)]$. Thus, we conclude that the impulse response of a system which is a cascade connection of two subsystems is equal to the convolution of the impulse responses of the subsystems, as illustrated in Fig. 4-12*b*.

As an example, let us consider the problem of building a system whose impulse response $h(t)$ is that shown in Fig. 4-13*a*, using integrators, amplifiers, and delay units. Note that

$$h(t) = h_1(t) * h_1(t)$$

where
$$h_1(t) = \begin{cases} 1 & 0 < t < 1 \\ 0 & \text{otherwise} \end{cases}$$

as shown in Fig. 4-13*b*. Since the impulse response of the system shown in Fig. 4-13*c* is $h_1(t)$, the impulse response of the system shown in Fig. 4-13*d* is $h(t)$.

It is clear that results similar to those in Eqs. (4-10) to (4-12) can be proved for discrete systems. We shall not repeat the proofs here.

We now examine how the signals $dy(t)/dt$ and $\int_{-\infty}^{t} y(\tau)\, d\tau$ are related to $x(t)$ and $h(t)$, where $y(t) = x(t) * h(t)$. Since

$$\frac{dy(t)}{dt} = \frac{d}{dt}\int_{-\infty}^{\infty} x(\tau)h(t-\tau)\, d\tau = \int_{-\infty}^{\infty} x(\tau)h'(t-\tau)\, d\tau$$

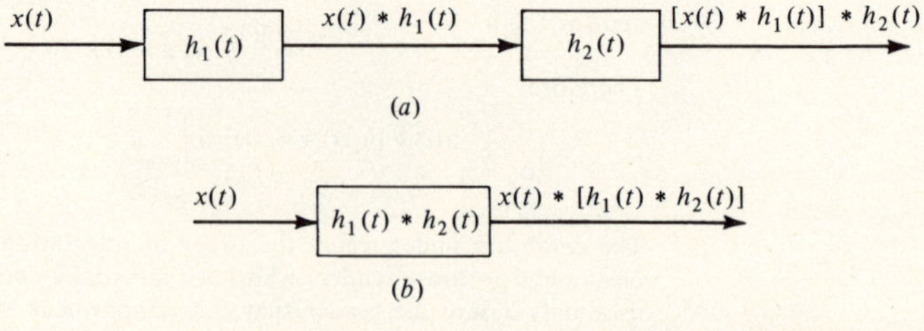

FIGURE 4-12

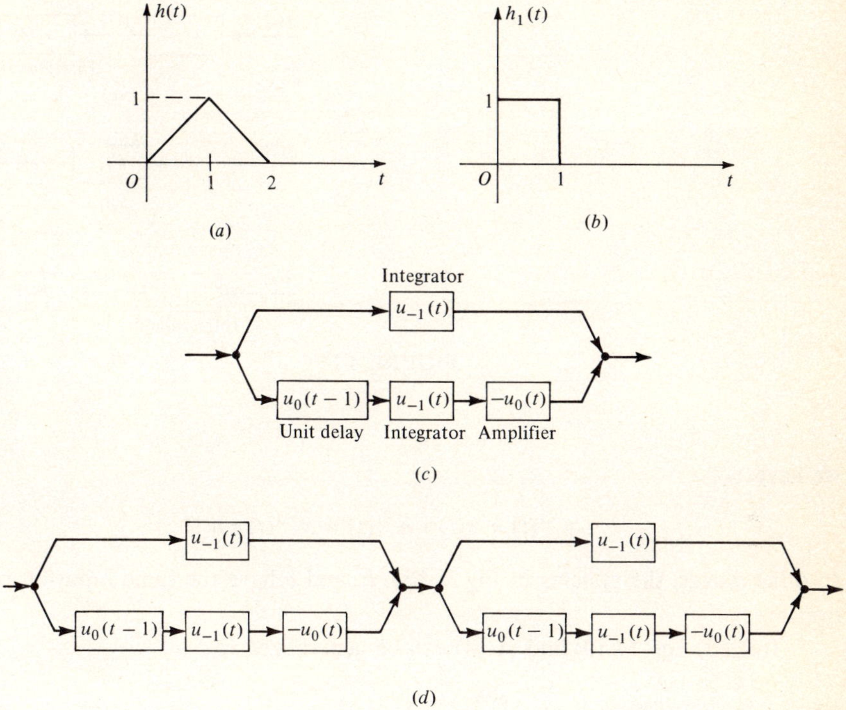

FIGURE 4-13

and

$$\frac{dy(t)}{dt} = \frac{d}{dt}\int_{-\infty}^{\infty} h(\tau)x(t-\tau)\,d\tau = \int_{-\infty}^{\infty} h(\tau)x'(t-\tau)\,d\tau$$

we have

$$y'(t) = x(t) * h'(t) = x'(t) * h(t) \tag{4-13}$$

In other words, the systems in Fig. 4-14*a*, *b*, and *c* have the same input-output relationship. Also, since

$$\int_{-\infty}^{\lambda} y(t)\,dt = \int_{-\infty}^{\lambda}\left[\int_{-\infty}^{\infty} x(\tau)h(t-\tau)\,d\tau\right]dt$$
$$= \int_{-\infty}^{\infty} x(\tau)\left[\int_{-\infty}^{\lambda} h(t-\tau)\,dt\right]d\tau$$

and

$$\int_{-\infty}^{\lambda} y(t)\,dt = \int_{-\infty}^{\lambda}\left[\int_{-\infty}^{\infty} h(\tau)x(t-\tau)\,d\tau\right]dt$$
$$= \int_{-\infty}^{\infty} h(\tau)\left[\int_{-\infty}^{\lambda} x(t-\tau)\,dt\right]d\tau$$

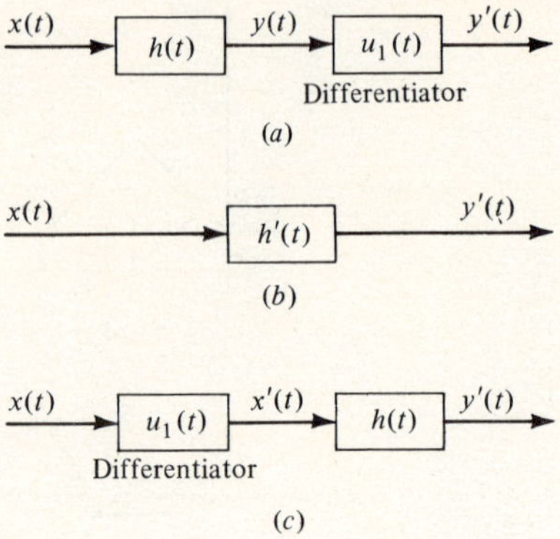

FIGURE 4-14

we have

$$y^{-1}(t) = x(t) * h^{-1}(t) = x^{-1}(t) * h(t) \tag{4-14}$$

In other words, the systems in Fig. 4-15*a*, *b*, and *c* have the same input-output relationship.

Indeed, Eqs. (4-13) and (4-14) can be generalized to

$$y^{(i)}(t) = x^{(j)}(t) * h^{(i-j)}(t) \tag{4-15}$$

for positive as well as negative i and j. We shall leave the proof to the interested reader.

Similarly, for discrete time systems, given that

$$y(n) = x(n) * h(n)$$

then, corresponding to Eq. (4-13), we have

$$\Delta y(n) = \Delta x(n) * h(n) = x(n) * \Delta h(n)$$
$$\nabla y(n) = \nabla x(n) * h(n) = x(n) * \nabla h(n)$$

and, corresponding to Eq. (4-14), we have

$$\sum_{i=-\infty}^{n} y(i) = \left[\sum_{i=-\infty}^{n} x(i)\right] * h(n) = x(n) * \sum_{i=-\infty}^{n} h(i)$$

Let us denote the kth backward difference and the kth forward difference of a discrete function $f(n)$ by $\Delta^{(k)}f(n)$ and $\nabla^{(k)}f(n)$, respectively. Then, corresponding to Eq. (4-15), we have

$$\Delta^{(k)}y(n) = \Delta^{(k)}x(n) * h(n) = x(n) * \Delta^{(k)}h(n)$$
$$\nabla^{(k)}y(n) = \nabla^{(k)}x(n) * h(n) = x(n) * \nabla^{(k)}h(n)$$

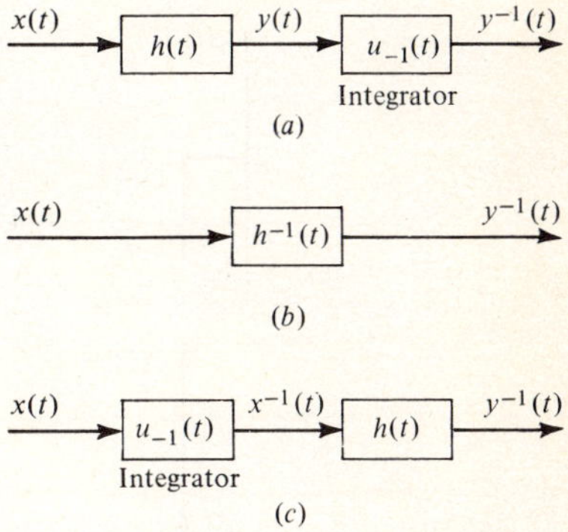

FIGURE 4-15

It can be shown that the results in Eqs. (4-10) to (4-15) are still valid when $x(t)$ and $h(t)$ and their derivatives contain singularity functions. We present now several illustrative examples.

Let $x(t)$ and $h(t)$ be as shown in Fig. 4-16*a*. When $x(t)$ is convolved with each of the singularity functions in $h(t)$, $x(t) * h(t)$ is that in Fig. 4-16*b* which simplifies to that in Fig. 4-16*c*.

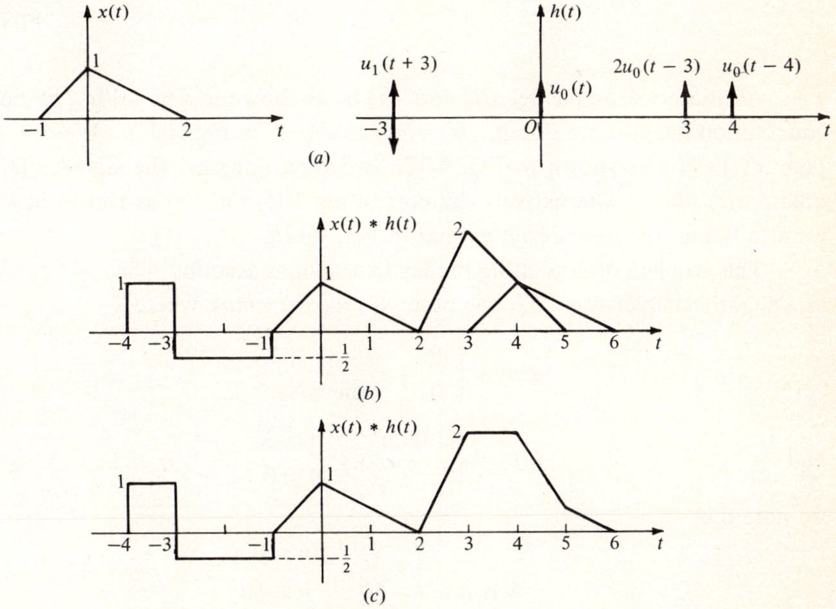

FIGURE 4-16

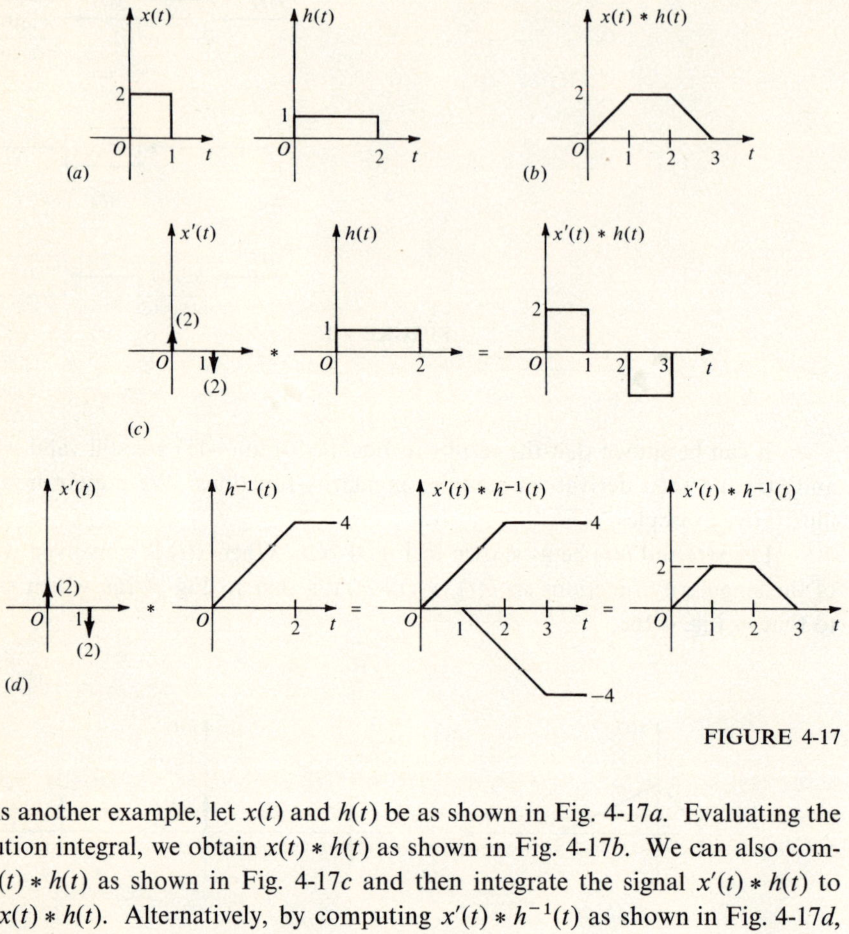

FIGURE 4-17

As another example, let $x(t)$ and $h(t)$ be as shown in Fig. 4-17*a*. Evaluating the convolution integral, we obtain $x(t) * h(t)$ as shown in Fig. 4-17*b*. We can also compute $x'(t) * h(t)$ as shown in Fig. 4-17*c* and then integrate the signal $x'(t) * h(t)$ to obtain $x(t) * h(t)$. Alternatively, by computing $x'(t) * h^{-1}(t)$ as shown in Fig. 4-17*d*, we also obtain the same result as that in Fig. 4-17*b*.

The problem of depositing money in a savings account in Sec. 4-1 can be solved in a slightly simpler way. To compute $y(n) = x(n) * h(n)$, where

$$x(n) = \begin{cases} 50 & n = 0, 1, 2, \ldots, 59 \\ 0 & \text{otherwise} \end{cases}$$

and

$$h(n) = \begin{cases} (1.005)^n & n \geq 0 \\ 0 & n < 0 \end{cases}$$

we note that

$$\nabla x(n) = \begin{cases} 50 & n = 0 \\ -50 & n = 60 \\ 0 & \text{otherwise} \end{cases}$$

Thus $\nabla y(n) = \nabla x(n) * h(n)$

$$= \begin{cases} 0 & n < 0 \\ 50(1.005)^n & 0 \le n \le 59 \\ 50[(1.005)^n - (1.005)^{n-60}] & 60 \le n \end{cases}$$

It follows that

$$y(n) = \begin{cases} 0 & n < 0 \\ \sum_{i=0}^{n} 50(1.005)^i & 0 \le n \le 59 \\ \sum_{i=0}^{n} 50(1.005)^i - \sum_{i=60}^{n} 50(1.005)^{i-60} & 60 \le n \end{cases}$$

4-5 DETERMINATION OF THE UNIT RESPONSE AND IMPULSE RESPONSE

So far, we have discussed how to determine the output signal of a system when the input signal and the unit response or impulse response of the system are given. We now investigate how the unit response or the impulse response of a system can be determined, when an input signal together with its corresponding output signal are known. Mathematically, determining the unit response $h(n)$ amounts to solving the *summation equation*

$$y(n) = \sum_{i=-\infty}^{\infty} x(i)h(n-i) \tag{4-16}$$

for the unknown function $h(n)$, and determining the impulse response $h(t)$ amounts to solving the *integral equation*

$$y(t) = \int_{-\infty}^{\infty} x(\tau)h(t-\tau)\,d\tau \tag{4-17}$$

for the unknown function $h(t)$. To be able to solve such equations is one of the motivations of introducing transformation representations of signals in later chapters. In this section, we are not ready to talk about a general procedure for solving these equations. We shall only illustrate the possibility by some examples.

Let us consider an example in which

$$x(n) = \begin{cases} 1 & n = 0 \\ 1 & n = 1 \\ 2 & n = 2 \\ 0 & \text{otherwise} \end{cases}$$

and

$$y(n) = \begin{cases} 1 & n = 0 \\ -1 & n = 1 \\ 3 & n = 2 \\ -1 & n = 3 \\ 6 & n = 4 \\ 0 & \text{otherwise} \end{cases}$$

are a pair of input and output signals of a *causal* system. According to Eq. (4-16), the unit response of the system can be determined by solving the infinite set of simultaneous equations

$$y(n) = x(0)h(n) + x(1)h(n-1) + x(2)h(n-2) \tag{4-18}$$

for $n = 0, \pm 1, \pm 2, \ldots$. That the system is causal yields

$$h(n) = 0 \qquad n < 0$$

Furthermore, for $n = 0$, Eq. (4-18) becomes

$$y(0) = x(0)h(0)$$

that is,

$$h(0) = 1$$

For $n = 1$, Eq. (4-18) becomes

$$y(1) = x(0)h(1) + x(1)h(0)$$

that is,

$$-1 = h(1) + 1(1)$$

or

$$h(1) = -2$$

Similarly, for $n = 2$, we obtain

$$y(2) = x(0)h(2) + x(1)h(1) + x(2)h(0)$$

or

$$h(2) = 3$$

For $n = 3$, we obtain

$$y(3) = x(0)h(3) + x(1)h(2) + x(2)h(1)$$

or

$$h(3) = 0$$

and for $n = 4$, we obtain

$$y(4) = x(0)h(4) + x(1)h(3) + x(2)h(2)$$

or

$$h(4) = 0$$

Finally, for $n \geq 5$, we obtain

$$h(n) = 0$$

Summing up, we have

$$h(n) = \begin{cases} 1 & n = 0 \\ -2 & n = 1 \\ 3 & n = 2 \\ 0 & \text{otherwise} \end{cases}$$

as the unit response of the system.

As an example of determining the impulse response of a continuous system, let $x(t)$ be an input to a causal system with $y(t)$ being the corresponding output, as shown in Fig. 4-18*a*. According to Eq. (4-13), the output of the system is $y'(t)$ when the input is $x'(t)$, as shown in Fig. 4-18*b*. From the input-output pair $x'(t)$ and $y'(t)$, the impulse response of the system, $h(t)$, which is shown in Fig. 4-18*c*, can be determined in a step-by-step manner:

1 Since the system is causal, within the range $0 < t < 1$, the output $y'(t)$ is due to the impulse $u_0(t)$ in $x'(t)$ alone. Thus, for $0 < t < 1$, $h(t) = 2$.

2 Within the range $1 < t < 2$, the output $y'(t)$ is due to both of the impulses in $x'(t)$. Thus, we have

$$h(t) - h(t-1) = -1$$

However, from step 1, $h(t-1) = 2$ for $1 < t < 2$. Therefore, for $1 < t < 2$,

$$h(t) = 1$$

3 Using the same argument as in step 2, we have, within the range $2 < t < 3$,

$$h(t) - h(t-1) = -1$$

As was found in step 2, $h(t-1) = 1$ for $2 < t < 3$; thus, for $2 < t < 3$

$$h(t) = 0$$

4 Similarly, within the range $3 < t < 4$, we have

$$h(t) - h(t-1) = 0$$

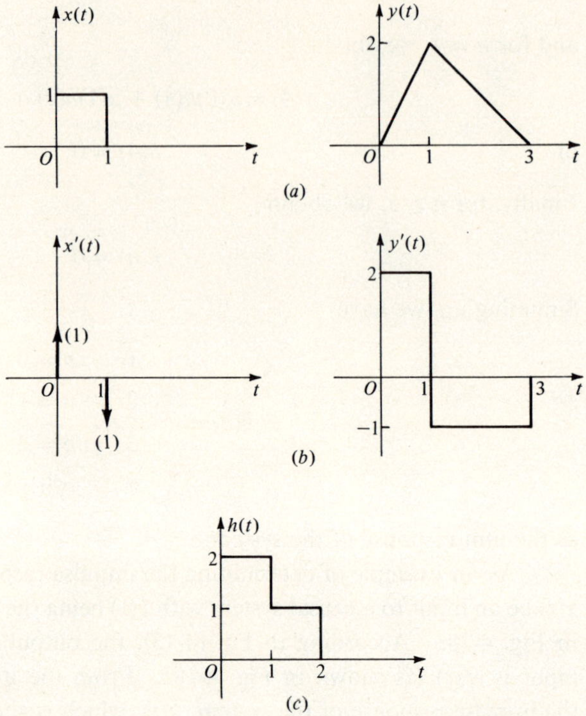

FIGURE 4-18

and thus $$h(t) = 0$$

5 The argument in step 4 can be repeated successively for $i < t < i + 1$, $4 \leq i$. We thus obtain $h(t) = 0$ for $t > 4$.

Note that in this example the result obtained in step 1 is not valid if the system is not known to be causal, and all subsequent steps do not follow. The reader may verify for himself that the impulse response $h(t)$ in Fig. 4-19 is also a solution to the integral

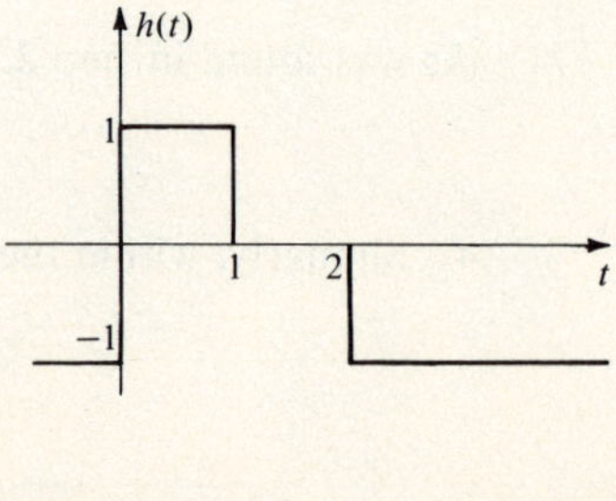

FIGURE 4-19

equation (4-17) with $x(t)$ and $y(t)$ being that shown in Fig. 4-18*a*. Note that the system is noncausal in this case.†

As another example, let $x(t)$ and $y(t)$ in Fig. 4-20*a* be a pair of input and output signals of a causal system. Then, $x'(t)$ and $y'(t)$ in Fig. 4-20*b* are also a pair of input and output signals of the system, and so are $x(t) + x'(t)$ and $y(t) + y'(t)$ in Fig. 4-20*c*. Using an argument similar to that in the preceding example, from $x(t) + x'(t)$ and $y(t) + y'(t)$ we determine $h(t)$ as that shown in Fig. 4-20*d*.

At this point, there are several questions one might raise: What is the relationship between the causality of a system and the uniqueness of the solution of Eqs. (4-16) and (4-17)? Does any pair of input and output signals determine uniquely the unit response or the impulse response of a system? Or are there exceptions? We shall answer these questions in Chap. 9.

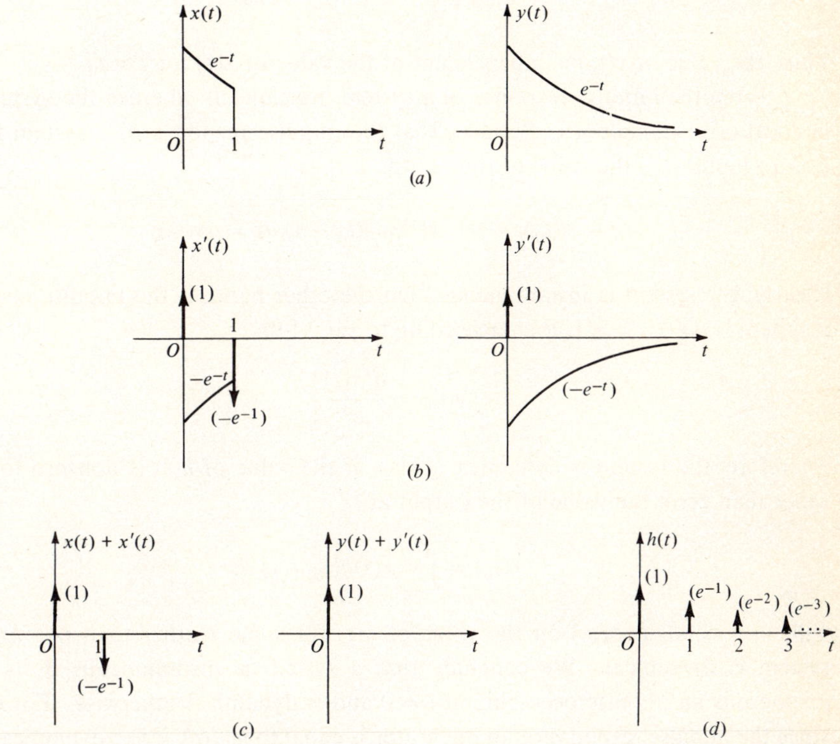

FIGURE 4-20

† As will be discussed further in Sec. 4-6, the causality of the system can be observed from its impulse response.

4-6 SYSTEM CHARACTERIZATION BY UNIT RESPONSE OR IMPULSE RESPONSE

We see in this chapter that the input-output relationship of a linear time-invariant system can be described by its unit response or impulse response. As a matter of fact, many of the properties of a system can also be derived directly from its unit response or impulse response. Our discussion will be in terms of continuous systems and signals. A parallel discussion in terms of discrete systems and signals can easily be supplied by the reader.

A system is causal if and only if its impulse response $h(t)$ is equal to 0 for $t < 0$. Clearly, if $h(t)$ is not equal to 0 for $t < 0$, it means that the system is able to *predict* the arrival of an impulse. Therefore, the system is not causal. On the other hand, if $h(t)$ is equal to 0 for $t < 0$, then according to Eq. (4-8) the output of the system at any time instant t_0 is

$$y(t_0) = \int_{-\infty}^{t_0} x(\tau)h(t_0 - \tau)\, d\tau$$

Thus, the value of $y(t_0)$ is independent of the value of $x(t)$ for $t > t_0$.

From the impulse response of a system, we can tell whether the system is dynamical or instantaneous. Suppose that the impulse response of a system is $au_0(t)$. For an input $x(t)$, the value of the output at t_0 is

$$y(t_0) = \int_{-\infty}^{\infty} x(\tau)au_0(t_0 - \tau)\, d\tau = ax(t_0)$$

Clearly, the system is instantaneous. On the other hand, if the impulse response of a system is $au_k(t)$, $k \geq 1$, then according to Eq. (4-9)

$$y(t_0) = \left.\frac{d^k x(t)}{dt^k}\right|_{t=t_0}$$

Therefore, the system is dynamical. Also, if the value of $h(t)$ is nonzero for some t other than zero, the value of the output at t_0

$$y(t_0) = \int_{-\infty}^{\infty} x(\tau)h(t_0 - \tau)\, d\tau$$

will, in general, depend on the value of $x(t)$ for some t other than t_0. Again, the system is dynamical. We conclude that a system is instantaneous if its impulse response is an impulse occurring at $t = 0$ and is dynamical otherwise. For example, since the impulse response of an integrator is equal to $u_{-1}(t)$, it is a dynamical system.

A system is stable in the bounded-input–bounded-output sense if and only if

$$\int_{-\infty}^{\infty} |h(t)|\, dt < \infty$$

Since for a bounded input $x(t)$, $|x(t-\tau)| \leq M$ for some real number M for all $t-\tau$,

$$|y(t)| = \left|\int_{-\infty}^{\infty} x(t-\tau)h(\tau)\,d\tau\right| \leq \int_{-\infty}^{\infty} |x(t-\tau)|\,|h(\tau)|\,d\tau$$

$$\leq M\int_{-\infty}^{\infty} |h(t)|\,dt < \infty$$

That is, the output of the system is also bounded. On the other hand, suppose that $\int_{-\infty}^{\infty}|h(t)|\,dt$ is infinite. Then for the bounded input

$$x(t) = \begin{cases} 1 & \text{if } h(-t) \text{ is positive} \\ 0 & \text{if } h(-t) \text{ is zero} \\ -1 & \text{if } h(-t) \text{ is negative} \end{cases}$$

the value of the output at $t = 0$ is

$$y(0) = \int_{-\infty}^{\infty} x(\tau)h(-\tau)\,dt = \int_{-\infty}^{\infty} |h(-\tau)|\,d\tau$$

Thus, the system under consideration is not stable in the bounded-input–bounded-output sense. For example, a delay element, whose impulse response is a delayed impulse, is a stable system. On the other hand, an integrator, whose impulse response is a unit step, is not stable in the bounded-input–bounded-output sense.

4-7 IMPULSE RESPONSE, DIFFERENTIAL EQUATION, AND STATE SPACE DESCRIPTIONS OF SYSTEMS

Thus far, we have studied three different ways of describing system behavior. The difference or differential equation representation and state space representation can be used to describe nonlinear systems as well as time-varying systems, although the solution of nonlinear time-varying difference or differential equations is not always straightforward. On the other hand, the unit response or impulse response representation is limited to linear systems. The unit response and impulse response representation is convenient when the system is initially at rest and the input $x(t)$ is known for $-\infty \leq t \leq \infty$. The difference or differential equation and state space representations are more suitable when the system is not initially at rest and the input $x(t)$ is known only for $t_0 \leq t \leq \infty$.

Given the difference equation description of a discrete system, the unit response of the system can be determined by solving the difference equation with the input signal being the unit function. For example, let

$$y(n) + 7y(n-1) + 6y(n-2) = 6x(n)$$

be a difference equation relating the input and output signals of a discrete system. Let

$$x(n) = \begin{cases} 1 & n = 0 \\ 0 & \text{otherwise} \end{cases}$$

Because the system is initially at rest, for $n < 0$

$$y(n) = 0$$

For $n \geq 0$, the total solution is

$$y(n) = A_1(-1)^n + A_2(-6)^n$$

because the particular solution is equal to 0. From the boundary conditions

$$y(0) = 6$$
$$y(1) = -42$$

(How do we determine these boundary conditions?), we obtain

$$y(n) = -\tfrac{6}{5}(-1)^n + \tfrac{36}{5}(-6)^n$$

Thus, the unit response of the system is

$$h(n) = \begin{cases} 0 & n < 0 \\ -\frac{6}{5}(-1)^n + \frac{36}{5}(-6)^n & n \geq 0 \end{cases}$$

The impulse response of a continuous system can be determined in a similar manner. For example, the input-output relationship of the electric circuit in Fig. 2-5 can be described by the differential equation

$$\frac{d^2y(t)}{dt^2} + 7\frac{dy(t)}{dt} + 6y(t) = 6x(t) \tag{4-19}$$

Let $x(t)$ be $u_0(t)$. Solving Eq. (4-19), we obtain the impulse response of the system:

$$h(t) = (\tfrac{6}{5}e^{-t} - \tfrac{6}{5}e^{-6t})u_{-1}(t)$$

The unit response or impulse response of a system can also be obtained from its state space description by solving the state equations. We leave the details to the reader.

Unfortunately, it is not always possible to obtain a difference or differential equation description of a system from its unit or impulse response (because the input-output relationship of some linear time-invariant systems cannot be described by linear difference or differential equations with constant coefficients). However, if the impulse response of a linear time-invariant continuous system is of the form

$$h(t) = (A_1 e^{\alpha_1 t} + A_2 e^{\alpha_2 t} + \cdots + A_r e^{\alpha_r t})u_{-1}(t) \tag{4-20}$$

it is possible to determine the corresponding differential equation description of the system. Since all the exponential terms in Eq. (4-20) are terms in the homogeneous solution, the characteristic equation of the differential equation is

$$(\alpha - \alpha_1)(\alpha - \alpha_2) \cdots (\alpha - \alpha_r) = 0 \tag{4-21}$$

The left-hand side of Eq. (4-21) is a polynomial of degree r and can be written

$$\alpha^r + C_1\alpha^{r-1} + C_2\alpha^{r-2} + \cdots + C_{r-1}\alpha + C_r$$

It follows that the left-hand side of the differential equation is

$$\frac{d^r y(t)}{dt^r} + C_1 \frac{d^{r-1}y(t)}{dt^{r-1}} + C_2 \frac{d^{r-2}y(t)}{dt^{r-2}} + \cdots + C_{r-1}\frac{dy(t)}{dt} + C_r y(t)$$

To determine the right-hand side of the differential equation, we make use of the fact that response of the system to the stimulus $u_0(t)$ is $h(t)$. Specifically, the right-hand side of the differential equation must be such that when the input signal is $u_0(t)$ the singularity functions at $t = 0$ in the right-hand side must match the singularity functions at $t = 0$ in

$$\frac{d^r h(t)}{dt^r} + C_1 \frac{d^{r-1}h(t)}{dt^r} + C_2 \frac{d^{r-2}h(t)}{dt^{r-2}} + \cdots + C_{r-1}\frac{dh(t)}{dt} + C_r h(t)$$

We illustrate the procedure by an example. Suppose that the impulse response of a linear time-invariant system is

$$h(t) = (-e^{-t} + 3e^{-2t})u_{-1}(t)$$

Since the terms $-e^{-t}$ and $3e^{-2t}$ are both terms in the homogeneous solution of the differential equation describing the input-output relationship of the system, the characteristic roots of the differential equation are -1 and -2. It follows that the left-hand side of the differential equation is

$$\frac{d^2y(t)}{dt^2} + 3\frac{dy(t)}{dt} + 2y(t)$$

Since

$$\frac{d^2h(t)}{dt^2} + 3\frac{dh(t)}{dt} + 2h(t) = u_0(t) + 2u_1(t)$$

the right-hand side of the differential equation is

$$2\frac{dx(t)}{dt} + x(t)$$

and the differential equation characterizing the input-output relationship of the system is

$$\frac{d^2y(t)}{dt^2} + 3\frac{dy(t)}{dt} + 2y(t) = 2\frac{dx(t)}{dt} + x(t)$$

Note that such a differential equation is not unique in the sense that there are higher-order differential equations that have the same response when the stimulus is $u_0(t)$. For example,

$$\frac{d^3y(t)}{dt^3} + 2\frac{d^2y(t)}{dt^2} - \frac{dy(t)}{dt} - 2y(t) = 2\frac{d^2x(t)}{dt^2} - \frac{dx(t)}{dt} - x(t)$$

is one such equation. At this point it is not obvious how one goes about finding such higher-order differential equations. We shall, however, come back to this topic in Chap. 9.

A parallel discussion can be carried out for discrete systems whose unit responses are of the form

$$h(n) = \begin{cases} 0 & n < 0 \\ A_1(\alpha_1)^n + A_2(\alpha_2)^n + \cdots + A_r(\alpha_r)^n & n \geq 0 \end{cases}$$

We present here an illustrative example. Let

$$h(n) = \begin{cases} 0 & n < 0 \\ (-1)^n + (-2)^n & n \geq 0 \end{cases}$$

be the unit response of a discrete system. Since -1 and -2 are the characteristic roots, the left-hand side of the difference equation is

$$y(n) + 3y(n-1) + 2y(n-2) \tag{4-22}$$

Substituting $h(n)$ into (4-22), we obtain

$$h(n) + 3h(n-1) + 2h(n-2) = \begin{cases} 2 & n = 0 \\ 3 & n = 1 \\ 0 & \text{otherwise} \end{cases}$$

We thus conclude that the right-hand side of the difference equation is

$$2x(n) + 3x(n-1)$$

and the difference equation characterizing the input-output relationship of the system is

$$y(n) + 3y(n-1) + 2y(n-2) = 2x(n) + 3x(n-1)$$

Although a state space description of a linear time-invariant system can be obtained from the differential equation derived from the impulse response as shown above, it can also be obtained directly from the impulse response, as we shall show. Let $h(t)$ be that in Eq. (4-20). For any input $x(t)$, the corresponding output is

$$\begin{aligned} y(t) &= \int_{-\infty}^{\infty} x(\tau)h(t-\tau)\,d\tau \\ &= \int_{-\infty}^{t} x(\tau)(A_1 e^{-\alpha_1(t-\tau)} + A_2 e^{-\alpha_2(t-\tau)} + \cdots + A_r e^{-\alpha_r(t-\tau)})\,d\tau \end{aligned} \tag{4-23}$$

Let

$$\begin{aligned} \lambda_1(t) &= \int_{-\infty}^{t} A_1 x(\tau) e^{-\alpha_1(t-\tau)}\,d\tau \\ \lambda_2(t) &= \int_{-\infty}^{t} A_2 x(\tau) e^{-\alpha_2(t-\tau)}\,d\tau \\ &\cdots\cdots\cdots\cdots\cdots\cdots \\ \lambda_r(t) &= \int_{-\infty}^{t} A_r x(\tau) e^{-\alpha_r(t-\tau)}\,d\tau \end{aligned} \tag{4-24}$$

be a set of state variables. Directly from Eq. (4-23), we have as an output equation

$$y(t) = \lambda_1(t) + \lambda_2(t) + \cdots + \lambda_r(t)$$

Differentiating Eqs. (4-24) we have as state equations

$$\begin{aligned} \frac{d\lambda_1(t)}{dt} &= -\alpha_1 \int_{-\infty}^{t} A_1 x(\tau) e^{-\alpha_1(t-\tau)}\,d\tau + A_1 x(t) \\ &= -\alpha_1 \lambda_1(t) + A_1 x(t) \\ \frac{d\lambda_2(t)}{dt} &= -\alpha_2 \lambda_2(t) + A_2 x(t) \\ &\cdots\cdots\cdots\cdots\cdots\cdots \\ \frac{d\lambda_r(t)}{dt} &= -\alpha_r \lambda_r(t) + A_r x(t) \end{aligned}$$

For the example above in which

$$h(t) = (-e^{-t} + 3e^{-2t})u_{-1}(t)$$

we have as state equations

$$\begin{aligned} \frac{d\lambda_1(t)}{dt} &= \lambda_1(t) - x(t) \\ \frac{d\lambda_2(t)}{dt} &= 2\lambda_2(t) + 3x(t) \end{aligned}$$

and as output equation

$$y(t) = \lambda_1(t) + \lambda_2(t)$$

4-8 REMARKS AND REFERENCES

In Chaps. 2, 3, and this chapter, we have studied three different ways of describing and analyzing the behavior of systems. It is hoped that the reader has not only learned the various techniques of system analysis but also gained insight into system behavior through looking at it from different points of view. In Chaps. 5, 6, 7, 9, and 11, we shall study these methods of analysis again from the point of view of "transformation representations."

For an introductory discussion of the convolution of signals, see either Cooper and McGillem [1, chap. 4] or Schwarz and Friedland [2, chap. 3]. See Zemanian [3, chap. 5] for a discussion of convolution of generalized functions.

[1] COOPER, G. R., and C. D. MCGILLEM: "Methods of Signal and System Analysis," Holt, Rinehart and Winston, Inc., New York, 1967.

[2] SCHWARZ, R. J., and B. FRIEDLAND: "Linear Systems," McGraw-Hill Book Company, New York, 1965.

[3] ZEMANIAN, A. H.: "Distribution Theory and Transform Analysis," McGraw-Hill Book Company, New York, 1965.

PROBLEMS

4-1 In Fig. 4P-1, let $x(n)$ be the input signal to a discrete linear time-invariant system with unit response $h(n)$. Sketch the corresponding output signal.

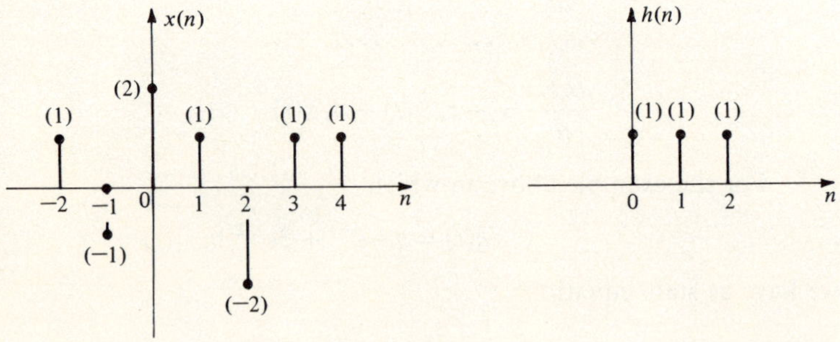

FIGURE 4P-1

4-2 In the following, let $h(n)$ denote the unit response of a linear time-invariant discrete system, and let $x(n)$ denote an input signal to the system. Determine the corresponding output signal $y(n)$. [Give either a sketch or a closed-form expression of $y(n)$, whichever is more convenient.]

(a) $h(n) = \begin{cases} 1 & n = 0, 1, 2 \\ 0 & \text{otherwise} \end{cases}$ $\quad x(n) = \begin{cases} 2 & n = 0, 1, 2 \\ 0 & \text{otherwise} \end{cases}$

(b) $h(n) = \begin{cases} 1 & n \geq 0 \\ 0 & \text{otherwise} \end{cases}$ $\quad x(n) = \begin{cases} 1 & n = 1 \\ 2 & n = 3 \\ 3 & n = 5 \\ -6 & n = 7 \\ 0 & \text{otherwise} \end{cases}$

(c) $h(n) = \begin{cases} 2^n & n \geq 0 \\ 0 & \text{otherwise} \end{cases}$ $\quad x(n) = \begin{cases} 2^n & n \geq -1 \\ 0 & \text{otherwise} \end{cases}$

(d) $h(n) = \begin{cases} 0 & n < 0 \\ 1 & n \geq 0 \end{cases}$ $\quad x(n) = \begin{cases} 0 & n < 0 \\ 2^{-n} & n \geq 0 \end{cases}$

(e) $h(n) = \begin{cases} 0 & n < 0 \\ e^{-n} & n \geq 0 \end{cases}$ $\quad x(n) = \begin{cases} 2^n & n \leq 0 \\ 0 & n > 0 \end{cases}$

4-3 (a) Express the discrete signal

$$x(n) = \begin{cases} 2 & n = 0 \\ 0 & n = 1 \\ -3 & n = 2 \\ 4 & n = 3 \\ 0 & \text{otherwise} \end{cases}$$

as a sum

$$x(n) = \sum_{i=-\infty}^{\infty} a_i u_{-1}(n - i)$$

(b) The unit response of a discrete linear time-invariant system is

$$h(n) = \begin{cases} 1 & n = -1 \\ 2 & n = 0 \\ 2 & n = 1 \\ 0 & \text{otherwise} \end{cases}$$

Determine the unit step response of the system, $h_{-1}(n)$, that is, the response of the system to the input signal $u_{-1}(n)$.

(c) Determine the output signal $y(n)$ by evaluating the convolution sum

$$y(n) = \sum_{i=-\infty}^{\infty} a_i h_{-1}(n - i)$$

4-4 (a) Every particle inside a nuclear reactor splits into two particles in each second. Suppose one particle is injected into the reactor every second beginning at $t = 0$. How many particles are there in the reactor at the nth second?

(*b*) When a particle splits into two, one of them is actually the original particle and the other is a newly created particle. Suppose a particle has a lifetime of 10 s. That is, a particle created in the 0th second will vanish after the 10th second. Repeat part (*a*), assuming that all injected particles are newly created.

4-5 Consider an air-traffic-control system modeled as a discrete linear time-invariant system. The input signal $x(n)$ of the system is the desired altitude (measured in thousands of feet) of an aircraft, and the output signal $y(n)$ is the actual altitude (also measured in thousands of feet) of the aircraft. Suppose that the unit response of the system is as shown in Fig. 4P-2*a*.

(*a*) Find the actual altitude, $y(n)$, of the aircraft when

$$x(n) = \begin{cases} 1 & n > 0 \\ 0 & n \leq 0 \end{cases}$$

(*b*) Sketch $y(n)$ when $x(n)$ is as shown in Fig. 4P-2*b*.

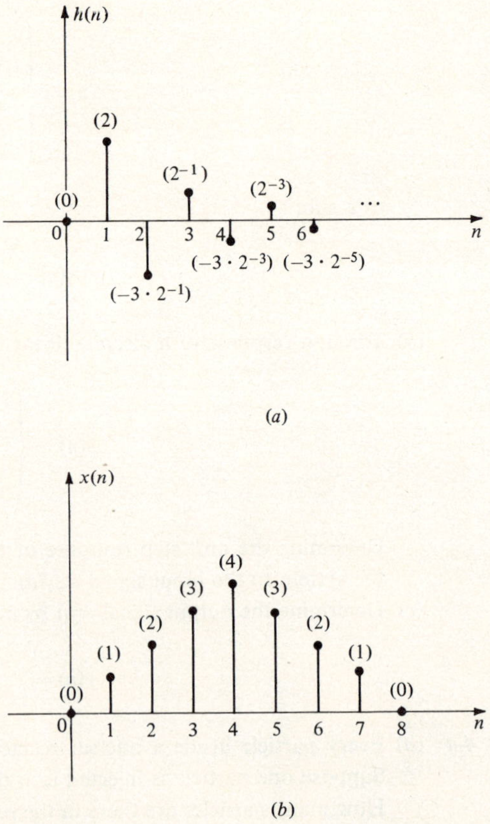

FIGURE 4P-2

4-6 The impulse response of a linear time-invariant system is

$$h(t) = \begin{cases} 1 & -1 < t < 3 \\ 0 & \text{otherwise} \end{cases}$$

Determine the response of the system to the input

$$x(t) = e^{2t} - 5e^{-t}$$

4-7 Sketch $x(t) * h(t)$ for the pairs of $x(t)$ and $h(t)$ shown in Fig. 4P-3.

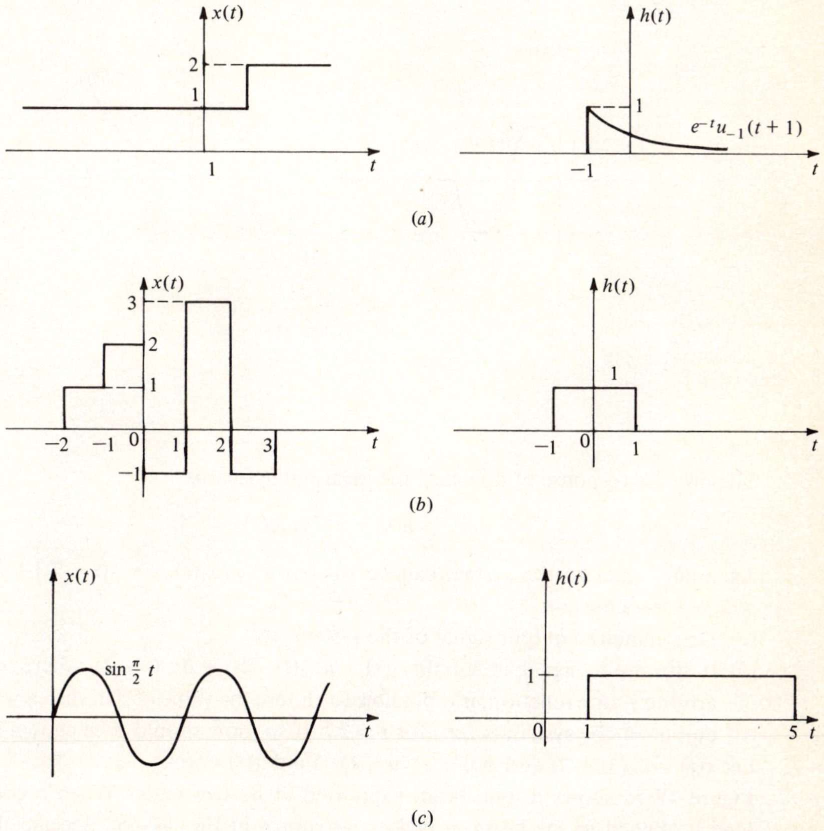

FIGURE 4P-3

4-8 Sketch $x(t) * h(t)$ for the pairs of $x(t)$ and $h(t)$ shown in Fig. 4P-4.

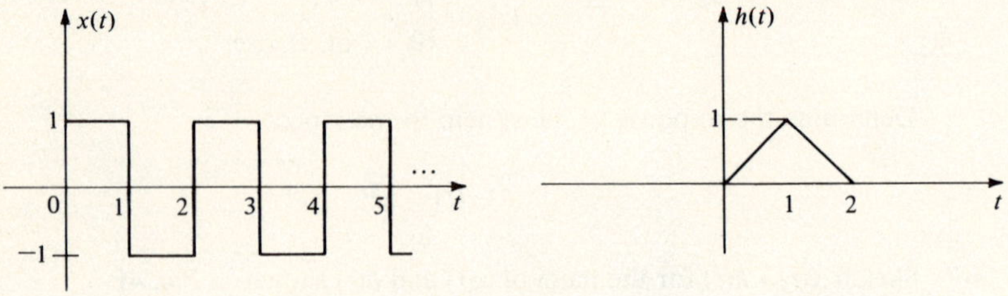

(*a*)

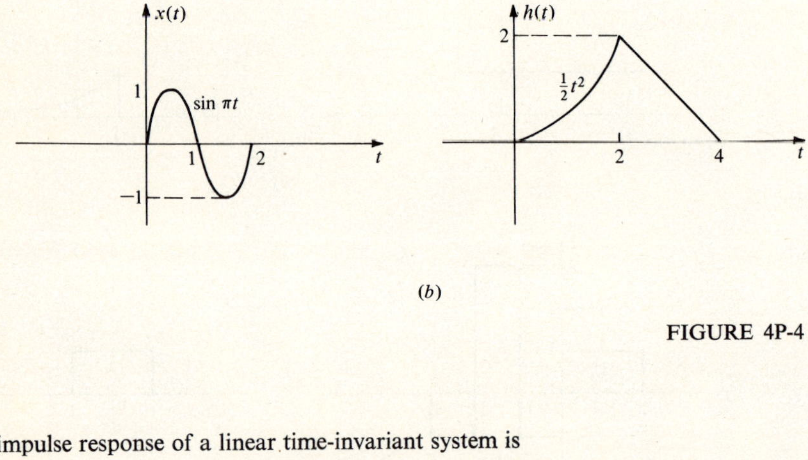

(*b*)

FIGURE 4P-4

4-9 The impulse response of a linear time-invariant system is

$$h(t) = e^{-2t}u_{-1}(t)$$

An input signal to the system can be written $e^{-t}[u_{-1}(t) - u_{-1}(t-2)] + \beta u_0(t-2)$, where β is a constant.

(*a*) Determine the output signal of the system, $y(t)$.

(*b*) If the input signal is $x_0(t)[u_{-1}(t) - u_{-1}(t-2)] + \beta u_0(t-2)$ where $x_0(t)$ is an arbitrary time function, is it possible to choose the value of β in such a way that the output of the system is zero for $t > 2$? If so, how should β be chosen?

4-10 Let $x(t) = u_0(3t+2)$ and $h(t) = e^{-t}u_{-1}(t)$. Find $x(t) * h(t)$.

4-11 Figure 4P-5*a* shows a long beam supported at its two ends. When a concentrated load is applied to the beam, it deflects as shown in Fig. 4P-5*b*. Specifically, when a load of unit weight is applied at the point *A* the deflection at a point *n* unit distance from A is 2^{-n} for $n \leq 10$ and is zero for $n > 10$. Determine the deflection of the beam for the load distribution shown in Fig. 4P-5c, where the distance between adjacent loads is 1.

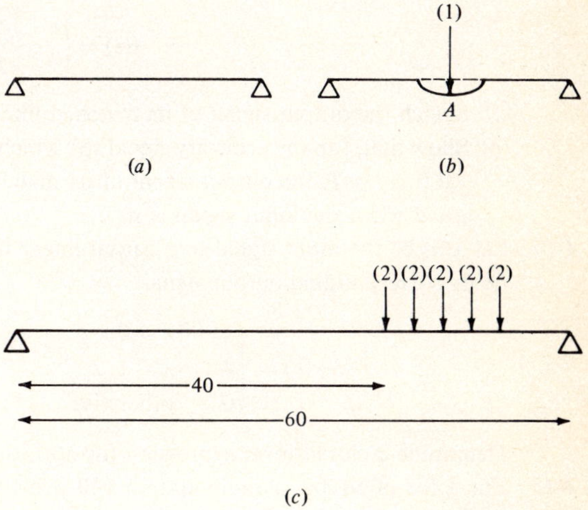

FIGURE 4P-5

4-12 Consider a signal $x(t)$ which is unspecified for $t < 0$ and is equal to that shown in Fig. 4P-6a for $t \geq 0$.

(*a*) Suppose that $x(t)$ is the input signal to a linear time-invariant system whose impulse response $h(t)$ is that shown in Fig. 4P-6b. Within what time interval can the output signal $y(t)$ be determined? Sketch $y(t)$ for this time interval.

(*b*) Suppose that $x(t)$ is the input signal to a linear time-invariant system whose impulse response $h(t)$ is as shown in Fig. 4P-6c. What additional information about $x(t)$ short of a complete specification of $x(t)$ for $t \leq 0$ is needed to determine the output signal $y(t)$ for $t \geq 0$?

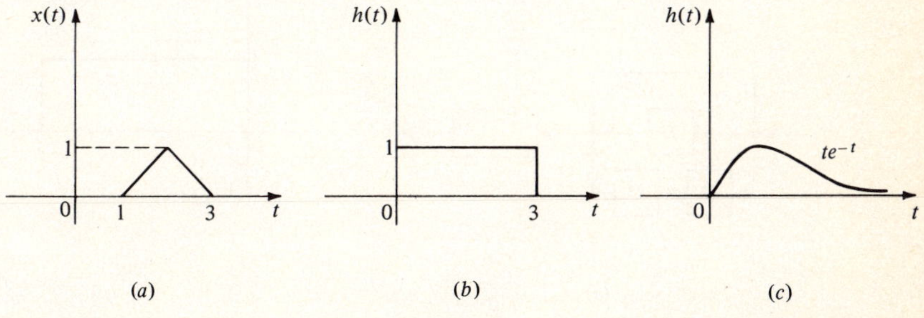

FIGURE 4P-6

4-13 Let $s(t)$ be a signal which is equal to zero outside the time interval $0 \leq t \leq T$. A linear time-invariant system is called a matched filter of the signal $s(t)$ if its impulse response equals $s(T-t)$.

(*a*) Suppose that

$$s(t) = \begin{cases} t & 0 \leq t \leq T \\ 0 & \text{elsewhere} \end{cases}$$

Sketch the output signal of its matched filter when the input signal is $s(t)$.

(*b*) Show that, for any arbitrary signal $s(t)$ which is equal to zero outside the time interval $0 \leq t \leq T$, the output signal of its matched filter attains the maximal value at $t = T$ when the input signal is $s(t)$.

4-14 Let $x(n)$ be the input signal to a causal linear time-invariant discrete system and $y(n)$ be its corresponding output signal.

$$x(n) = \begin{cases} 1 & n = 0 \\ 4 & n = 1, 2 \\ 0 & \text{otherwise} \end{cases} \qquad y(n) = \begin{cases} 0 & n < 0 \\ 3^n & n \geq 0 \end{cases}$$

Determine a closed-form expression for $h(n)$, the unit response of the system.

4-15 For a fee of \$1000, a radio station will provide a store with 10 min of commercial time each day for five consecutive days and then 2 min of commercial time each day for the next five consecutive days. Suppose that every minute of commercial time on a particular day will produce a total sale of \$500 on that day and a total sale of \$500(2^{-n}) on the nth day from that day.

(*a*) Describe a discrete system model with the fees paid by a store on successive days as input signal and the total sales on successive days as output signal.

(*b*) Determine the unit response of the system. (Use \$1000 as a unit.)

(*c*) If the store pays the radio station \$2000 daily for 10 consecutive days starting from January 1, 1974, determine the total sales on the nth day of that year.

4-16 Determine the step response of the system shown in Fig. 4P-7*a*, where D denotes a unit delay. The impulse response of the subsystem A, $h_A(t)$, is shown in Fig. 4P-7*b*.

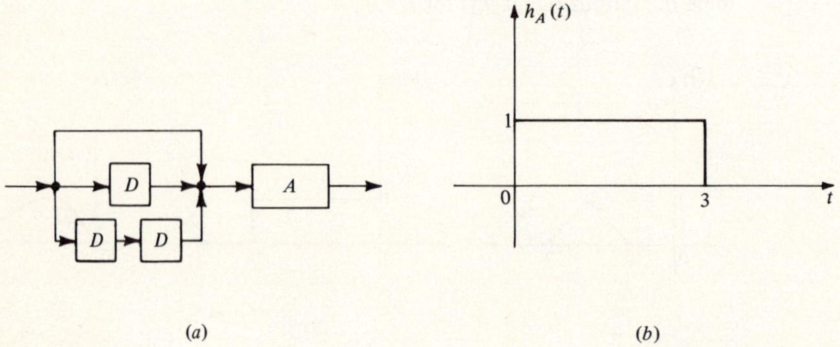

FIGURE 4P-7

4-17 Let $x_1(t)$ and $y_1(t)$ in Fig. 4P-8*a* be a pair of input and output signals of a linear time-invariant system. Determine the output signal of the system corresponding to the input $x_2(t)$ in Fig. 4P-8*b*.

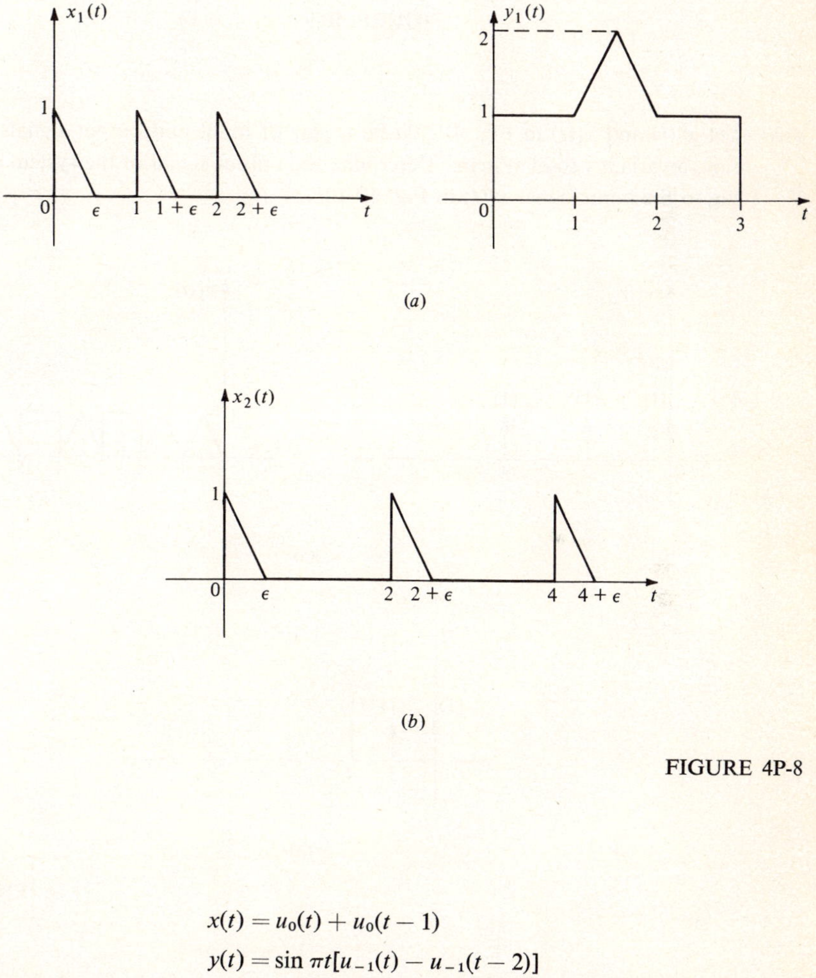

FIGURE 4P-8

4-18 Let

$$x(t) = u_0(t) + u_0(t-1)$$

and
$$y(t) = \sin \pi t[u_{-1}(t) - u_{-1}(t-2)]$$

be the input and output signals, respectively, of a linear time-invariant system. Determine the impulse response of the system.

4-19 The output signal of a linear time-invariant system to the input signal $\sin tu_{-1}(t)$ is shown in Fig. 4P-9. Determine the impulse response of the system.

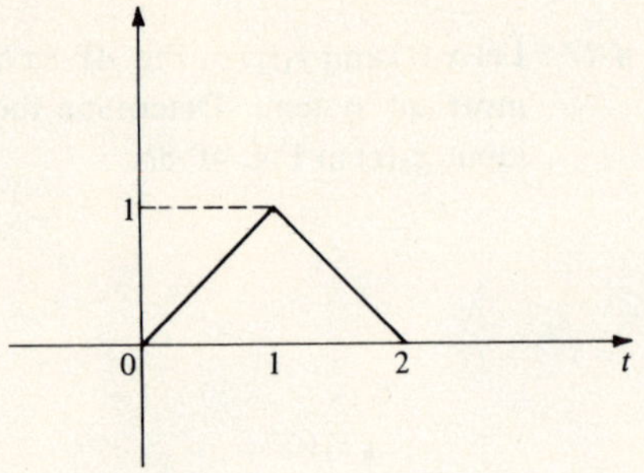

FIGURE 4P-9

4-20 Let $x_1(t)$ and $y_1(t)$ in Fig. 4P-10*a* be a pair of input and output signals of a linear time-invariant causal system. Determine the output signal of the system corresponding to the input signal $x_2(t)$ in Fig. 4P-10*b*.

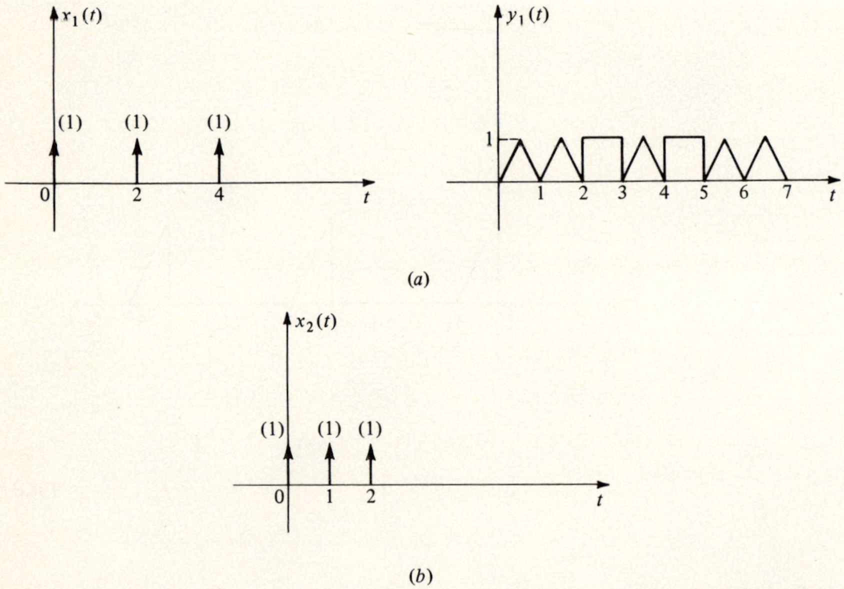

FIGURE 4P-10

4-21 The impulse response of the system in Fig. 4P-11*a* is shown in Fig. 4P-11*b*. Given that the impulse responses of the subsystems *A* and *B*, $h_A(t)$ and $h_B(t)$, are as shown in Fig. 4P-11*c*, determine the impulse response of the subsystem *C*, if it is known that *C* is a causal system.

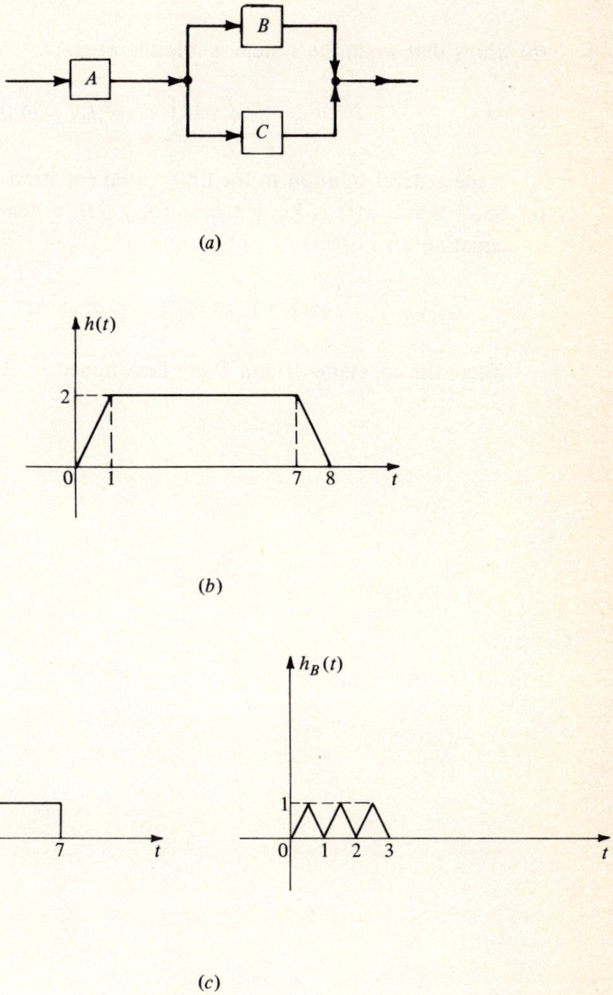

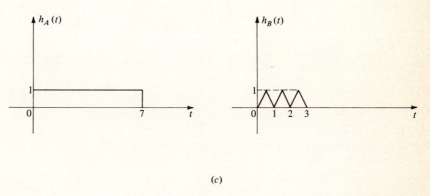

FIGURE 4P-11

4-22 Consider a linear time-invariant system whose input-output relationship is described by the differential equation

$$\frac{d^2y(t)}{dt^2} + 5\frac{dy(t)}{dt} + 6y(t) = 2x(t)$$

(*a*) Find its impulse response $h(t)$, assuming that $y(-\infty)$ and $\left.\frac{dy}{dt}\right|_{t=-\infty}$ are zero, i.e., the system is initially at rest.

(*b*) Show that when the system is initially at rest

$$y(t) = \int_{-\infty}^{\infty} x(\tau)h(t-\tau)\,d\tau$$

is the general solution to the differential equation.

(*c*) Show that if $x(t)$ is known only for $t \geq 0$, a general solution to the differential equation for $t \geq 0$ is

$$y(t) = \int_{0}^{\infty} x(\tau)h(t-\tau)\,d\tau + Ae^{-3t} + Be^{-2t}$$

where the constants A and B are determined by the boundary conditions at $t = 0$.

5

THE z TRANSFORMATION

5-1 INTRODUCTION

In Chaps. 2 to 4 we studied three different techniques for describing and analyzing the behavior of linear time-invariant systems. These techniques are known as *time domain analysis techniques* because, throughout the course of analysis, signals are represented explicitly as functions of time. In the remaining chapters of this book, we shall study techniques known as *frequency domain analysis* or *transformation analysis techniques*. In a transformation analysis of linear time-invariant systems, "transformation representations" of signals are used. We shall present in this chapter a transformation representation of discrete signals and then carry out a parallel development for continuous signals in later chapters.

A transformation theory of signals can be developed and introduced in two different ways. From a mathematical point of view, a transformation of a signal is simply an alternative representation of the signal. From a physical point of view, a transformation of a signal leads to a decomposition of the signal into components. (We must admit that these statements may seem rather vague. They will become clear after our discussion in this chapter and Chaps. 6, 7, and 9.) We shall introduce the z transformation of discrete signals following the first point of view and then introduce the Fourier transformation and the Laplace transformation of continuous signals following the second point of view. We hope that such an approach will aid the reader to gain further insight into the topic of transformation of signals.

5-2 THE *z* TRANSFORM

Consider the discrete signal

$$x(n) = \begin{cases} (\frac{1}{3})^n & n \geq 0 \\ (\frac{1}{2})^{-n} & n < 0 \end{cases} \tag{5-1}$$

It is clear that the signal $x(n)$ can also be described by the infinite sequence of values $\{\ldots, (\frac{1}{2})^2, \frac{1}{2}, 1, \frac{1}{3}, (\frac{1}{3})^2, (\frac{1}{3})^3, \ldots, (\frac{1}{3})^k, \ldots\}$. Alternatively, this infinite sequence of values can also be represented as an infinite series

$$\cdots + (\tfrac{1}{2})^2 z^2 + \tfrac{1}{2}z + 1 + \tfrac{1}{3}z^{-1} + (\tfrac{1}{3})^2 z^{-2} + (\tfrac{1}{3})^3 z^{-3} + \cdots + (\tfrac{1}{3})^k z^{-k} + \cdots \tag{5-2}$$

where the coefficient of z^{-k} is the value of $x(k)$. Conceptually, one can view z as merely a formal variable and the powers of z as "indicators" which are used to identify the values of $x(n)$ in the infinite series. However, so that we can express the sum of the infinite series in closed form and, thus, obtain a compact alternative representation of the signal $x(n)$, we should know for what values of z the infinite series converges. For reasons that will become clear later on, we shall let z assume complex values instead of real values only. For example, the infinite series (5-2) converges when the absolute value of z is within the range $\frac{1}{3} < |z| < 2$. For these values of z, the infinite series can be written

$$\frac{1}{1 - \frac{1}{2}z} + \frac{1}{1 - \frac{1}{3}z^{-1}} - 1 = \frac{2}{2 - z} + \frac{3z}{3z - 1} - 1 = \frac{5z}{(2 - z)(3z - 1)} \tag{5-3}$$

Thus, the expression (5-3) can be used as an alternative representation of the discrete signal $x(n)$ in Eq. (5-1).

Formally, we define the *z transform* of a discrete signal $x(n)$, denoted $X(z)$, to be

$$\begin{aligned} X(z) &= \cdots + x(-1)z + x(0) + x(1)z^{-1} + x(2)z^{-2} + \cdots + x(k)z^{-k} + \cdots \\ &= \sum_{n=-\infty}^{\infty} x(n)z^{-n} \end{aligned}$$

where z is a *complex variable*. We also define the *region of absolute convergence* of the signal $x(n)$ to be the area in the z plane (the complex plane) containing all z such that the infinite series $X(z)$ converges absolutely, that is,

$$\sum_{n=-\infty}^{\infty} |x(n)z^{-n}| < \infty$$

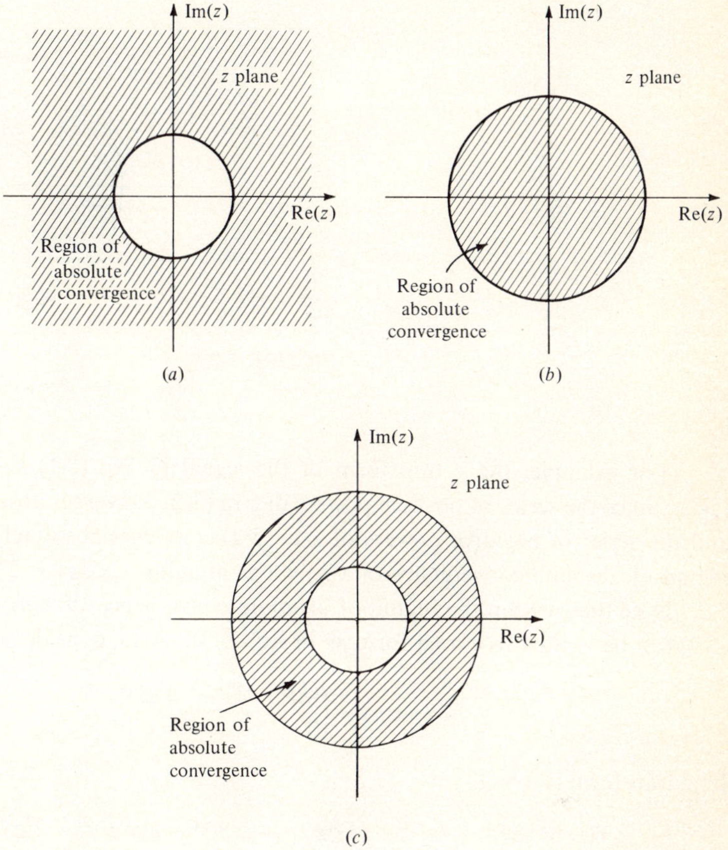

FIGURE 5-1

For convenience, we shall also refer to the region of absolute convergence of $x(n)$ as the region of absolute convergence of $X(z)$. Let us note that if $x(n)$ is a signal that is equal to zero for $n < 0$, then $X(z)$ is a series of negative powers of z only. Consequently, the region of absolute convergence of $x(n)$ is the area outside a circle centered at the origin of the z plane, as illustrated in Fig. 5-1a [because if $X(z)$ converges absolutely for $|z| = \rho_0$, then $x(z)$ also converges absolutely for all z, $|z| > \rho_0$]. Similarly, if $x(n)$ is a signal that is equal to zero for $n > 0$, then $X(z)$ is a series of positive powers of z only, and the region of absolute convergence of $x(n)$ is the area inside a circle centered at the origin of the z plane, as illustrated in Fig. 5-1b. Combining these two observations, we note that in the general case when $X(z)$ contains both positive and negative powers of z, its region of absolute convergence is an annulus centered at the origin of the z plane, as illustrated in Fig. 5-1c.

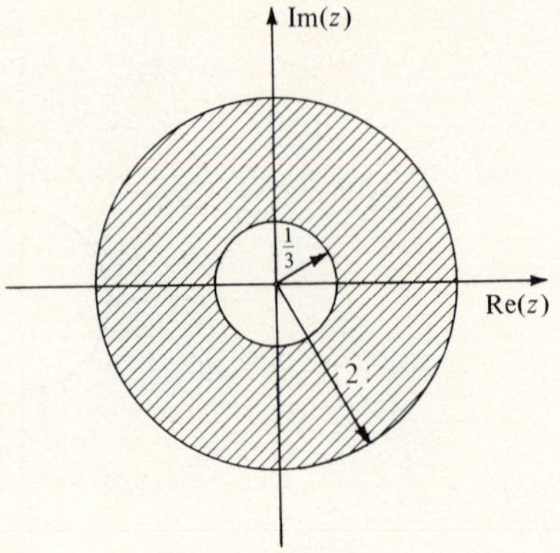

FIGURE 5-2

For example, the z transform of the signal in Eq. (5-1) is the infinite series (5-2). Since the series of positive powers of z in (5-2) converges absolutely for $|z| < 2$ and the series of negative powers of z in (5-2) converges absolutely for $|z| > \frac{1}{3}$, the region of absolute convergence of $x(n)$ is the annulus $\frac{1}{3} < |z| < 2$ as shown in Fig. 5-2. Note that, within the region of absolute convergence of $x(n)$, the infinite series (5-2) can be written in closed form as in Eq. (5-3). Also, consider the signal

$$x(n) = \begin{cases} (\frac{1}{3})^n - 2^n & n \geq 0 \\ 0 & n < 0 \end{cases} \tag{5-4}$$

Its z transform is

$$\begin{aligned} X(z) &= \tfrac{1}{3}z^{-1} - 2z^{-1} + (\tfrac{1}{3})^2 z^{-2} - 2^2 z^{-2} + (\tfrac{1}{3})^3 z^{-3} - 2^3 z^{-3} + \cdots \\ &\quad + (\tfrac{1}{3})^k z^{-k} - 2^k z^{-k} + \cdots \\ &= \frac{1}{1 - \frac{1}{3}z^{-1}} - \frac{1}{1 - 2z^{-1}} \\ &= \frac{5z}{(2 - z)(3z - 1)} \end{aligned}$$

Because the series

$$\tfrac{1}{3}z^{-1} + (\tfrac{1}{3})^2 z^{-2} + (\tfrac{1}{3})^3 z^{-3} + \cdots + (\tfrac{1}{3})^k z^{-k} + \cdots$$

converges absolutely for $|z| > \frac{1}{3}$, and the series

$$-2z^{-1} - 2^2 z^{-2} - 2^3 z^{-3} - \cdots - 2^k z^{-k} - \cdots$$

converges absolutely for $|z| > 2$, the region of absolute convergence of $X(z)$ is $|z| > 2$. Note that the z transforms of the two signals in Eqs. (5-1) and (5-4) have the same closed-form expression but have different regions of absolute convergence. Indeed, it should be borne in mind that although a z transform expressed as an infinite series of positive powers and negative powers of z corresponds to a unique discrete time signal $x(n)$, a z transform in a closed-form expression does not specify uniquely a discrete time signal. Rather, the region of absolute convergence of the signal must also be known. What probably is not clear at this point is whether a closed-form expression of $X(z)$ together with a region of absolute convergence will specify uniquely a discrete signal $x(n)$. The answer to this question is affirmative as we shall see in the next section.

As another example, let

$$x(n) = \begin{cases} \dfrac{a^n}{n!} & n \geq 0 \\ 0 & n < 0 \end{cases}$$

Then

$$X(z) = \sum_{n=0}^{\infty} \frac{a^n}{n!} z^{-n} = \sum_{n=0}^{\infty} \frac{1}{n!} \left(\frac{a}{z}\right)^n = e^{a/z}$$

with the region of absolute convergence being $|z| > 0$, that is, the entire z plane with the origin $z = 0$ excluded.

Table 5-1

$x(n)$		$X(z)$	Region of absolute convergence
1 0	$n = k$ otherwise	z^{-k}	The entire z plane for negative k The entire z plane except the origin for positive k
1 0	$n \geq 0$ $n < 0$	$\dfrac{1}{1 - z^{-1}}$	$\lvert z\rvert > 1$
0 1	$n > 0$ $n \leq 0$	$\dfrac{1}{1 - z}$	$\lvert z\rvert < 1$
a^n 0	$n \geq 0$ $n < 0$	$\dfrac{1}{1 - az^{-1}}$	$\lvert z\rvert > a$
a^n b^n	$n \geq 0$ $n < 0$	$\dfrac{1}{1 - b^{-1}z} + \dfrac{1}{1 - az^{-1}} - 1$	$b > \lvert z\rvert > a$
n 0	$n \geq 0$ $n < 0$	$\dfrac{1}{(1 - z^{-1})^2}$	$\lvert z\rvert > 1$
0 n	$n > 0$ $n \leq 0$	$\dfrac{z}{(1 - z)^2}$	$\lvert z\rvert < 1$

Note that the signal

$$x(n) = \begin{cases} 1 & n \geq 0 \\ 2^{-n} & n < 0 \end{cases}$$

does not have a region of absolute convergence. This conclusion follows from the observation that in the infinite series

$$\cdots + 2^k z^k + \cdots + 2^2 z^2 + 2z + 1 + z^{-1} + z^{-2} + z^{-3} + \cdots + z^{-k} + \cdots$$

the series of negative powers of z converges for $|z| > 1$ and the series of positive powers of z converges for $|z| < \frac{1}{2}$. When a signal does not have a region of absolute convergence, we say that the z transform of the signal does not exist.

The z transforms of some frequently encountered discrete signals are listed in Table 5-1.

5-3 THE INVERSE TRANSFORMATION

The technique of z transformation is most useful in analyzing the behavior of discrete systems. However, before we discuss the application of such a technique to system analysis problems, let us study first the properties of the z transforms of discrete signals.

We saw in Sec. 5-2 how a discrete signal $x(n)$ can be represented by its z transform $X(z)$. Conversely, we can recover $x(n)$ from $X(z)$ by expanding $X(z)$ into an infinite series of positive and negative powers of z. Note, however, that the expansion must be such that the infinite series indeed converges within the given region of absolute convergence of $x(n)$. As an example, let

$$X(z) = \frac{1}{1 - z}$$

with its region of absolute convergence being $|z| > 1$. Although $X(z)$ can be expanded either as

$$X(z) = -z^{-1} - z^{-2} - z^{-3} - \cdots - z^{-k} - \cdots \tag{5-5}$$

or as

$$X(z) = 1 + z + z^2 + z^3 + \cdots + z^k + \cdots \tag{5-6}$$

only the infinite series in Eq. (5-5) converges for $|z| > 1$. Thus, we obtain

$$x(n) = \begin{cases} -1 & n > 0 \\ 0 & n \leq 0 \end{cases}$$

If, on the other hand, the region of absolute convergence of $X(z)$ is $|z| < 1$ instead, then only the infinite series in Eq. (5-6) converges and we obtain

$$x(n) = \begin{cases} 0 & n > 0 \\ 1 & n \leq 0 \end{cases}$$

As another example, let

$$X(z) = \frac{5z}{(2 - z)(3z - 1)}$$

with its region of absolute convergence being $\frac{1}{3} < |z| < 2$. By the method of partial fraction, we write $X(z)$ as

$$X(z) = \frac{2}{2 - z} + \frac{1}{3z - 1}$$

Expanding $X(z)$ as an infinite series that converges within the region $\frac{1}{3} < |z| < 2$, we obtain

$$X(z) = 1 + \tfrac{1}{2}z + (\tfrac{1}{2})^2z^2 + \cdots + (\tfrac{1}{2})^kz^k + \cdots + \tfrac{1}{3}z^{-1} + (\tfrac{1}{3})^2z^{-2} + \cdots + (\tfrac{1}{3})^kz^{-k} + \cdots$$

That is,

$$x(n) = \begin{cases} (\frac{1}{3})^n & n \geq 0 \\ (\frac{1}{2})^{-n} & n < 0 \end{cases}$$

We leave it to the reader to show that if the region of absolute convergence is $|z| > 2$ then

$$x(n) = \begin{cases} (\frac{1}{3})^n - 2^n & n \geq 0 \\ 0 & n < 0 \end{cases}$$

and if the region of absolute convergence is $|z| < \frac{1}{3}$ then

$$x(n) = \begin{cases} 0 & n > 0 \\ (\frac{1}{2})^{-n} - 3^{-n} & n \leq 0 \end{cases}$$

As a further example, let

$$X(z) = \sqrt{1 - \frac{1}{z^2}}$$

with its region of absolute convergence being $|z| > 0$. We can write

$$X(z) = 1 + \frac{1}{2}z^{-2} + \frac{1(3)}{2(4)}z^{-4} + \frac{1(3)(5)}{2(4)(6)}z^{-6} + \frac{1(3)(5)(7)}{2(4)(6)(8)}z^{-8} + \cdots$$

It follows that

$$x(n) = \begin{cases} \dfrac{1(3)(5)\cdots(2n-1)}{2(4)(6)\cdots n} & n > 0 \text{ and } n \text{ is even} \\ 1 & n = 0 \\ 0 & \text{otherwise} \end{cases}$$

There is another way to determine $x(n)$ from its z transform $X(z)$. In the terminologies used in the theory of functions of a complex variable $x(n)$ are the coefficients of the Laurent's expansion of $X(z)$ at $z = 0$ which can be computed according to the formula†

$$x(n) = \frac{1}{2\pi j} \oint_C X(z) z^{n-1}\, dz \tag{5-7}$$

In this equation C is a closed path in the region of absolute convergence of $X(z)$ and $\oint_C$ means a contour integration along C in the counterclockwise direction. Equation (5-7) is known as the *inverse transformation formula*.

Furthermore, the contour integral in Eq. (5-7) can be evaluated by the method of residues. It can be shown that

$$x(n) = \begin{cases} \text{sum of residues of } X(z)z^{n-1} \text{ at poles inside contour } C & n \geq 0 \\ -(\text{sum of residues of } X(z)z^{n-1} \text{ at poles outside contour } C) & n < 0 \end{cases}$$

where the residue of the function $X(z)z^{n-1}$ at a simple pole z_0 is equal to

$$(z - z_0)X(z)z^{n-1} \big|_{z=z_0}$$

and the residue of the function $X(z)z^{n-1}$ at a kth-order pole z_0 is equal to

$$\frac{1}{(k-1)!} \frac{d^{k-1}}{dz^{k-1}} [(z - z_0)^k X(z) z^{n-1}] \bigg|_{z=z_0}$$

Let us now use the inverse transformation formula (5-7) to work out some of the examples presented earlier in this section. Let

$$X(z) = \frac{5z}{(2 - z)(3z - 1)}$$

with its region of absolute convergence being $\frac{1}{3} < |z| < 2$. Let C be a closed path in the region $\frac{1}{3} < |z| < 2$ as shown in Fig. 5-3*a*. Since for $n \geq 0$, the function

$$\frac{5z}{(2 - z)(3z - 1)} z^{n-1}$$

† We assume that the reader has been introduced to the basic ideas in the theory of functions of a complex variable. On the other hand, the detailed steps in evaluating the contour integral in Eq. (5-7) discussed in the remainder of this section can be omitted without causing any discontinuity in the presentation.

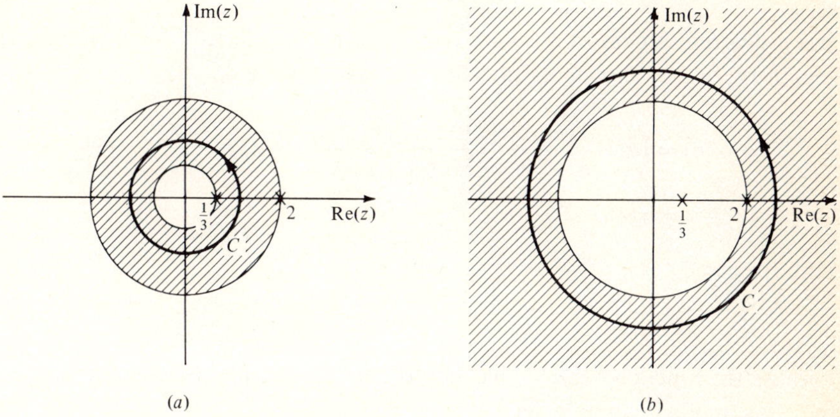

FIGURE 5-3

has a pole at $z = \frac{1}{3}$ that is inside the contour C. We have, for $n \geq 0$,

$$\begin{aligned} x(n) &= \text{residue of } \frac{5z(z^{n-1})}{(2-z)(3z-1)} \quad \text{at} \quad z = \tfrac{1}{3} \\ &= \left(z - \frac{1}{3}\right) \frac{5z(z^{n-1})}{(2-z)(3z-1)}\bigg|_{z=\frac{1}{3}} \\ &= (\tfrac{1}{3})^n \end{aligned}$$

Similarly, since for $n < 0$ the function

$$\frac{5z}{(2-z)(3z-1)} z^{n-1}$$

has a pole at $z = 2$ that is outside the contour C, we have, for $n < 0$,

$$\begin{aligned} x(n) &= -\text{residue of } \frac{5z(z^{n-1})}{(2-z)(3z-1)} \quad \text{at} \quad z = 2 \\ &= -(z-2) \frac{5z(z^{n-1})}{(2-z)(3z-1)}\bigg|_{z=2} \\ &= 2^n \end{aligned}$$

For the same expression

$$X(z) = \frac{5z}{(2-z)(3z-1)}$$

but a different region of absolute convergence $|z| > 2$, let C be a closed path in the region $|z| > 2$, as shown in Fig. 5-3b. We have, for $n \geq 0$,

$$x(n) = \left(z - \frac{1}{3}\right) \frac{5z(z^{n-1})}{(2 - z)(3z - 1)}\bigg|_{z=\frac{1}{3}} + (z - 2) \frac{5z(z^{n-1})}{(2 - z)(3z - 1)}\bigg|_{z=2}$$
$$= (\tfrac{1}{3})^n - 2^n$$

and, for $n < 0$,

$$x(n) = 0$$

because the function $x(z)z^{n-1}$ has no pole that is outside the contour C.

As another example, let

$$X(z) = \frac{1}{1 - z}$$

with its region of absolute convergence being $|z| < 1$. Let C be a closed path in the region of absolute convergence, as shown in Fig. 5-4. For $n > 0$, the function

$$\frac{1}{1 - z} z^{n-1}$$

has no pole that is inside the contour C. Therefore, for $n > 0$,

$$x(n) = 0$$

For $n = 0$, the function

$$\frac{1}{1 - z} z^{n-1} = \frac{1}{(1 - z)z}$$

has a pole $z = 0$ that is inside the contour C. Therefore, for $n = 0$,

$$x(n) = z \frac{1}{(1 - z)z}\bigg|_{z=0} = 1$$

Similarly, for $n < 0$,

$$x(n) = -(z - 1) \frac{1}{1 - z} z^{n-1}\bigg|_{z=1} = 1$$

Finally, let

$$x(z) = \frac{z^3}{(z - 1)(z - 2)^2}$$

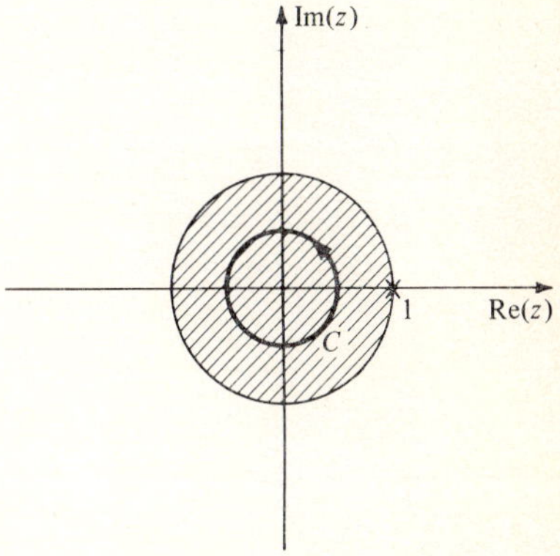

FIGURE 5-4

with its region of absolute convergence being $1 < |z| < 2$. Let C be a closed path in the region of absolute convergence, as shown in Fig. 5-5. For $n \geq 0$, the function

$$\frac{z^3}{(z-1)(z-2)^2} z^{n-1}$$

has a pole $z = 1$ inside the contour C. Therefore, for $n \geq 0$,

$$x(n) = \left. \frac{z^{n+2}}{(z-2)^2} \right|_{z=1} = 1$$

For $n < 0$, the function

$$\frac{z^3}{(z-1)(z-2)^2} z^{n-1}$$

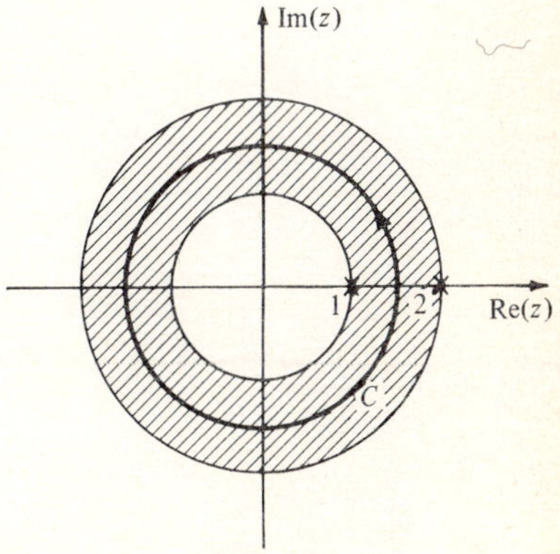

FIGURE 5-5

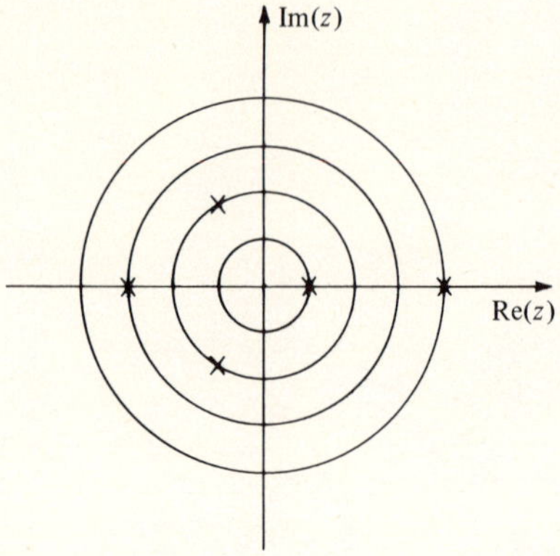

FIGURE 5-6

has a second-order pole $z = 2$ outside the contour C. Therefore, for $n < 0$,

$$x(n) = -\frac{d}{dz}\frac{z^{n+2}}{z-1}\bigg|_{z=2} = -n(2^{n+1})$$

In general, the poles of $X(z)$ divide the z plane into a number of regions, as illustrated in Fig. 5-6, where the poles are marked as crosses. The innermost region is a circle of finite radius, the outermost region is the exterior of a circle of finite radius, and all other regions are annuli. Each of these regions is a possible region of absolute convergence of $X(z)$. [Recall our discussion in Sec. 5-2 that the region of absolute convergence of $X(z)$ is always the interior of a circle, the exterior of a circle, or an annulus. Therefore, a region such as the shaded area in Fig. 5-7 is *not* a possible region of absolute convergence.] For a given region of absolute convergence, we can recover $x(n)$ from $X(z)$ by using the inverse transformation formula

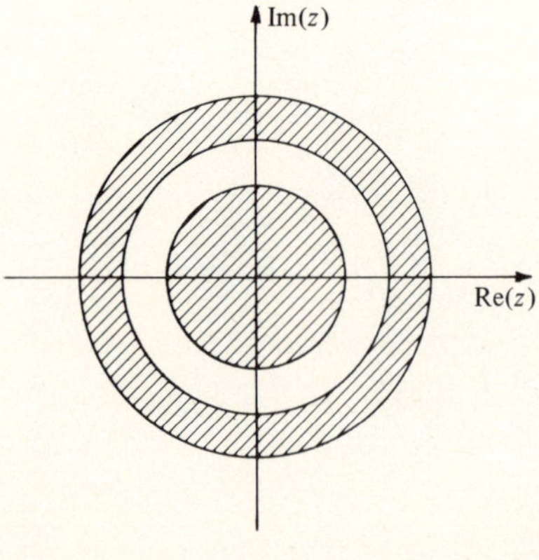

FIGURE 5-7

(5-7). Since two different regions of absolute convergence will yield two different contour integrals, we can conclude that a closed-form expression of $X(z)$ together with its region of absolute convergence will indeed determine $x(n)$ uniquely.

5-4 MANIPULATION OF SIGNALS

Let $X(z)$ be the z transform of a discrete signal $x(n)$ with its region of absolute convergence being $\rho_1 < |z| < \rho_2$. The z transform of the transposed signal $x(-n)$ is

$$\sum_{n=-\infty}^{\infty} x(-n)z^{-n} = \sum_{n=-\infty}^{\infty} x(n)\left(\frac{1}{z}\right)^{-n} = X\left(\frac{1}{z}\right)$$

with its region of absolute convergence being $1/\rho_2 < |z| < 1/\rho_1$. For example, let

$$x(n) = \begin{cases} (\frac{1}{3})^n & n \geq 0 \\ (\frac{1}{2})^{-n} & n < 0 \end{cases} \tag{5-8}$$

Its z transform is

$$\begin{aligned} X(z) &= \frac{1}{1 - \frac{1}{2}z} + \frac{1}{1 - \frac{1}{3}z^{-1}} - 1 \\ &= \frac{5z}{(2 - z)(3z - 1)} \end{aligned}$$

with its region of absolute convergence being $\frac{1}{3} < |z| < 2$. Thus,

$$x(-n) = \begin{cases} (\frac{1}{2})^n & n > 0 \\ (\frac{1}{3})^{-n} & n \leq 0 \end{cases}$$

and its z transform is

$$\frac{5z^{-1}}{(2 - z^{-1})(3z^{-1} - 1)} = \frac{5z}{(2z - 1)(3 - z)} = \frac{1}{1 - \frac{1}{3}z} + \frac{1}{1 - \frac{1}{2}z^{-1}} - 1$$

with the region of absolute convergence being $\frac{1}{2} < |z| < 3$.

The z transform of the signal $x(n + k)$ is $z^k X(z)$ for positive as well as negative k. Moreover, the region of absolute convergence of $z^k X(z)$ is the same as that of $X(z)$ for positive k and is at least the intersection of that of $X(z)$ and the region $|z| > 0$ for negative k. Thus, for the discrete signal in Eq. (5-8),

$$x(n + 5) = \begin{cases} (\frac{1}{3})^{n+5} & n \geq -5 \\ (\frac{1}{2})^{-(n+5)} & n < -5 \end{cases}$$

and its z transform is

$$\cdots + (\tfrac{1}{2})^2 z^7 + \tfrac{1}{2}z^6 + z^5 + \tfrac{1}{3}z^4 + (\tfrac{1}{3})^2 z^3 + (\tfrac{1}{3})^3 z^2 + (\tfrac{1}{3})^4 z + (\tfrac{1}{3})^5 + (\tfrac{1}{3})^6 z^{-1} + (\tfrac{1}{3})^7 z^{-2} + \cdots$$

which can be written

$$\frac{z^5}{1 - \frac{1}{2}z} + \frac{z^5}{1 - \frac{1}{3}z^{-1}} - z^5$$

Also,

$$x(n-5) = \begin{cases} (\frac{1}{3})^{n-5} & n \geq 5 \\ (\frac{1}{2})^{-(n-5)} & n < 5 \end{cases}$$

and its z transform is

$$\cdots (\tfrac{1}{2})^5 + (\tfrac{1}{2})^4 z^{-1} + (\tfrac{1}{2})^3 z^{-2} + (\tfrac{1}{2})^2 z^{-3} + (\tfrac{1}{2}) z^{-4} + z^{-5} + \tfrac{1}{3}z^{-6} + (\tfrac{1}{3})^2 z^{-7} + (\tfrac{1}{3})^3 z^{-8} + \cdots$$

which can be written

$$\frac{z^{-5}}{1 - \frac{1}{2}z} + \frac{z^{-5}}{1 - \frac{1}{3}z^{-1}} - z^{-5}$$

Let $x(n)$ and $y(n)$ be two discrete signals, and $X(z)$ and $Y(z)$ be their z trans. forms. Let a and b be two arbitrary constants. The z transform of the signal $ax(n) + by(n)$ is $aX(z) + bY(z)$ with its region of absolute convergence being *at least* the intersection of that of $x(n)$ and $y(n)$. For example, let

$$x(n) = \begin{cases} 1 & n \geq 0 \\ 3^n & n < 0 \end{cases} \tag{5-9}$$

and

$$y(n) = \begin{cases} 2^n & n \geq 0 \\ 4^n & n < 0 \end{cases}$$

Then

$$X(z) = \frac{1}{1 - z^{-1}} + \frac{1}{1 - \frac{1}{3}z} - 1 = \frac{2z}{(z-1)(3-z)}$$

with its region of absolute convergence being $1 < |z| < 3$, and

$$Y(z) = \frac{1}{1 - 2z^{-1}} + \frac{1}{1 - \frac{1}{4}z} - 1 = \frac{2z}{(z-2)(4-z)}$$

with its region of absolute convergence being $2 < |z| < 4$. It follows that the z transform of the signal

$$x(n) - y(n) = \begin{cases} 1 - 2^n & n \geq 0 \\ 3^n - 4^n & n < 0 \end{cases}$$

is

$$X(z) - Y(z) = \frac{2z}{(z-1)(3-z)} - \frac{2z}{(z-2)(4-z)} = \frac{2z(2z-5)}{(z-1)(z-2)(3-z)(4-z)}$$

with its region of absolute convergence being $2 < |z| < 3$.

It follows from the last two properties of z transforms that the z transform of the forward difference of $x(n)$

$$\Delta x(n) = x(n+1) - x(n)$$

is $(z-1)X(z)$ with its region of absolute convergence being at least that of $X(z)$. Also, the z transform of the backward difference of $x(n)$

$$\nabla x(n) = x(n) - x(n-1)$$

is $(1 - z^{-1})X(z)$ with its region of absolute convergence being at least the intersection of that of $X(z)$ and the region $|z| > 0$. For example, the backward difference of the signal $x(n)$ in Eq. (5-9) is

$$\nabla x(n) = x(n) - x(n-1) = \begin{cases} 0 & n > 0 \\ \frac{2}{3}(3^n) & n \leq 0 \end{cases}$$

Its z transform is

$$(1 - z^{-1})X(z) = \frac{2z(1 - z^{-1})}{(z-1)(3-z)} = \frac{2}{3-z}$$

with its region of absolute convergence being $|z| < 3$. Note that in this example, because of the cancellation of the pole $z = 1$ in $X(z)$, the region of absolute convergence of $\nabla x(n)$ is larger than the intersection of that of $x(n)$ and the region $|z| > 0$.

5-5 THE CONVOLUTION PROPERTY

We are now ready to study how the technique z transformation can be applied to system analysis problems. Let $h(n)$ denote the unit response of a linear time-invariant discrete system. Let $x(n)$ be an input signal to the system and $y(n)$ the corresponding output signal. According to our discussion in Chap. 4,

$$y(n) = x(n) * h(n) = \sum_{i=-\infty}^{\infty} x(i)h(n-i)$$

It follows that

$$\begin{aligned} Y(z) &= \sum_{n=-\infty}^{\infty} y(n)z^{-n} \\ &= \sum_{n=-\infty}^{\infty} \sum_{i=-\infty}^{\infty} x(i)h(n-i)z^{-n} \\ &= \sum_{n=-\infty}^{\infty} \sum_{i=-\infty}^{\infty} x(i)z^{-i}h(n-i)z^{-(n-i)} \\ &= \left[\sum_{i=-\infty}^{\infty} x(i)z^{-i}\right]\left[\sum_{j=-\infty}^{\infty} h(j)z^{-j}\right] \\ &= X(z)H(z) \end{aligned} \tag{5-10}$$

Observe that if both $X(z)$ and $H(z)$ converge absolutely for $z = z_0$, then $Y(z)$ will also converge absolutely for $z = z_0$. Therefore, the region of absolute convergence of $y(n)$ is at least the intersection of that of $x(n)$ and $h(n)$. $H(z)$, the z transform of the unit response of the system, is known as the *transfer function* (or the *system function*) of the system.

The result in Eq. (5-10) suggests immediately an alternative way to compute the response of a linear time-invariant discrete system to a given stimulus. Instead of computing the convolution sum of the stimulus and the unit response of the system, we can compute the z transform of the response $y(n)$ first and then carry out an inverse transformation to determine $y(n)$. As will be demonstrated in the following, there are many occasions in which it is computationally easier to determine $Y(z)$ according to Eq. (5-10) and then carry out an inverse transformation than to determine $y(n)$ directly by evaluating the convolution sum of $x(n)$ and $h(n)$. It should also be pointed out that application of the result in Eq. (5-10) is not restricted to analyzing the input-output relationship of discrete systems. For two arbitrary discrete signals $x(n)$ and $h(n)$, the z transform of the convolution sum of these two signals is equal to the product of their z transforms.

As an example, let

$$h(n) = \begin{cases} (\frac{1}{2})^n & n \geq 0 \\ 0 & n < 0 \end{cases}$$

be the unit response of a linear time-invariant system. Let

$$x(n) = \begin{cases} (\frac{1}{3})^n & n \geq 0 \\ (\frac{1}{2})^{-n} & n < 0 \end{cases}$$

be an input signal to the system. The transfer function of the system is

$$H(z) = \frac{1}{1 - \frac{1}{2}z^{-1}} = \frac{2z}{2z - 1}$$

with its region of absolute convergence being $|z| > \frac{1}{2}$. Also,

$$X(z) = \frac{1}{1 - \frac{1}{2}z} + \frac{1}{1 - \frac{1}{3}z^{-1}} - 1 = \frac{5z}{(2 - z)(3z - 1)}$$

with its region of absolute convergence being $\frac{1}{3} < |z| < 2$. It follows that the z transform of the output signal is

$$Y(z) = X(z)H(z) = \frac{10z^2}{(2 - z)(2z - 1)(3z - 1)}$$

with its region of absolute convergence being $\frac{1}{2} < |z| < 2$. Writing $Y(z)$ as

$$Y(z) = \frac{\frac{8}{3}}{2 - z} + \frac{\frac{10}{3}}{2z - 1} - \frac{2}{3z - 1}$$

we obtain

$$y(n) = \begin{cases} \frac{10}{3}(\frac{1}{2})^n - 2(\frac{1}{3})^n & n > 0 \\ \frac{4}{3}(2)^n & n \leq 0 \end{cases}$$

The reader might wish to check the result by computing directly the convolution sum of $x(n)$ and $h(n)$.

We studied in Sec. 4-5 the problem of determining the unit response of a system from a pair of input and output signals. Such a problem can be handled most easily by the technique of z transformation. Given $x(n)$ and $y(n)$, we can determine $X(z)$ and $Y(z)$ and then $Y(z)/X(z)$ which, according to Eq. (5-10), is the transfer function of the system. Although we have already developed all the necessary mathematical concepts, for pedagogical reasons we shall defer the discussion of this problem to Chap. 9,† after we have developed the counterpart of the z transformation for continuous signals, the Laplace transformation.

We present now more illustrative examples of the application of the formula in Eq. (5-10), which, as was pointed out earlier, is not limited to analyzing the input-output behavior of discrete systems.

Suppose that we want to determine the z transform of the signal $\sum_{i=-\infty}^{n} x(i)$. Let

$$h(n) = \begin{cases} 1 & n \geq 0 \\ 0 & n < 0 \end{cases}$$

Then

$$\sum_{i=-\infty}^{n} x(i) = x(n) * h(n)$$

Consequently, the z transform of the signal $\sum_{i=-\infty}^{n} x(i)$ is equal to

$$X(z)H(z) = \frac{X(z)}{1 - z^{-1}}$$

with its region of absolute convergence being at least the intersection of that of $X(z)$ and the region $|z| > 1$. As an example, let us determine the z transform of the signal

$$y(n) = \begin{cases} n + 1 & n \geq 0 \\ 0 & n < 0 \end{cases}$$

Let

$$x(n) = \begin{cases} 1 & n \geq 0 \\ 0 & n < 0 \end{cases}$$

† There is a question concerning the uniqueness of the unit response of the system because the region of absolute convergence of $H(z)$, which is computed according to Eq. (5-10), might not be unique. To discuss this question now might cause some confusion later on when we discuss the solution of the same problem for continuous systems, using the method of Fourier transformation in Chap. 7 where the question of the uniqueness of the impulse response does not arise.

We have

$$X(z) = \frac{1}{1 - z^{-1}}$$

with its region of absolute convergence being $|z| > 1$. Since

$$y(n) = \sum_{i=-\infty}^{n} x(i) = x(n) * h(n)$$

we have

$$Y(z) = \frac{1}{1 - z^{-1}} \frac{1}{1 - z^{-1}} = \frac{1}{(1 - z^{-1})^2}$$

with its region of absolute convergence being $|z| > 1$.

Let us consider a problem of the multiplication of rabbits. Suppose that the life span of a rabbit is exactly 10 years. Furthermore, suppose that at the beginning there were two newborn rabbits, and the number of newborn rabbits in each year is two times that in the previous year. We want to determine the number of rabbits there are in the nth year. Let

$$x(n) = \begin{cases} 2^{n+1} & n \geq 0 \\ 0 & n < 0 \end{cases}$$

and

$$h(n) = \begin{cases} 1 & 0 \leq n \leq 10 \\ 0 & \text{otherwise} \end{cases}$$

Then

$$y(n) = x(n) * h(n)$$

will give the number of rabbits there are in the nth year. Since

$$X(z) = \frac{2}{1 - 2z^{-1}}$$

with its region of absolute convergence being $|z| > 2$, and

$$\begin{aligned} H(z) &= 1 + z^{-1} + z^{-2} + \cdots + z^{-10} \\ &= \frac{1 - z^{-11}}{1 - z^{-1}} \end{aligned}$$

with its region of absolute convergence being $|z| > 0$, we obtain

$$Y(z) = X(z)H(z) = \frac{2(1 - z^{-11})}{(1 - z^{-1})(1 - 2z^{-1})} = (1 - z^{-11})\left(\frac{-2}{1 - z^{-1}} + \frac{4}{1 - 2z^{-1}}\right)$$

with its region of absolute convergence being $|z| > 2$. To recover $y(n)$ from $Y(z)$, we note that corresponding to the z transform

$$W(z) = \frac{-2}{1 - z^{-1}} + \frac{4}{1 - 2z^{-1}}$$

the discrete signal is

$$w(n) = \begin{cases} 2^{n+2} - 2 & n \geq 0 \\ 0 & n < 0 \end{cases}$$

Thus,

$$y(n) = w(n) - w(n - 11) = \begin{cases} 2^{n+2} - 2^{n-9} & n \geq 11 \\ 2^{n+2} - 2 & 0 \leq n < 11 \\ 0 & n < 0 \end{cases}$$

As another example of the application of Eq. (5-10), we consider the movement of a particle in a plane. The particle starts at the origin and makes a sequence of left and right moves and then a sequence of up and down moves. Specifically, during its left and right moves the particle either moves to its left or moves to its right for a unit distance in each second; during its up and down moves, the particle either moves upward for a distance of two units, or moves upward for a unit distance, or moves downward for a unit distance in each second. Therefore, a path taken by the particle in n seconds is always a sequence of k left and right moves followed by a sequence of $n - k$ up and down moves for some k, $0 \leq k \leq n$. Since the number of sequences of k left and right moves is 2^k and the number of sequences of $n - k$ up and down moves is 3^{n-k}, the number of different paths the particle can take in n seconds is

$$3^n + 2(3^{n-1}) + 2^2(3^{n-2}) + \cdots + 2^k(3^{n-k}) + \cdots + 2^{n-1}(3) + 2^n \tag{5-11}$$

To evaluate this sum, we let

$$x(n) = \begin{cases} 2^n & n \geq 0 \\ 0 & n < 0 \end{cases}$$

and

$$y(n) = \begin{cases} 3^n & n \geq 0 \\ 0 & n < 0 \end{cases}$$

Then

$$w(n) = x(n) * y(n)$$

is equal to the number of paths the particle can take in n seconds given by expression (5-11). Since

$$X(z) = \frac{1}{1 - 2z^{-1}}$$

and

$$Y(z) = \frac{1}{1 - 3z^{-1}}$$

we have

$$W(z) = X(z)Y(z) = \frac{1}{1-2z^{-1}}\frac{1}{1-3z^{-1}} = \frac{-2}{1-2z^{-1}} + \frac{3}{1-3z^{-1}}$$

or

$$w(n) = \begin{cases} 3^{n+1} - 2^{n+1} & n \geq 0 \\ 0 & n < 0 \end{cases}$$

We now present examples from probability theory in which z transformation is a very important tool. Suppose we have a "crooked" die. When the die is rolled, the probability that a 1 will show is $\frac{1}{12}$, that a 2 will show is $\frac{1}{12}$, that a 3 will show is $\frac{1}{6}$, that a 4 will show is $\frac{1}{6}$, that a 5 will show is $\frac{1}{4}$, and that a 6 will show is $\frac{1}{4}$. We want to determine the probability that the sum of the scores is 10 when the die is rolled twice. Let

$$p(n) = \begin{cases} \frac{1}{12} & n = 1, 2 \\ \frac{1}{6} & n = 3, 4 \\ \frac{1}{4} & n = 5, 6 \\ 0 & \text{otherwise} \end{cases}$$

The probability that the total score is 10 when the die is rolled twice is given by

$$\begin{aligned} p(6)p(4) + p(5)p(5) + p(4)p(6) &= \tfrac{1}{4}\tfrac{1}{6} + \tfrac{1}{4}\tfrac{1}{4} + \tfrac{1}{6}\tfrac{1}{4} \\ &= \tfrac{7}{48} \end{aligned}$$

In general, the probability that the total score is n when the die is rolled twice is given by $p(n) * p(n)$. Since

$$\begin{aligned} P(z) &= \tfrac{1}{12}z^{-1} + \tfrac{1}{12}z^{-2} + \tfrac{1}{6}z^{-3} + \tfrac{1}{6}z^{-4} + \tfrac{1}{4}z^{-5} + \tfrac{1}{4}z^{-6} \\ &= (\tfrac{1}{12}z^{-1} + \tfrac{1}{6}z^{-3} + \tfrac{1}{4}z^{-5})(1 + z^{-1}) \end{aligned}$$

we obtain

$$\begin{aligned} [P(z)]^2 = {} & \tfrac{1}{144}z^{-2} + \tfrac{1}{72}z^{-3} + \tfrac{5}{144}z^{-4} + \tfrac{1}{18}z^{-5} + \tfrac{7}{72}z^{-6} + \tfrac{5}{36}z^{-7} \\ & + \tfrac{11}{72}z^{-8} + \tfrac{1}{6}z^{-9} + \tfrac{7}{48}z^{-10} + \tfrac{1}{8}z^{-11} + \tfrac{1}{16}z^{-12} \end{aligned}$$

Because the coefficient of z^{-10} in $[P(z)]^2$ is $\frac{7}{48}$, we confirm that the probability of the sum of the scores being 10 when the die is rolled twice is indeed $\frac{7}{48}$, as was determined above. It follows immediately from this discussion that the probability of the sum of the scores being n when the die is rolled k times is equal to the coefficient of z^{-n} in $[P(z)]^k$.

As another example, we consider a problem concerning the number of descendants of African elephants. The probability that an African elephant has n offspring is $(\frac{1}{2})^{n+1}$; that is, the probability that it has no offspring is $\frac{1}{2}$, that it has one offspring is $\frac{1}{4}$, that it has two offspring is $\frac{1}{8}$, and so on. We want to know the probability that the total number of offspring of k elephants is equal to n. Let

$$x(n) = \begin{cases} \dfrac{1}{2^{n+1}} & n \geq 0 \\ 0 & n < 0 \end{cases}$$

We have

$$X(z) = \frac{1}{2 - z^{-1}}$$

with its region of absolute convergence being $|z| > \frac{1}{2}$. According to our discussion above on the crooked die, the probability of the total number of offspring of k elephants being n is the coefficient of z^{-n} in $[X(z)]^k$. We now ask for the probability that an elephant will have exactly n grandchildren, $n \geq 0$. Let $y(n)$ denote such probability. Then

$$y(n) = \sum_{k=0}^{\infty} \begin{pmatrix}\text{probability that an} \\ \text{elephant has } k \text{ offspring}\end{pmatrix} \begin{pmatrix}\text{probability that these } k \text{ elephants} \\ \text{will have all together } n \text{ offspring}\end{pmatrix}$$

That is,

$$y(n) = \sum_{k=0}^{\infty} \frac{1}{2^{k+1}} \{\text{coefficient of } z^{-n} \text{ in } [X(z)]^k\}$$

It follows that

$$Y(z) = \sum_{k=0}^{\infty} \frac{1}{2^{k+1}} [X(z)]^k = X\left[\frac{1}{X(z)}\right]^\dagger$$

That is,

$$Y(z) = \frac{1}{2 - 1/(2 - z^{-1})} = \frac{2 - z^{-1}}{3 - 2z^{-1}} = \frac{\frac{2}{3}}{1 - \frac{2}{3}z^{-1}} - \frac{\frac{1}{3}z^{-1}}{1 - \frac{2}{3}z^{-1}}$$

Therefore, for $n = 0$

$$y(n) = \tfrac{2}{3}$$

and for $n > 0$

$$y(n) = \tfrac{2}{3}(\tfrac{2}{3})^n - \tfrac{1}{3}(\tfrac{2}{3})^{n-1} = \tfrac{1}{9}(\tfrac{2}{3})^{n-1}$$

† $X[1/X(z)]$ means replace z by $1/X(z)$ in $X(z)$.

5-6 CAUSALITY AND STABILITY

As was pointed out in Chap. 4, many of the properties of a linear time-invariant system can be determined from its unit response $h(n)$. Consequently, such properties can also be determined from the transfer function of the system. We define the region of absolute convergence of a linear time-invariant system to be the region of absolute convergence of $h(n)$.

According to the inverse transformation formula (5-7), $h(n) = 0$ for $n < 0$ if and only if the region of absolute convergence of $h(n)$ is the area outside a circle of finite radius in the z plane. Consequently, a linear time-invariant discrete system is causal if and only if its region of absolute convergence is the area outside a circle of finite radius in the z plane. For example, let

$$H(z) = \frac{z^{-2}}{1 - 3z^{-1}}$$

be the transfer function of a linear time-invariant system with its region of absolute convergence being $|z| > 3$. Because the region of absolute convergence is the exterior of a circle of finite radius, we can conclude that the system is causal. Indeed, the unit response of the system is

$$h(n) = \begin{cases} 3^{n-2} & n \geq 2 \\ 0 & n < 2 \end{cases}$$

We show now that a linear time-invariant discrete system is stable if and only if its region of absolute convergence includes the unit circle centered at the origin of the z plane. According to our discussion in Chap. 4, a system is stable if and only if

$$\sum_{n=-\infty}^{\infty} |h(n)| < \infty$$

Since

$$H(z) = \sum_{n=-\infty}^{\infty} h(n)z^{-n}$$

the summation $\sum_{n=-\infty}^{\infty} |h(n)|$ is finite if the region of absolute convergence of $H(z)$ includes the unit circle. Thus, the system in the example shown above is not stable.

Combining the two observations above, we conclude that a causal system is stable if and only if its region of absolute convergence is the exterior of a circle of radius less than 1 in the z plane. For example, let

$$H(z) = \frac{z(12z - 5)}{(2z - 1)(3z - 1)}$$

be the transfer function of a causal system. The two poles of $H(z)$ are at $z = \frac{1}{2}$ and $z = \frac{1}{3}$. Since the region of absolute convergence of $H(z)$ must be the exterior of a circle of finite radius, we conclude that the region of absolute convergence of $H(z)$ is $|z| > \frac{1}{2}$. Consequently, the system is a stable one. On the other hand, let

$$H(z) = \frac{z(4z - 7)}{(2z - 1)(z - 3)}$$

be the transfer function of a causal system. Because the region of absolute convergence is now $|z| > 3$, we can conclude that the system is not stable.

5-7 REMARKS AND REFERENCES

Besides the general references cited in Chap. 1, we also recommend Freeman [2] and Kuo [3] as references to the material presented in this chapter.

In some textbooks, the z transform of a discrete signal is defined as

$$X(z) = \sum_{n=-\infty}^{\infty} x(n)z^n$$

Such a definition is simply a matter of convenience and preference and is conceptually equivalent to the definition used in this chapter.

The method of z transformation is not only very useful in system analysis but is also an important tool in probability theory and combinatorial analysis, where the z transform of a discrete function is usually known as the *generating function* of the function. For a further discussion on the application of z transformation in probability theory, see Feller [1], and in combinatorial analysis, see Liu [4].

[1] FELLER, W.: "An Introduction to Probability Theory and Its Applications," 2d ed., vol. I, John Wiley & Sons, Inc., New York, 1957.

[2] FREEMAN, H.: "Discrete-Time Systems," John Wiley & Sons, Inc., New York, 1965.

[3] KUO, B. C.: "Discrete-Data Control Systems," Prentice-Hall, Inc., Englewood Cliffs, N.J., 1970.

[4] LIU, C. L.: "Introduction to Applied Combinatorial Mathematics," McGraw-Hill Book Company, New York, 1968.

PROBLEMS

5-1 Determine the z transform and the corresponding region of absolute convergence of each of the following signals:

(*a*) $x(n) = \begin{cases} n(n+1) & n \geq 0 \\ 0 & n < 0 \end{cases}$

(*b*) $x(n) = \begin{cases} n^2 & n \geq 0 \\ 0 & n < 0 \end{cases}$

(*c*) $x(n) = \begin{cases} 0 & n > 0 \\ 0 & n = -1, -3, -5, \ldots \\ \dfrac{2^n}{n!} & n = 0, -2, -4, -6, \ldots \end{cases}$

(*d*) $x(n) = \begin{cases} 0 & n \geq 0 \\ \dfrac{1}{n}(-2)^n & n < 0 \end{cases}$

5-2 (*a*) Find the z transform and the region of absolute convergence of the discrete signal

$$x(n) = \begin{cases} \dbinom{m}{n} 2^n & 0 \leq n \leq m \\ 0 & \text{otherwise} \end{cases}$$

where m is a positive integer.

(*b*) Evaluate the sum

$$\binom{m}{0} + 2\binom{m}{1} + 2^2\binom{m}{2} + \cdots + 2^m\binom{m}{m}$$

(*c*) Evaluate the sum

$$1 - \frac{1}{2} + \frac{1(3)}{2(4)} - \frac{1(3)(5)}{2(4)(6)} + \frac{1(3)(5)(7)}{2(4)(6)(8)} + \cdots + (-1)^n \frac{1(3)(5)\cdots(2n-1)}{2(4)(6)\cdots 2n} + \cdots$$

5-3 (*a*) Show that the z transform of the signal

$$x(n) = \begin{cases} \dbinom{m}{n} + \dbinom{m-1}{n-1} + \dbinom{m-2}{n-2} + \cdots + \dbinom{m-n+1}{1} + \dbinom{m-n}{0} & n \geq 0 \\ 0 & n < 0 \end{cases} \tag{5P-1}$$

where m is a positive integer, is

$$X(z) = (1 + z^{-1})^m + z^{-1}(1 + z^{-1})^{m-1} + z^{-2}(1 + z^{-1})^{m-2} + \cdots + z^{-n}(1 + z^{-1})^{m-n}$$
$$= (1 + z^{-1})^{m+1} - z^{-(n+1)}(1 + z^{-1})^{m-n} \tag{5P-2}$$

What is its region of absolute convergence? Determine $x(n)$ from Eq. (5P-2) and thus find a closed-form expression of the sum in Eq. (5P-1).

(*b*) Find the z transform and the region of absolute convergence of the signal

$$y(n) = \begin{cases} \dbinom{2n}{n} + 2\dbinom{2n-1}{n} + 2^2\dbinom{2n-2}{n} + \cdots + 2^n\dbinom{n}{n} & n \geq 0 \\ 0 & n < 0 \end{cases}$$

and show that

$$y(n) = \begin{cases} 2^{2n} & n \geq 0 \\ 0 & n < 0 \end{cases}$$

5-4 (*a*) Determine the z transform and the region of absolute convergence of the discrete signal

$$x(n) = \begin{cases} \binom{m}{n+1} + \sum_{j=0}^{k-1} \binom{m+j}{n} & n \geq 0 \\ 0 & n < 0 \end{cases}$$

where m and k are positive integers.

(*b*) Show that

$$\binom{m}{n+1} + \sum_{j=0}^{k-1} \binom{m+j}{n} = \binom{m+k}{n+1} \qquad n \geq 0$$

(*c*) Show that for $n \geq 0$

$$\sum_{j=0}^{n-1} \binom{2+j}{2} = \binom{2+n}{3}$$

and thus

$$1(2) + 2(3) + 3(4) + \cdots + n(n+1) = \tfrac{1}{3}n(n+1)(n+2)$$

5-5 Determine the z transform and the region of absolute convergence of each of the time signals shown in Fig. 5P-1.

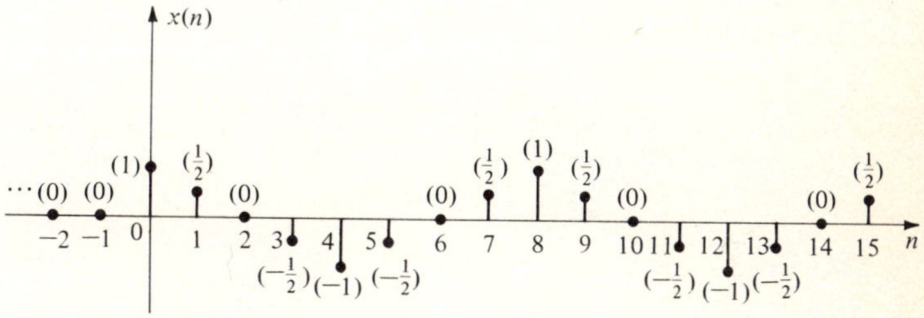

(*a*)

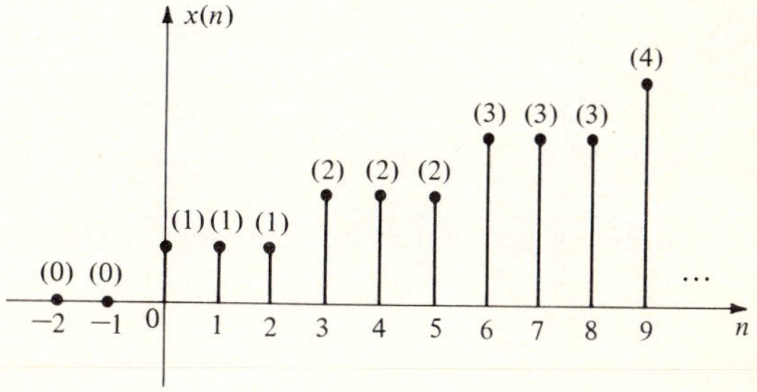

(*b*)

FIGURE 5P-1

5-6 Determine the time signals corresponding to the following z transforms:

$$(a)\quad X(z)=\frac{z^5}{z^2+3z+2}\qquad |z|>2$$

$$(b)\quad X(z)=\frac{3z^2-6z-1}{z^3-3z^2-z+3}\qquad \tfrac{1}{3}<|z|<1$$

$$(c)\quad X(z)=\frac{1}{(1-z)(1-z^2)(1-z^3)}\qquad |z|<1$$

$$(d)\quad X(z)=\tan^{-1} z\qquad |z|<1$$

5-7 Determine the time signals corresponding to the following z transforms:

$$(a)\quad X(z)=\frac{a^k}{z(z-a)^{k+1}}\qquad |z|>a$$

where k is a positive integer.

$$(b)\quad X(z)=\frac{a(z+a)}{(1-az^{-1})^3}\qquad |z|>a$$

5-8 (*a*) Find the z transform and the region of absolute convergence of the signal

$$x(n)=\begin{cases}\binom{m}{n}+\binom{m+1}{n}+\binom{m+2}{n}+\cdots+\binom{m+k}{n} & n\geq 0\\ 0 & n<0\end{cases}$$

(*b*) Let

$$y(n)=\sum_{i=n}^{\infty} x(i)$$

Determine the z transform of $y(n)$ and its region of absolute convergence.

5-9 The impulse response of a linear time-invariant system is

$$h(n)=\begin{cases}ne^{-n} & n\geq 0\\ 0 & n<0\end{cases}$$

Find the input signal $x(n)$ so that the output signal of the system is

$$y(n)=\begin{cases}1 & n=2\\ 0 & \text{otherwise}\end{cases}$$

5-10 In Fig. 5P-2, $x(n)$ is the input signal to a discrete linear time-invariant system with unit response $h(n)$. Find the corresponding output signal $y(n)$, its z transform, and the region of absolute convergence.

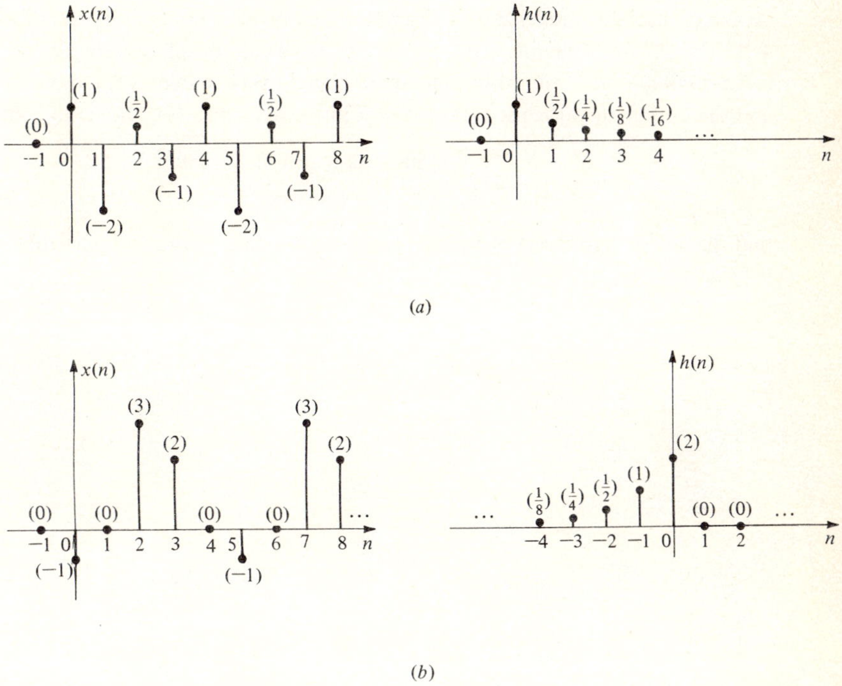

FIGURE 5P-2

5-11 In the following, let $h(n)$ denote the unit response of a linear time-invariant system, and let $x(n)$ denote an input signal to the system. Determine the z transform and the region of absolute convergence of the corresponding output signal $y(n)$. Also determine $y(n)$.

(a) $$h(n) = \begin{cases} (\frac{1}{3})^n & n \geq 0 \\ 0 & n < 0 \end{cases}$$

$$x(n) = \begin{cases} 2^n & n \geq 0 \\ 0 & n < 0 \end{cases}$$

(b) $$h(n) = \begin{cases} (\frac{1}{3})^n & n \geq 0 \\ 0 & n < 0 \end{cases}$$

$$x(n) = \begin{cases} 0 & n > 0 \\ 2^n & n \geq 0 \end{cases}$$

5-12 A savings account modeled as a discrete time-invariant system was described in Sec. 4-1. The savings account pays an interest at the rate of 0.5 percent per month with interest compounded monthly. The input signal to the system is the monthly deposit, and the output signal is the total amount in the account. For the input signal

$$x(n) = \begin{cases} 50 & n = 0, 1, 2, \ldots, 59 \\ 0 & \text{otherwise} \end{cases}$$

find the output signal $y(n)$ by the method of z transformation. In particular, evaluate $y(47)$ and $y(239)$.

6

FOURIER SERIES

6-1 INTRODUCTION

After developing a transformation theory for discrete signals, one naturally wants to see a corresponding theory for continuous signals. As a matter of fact, the transformation theory for continuous signals is richer in contents and has a wider range of application. We shall devote this chapter and Chaps. 7 and 9 to its development. The counterpart of z transformation for continuous signals is bilateral Laplace transformation which will be discussed in Chap. 9. The topics of Fourier series and Fourier transformation, discussed in this chapter and Chap. 7, are special cases of bilateral Laplace transformation. Although it was possible to do so, we did not discuss the corresponding special cases in Chap. 5 because these special cases are not particularly useful for discrete signals.

As was promised in Chap. 5, we shall discuss a transformation theory for continuous signals from a slightly different point of view, that of decomposing a signal into components. We suggested in Chap. 4 the idea of expressing a signal $x(t)$ as a weighted sum of a chosen set of signals:

$$x(t) = \cdots + a_{-1}\xi_{-1}(t) + a_0\,\xi_0(t) + a_1\xi_1(t) + a_2\,\xi_2(t) + \cdots + a_k\,\xi_k(t) + \cdots$$

We also recall the intention of expressing a signal in this way: If the responses of a linear system to the signals $\ldots, \xi_{-1}(t), \xi_0(t), \xi_1(t), \xi_2(t), \ldots, \xi_k(t), \ldots$ are known, the response of the system to any signal $x(t)$ that is expressible as a weighted sum of $\ldots, \xi_{-1}(t), \xi_0(t), \xi_2(t), \ldots, \xi_k(t), \ldots$ can be determined. In Chap. 4, $\ldots, \xi_{-1}(t), \xi_1(t), \xi_2(t), \ldots, \xi_k(t), \ldots$ were chosen to be advanced and delayed versions of $\xi_0(t)$. As we have seen in Chap. 4, such a choice is a good one for linear time-invariant systems, because the input-output relationship of a system can then be described completely by the response of the system to the signal $\xi_0(t)$. Another interesting possibility is to choose $\ldots, \xi_{-1}(t), \xi_0(t), \xi_1(t), \xi_2(t), \ldots, \xi_k(t), \ldots$ such that for each $\xi_i(t)$ the response of a linear system to $\xi_i(t)$ is $\lambda_i \xi_i(t)$. That is, the response is equal to the stimulus $\xi_i(t)$ scaled by a finite multiplying constant λ_i. Such a signal $\xi_i(t)$ is said to be an *eigenfunction* of the linear system and λ_i is said to be the corresponding *eigenvalue*. If such a choice is indeed possible, then for the input signal

$$x(t) = \cdots + a_{-1}\xi_{-1}(t) + a_0\xi_0(t) + a_1\xi_1(t) + a_2\xi_2(t) + \cdots + a_k\xi_k(t) + \cdots$$

the corresponding output signal will be

$$y(t) = \cdots + \lambda_{-1}a_{-1}\xi_{-1}(t) + \lambda_0 a_0\xi_0(t) + \lambda_1 a_1\xi_1(t) + \lambda_2 a_2\xi_2(t) + \cdots + \lambda_k a_k\xi_k(t) + \cdots$$

Consequently, the input-output relationship of a system can be specified by its eigenvalues $\ldots, \lambda_{-1}, \lambda_0, \lambda_1, \lambda_2, \ldots, \lambda_k, \ldots$.

As it turns out, for most linear time-invariant systems of practical interest, $\xi_i(t)$ can be chosen to be a complex exponential function e^{st} for some complex number s. Let $y(t)$ denote the response of a linear time-invariant system to the input e^{st}. Since the system is time-invariant, its response to $e^{s(t+\tau)}$ is $y(t+\tau)$ for any constant τ. On the other hand, since $e^{s(t+\tau)}$ can be written $e^{s\tau}e^{st}$ and since the system is linear, its response to $e^{s(t+\tau)}$ is equal to $e^{s\tau}y(t)$. We thus have

$$y(t+\tau) = e^{s\tau}y(t)$$

Setting $t = 0$, we obtain

$$y(\tau) = y(0)e^{s\tau}$$

for all τ. That is, the output $y(t)$ of the system is equal to the input e^{st} scaled by a constant $y(0)$. Thus the signal e^{st} is an eigenfunction of the linear time-invariant system if the constant $y(0)$ is finite. And if so, $y(0)$ is the corresponding eigenvalue. We now determine the condition under which $y(0)$ is finite. Let $h(t)$ denote the impulse response of the system under consideration. Since

$$y(t) = e^{st} * h(t) = \int_{-\infty}^{\infty} e^{s(t-\tau)}h(\tau)\,d\tau$$

we have

$$y(0) = \int_{-\infty}^{\infty} e^{-s\tau}h(\tau)\,d\tau \tag{6-1}$$

Therefore, the signal e^{st} is an eigenfunction of the system if the value of the integral $\int_{-\infty}^{\infty} e^{-s\tau} h(\tau)\, d\tau$ is finite.

For a given system we can determine those s such that the functions e^{st} are eigenfunctions of the system. Let us write $s = \sigma + j\omega$. Since

$$\begin{aligned} |y(0)| &= \left| \int_{-\infty}^{\infty} e^{-s\tau} h(\tau)\, d\tau \right| \\ &\le \int_{-\infty}^{\infty} |e^{-s\tau}|\, |h(\tau)|\, d\tau \\ &= \int_{-\infty}^{\infty} |e^{-(\sigma + j\omega)\tau}|\, |h(\tau)|\, d\tau \\ &= \int_{-\infty}^{\infty} e^{-\sigma\tau} |h(\tau)|\, d\tau \end{aligned}$$

we conclude that if

$$\int_{-\infty}^{\infty} e^{-\sigma\tau} |h(\tau)|\, d\tau < \infty \tag{6-2}$$

then e^{st} is an eigenfunction of the system under consideration. [Clearly, the condition in (6-2) is sufficient, but not necessary, to assure the finiteness of $|y(0)|$.]†

For a given linear time-invariant system, the region in the s plane (the complex plane) consisting of all s such that the condition in (6-2) is satisfied is called the *region of absolute convergence* of the system. We note that if a point $s_0 = \sigma_0 + j\omega_0$ is in the region of absolute convergence, the points $\sigma_0 + j\omega$ for all ω are also in the region of absolute convergence. Thus, the region of absolute convergence consists of points along vertical lines in the s plane. We can further show that the region of absolute convergence must always be a continuous vertical strip in the s plane. As examples, the shaded areas in Fig. 6-1*a*, *b*, and *c* are possible regions of absolute convergence and the shaded area in Fig. 6-1*d* is an impossibility. To show this, we want to show first that for $\sigma_1 < \sigma_2 < \sigma_3$ and for any value of ω, if the point $s_1 = \sigma_1 + j\omega$ is in the region of absolute convergence but the point $s_2 = \sigma_2 + j\omega$ is not, then the point $s_3 = \sigma_3 + j\omega$ cannot be in the region of absolute convergence. That is, we want to show if

$$\int_{-\infty}^{\infty} e^{-\sigma_1 t} |h(t)|\, dt < \infty$$

and

$$\int_{-\infty}^{\infty} e^{-\sigma_2 t} |h(t)|\, dt = \infty$$

then

$$\int_{-\infty}^{\infty} e^{-\sigma_3 t} |h(t)|\, dt = \infty$$

† An entirely parallel derivation can be used to show that $\xi(n) = z^{-n}$ for some complex numbers z are eigenfunctions of linear time-invariant discrete systems. We leave the derivation to the reader.

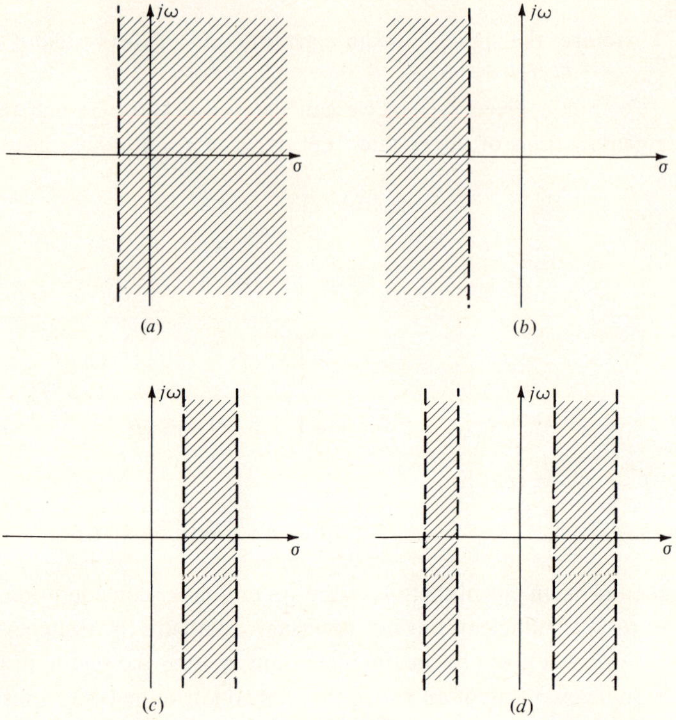

FIGURE 6-1

We write

$$\int_{-\infty}^{\infty} e^{-\sigma_1 t}|h(t)|\ dt = \int_{-\infty}^{0} e^{-\sigma_1 t}|h(t)|\ dt + \int_{0}^{\infty} e^{-\sigma_1 t}|h(t)|\ dt$$

and

$$\int_{-\infty}^{\infty} e^{-\sigma_2 t}|h(t)|\ dt = \int_{\infty}^{0} e^{-\sigma_2 t}|h(t)|\ dt + \int_{0}^{\infty} e^{-\sigma_2 t}|h(t)|\ dt$$

Since the values of

$$\int_{-\infty}^{0} e^{-\sigma_1 t}|h(t)|\ dt$$

and

$$\int_{0}^{\infty} e^{-\sigma_1 t}|h(t)|\ dt$$

are nonnegative and their sum is finite, the value of each of these integrals must be finite. That $\sigma_1 < \sigma_2$ implies

$$\int_{0}^{\infty} e^{-\sigma_2 t}|h(t)|\ dt \leq \int_{0}^{\infty} e^{-\sigma_1 t}|h(t)|\ dt$$

Since

$$\int_{-\infty}^{\infty} e^{-\sigma_2 t}|h(t)|\, dt = \infty$$

it follows that

$$\int_{-\infty}^{0} e^{-\sigma_2 t}|h(t)|\, dt = \infty$$

Since $\sigma_2 < \sigma_3$ implies

$$\int_{-\infty}^{0} e^{-\sigma_3 t}|h(t)|\, dt \geq \int_{-\infty}^{0} e^{-\sigma_2 t}|h(t)|\, dt$$

and since the value of the integral

$$\int_{0}^{\infty} e^{-\sigma_3 t}|h(t)|\, dt$$

is nonnegative, we conclude that

$$\int_{-\infty}^{\infty} e^{-\sigma_3 t}|h(t)|\, dt = \infty$$

In a similar manner, we can show that for $\sigma_1 > \sigma_2 > \sigma_3$, if

$$\int_{-\infty}^{\infty} e^{-\sigma_1 t}|h(t)|\, dt < \infty$$

and

$$\int_{-\infty}^{\infty} e^{-\sigma_2 t}|h(t)|\, dt = \infty$$

then

$$\int_{-\infty}^{\infty} e^{-\sigma_3 t}|h(t)|\, dt = \infty$$

Many properties of a linear time-invariant system are revealed by its region of absolute convergence. We note that a linear time-invariant system is stable if and only if its region of absolute convergence includes the $j\omega$ axis. Such a conclusion comes from an earlier observation in Sec. 4-6 that a linear time-invariant system is stable if and only if

$$\int_{-\infty}^{\infty} |h(t)|\, dt < \infty$$

which, in turn, implies the inclusion of the $j\omega$ axis in the region of absolute convergence.

The region of absolute convergence of a causal system is always a half plane to the right of some vertical line in the s plane. This conclusion comes from the following observation: If $h(t) = 0$ for $t < 0$, then

$$\int_{-\infty}^{\infty} e^{-\sigma_2 t}|h(t)|\,dt = \int_{0}^{\infty} e^{-\sigma_2 t}|h(t)|\,dt$$

and
$$\int_{-\infty}^{\infty} e^{-\sigma_1 t}|h(t)|\,dt = \int_{0}^{\infty} e^{-\sigma_1 t}|h(t)|\,dt$$

Consequently, for $\sigma_1 > \sigma_2$,

$$\int_{-\infty}^{\infty} e^{-\sigma_2 t}|h(t)|\,dt < \infty$$

implies that

$$\int_{-\infty}^{\infty} e^{-\sigma_1 t}|h(t)|\,dt < \infty$$

It can be shown in a similar way that for some finite t_0, if $h(t) = 0$ for $t < t_0$, the region of absolute convergence is a half plane to the right of a vertical line; and if $h(t) = 0$ for $t > t_0$, the region of absolute convergence is a half plane to the left of a vertical line. This is illustrated by the examples in Fig. 6-2*a* and *b*, where the shaded areas are the regions of absolute convergence.

Let us clarify a point before we conclude our discussion in this section. According to our discussion in Chap. 2, a system whose input-output relationship can be described by a linear differential equation with constant coefficients is a linear time-invariant system if the system is initially at rest. Furthermore, if the input to the system is e^{st}, the particular solution of the output is of the form Ke^{st}. But then one recalls that the output also contains a homogeneous solution, and one might wonder why the homogeneous solution was not included in our discussion in the preceding paragraphs. Let us use a simple example to illustrate how this discrepancy can be explained. Consider a linear time-invariant system the input-output relationship of which is described by the differential equation

$$\frac{d^2y(t)}{dt^2} + 4\frac{dy(t)}{dt} + 3y(t) = 3x(t) \tag{6-3}$$

The impulse response of the system is

$$h(t) = \tfrac{3}{2}(e^{-t} - e^{-3t})u_{-1}(t)$$

Since the integral $\int_{-\infty}^{\infty} e^{-\sigma\tau}|h(\tau)|\,d\tau$ is finite for $\sigma > -1$, the region of absolute convergence of the system is the half plane to the right of the vertical line $s = -1$, as shown

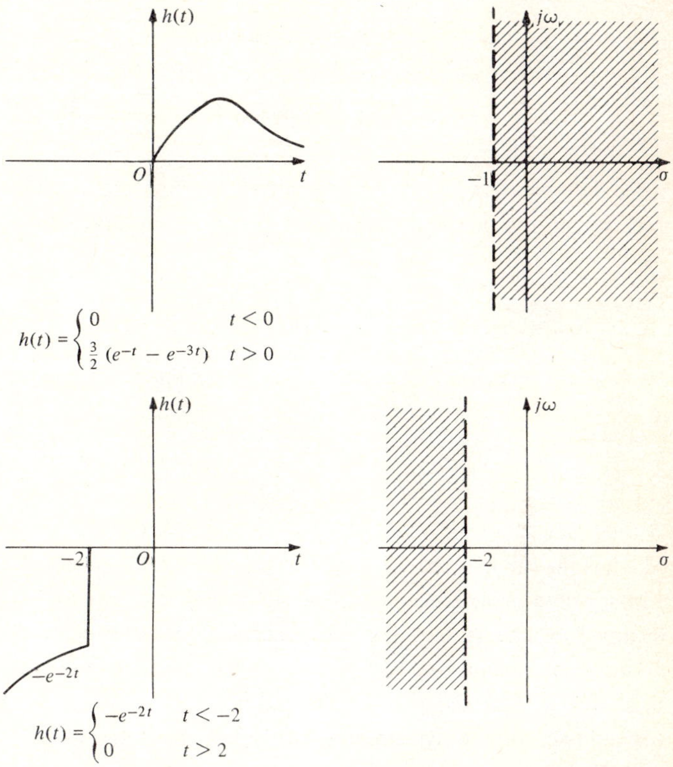

FIGURE 6-2

in Fig. 6-2*a*. Therefore, according to our discussion in preceding paragraphs, the response of the system to the stimulus e^{st} is

$$\left[\int_{-\infty}^{\infty} e^{-s\tau} h(\tau)\, d\tau\right] e^{st} = \frac{3}{s^2 + 4s + 3} e^{st} \tag{6-4}$$

provided that the real part of the complex number s is larger than -1.

Let us also determine the response of the system by solving Eq. (6-3) with $x(t) = e^{st}$ for $-\infty \leq t \leq \infty$. We assume first that the input is applied to the system at $t = -T$, that is, $x(t) = e^{st}u_{-1}(t + T)$, and shall let $T \to \infty$ to obtain the desired solution later on. The particular solution of the differential equation is

$$\frac{3e^{st}}{s^2 + 4s + 3}$$

The total solution of the differential equation is

$$y(t) = \left(Ae^{-t} + Be^{-3t} + \frac{3e^{st}}{s^2 + 4s + 3}\right)u_{-1}(t + T)$$

Since the system is assumed to be initially at rest, the constants A and B can be determined from the boundary condition $y(-T-) = 0$ and $y'(-T-) = 0$. We obtain

$$y(t) = \frac{3}{2}\left[\left(e^{-T} - \frac{e^{-(s+1)T}}{s+1}\right)e^{-t} - \left(e^{-3T} - \frac{e^{-(s+3)T}}{s+3}\right)e^{-3t} + \frac{2e^{st}}{s^2 + 4s + 3}\right]u_{-1}(t + T)$$

As $T \to \infty$, because the real part of s is larger than -1, $y(t)$ remains finite and is equal to

$$\frac{3e^{st}}{s^2 + 4s + 3}$$

which is the same as that obtained in Eq. (6-4).† From this example, we note that when a linear time-invariant system is excited by e^{st} starting at $t = t_0$ for some finite t_0, the response of the system will, in general, contain a particular solution as well as a homogeneous solution. On the other hand, when a linear time-invariant system is excited by e^{st} starting at $t = -\infty$, and if the value of s is within the region of absolute convergence of the system, the homogeneous solution in the response decays very rapidly and is equal to 0 for $t > -\infty$.

6-2 RESOLUTION OF SIGNALS

We turn now to the problem of decomposing a signal into a weighted sum or integral of complex exponential functions. Of course, for a given linear time-invariant system, such a decomposition is useful only when the complex exponential functions are all eigenfunctions of the system. Our approach is not to carry out a decomposition for each particular system but rather to study several possibilities of choosing a set of complex exponential functions e^{st} into weighted sums or integrals of which signals can be decomposed.

Suppose we choose a set of complex exponential functions e^{st} such that the s are equidistant discrete points on the $j\omega$ axis as illustrated in Fig. 6-3*a*. That is, the complex exponential functions are $\ldots, e^{-j\omega_0 t}, 1, e^{j\omega_0 t}, \ldots, e^{jk\omega_0 t}, \ldots,$ where ω_0 is a

† If the real part of s is less than or equal to -1, we have $y(t) \to \infty$ as $T \to \infty$.

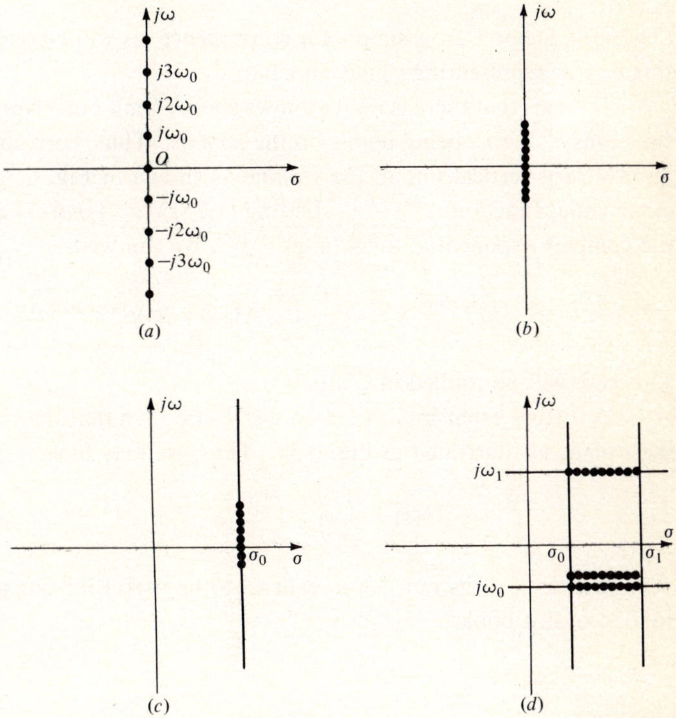

FIGURE 6-3

constant we have chosen. If a signal $x(t)$ can be decomposed as a weighted sum of these complex exponential functions, we can write

$$
\begin{aligned}
x(t) &= \cdots + x_{-2}e^{-j2\omega_0 t} + x_{-1}e^{-j\omega_0 t} + x_0 + x_1 e^{j\omega_0 t} + x_2 e^{j2\omega_0 t} + \cdots + x_k e^{jk\omega_0 t} + \cdots \\
&= \sum_{n=-\infty}^{\infty} x_n e^{jn\omega_0 t}
\end{aligned}
$$

where x_n is, in general, a complex number and can be viewed as the "amplitude" of the complex exponential function $e^{jn\omega_0 t}$.

An immediate extension is to choose a set of complex exponential functions e^{st} such that s are all the points along the $j\omega$ axis, as illustrated in Fig. 6-3b. A signal $x(t)$ can then be expressed as an integral of complex exponential functions of the form $e^{j\omega t}$. If we let $(1/2\pi)X(\omega)\,d\omega$ denote the "amplitude" of the complex exponential function $e^{j\omega t}$, we can write

$$x(t) = \frac{1}{2\pi}\int_{-\infty}^{\infty} X(\omega)e^{j\omega t}\,d\omega$$

(The scale factor $1/2\pi$ is simply for convenience, as will be seen later.) We shall study this way of representing signals in Chap. 7.

It seems that there is no reason we should limit ourselves to complex exponential functions e^{st} with s being points on the $j\omega$ axis. Thus, corresponding to the points on an arbitrary vertical line in the s plane as shown in Fig. 6-3c, we have the complex exponential functions $e^{(\sigma_0+j\omega)t}$. Letting $(1/2\pi)X(\sigma_0+j\omega)\,d\omega$ denote the amplitude of the complex exponential function $e^{(\sigma_0+j\omega)t}$, we can write

$$x(t) = \frac{1}{2\pi}\int_{-\infty}^{\infty} X(\sigma_0 + j\omega)e^{(\sigma_0+j\omega)t}\,d\omega$$

This case will be studied in Chap. 9.

A further extension is to choose e^{st} to be such that the s are points in a region in the s plane as illustrated in Fig. 6-3d. Thus, we may have

$$x(t) = \frac{1}{2\pi}\int_{\omega_0}^{\omega_1}\int_{\sigma_0}^{\sigma_1} X(\sigma + j\omega)e^{(\sigma+j\omega)t}\,d\sigma\,d\omega$$

This case, as it turns out, is too general to be useful for our purpose and will not be studied in this book.

6-3 THE FOURIER SERIES

We investigate now the possibility of decomposing a signal $x(t)$ into a sum of complex exponential functions e^{st}, where the s are equidistant discrete points on the $j\omega$ axis.†
That is,

$$x(t) = \sum_{n=-\infty}^{\infty} x_n e^{jn\omega_0 t} \tag{6-5}$$

We note first that not all continuous signals can be expressed as such a sum. Since, according to Eq. (6-5),

$$\begin{aligned} x\left(t + \frac{2\pi}{\omega_0}\right) &= \sum_{n=-\infty}^{\infty} x_n e^{jn\omega_0(t+2\pi/\omega_0)} \\ &= \sum_{n=-\infty}^{\infty} x_n e^{jn\omega_0 t} e^{j2\pi n} \\ &= \sum_{n=-\infty}^{\infty} x_n e^{jn\omega_0 t} \\ &= x(t) \end{aligned}$$

† Although most of the signals we shall encounter are real signals, our results are valid for the more general case of complex signals.

$x(t)$ must necessarily be a periodic signal of period $2\pi/\omega_0$, or $2\pi/m\omega_0$ for some positive integer m. [A continuous signal $x(t)$ is said to be periodic if $x(t + T) = x(t)$ for some constant T. Moreover, the smallest constant T such that $x(t + T) = x(t)$ is said to be the period of $x(t)$.] Without loss of generality, we shall assume that the period of $x(t)$ is $2\pi/\omega_0$, which will also be denoted T. As a matter of fact, the decomposition in Eq. (6-5) is possible if $x(t)$ is a periodic signal satisfying the following conditions:

1 $\int_{-T/2}^{T/2} |x(t)|\, dt < \infty$

2 $x(t)$ has a finite number of maxima and minima within a period.

3 $x(t)$ has a finite number of discontinuities within a period.

These conditions are known as the *Dirichlet conditions*.† Let us note that since the function $e^{jn\omega_0 t}$ is continuous for every t, the sum $\sum_{n=-\infty}^{\infty} x_n e^{jn\omega_0 t}$ is also continuous for every t. Thus one might wonder what value the sum $\sum_{n=\infty}^{\infty} x_n e^{jn\omega_0 t}$ will assume at $t = t_0$ if the signal $x(t)$ has a discontinuity at $t = t_0$ [because $\sum_{n=-\infty}^{\infty} x_n e^{jn\omega_0 t_0}$ assumes a unique value at $t = t_0$, yet the value of $x(t)$ can be considered $x(t_0-)$ or $x(t_0+)$ or any value in between]. As it turns out, it can be shown that, for a signal $x(t)$ satisfying the Dirichlet conditions,

$$\sum_{n=-\infty}^{\infty} x_n e^{jn\omega_0 t} = \tfrac{1}{2}[x(t-) + x(t+)]$$

for all t. That is, if $x(t)$ is continuous at t_0

$$\sum_{n=-\infty}^{\infty} x_n e^{jn\omega_0 t_0} = x(t_0)$$

and if $x(t)$ has a discontinuity at t_0

$$\sum_{n=-\infty}^{\infty} x_n e^{jn\omega_0 t_0} = \tfrac{1}{2}[x(t_0-) + x(t_0+)]$$

For a periodic signal satisfying the Dirichlet conditions, the sum in Eq. (6-5) is known as the *Fourier series representation* of the periodic signal. The constants x_n, known as the *Fourier coefficients* of the periodic signal, can be determined as follows: We multiply both sides of Eq. (6-5) by $e^{-jk\omega_0 t}$

$$x(t)e^{-jk\omega_0 t} = \sum_{n=-\infty}^{\infty} x_n e^{j(n-k)\omega_0 t}$$

† To prove that the Dirichlet conditions guarantee the possibility of decomposing a periodic signal as in Eq. (6-5) is beyond the scope of our discussion. Let us only point out that these conditions are essential in showing the convergence of the infinite series in the right-hand side of Eq. (6-5).

and integrate from $t = -T/2$ to $t = T/2$:

$$\int_{-T/2}^{T/2} x(t)e^{-jk\omega_0 t}\,dt = \int_{-T/2}^{T/2} \sum_{n=-\infty}^{n} x_n e^{j(n-k)\omega_0 t}\,dt \tag{6-6}$$

Since

$$\int_{-T/2}^{T/2} e^{j(n-k)\omega_0 t}\,dt = \begin{cases} T & n = k \\ 0 & n \neq k \end{cases}$$

the right-hand side of Eq. (6-6) can be written

$$\int_{-T/2}^{T/2} \sum_{n=-\infty}^{\infty} x_n e^{j(n-k)\omega_0 t}\,dt = \sum_{n=-\infty}^{\infty} \int_{-T/2}^{T/2} x_n e^{j(n-k)\omega_0 t}$$
$$= x_k T$$

Thus, Eq. (6-6) yields

$$\int_{-T/2}^{T/2} x(t)e^{-jk\omega_0 t}\,dt = x_k T$$

or

$$x_k = \frac{1}{T}\int_{-T/2}^{T/2} x(t)e^{-jk\omega_0 t}\,dt \tag{6-7}$$

Changing the subscript from k to n in Eq. (6-7), we obtain a formula for computing the Fourier coefficients:

$$x_n = \frac{1}{T}\int_{-T/2}^{T/2} x(t)e^{-jn\omega_0 t}\,dt = \frac{1}{T}\int_{-T/2}^{T/2} x(t)e^{-j2\pi nt/T}\,dt \qquad n = 0, \pm 1, \pm 2, \ldots \tag{6-8}$$

The formula for the Fourier coefficients in Eq. (6-8) can be put into a slightly more general form, namely,

$$x_n = \frac{1}{T}\int_{-T/2+\Delta}^{T/2+\Delta} x(t)e^{-jn\omega_0 t}\,dt$$
$$= \frac{1}{T}\int_{-T/2+\Delta}^{T/2+\Delta} x(t)e^{-j2\pi nt/T}\,dt \qquad n = 0, \pm 1, \pm 2, \ldots \tag{6-9}$$

for any constant Δ. We leave it to the reader to verify this result.†

† From a mathematical point of view, it is interesting to observe the possibility of specifying a continuous periodic signal by a discrete set of complex numbers, its Fourier coefficients. One can indeed view the Fourier coefficients as a discrete signal. In this way, we establish a correspondence between a continuous signal and a discrete signal. A reader who has been exposed to elementary set theory will recall that a continuous complex function (of a certain period) is specified by an uncountably infinite number ($2^{\aleph_0}$) of complex numbers, yet the set of Fourier coefficients contains only a countable infinite number ($\aleph_0$) of complex numbers. We thus conclude from this observation that not every periodic signal has a Fourier series representation. Indeed, not every signal satisfies the Dirichlet conditions.

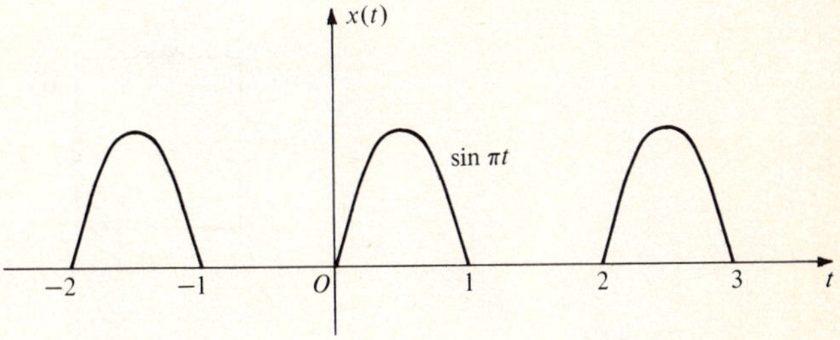

FIGURE 6-4

As an example, consider the periodic signal $x(t)$ shown in Fig. 6-4. The period of $x(t)$ is 2. Within the interval $0 \leq t \leq 2$

$$x(t) = \begin{cases} \sin \pi t & 0 \leq t \leq 1 \\ 0 & 1 \leq t \leq 2 \end{cases}$$

The Fourier coefficients of $x(t)$ can be computed as

$$\begin{aligned} x_n &= \frac{1}{2}\int_0^1 e^{-j\pi n t} \sin \pi t \, dt \\ &= \frac{1}{4j}\int_0^1 (e^{-j\pi(n-1)t} - e^{-j\pi(n+1)t})\, dt \\ &= \frac{1}{4j}\left[-\frac{e^{-j\pi(n-1)t}}{j\pi(n-1)} + \frac{e^{-j\pi(n+1)t}}{j\pi(n+1)} \right]\Bigg|_0^1 \\ &= \begin{cases} \dfrac{1}{\pi(1-n^2)} & n \text{ is even} \\ \mp\dfrac{j}{4} & n = \pm 1 \\ 0 & n \neq \pm 1, n \text{ is odd} \end{cases} \end{aligned} \tag{6-10}$$

Thus, we have

$$x(t) = \frac{1}{\pi} - \frac{j}{4} e^{j\pi t} + \frac{j}{4} e^{-j\pi t} + \sum_{n=\pm 2, \pm 4, \ldots}^{\infty} \frac{1}{\pi(1-n^2)} e^{jn\pi t}$$

As another example, let us determine the current $i(t)$ in the electric circuit in Fig. 6-5a when the voltage source $v(t)$ is a periodic signal as shown in Fig. 6-5b. From the differential equation

$$\frac{di(t)}{dt} + i(t) = v(t)$$

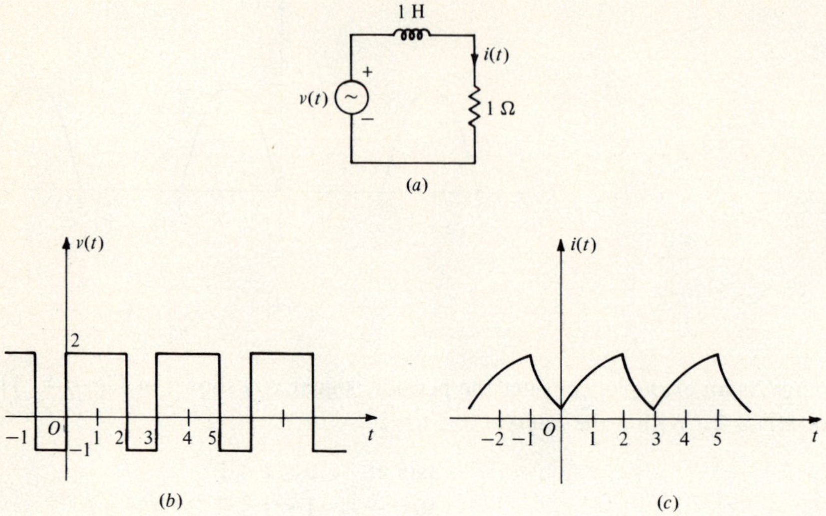

FIGURE 6-5

we determine the impulse response of the system to be

$$h(t) = e^{-t}u_{-1}(t)$$

Since, according to Eq. (6-1),

$$y(0) = \int_{-\infty}^{\infty} e^{-j\omega\tau}h(\tau)\,d\tau = \int_{0}^{\infty} e^{-j\omega\tau}e^{-\tau}\,d\tau$$

$$= \frac{1}{1 + j\omega}$$

we conclude that $e^{j\omega t}$ is an eigenfunction of the system for any ω and $1/(1 + j\omega)$ is the corresponding eigenvalue. According to Eq. (6-9), the Fourier coefficients of $v(t)$ can be computed as

$$v_n = \tfrac{1}{3}\int_{0}^{2} 2e^{-j2n\pi t/3}\,dt - \tfrac{1}{3}\int_{2}^{3} e^{-j2n\pi t/3}\,dt$$

$$= \begin{cases} 1 & n = 0 \\ \dfrac{j3}{2\pi n}\left(e^{-j4n\pi/3} - 1\right) & n \neq 0 \end{cases}$$

Consequently, the Fourier coefficients of $i(t)$ are

$$i_n = v_n \frac{1}{1 + j2n\pi/3}$$

and we have

$$i(t) = 1 + \sum_{n=\pm 1, \pm 2, \ldots}^{\infty} \frac{j3(e^{-j4n\pi/3} - 1)}{2\pi n(1 + j2n\pi/3)} e^{j2n\pi t/3}$$

The signal $i(t)$ is plotted in Fig. 6-5c.

When the periodic signal $x(t)$ is a real time function, an alternative form for the Fourier series representation is quite useful. Since

$$x_n = \frac{1}{T} \int_{-T/2}^{T/2} x(t) e^{-jn\omega_0 t}\, dt$$

we have

$$\begin{aligned} x_n^* &= \frac{1}{T} \int_{-T/2}^{T/2} x^*(t) e^{jn\omega_0 t}\, dt \\ &= \frac{1}{T} \int_{-T/2}^{T/2} x(t) e^{jn\omega_0 t}\, dt \\ &= x_{-n} \end{aligned}$$

Therefore,

$$\begin{aligned} x(t) &= x_0 + \sum_{n=1}^{\infty} (x_n e^{jn\omega_0 t} + x_{-n} e^{-jn\omega_0 t}) \\ &= x_0 + \sum_{n=1}^{\infty} (x_n e^{jn\omega_0 t} + x_n^* e^{-jn\omega_0 t}) \end{aligned} \tag{6-11}$$

Letting $x_n = a_n + jb_n$ and $x_n^* = a_n - jb_n$, we can write Eq. (6-11)

$$\begin{aligned} x(t) &= x_0 + \sum_{n=1}^{\infty} [(a_n + jb_n) e^{jn\omega_0 t} + (a_n - jb_n) e^{-jn\omega_0 t}] \\ &= x_0 + \sum_{n=1}^{\infty} 2a_n \cos n\omega_0 t - \sum_{n=1}^{\infty} 2b_n \sin n\omega_0 t \end{aligned} \tag{6-12}$$

That is, for a real time signal $x(t)$, we can write†

$$x(t) = A_0 + \sum_{n=1}^{\infty} A_n \cos n\omega_0 t + \sum_{n=1}^{\infty} B_n \sin n\omega_0 t \tag{6-13}$$

† See Prob. 6-1 where a condition for the sinusoids and cosinusoids to be eigenfunctions of a linear time-invariant system is discussed.

where $$A_0 = x_0 = \frac{1}{T}\int_{-T/2}^{T/2} x(t)\,dt$$

$$A_n = 2a_n = 2\,\mathrm{Re}\,(x_n) = 2\,\mathrm{Re}\left[\frac{1}{T}\int_{-T/2}^{T/2} x(t)e^{-jn\omega_0 t}\,dt\right]$$

$$= \frac{2}{T}\int_{-T/2}^{T/2} x(t)\cos n\omega_0 t\,dt \qquad n = 1, 2, \ldots$$

$$B_n = -2b_n = -2\,\mathrm{Im}\,(x_n) = -2\,\mathrm{Im}\left[\frac{1}{T}\int_{-T/2}^{T/2} x(t)e^{-jn\omega_0 t}\,dt\right]$$

$$= \frac{2}{T}\int_{-T/2}^{T/2} x(t)\sin n\omega_0 t\,dt \qquad n = 1, 2, \ldots$$

For example, the Fourier series of the periodic signal in Fig. 6-4 can be written alternatively:

$$x(t) = \frac{1}{\pi} + \frac{1}{2}\sin \pi t + \sum_{n=2,4,6,\ldots}^{\infty} \frac{2}{\pi(1-n^2)}\cos n\pi t$$

The terms of Eq. (6-13) are called the frequency components of $x(t)$. In particular, the term A_0 is said to be the *average value*, or the *dc component*, of $x(t)$; the terms $A_1 \cos \omega_0 t$ and $B_1 \sin \omega_0 t$ are said to be the *fundamental frequency components* of $x(t)$; and, for $n \geq 2$, the terms $A_n \cos n\omega t$ and $B_n \sin n\omega t$ are said to be the nth *harmonics* of $x(t)$. Extending such terminologies to the case of complex time signals, we also call the terms in Eq. (6-5) the frequency components of $x(t)$. In particular, we refer to the term $x_n e^{jn\omega_0 t}$ as the nth frequency component, and the term $x_{-n}e^{-jn\omega_0 t}$ as the $-n$th frequency component of $x(t)$. The *fundamental frequency* of $x(t)$ is defined to be $\omega_0/2\pi$ cycles per second.† As will be seen in Chap. 8, that a signal contains different frequency components is the most fundamental concept in the design and construction of many communication systems. Qualitatively, the fast-varying high-frequency components in the Fourier series are due to the fast variations in the signal $x(t)$. Therefore, the high-frequency components in a signal containing sharp discontinuities will be more prominent than those in a smooth, slow-varying signal. This point will be seen in a more quantitative manner in Secs. 6-5 and 6-6.

6-4 SYMMETRY PROPERTIES OF TIME SIGNALS

So that we can have a better understanding of the relationship between a periodic signal and its corresponding Fourier series representation, we shall investigate how some of the symmetry properties of a periodic signal are reflected in its Fourier series representation.

† We remind the reader that ω_0 is the angular velocity.

Recall that a function $x(t)$ is said to be even if $x(t) = x(-t)$ and to be odd if $x(t) = -x(-t)$. Clearly, a complex time function† is even if and only if both its real part and its imaginary part are even, and it is odd if and only if both its real part and imaginary part are odd. Thus, for the sake of simplicity, we shall limit our discussion to real time functions.

Since $\cos n\omega_0 t$ is an even time function, a sum of terms of the form $A_n \cos n\omega_0 t$ is an even time function. Similarly, since $\sin n\omega_0 t$ is an odd time function, a sum of terms of the form $B_n \sin n\omega_0 t$ is an odd time function. Conversely, we show now that the Fourier series of an even time signal contains only cosinusoidal terms and the Fourier series of an odd time signal contains only sinusoidal terms. We recall that

$$A_n = \frac{2}{T}\int_{-T/2}^{T/2} x(t) \cos n\omega_0 t \, dt$$

$$B_n = \frac{2}{T}\int_{-T/2}^{T/2} x(t) \sin n\omega_0 t \, dt$$

If $x(t)$ is an even function, $x(t) \sin n\omega_0 t$ is an odd function. Thus,

$$\begin{aligned} B_n &= \frac{2}{T}\int_{-T/2}^{0} x(t) \sin n\omega_0 t \, dt + \frac{2}{T}\int_{0}^{T/2} x(t) \sin n\omega_0 t \, dt \\ &= -\frac{2}{T}\int_{0}^{T/2} x(t) \sin n\omega_0 \, dt + \frac{2}{T}\int_{0}^{T/2} x(t) \sin n\omega_0 t \, dt \\ &= 0 \end{aligned}$$

On the other hand, if $x(t)$ is an odd function, $x(t) \cos n\omega_0 t$ is an odd function. Using the same argument, we obtain $A_n = 0$.

If $x(t)$ is neither even nor odd, the Fourier series representation of $x(t)$ will contain both sinusoidal and cosinusoidal terms. In this case, $x(t)$ can be expressed as the sum of an even function $x_e(t)$ and an odd function $x_0(t)$, whereas $x_e(t)$ corresponds to the cosinusoidal terms and $x_0(t)$ corresponds to the sinusoidal terms in the Fourier series representation of $x(t)$. It can be shown immediately that

$$x_e(t) = \tfrac{1}{2}[x(t) + x(-t)]$$

and

$$x_0(t) = \tfrac{1}{2}[x(t) - x(-t)]$$

As an example, the periodic signal $x(t)$ in Fig. 6-6*a* is equal to the sum of $x_e(t)$ and $x_0(t)$ shown in Fig. 6-6*b*. The reader can check the assertion that the Fourier series representation of $x_e(t)$ contains only cosinusoidal terms and the Fourier series

† A complex time function can be written $u(t) + jv(t)$, where $u(t)$ and $v(t)$ are real time functions. $u(t)$ is said to be the *real part* and $v(t)$ is said to be the *imaginary part* of the complex time function.

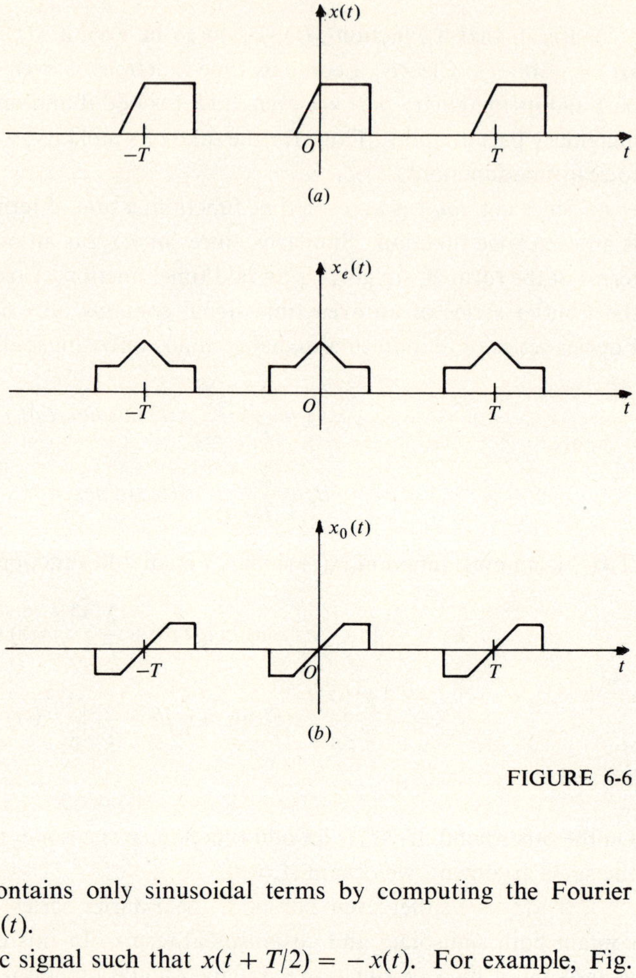

FIGURE 6-6

representation of $x_0(t)$ contains only sinusoidal terms by computing the Fourier coefficients of $x_e(t)$ and $x_0(t)$.

Let $x(t)$ be a periodic signal such that $x(t + T/2) = -x(t)$. For example, Fig. 6-7 shows one such signal. We compute the Fourier coefficients of $x(t)$:

$$x_n = \frac{1}{T}\int_{-T/2}^{T/2} x(t)e^{-jn\omega_0 t}\,dt$$

That is,

$$x_n = \frac{1}{T}\int_{-T/2}^{0} x(t)e^{-jn\omega_0 t}\,dt + \frac{1}{T}\int_{0}^{T/2} x(t)e^{-jn\omega_0 t}\,dt \tag{6-14}$$

Letting $t = \tau + T/2$ in the second integral of Eq. (6-14), we obtain

$$x_n = \frac{1}{T}\int_{-T/2}^{0} x(t)e^{-jn\omega_0 t}\,dt + \frac{1}{T}\int_{-T/2}^{0} x\left(\tau + \frac{T}{2}\right)e^{-jn\omega_0(\tau + T/2)}\,d\tau$$

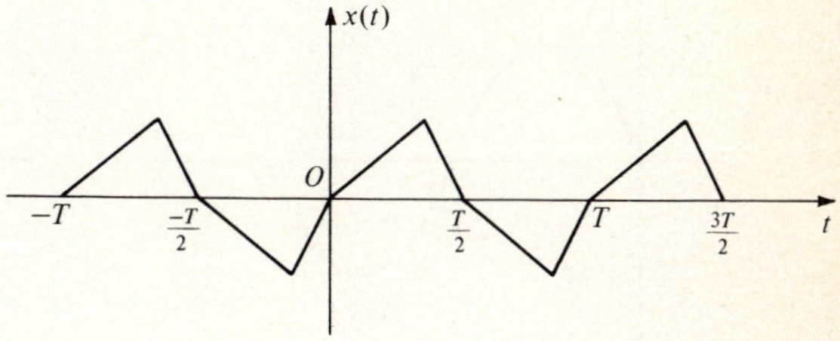

FIGURE 6-7

Since $e^{-jn\pi} = 1$ for even n, we have

$$x_n = \frac{1}{T}\int_{-T/2}^{0} x(t)e^{-jn\omega_0 t}\,dt - \frac{1}{T}e^{jn\pi}\int_{-T/2}^{0} x(\tau)e^{-jn\omega_0\tau}\,d\tau$$
$$= 0$$

when n is even. It follows that $A_n = B_n = 0$ for even n. That is, the Fourier series representation of $x(t)$ contains odd harmonics only. As a matter of fact, such a conclusion can be arrived at by observing, as illustrated in Fig. 6-8, that all odd harmonics satisfy the condition $x(t + T/2) = -x(t)$, whereas the even ones do not.

By using the same argument, it can be shown that if $x(t)$ is a periodic signal such that $x(t + T/2) = x(t)$, then the Fourier series representation of $x(t)$ will contain only even harmonics. However, it is clear that the period of $x(t)$, in this case, is actually $T/2$ instead of T. Consequently, the fundamental frequency of $x(t)$ is $2/T$ which accounts for the existence of even harmonics only.

We consider now an illustrative example: Let $x(t)$ be a periodic signal of period 4 that contains only odd cosinusoidal harmonics. We want to determine the signal $x(t)$, if it is given that, for $0 \le t \le 1$,

$$x(t) = t$$

as shown in Fig. 6-9*a*. Because $x(t)$ is an even time function, we have, for $-1 \le t \le 0$,

$$x(t) = -t$$

Thus, for $3 \le t \le 4$,

$$x(t) = -t + 4$$

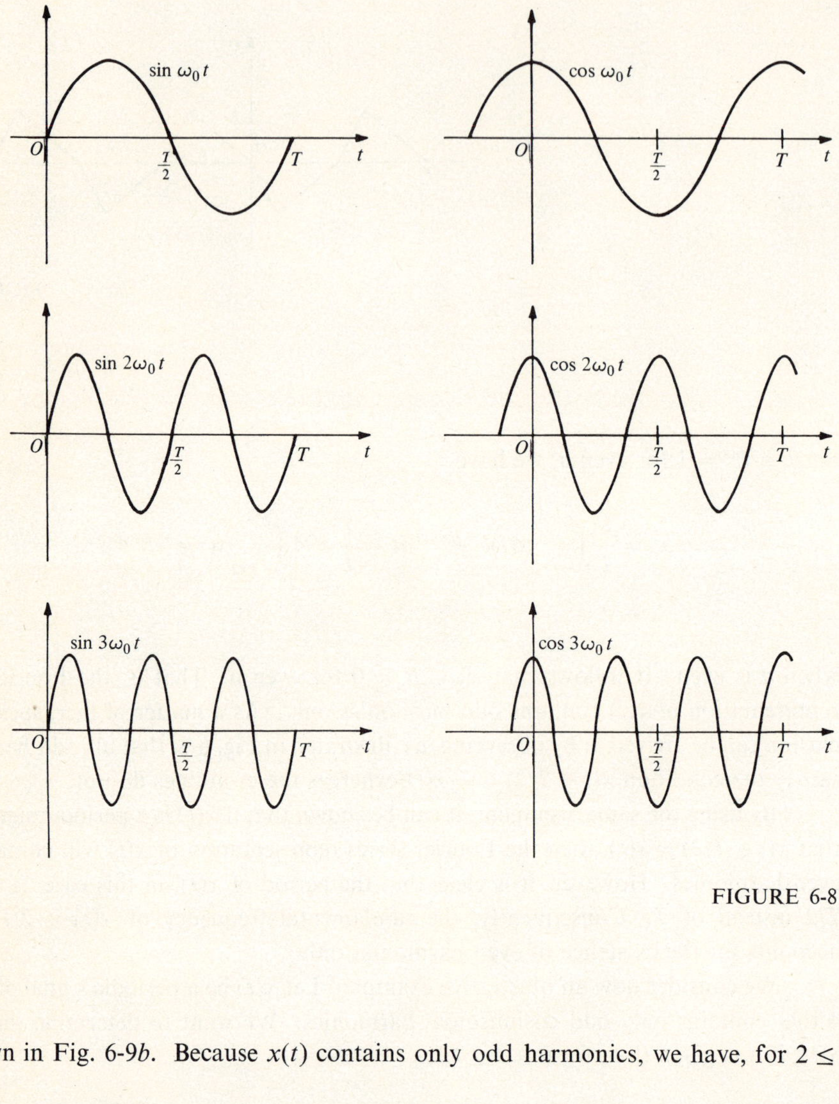

FIGURE 6-8

as shown in Fig. 6-9*b*. Because $x(t)$ contains only odd harmonics, we have, for $2 \le t \le 3$,

$$x(t) = -(t-2) = -t+2$$

and for $1 \le t \le 2$

$$x(t) = -(-t+4-2) = t-2$$

The signal $x(t)$ is shown in Fig. 6-9*c*.

As another example, we note that after decomposing the signal $x(t)$ in Fig. 6-6*a*

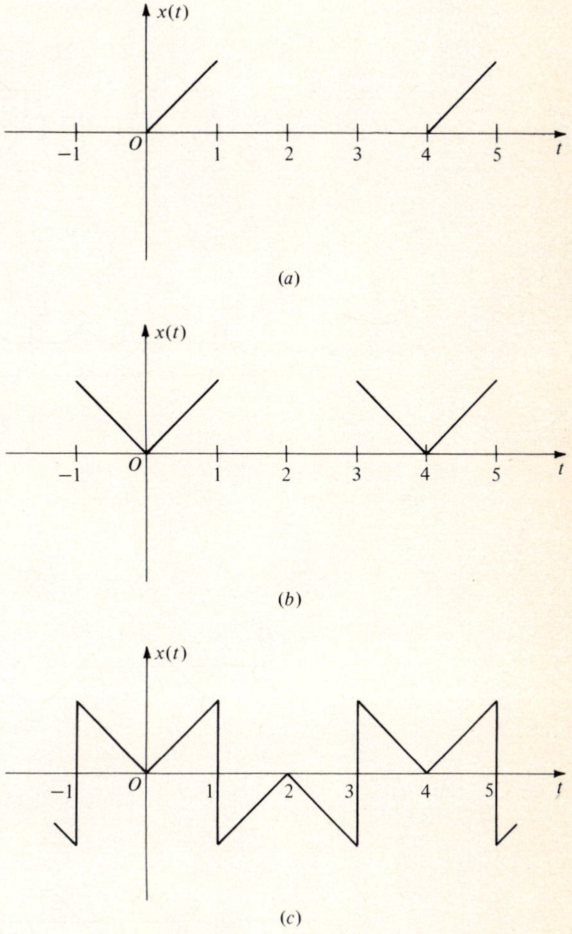

FIGURE 6-9

into two signals $x_e(t)$ and $x_0(t)$ as shown in Fig. 6-6*b*, we can further decompose $x_e(t)$ and $x_0(t)$ so that

$$x_e(t) = x_{e1}(t) + x_{e2}(t)$$
$$x_0(t) = x_{01}(t) + x_{02}(t)$$

as shown in Fig. 6-10*a* and *b*. Since

$$x_{e1}\left(t + \frac{T}{2}\right) = -x_{e1}(t)$$

$$x_{01}\left(t + \frac{T}{2}\right) = -x_{01}(t)$$

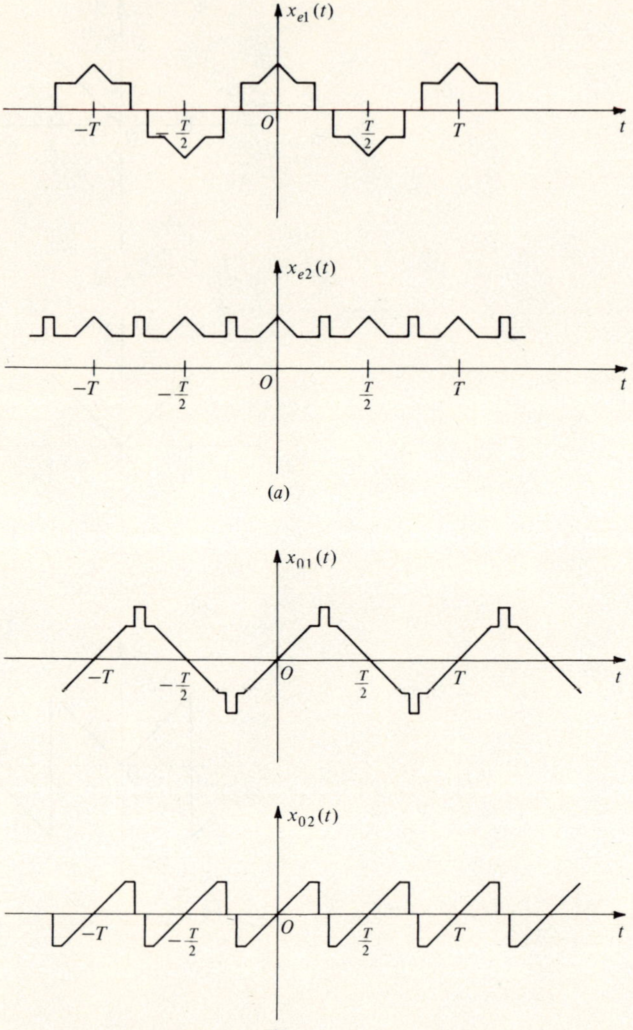

FIGURE 6-10

$x_{e1}(t)$ corresponds to the odd cosinusoidal terms and $x_{01}(t)$ corresponds to the odd sinusoidal terms in the Fourier series representation of $x(t)$. Similarly, since

$$x_{e2}\left(t + \frac{T}{2}\right) = x_{e2}(t)$$

$$x_{02}\left(t + \frac{T}{2}\right) = x_{02}(t)$$

$x_{e2}(t)$ corresponds to the even cosinusoidal terms and $x_{02}(t)$ corresponds to the even sinusoidal terms in the Fourier series representation of $x(t)$. [The reader might wish to know how we determine $x_{e1}(t)$ and $x_{e2}(t)$ from $x_e(t)$, and $x_{01}(t)$ and $x_{02}(t)$ from $x_0(t)$. See Prob. 6-9.]

6-5 MANIPULATION OF SIGNALS

When a periodic signal $x(t)$ is translated, transposed, or differentiated, the Fourier coefficients of the resultant periodic signals can be derived from the Fourier coefficients of $x(t)$. We shall show such derivations in this section to further illustrate the relationship between a periodic signal and its Fourier series representation.

Let $x(t)$ be a periodic signal of period T, and x_n be its Fourier coefficients. Clearly, $x(-t)$ is also a periodic signal of period T. The Fourier coefficients of $x(-t)$ can be computed as

$$\begin{aligned}\frac{1}{T}\int_{-T/2}^{T/2} x(-t)e^{-jn\omega_0 t}\,dt &= \frac{1}{T}\int_{-T/2}^{T/2} x(\tau)e^{jn\omega_0 \tau}\,d\tau \\ &= x_{-n}\end{aligned}$$

That is, the nth Fourier coefficient of $x(-t)$ is equal to the $-n$th Fourier coefficient of $x(t)$. This result can be seen quite easily when we write

$$\begin{aligned}e^{jn\omega_0 t} &= \cos n\omega_0 t + j\sin n\omega_0 t \\ e^{-jn\omega_0 t} &= \cos n\omega_0 t - j\sin n\omega_0 t\end{aligned}$$

Since $\cos(-n\omega_0 t) = \cos n\omega_0 t$ and $\sin(-n\omega_0 t) = -\sin n\omega_0 t$, transposing $x(t)$ will leave the cosinusoids as they were and will transpose the sinusoids. Consequently, the Fourier coefficients of the terms $e^{jn\omega_0 t}$ and $e^{-jn\omega_0 t}$, x_n and x_{-n}, will be interchanged.

The translated signal $x(t+t_0)$ is also of period T. Its Fourier coefficients can be computed as

$$\begin{aligned}\frac{1}{T}\int_{-T/2}^{T/2} x(t+t_0)e^{-jn\omega_0 t}\,dt &= \frac{1}{T}\int_{-T/2+t_0}^{T/2+t_0} x(\tau)e^{-jn\omega_0(\tau-t_0)}\,d\tau \\ &= \frac{e^{jn\omega_0 t_0}}{T}\int_{-T/2+t_0}^{T/2+t_0} x(\tau)e^{-jn\omega_0 \tau}\,d\tau \\ &= e^{jn\omega_0 t_0}x_n \qquad (6\text{-}15)\end{aligned}$$

That is, the nth Fourier coefficient of $x(t+t_0)$ is equal to the nth Fourier coefficient of $x(t)$ multiplied by $e^{jn\omega_0 t_0}$. The reader should note that if we express $x(t)$ as a sum of sinusoids and cosinusoids the magnitudes of both the term $\cos n\omega_0 t$ and the term $\sin n\omega_0 t$ will be changed when $x(t)$ is translated.

To illustrate the results we have just derived, let us consider a simple example. Let $y(t)$ be the periodic signal shown in Fig. 6-11, where within the interval $0 \leq t \leq 2$

$$y(t) = \begin{cases} 0 & 0 \leq t \leq 1 \\ -\sin \pi t & 1 \leq t \leq 2 \end{cases}$$

To determine the Fourier coefficients y_n, we make use of the results in Eq. (6-10) which are the Fourier coefficients of the signal $x(t)$ shown in Fig. 6-4. Since $y(t) = x(-t)$, we have

$$y_n = x_{-n} = \begin{cases} \dfrac{1}{\pi(1-n^2)} & n \text{ is even} \\ \pm\dfrac{j}{4} & n = \pm 1 \\ 0 & n \neq \pm 1,\ n \text{ is odd} \end{cases}$$

Alternatively, since $y(t) = x(t+1)$, we have, according to Eq. (6-15),

$$y_n = e^{jn(2\pi/2)(1)} x_n = \begin{cases} \dfrac{1}{\pi(1-n^2)} e^{jn\pi} & n \text{ is even} \\ \mp\dfrac{j}{4} e^{\pm j\pi} & n = \pm 1 \\ 0 & n \neq \pm 1,\ n \text{ is odd} \end{cases}$$

$$= \begin{cases} \dfrac{1}{\pi(1-n^2)} & n \text{ is even} \\ \pm\dfrac{j}{4} & n = \pm 1 \\ 0 & n \neq \pm 1,\ n \text{ is odd} \end{cases}$$

The signal $x(at)$ is a periodic signal of period T/a. Thus the fundamental frequency of $x(at)$ is $a\omega_0/2\pi$. The Fourier coefficients of $x(at)$ can be computed as

$$\frac{a}{T}\int_{-T/2a}^{T/2a} x(at)e^{-jna\omega_0 t}\,dt = \frac{a}{T}\int_{-T/2}^{T/2} x(\tau)e^{-jn\omega_0\tau}\,\frac{d\tau}{a}$$

$$= \frac{1}{T}\int_{-T/2}^{T/2} x(\tau)e^{-jn\omega_0\tau}\,d\tau$$

$$= x_n$$

That is, the Fourier coefficients of a time signal do not change when the time scale is contracted or extended. However, the frequencies of the harmonics change from $n\omega_0/2\pi$ to $na\omega_0/2\pi$.

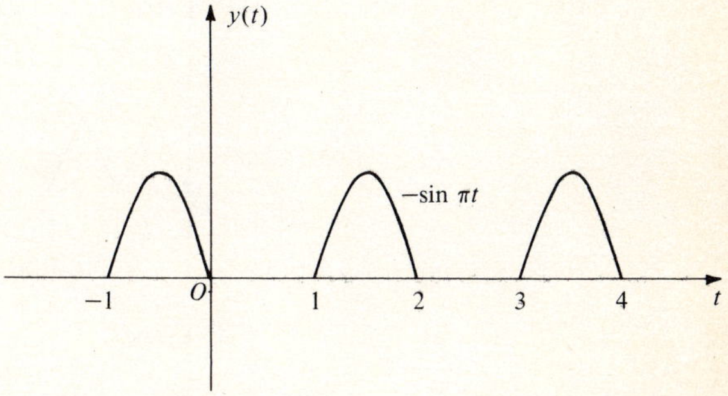

FIGURE 6-11

If $x(t)$ is a periodic signal of period T, then $dx(t)/dt$ is also a periodic signal of period T. [It is quite clear that $dx(t)/dt$ cannot have a period that is larger than T. The reader should also convince himself that it cannot have a period that is a fraction of T either.] For the moment, we confine our attention to signals that are continuous everywhere. For $n \neq 0$, the nth Fourier coefficients of $dx(t)/dt$ can be computed as

$$\frac{1}{T}\int_{-T/2}^{T/2} \frac{dx(t)}{dt} e^{-jn\omega_0 t}\, dt = \frac{1}{T}\left[x(t)e^{-jn\omega_0 t}\right]\Bigg|_{-T/2}^{T/2} + \frac{jn\omega_0}{T}\int_{-T/2}^{T/2} x(t)e^{-jn\omega_0 t}\, dt$$

$$= \frac{1}{T}\left[x\left(\frac{T}{2}\right)e^{-jn\pi} - x\left(-\frac{T}{2}\right)e^{jn\pi}\right] + jn\omega_0 x_n$$

$$= jn\omega_0 x_n$$

This result says that, for $n \neq 0$, the nth Fourier coefficient of $dx(t)/dt$ is equal to the nth Fourier coefficient of $x(t)$ multiplied by $jn\omega_0$. Let $x^{-1}(t)$ denote a periodic signal such that for $-T/2 \leq t \leq T/2$

$$x^{-1}(t) = \int_{-T/2}^{t} x(\tau)\, d\tau$$

It follows directly that for $n \neq 0$ the nth Fourier coefficients of $x^{-1}(t)$ is equal to the nth Fourier coefficients of $x(t)$ divided by $jn\omega_0$. We note that the high-frequency components in $dx(t)/dt$ are more prominent and those in $x^{-1}(t)$ are less prominent than the corresponding components in $x(t)$. Since differentiating a signal accentuates and integrating a signal reduces its fast variations, we confirm an earlier statement that the high-frequency components in a signal are related to the fast variations in the signal.

Let $x(t)$ and $y(t)$ be two periodic signals of period T. Let x_n and y_n be their corresponding Fourier coefficients. The sum of $x(t)$ and $y(t)$, $x(t) + y(t)$, is a signal whose period is T or T/m for some positive integer m. [To see that it is possible

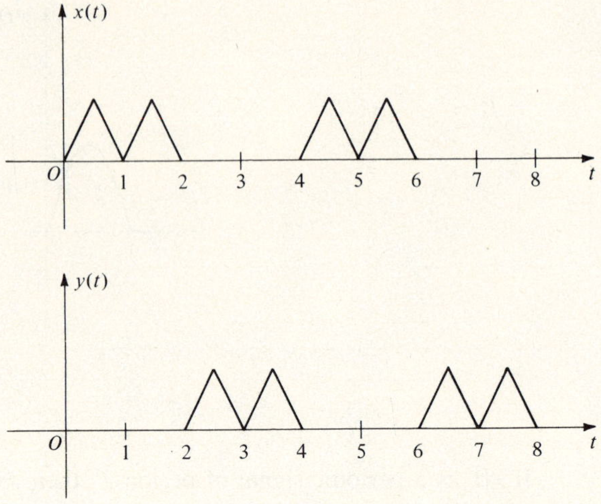

FIGURE 6-12

that the period of $x(t) + y(t)$ is a fraction of T, note that the periods of $x(t)$ and $y(t)$ in Fig. 6-12 are both 4 and the period of $x(t) + y(t)$ is 1.] The Fourier coefficients of $x(t) + y(t)$ are

$$\frac{1}{T}\int_0^T [x(t) + y(t)]e^{-jn\omega_0 t}\,dt = \frac{1}{T}\int_0^T x(t)e^{-jn\omega_0 t}\,dt + \frac{1}{T}\int_0^T y(t)e^{-jn\omega_0 t}\,dt$$

$$= x_n + y_n$$

That is, the nth Fourier coefficient of $x(t) + y(t)$ is equal to the sum of the nth Fourier coefficient of $x(t)$ and the nth Fourier coefficient $y(t)$. Note that in computing the Fourier coefficients of $x(t) + y(t)$ there is no need to determine exactly whether the period of $x(t) + y(t)$ is T or T/m. We can always assume that the period of $x(t) + y(t)$ is T. If it turns out that the period of $x(t) + y(t)$ is equal to T/m, the nth Fourier coefficient will equal zero if n is not a multiple of m.

Let us consider the following illustrative example in which we are to compute the Fourier coefficients of the signal $w(t)$ in Fig. 6-13a and those of the signals $x(t)$ and $y(t)$ in Fig. 6-12. First, we differentiate $w(t)$ to obtain $w'(t)$ as shown in Fig. 6-13b. We then determine the nth Fourier coefficient of $w'(t)$ which is

$$\frac{1}{8}\int_0^1 e^{-jn\pi t/4}\,dt - \frac{1}{8}\int_1^2 e^{-jn\pi t/4}\,dt = \frac{j}{2n\pi}(2e^{-jn\pi/4} - 1 - e^{-jn\pi/2})$$

$$= \frac{-j}{2n\pi}(1 - e^{-jn\pi/4})^2$$

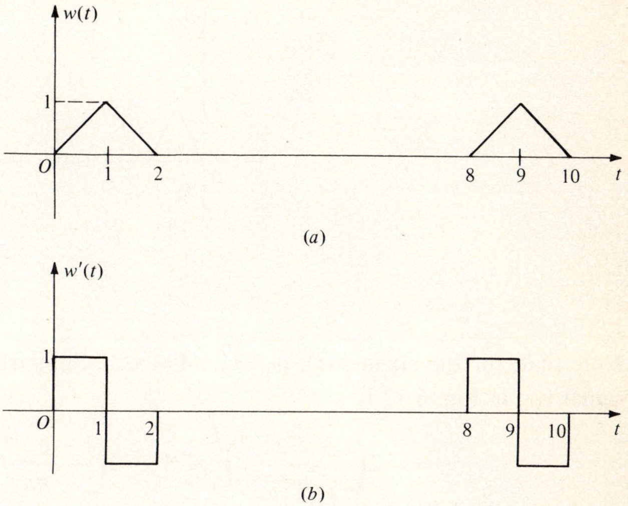

FIGURE 6-13

For $n \neq 0$ the nth Fourier coefficient of $w(t)$, w_n, is equal to $1/jn\omega_0$ times the nth Fourier coefficient of $w'(t)$. That is,

$$w_n = \frac{1}{jn(2\pi/8)} \frac{-j}{2n\pi} (1 - e^{-jn\pi/4})^2 = -2\left(\frac{1 - e^{-jn\pi/4}}{n\pi}\right)^2 \qquad n \neq 0$$

The average value of $w(t)$ is

$$w_0 = \tfrac{1}{8}$$

as the reader can check immediately.

Since the signal $v(t)$ in Fig. 6-14 is equal to $w(2t)$, the nth Fourier coefficient of $v(t)$ is

$$v_n = -2\left(\frac{1 - e^{-jn\pi/4}}{n\pi}\right)^2 \qquad n \neq 0$$

For the signal $x(t)$ in Fig. 6-12, since $x(t) = v(t) + v(t-1)$, the nth Fourier coefficient of $x(t)$ is

$$\begin{aligned} x_n &= -2\left(\frac{1 - e^{-jn\pi/4}}{n\pi}\right)^2 - 2\left(\frac{1 - e^{-jn\pi/4}}{n\pi}\right)^2 e^{-jn\pi/2} \\ &= -2\left(\frac{1 - e^{-jn\pi/4}}{n\pi}\right)^2 (1 + e^{-jn\pi/2}) \qquad n \neq 0 \end{aligned}$$

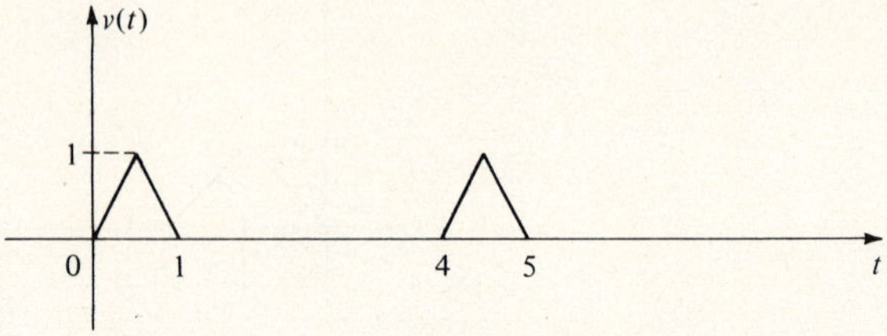

FIGURE 6-14

Note that, for the signal $v(t)$, $\omega_0 = 2\pi/4 = \pi/2$. Similarly, the nth coefficient of the signal $y(t)$ in Fig. 6-12 is

$$y_n = -2\left(\frac{1 - e^{-jn\pi/4}}{n\pi}\right)^2 e^{-jn\pi} - 2\left(\frac{1 - e^{-jn\pi/4}}{n\pi}\right)^2 e^{-j3n\pi/2}$$

$$= -2\left(\frac{1 - e^{-jn\pi/4}}{n\pi}\right)^2 (e^{-jn\pi} + e^{-j3n\pi/2}) \qquad n \neq 0$$

The nth Fourier coefficient of $x(t) + y(t)$ is

$$-2\left(\frac{1 - e^{-jn\pi/4}}{n\pi}\right)^2 (1 + e^{-jn\pi/2} + e^{-jn\pi} + e^{-j3n\pi/2}) \qquad n \neq 0$$

The reader can check that the value of the expression within the second parentheses is 4 for n being a multiple of 4 and is 0 for n not being a multiple of 4. This, of course, follows from the fact that the period of $x(t) + y(t)$ is 1 instead of 4. Also, the average value of $x(t) + y(t)$ is $\frac{1}{2}$.

6-6 FOURIER SERIES OF SIGNALS CONTAINING SINGULARITY FUNCTIONS

Clearly, periodic signals containing singularity functions do not satisfy the Dirichlet conditions and hence might not have Fourier series representation. In order to define the Fourier series representation for these signals it is necessary to go back to the theory of distributions. Here, we shall ignore mathematical rigor and extend the results we obtained in preceding sections to include periodic signals containing singularity functions. Let us consider several illustrative examples.

Consider the periodic signal $x(t) = \sum_{n=-\infty}^{\infty} u_0(t - nT)$ shown in Fig. 6-15. The Fourier coefficients of $x(t)$ can be computed as

$$x_n = \frac{1}{T}\int_{-T/2}^{T/2} u_0(t)e^{-jn\omega_0 t}\, dt$$

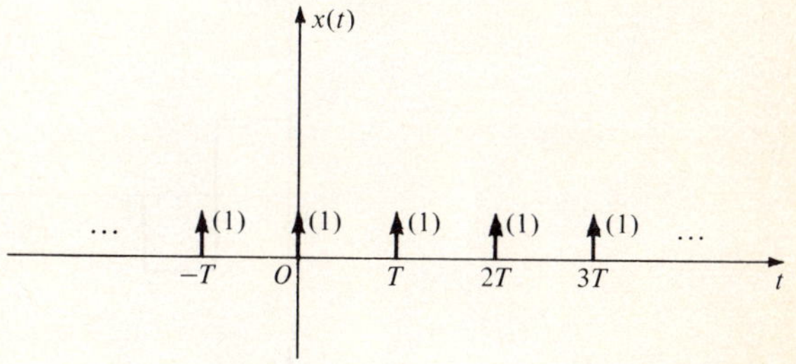

FIGURE 6-15

From Eq. (1-9), we obtain

$$x_n = \frac{1}{T}$$

That is,

$$x(t) = \frac{1}{T}\sum_{n=-\infty}^{\infty} e^{j2\pi nt/T} = 1 + \frac{2}{T}\sum_{n=1}^{\infty} \cos\frac{2\pi nt}{T} \tag{6-16}$$

Observe that according to Eq. (6-16), at $t = 0, \pm T, \pm 2T, \ldots$, the value of $x(t)$ indeed goes to infinity.

Similarly, the Fourier coefficients of the signal $\sum_{n=-\infty}^{\infty} u_1(t - nT)$ are

$$\frac{1}{T}\int_{-T/2}^{T/2} u_1(t)e^{-jn\omega_0 t}at = \frac{1}{T}(-jn\omega_0)$$

In general, the Fourier coefficients of the signal $\sum_{n=-\infty}^{\infty} u_k(t - nT)$ are

$$\frac{1}{T}\int_{-T/2}^{T/2} u_k(t)e^{-jn\omega_0 t}\,dt = \frac{1}{T}(-jn\omega_0)^k$$

As another example, let us determine the Fourier coefficients of the signal $w(t)$ in Fig. 6-13*a* in an alternative way. We differentiate $w(t)$ twice to obtain $w'(t)$ and then $w''(t)$ as shown in Fig. 6-16*a* and *b*. The nth Fourier coefficient of $w''(t)$ is equal to

$$\frac{1}{8}\int_0^8 [u_0(t) - 2u_0(t-1) + u_0(t-2)]e^{-jn\pi t/4}\,dt = \tfrac{1}{8}(1 - 2e^{-jn\pi/4} + e^{-jn\pi/2})$$
$$= \tfrac{1}{8}(1 - e^{-jn\pi/4})^2$$

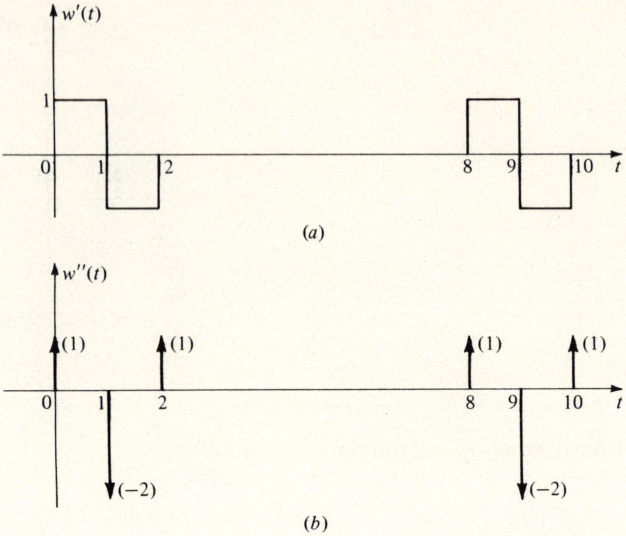

FIGURE 6-16

Thus, the nth Fourier coefficient of $w(t)$, $n \neq 0$, is equal to

$$\frac{1}{[jn(2\pi/8)]^2}\frac{1}{8}(1 - e^{-jn\pi/4})^2 = -2\left(\frac{1 - e^{-jn\pi/4}}{n\pi}\right)^2 \qquad n \neq 0$$

which was also computed in Sec. 6-5.

The Fourier coefficients of the signal $x(t)$ shown in Fig. 6-4 can be computed in still another way. We compute $x'(t)$ and $x''(t)$ as shown in Fig. 6-17a and b. Since

$$x(t) + \frac{1}{\pi^2}x''(t) = \sum_{n=-\infty}^{\infty} \frac{1}{\pi} u_0(t - n)$$

as shown in Fig. 6-17c, taking the period of $x(t) + (1/\pi^2)x''(t)$ to be 2, the period of $x(t)$, we find that its Fourier coefficients equal $1/\pi$ for all even n and equal zero for all odd n. That is

$$x_n + \frac{(jn\pi)^2}{\pi^2} x_n = \begin{cases} \dfrac{1}{\pi} & n \text{ is even} \\ 0 & n \text{ is odd} \end{cases}$$

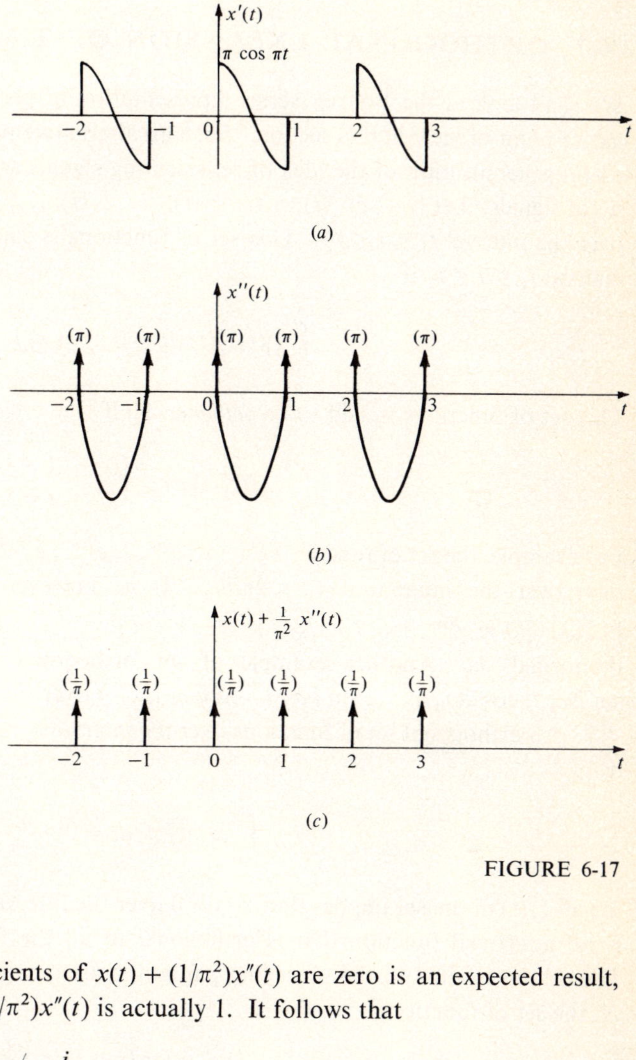

FIGURE 6-17

That all odd Fourier coefficients of $x(t) + (1/\pi^2)x''(t)$ are zero is an expected result, since the period of $x(t) + (1/\pi^2)x''(t)$ is actually 1. It follows that

$$x_n = \begin{cases} \mp \dfrac{j}{4} & n = \pm 1 \\ 0 & n \neq \pm 1,\ n \text{ is odd} \\ \dfrac{1}{\pi(1 - n^2)} & n \text{ is even} \end{cases}$$

In closing, we state that the results derived in the preceding sections can be extended to include periodic signals that contain singularity functions. Of course, computation and manipulation involving singularity functions follow the properties of these functions defined in Chaps. 1 and 4.

*6-7 ORTHOGONAL EXPANSION OF TIME SIGNALS

We shall look at the Fourier series representation of periodic signals from an alternative point of view in this section. Such an alternative point of view leads to a most useful generalization of the idea of representing signals as weighted sums of a chosen set of signals. Let $\{\ldots, \xi_{-1}(t), \xi_0(t), \xi_1(t), \ldots, \xi_k(t), \ldots\}$ be a set of functions defined over the interval $t_1 \le t \le t_2$. This set of functions is said to be *orthogonal* over the interval $t_1 \le t \le t_2$ if

$$\int_{t_1}^{t_2} \xi_i(t)\xi_j^*(t)\,dt = 0 \qquad i \ne j$$

This set of functions is said to be *orthonormal* if

$$\int_{t_1}^{t_2} \xi_i(t)\xi_j^*(t)\,dt = \begin{cases} 0 & i \ne j \\ 1 & i = j \end{cases}$$

For example, the set of functions $\{\ldots, e^{-j\omega_0 t}, 1, e^{j\omega_0 t}, e^{j2\omega_0 t}, \ldots, e^{jk\omega_0 t}, \ldots\}$ is orthogonal over the interval $0 \le t \le 2\pi/\omega_0$. It is, however, not orthonormal, whereas $\{\ldots, (1/\sqrt{T})e^{-j\omega_0 t}, 1/\sqrt{T}, (1/\sqrt{T})e^{j\omega_0 t}, (1/\sqrt{T})e^{j2\omega_0 t}, \ldots, (1/\sqrt{T})e^{jk\omega_0 t}, \ldots\}$ is an orthonormal set. Another example of an orthogonal set is $\{1, \sin\omega_0 t, \cos\omega_0 t, \sin 2\omega_0 t, \cos 2\omega_0 t, \ldots, \sin k\omega_0 t, \cos k\omega_0 t, \ldots\}$ over the interval $0 \le t \le 2\pi/\omega_0$.

An orthogonal set of functions over the interval $t_1 \le t \le t_2$ is said to be *complete* if

$$\int_{t_1}^{t_2} x(t)\xi_i^*(t)\,dt = 0$$

for all $\xi_i(t)$ in the set implies that $x(t) = 0$ over the interval $t_1 \le t \le t_2$. That is, there is no nontrivial function that is orthogonal to all the functions in a complete set. The examples cited in the preceding paragraph are complete sets. For a fixed integer N, the set of functions

$$\xi_i(t) = \begin{cases} 1 & \text{for } (i-1) < t < i \\ 0 & \text{elsewhere} \end{cases}$$

for $i = 1, 2, \ldots, N$ are orthogonal over the interval $0 \le t \le N$. However, this set is not complete since the function

$$x(t) = \begin{cases} 1 & \text{for } 0 \le t < \frac{1}{2} \\ -1 & \text{for } \frac{1}{2} < t < 1 \\ 0 & \text{for } 1 < t \le N \end{cases}$$

is orthogonal to all $\xi_i(t)$ in the set.

Let $\{\xi_0(t), \xi_1(t), \ldots, \xi_k(t), \ldots\}$ be a complete orthogonal set of functions over the interval $t_1 \le t \le t_2$. Suppose a function $x(t)$ defined over the interval $t_1 \le t \le t_2$ is to be expressed as a weighted sum of the functions in the set. That is,

$$x(t) = a_0 \xi_0(t) + a_1\xi_1(t) + \cdots + a_k \xi_k(t) + \cdots \tag{6-17}$$

To determine the coefficient a_k, we multiply both sides of Eq. (6-17) by $\xi_k^*(t)$ and integrate from t_1 to t_2. We obtain

$$\int_{t_1}^{t_2} x(t)\xi_k^*(t)\,dt = \int_{t_1}^{t_2} a_0 \xi_0(t)\xi_k^*(t)\,dt + \int_{t_1}^{t_2} a_1\xi_1(t)\xi_k^*(t)\,dt + \cdots + \int_{t_1}^{t_2} a_k \xi_k(t)\xi_k^*(t)\,dt + \cdots$$

or

$$a_k = \frac{\int_{t_1}^{t_2} x(t)\xi_k^*(t)\,dt}{\int_{t_1}^{t_2} \xi_k(t)\xi_k^*(t)\,dt} \tag{6-18}$$

The relation in Eq. (6-18) is called the *Euler-Fourier formula;* the coefficients a_0, a_1, $\ldots$, a_k, $\ldots$ are called the Fourier coefficients of $x(t)$ with respect to the set of functions $\{\xi_0(t), \xi_1(t), \ldots, \xi_k(t), \ldots\}$. As the reader will have recognized by now, our discussion on Fourier series in the previous sections is one special case in which we choose $\xi_k(t)$ to be the complex exponential function $e^{jk\omega_0 t}$.

Suppose we are to approximate a function $x(t)$, within the interval $t_1 \le t \le t_2$, by another function, $x_N(t)$, which is a linear combination of $\xi_0(t), \xi_1(t), \ldots, \xi_N(t)$ for a certain finite N. That is,

$$x_N(t) = b_0 \xi_0(t) + b_1\xi_1(t) + \cdots + b_N \xi_N(t)$$

where the coefficients b_k are multiplying constants we have chosen. The mean-square error of such an approximation is defined to be

$$E = \int_{t_1}^{t_2} |x(t) - x_N(t)|^2\,dt$$

Thus,

$$E = \int_{t_1}^{t_2} |x(t) - b_0 \xi_0(t) - b_1\xi_1(t) - \cdots - b_N \xi_N(t)|^2\,dt$$

or

$$E = \int_{t_1}^{t_2} x(t)x^*(t)\,dt + \sum_{i=0}^{N} b_i b_i^* \int_{t_1}^{t_2} \xi_i(t)\xi_i^*(t)\,dt - \sum_{i=0}^{N} b_i^* \int_{t_1}^{t_2} x(t)\xi_i^*(t)\,dt - \sum_{i=0}^{N} b_i \int_{t_1}^{t_2} x^*(t)\xi_i(t)\,dt \tag{6-19}$$

Equation (6-19) can be written

$$E = \int_{t_1}^{t_2} |x(t)|^2\,dt + \sum_{i=0}^{N} |b_i|^2 - \sum_{i=0}^{N} (a_i b_i^* + a_i^* b_i) \tag{6-20}$$

where a_i are the Fourier coefficients of $x(t)$ with respect to the set of functions $\{\xi_0(t), \xi_1(t), \ldots, \xi_k(t), \ldots\}$. Since

$$\begin{aligned} |a_i - b_i|^2 &= (a_i - b_i)(a_i - b_i)^* \\ &= a_i a_i^* + b_i b_i^* - a_i b_i^* - a_i^* b_i \\ &= |a_i|^2 + |b_i|^2 - a_i b_i^* - a_i^* b_i \end{aligned}$$

Eq. (6-20) can be written

$$E = \int_{t_1}^{t_2} |x(t)|^2\,dt + \sum_{i=0}^{N} |a_i - b_i|^2 - \sum_{i=0}^{N} |a_i|^2 \tag{6-21}$$

Equation (6-21) leads to some interesting observations. First, when we approximate $x(t)$ by a linear combination of the functions $\xi_0(t), \xi_1(t), \ldots, \xi_N(t)$, the mean-square error is minimum if the b_i are chosen to be the Fourier coefficients of $x(t)$ with respect to the set of functions $\{\xi_0(t), \xi_1(t), \ldots, \xi_N(t)\}$. Second, if we approximate $x(t)$ by a finite number of the terms in the representation in Eq. (6-17), then the more terms we include in the summation, the smaller the mean-square error will be. Third, since the mean-square error is nonnegative, we obtain

$$\sum_{i=0}^{N} |a_i|^2 \le \int_{t_1}^{t_2} |x(t)|^2\,dt$$

which is known as *Bessel's inequality*.

*6-8 POWER DISTRIBUTION IN PERIODIC SIGNALS

It is well known that if $i(t)$ is the current in an r-Ω resistor the total energy dissipated in the resistor from t_1 to t_2 can be computed as

$$\int_{t_1}^{t_2} r[i(t)]^2\,dt$$

and the average power dissipated in the resistor within this time interval is

$$\frac{1}{t_2 - t_1} \int_{t_1}^{t_2} r[i(t)]^2\,dt$$

Extending such an observation, we define the total energy in a signal $x(t)$ within the time interval $t_1 \leq t \leq t_2$ to be

$$\int_{t_1}^{t_2} |x(t)|^2 \, dt \tag{6-22}$$

Note that $x(t)$, in general, is a complex time function. If $x(t)$ is a real time function, (6-22) is reduced to

$$\int_{t_1}^{t_2} [x(t)]^2 \, dt$$

We also define the average power in $x(t)$, within the time interval $t_1 \leq t \leq t_2$, to be

$$\frac{1}{t_2 - t_1} \int_{t_1}^{t_2} |x(t)|^2 \, dt$$

The energy or the power in a signal is a quantity of important physical significance. In systems built to generate or to process signals, the physical parameters of the systems place upper or lower limits on the total energy or the average power in these signals. For example, the size of the field windings and the strength of the insulation material in an electric generator place a limit on the maximum power in the generated signal, and the gain of the output amplifier of a radar system affects directly the maximum power of the radar signal to be transmitted. On the other hand, a radio receiver will not be able to detect a radio signal if the power in the signal is below a certain minimum level.

For a periodic signal $x(t)$ of period T, we define the average power in $x(t)$ to be

$$\frac{1}{T} \int_{-T/2}^{T/2} |x(t)|^2 \, dt \tag{6-23}$$

The integral in (6-23) can be computed as

$$\begin{aligned} \frac{1}{T} \int_{-T/2}^{T/2} |x(t)|^2 \, dt &= \frac{1}{T} \int_{-T/2}^{T/2} x(t) x^*(t) \, dt \\ &= \frac{1}{T} \int_{-T/2}^{T/2} x(t) \left(\sum_{n=-\infty}^{\infty} x_n^* e^{-j2\pi nt/T} \right) dt \end{aligned}$$

Interchanging the order of integration and summation, we obtain

$$\begin{aligned} \frac{1}{T} \int_{-T/2}^{T/2} |x(t)|^2 \, dt &= \sum_{n=-\infty}^{\infty} x_n^* \left[\frac{1}{T} \int_{-T/2}^{T/2} x(t) e^{-j2\pi nt/T} \, dt \right] \\ &= \sum_{n=-\infty}^{\infty} x_n^* x_n \\ &= \sum_{n=-\infty}^{\infty} |x_n|^2 \end{aligned} \tag{6-24}$$

We note that the average power in the nth frequency component $x_n e^{j2\pi nt/T}$, within the time interval $-T/2 \le t \le T/2$, is equal to

$$\frac{1}{T}\int_{-T/2}^{T/2} (x_n e^{j2\pi nt/T})(x_n^* e^{-j2\pi nt/T})\, dt = |x_n|^2$$

Thus, according to Eq. (6-24), the average power in $x(t)$ is equal to the sum of the average powers in its frequency components. The reader is reminded that, in general, the average power in a signal does not equal the sum of the average powers of its components. However, in the Fourier series representation, the "cross power" between them is equal to zero because the frequency components are orthogonal functions.

The result will be brought out in further evidence, if we consider the case in which $x(t)$ is a real signal. Writing

$$x(t) = x_0 + \sum_{n=1}^{\infty} A_n \cos n\omega_0 t + \sum_{n=1}^{\infty} B_n \sin n\omega_0 t$$

where $A_n = 2\ \mathrm{Re}\ [x_n]$ and $B_n = -2\ \mathrm{Im}\ [x_n]$, we have

$$\frac{1}{T}\int_{-T/2}^{T/2} |x(t)|^2\, dt = x_0^2 + \sum_{n=1}^{\infty}\left(\frac{A_n^2}{2} + \frac{B_n^2}{2}\right)$$

Since the average power in a sinusoid or cosinusoid of amplitude A is equal to $A^2/2$, the average power in the signal is equal to the sum of the power in the dc component and that in the sinusoidal and cosinusoidal components.

As an example, consider the periodic signal $x(t)$ in Fig. 6-4. A direct computation of the average power in $x(t)$ yields

$$\frac{1}{2}\int_0^1 (\sin \pi t)^2\, dt = \frac{1}{4} \tag{6-25}$$

According to Eq. (6-10), the average power in $x(t)$ is also equal to

$$\left(\frac{1}{\pi}\right)^2 + \left(\frac{1}{4}\right)^2 + \left(\frac{1}{4}\right)^2 + \sum_{n=\pm 2, \pm 4, \ldots}^{\infty} \frac{1}{\pi^2(1-n^2)^2} = \left(\frac{1}{\pi}\right)^2 + \frac{1}{8} + \sum_{n=2,4,\ldots}^{\infty} \frac{2}{\pi^2(1-n^2)^2}\,\dagger \tag{6-26}$$

† Combining the results in Eqs. (6-25) and (6-26), we obtain an interesting infinite sum

$$\left(\frac{1}{\pi}\right)^2 + \frac{1}{8} + \sum_{n=2,4,\ldots}^{\infty} \frac{2}{\pi^2(1-n^2)^2} = \frac{1}{4}$$

which is simplified to

$$\sum_{n=2,4,\ldots}^{\infty} \frac{2}{(1-n^2)^2} = \frac{\pi^2}{8} - 1$$

Consider now the sequence of numbers $\ldots, |x_{-1}|^2, |x_0|^2, |x_1|^2, |x_2|^2, \ldots, |x_k|^2, \ldots$. We ask whether they are the Fourier coefficients of some periodic signal. The answer is an affirmative one. As a matter of fact, they are the Fourier coefficients of the time function

$$r(t) = \frac{1}{T}\int_{-T/2}^{T/2} x^*(\tau)x(\tau + t)\,d\tau$$

The function $r(t)$ is called the *autocorrelation function* of the periodic signal $x(t)$. It is easy to show that $r(t)$ is periodic with a period equal to T:

$$\begin{aligned} r(t+T) &= \frac{1}{T}\int_{-T/2}^{T/2} x^*(\tau)x(\tau + t + T)\,d\tau \\ &= \frac{1}{T}\int_{-T/2}^{T/2} x^*(\tau)x(\tau + t)\,d\tau \\ &= r(t) \end{aligned}$$

The Fourier coefficients of $r(t)$ can be computed as

$$r_n = \frac{1}{T}\left\{\int_{-T/2}^{T/2}\left[\frac{1}{T}\int_{-T/2}^{T/2} x^*(\tau)x(\tau + t)\,d\tau\right]e^{-j2\pi nt/T}\,dt\right\}$$

Letting $t + \tau = \lambda$, we obtain

$$\begin{aligned} r_n &= \frac{1}{T^2}\int_{-T/2+\tau}^{T/2+\tau}\left[\int_{-T/2}^{T/2} x^*(\tau)x(\lambda)\,d\tau\right]e^{-j2\pi n(\lambda-\tau)/T}\,d\lambda \\ &= \left[\frac{1}{T}\int_{-T/2+\tau}^{T/2+\tau} x(\lambda)e^{-j2\pi n\lambda/T}\,d\lambda\right]\left[\frac{1}{T}\int_{-T/2}^{T/2} x^*(\tau)e^{j2\pi n\tau/T}\,d\tau\right] \\ &= |x_n|^2 \end{aligned}$$

The sequence of numbers $\ldots, |x_{-1}|^2, |x_0|^2, |x_1|^2, |x_2|^2, \ldots, |x_k|^2, \ldots$ is called the *power spectrum* of $x(t)$, because they equal the average powers of the frequency components of $x(t)$.

For example, the autocorrelation function of the signal $x(t)$ in Fig. 6-4 can be computed as follows: For $-1 \le t \le 0$

$$\begin{aligned} r(t) &= \frac{1}{2}\int_{-t}^{1} \sin \pi\tau \sin \pi(\tau + t)\,d\tau \\ &= \frac{1}{4}(1 + t)\cos \pi t - \frac{1}{4\pi}\sin \pi t \qquad -1 \le t \le 0 \end{aligned}$$

and

$$\begin{aligned} r(t) &= \frac{1}{2}\int_{0}^{1-t} \sin \pi\tau \sin \pi(\tau + t)\,dt \\ &= \frac{1}{4}(1 - t)\cos \pi t + \frac{1}{4\pi}\sin \pi t \qquad 0 \le t \le 1 \end{aligned}$$

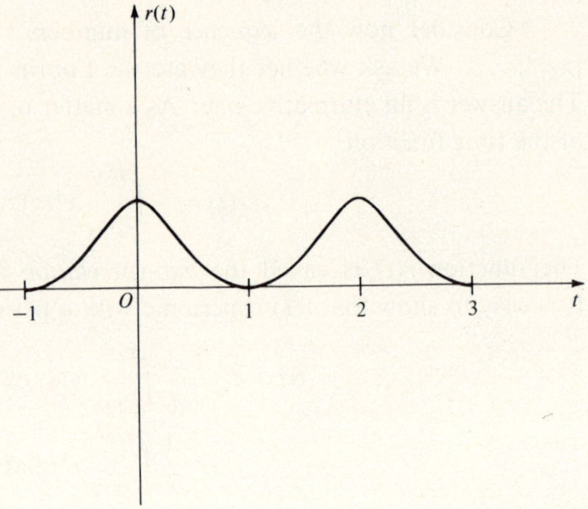

FIGURE 6-18

The signal $r(t)$ is plotted in Fig. 6-18. According to Eq. (6-10), the Fourier coefficients of $r(t)$ are

$$r_n = \begin{cases} \dfrac{1}{\pi^2(1-n^2)^2} & n \text{ is even} \\ (\frac{1}{4})^2 & n = \pm 1 \\ 0 & n \neq \pm 1, n \text{ is odd} \end{cases}$$

6-9 REMARKS AND REFERENCES

We must admit that the possibility of expressing a large class of periodic time signals as sums of sinusoidal and cosinusoidal terms is not at all intuitively obvious. Indeed, reactions ranged from surprise to disbelief when Fourier first presented his result around one hundred years ago. A brief historical account of the development of the subject can be found in Carslaw [1]. Today, the Fourier series is studied in many courses in engineering and applied mathematics. Most books on advanced calculus for engineers are possible references. See, for example, Sokolnikoff and Redheffer [4].

We should also mention that there are some very deep questions concerning the Fourier series representation of signals which are of significant mathematical interest. However, a discussion of these questions is beyond the scope of this book. A reader who wishes to have some general idea of a rigorous mathematical treatment of the subject might want to consult Churchill [2], Kaplan [3], or Carslaw [1].

[1] CARSLAW, H. S.: "Fourier's Series and Integrals," 3d ed., Macmillan & Co., Ltd., London, 1930.

[2] CHURCHILL, R. V.: "Fourier Series and Boundary Value Problems," 2d ed., McGraw-Hill Book Company, New York, 1963.

[3] KAPLAN, W.: "Operational Methods for Linear Systems," Addison-Wesley Publishing Company, Inc., Reading, Mass., 1962.

[4] SOKOLNIKOFF, I. S., and R. M. REDHEFFER: "Mathematics of Physics and Modern Engineering," McGraw-Hill Book Company, New York, 1958.

PROBLEMS

6-1 Show that $\sin \omega t$ and $\cos \omega t$ are eigenfunctions of a linear time-invariant system whose impulse response $h(t)$ is an even time function and which is stable in the bounded-input–bounded-output sense.

6-2 (*a*) Let $\sin \omega t$ be the input signal to an arbitrary linear time-invariant system. Show that $A \sin (\omega t + \theta)$ is the corresponding output signal, where A and θ are constants. Under what condition will the constant A be finite?

(*b*) What is the output signal corresponding to the input signal $\cos \omega t$?

6-3 (*a*) Show that z^{-n} are eigenfunctions of a linear time-invariant discrete system for some complex numbers z.

(*b*) Determine the condition which the complex numbers z must satisfy so that z^{-n} are eigenfunctions of a discrete linear time-invariant system with unit response $h(n)$.

6-4 The impulse response $h(t)$ of a linear time-invariant system is such that

$$\int_{-\infty}^{\infty} e^{-s\tau} |h(\tau)| \, d\tau$$

is finite for all s except at the points marked with crosses in the s plane shown in Fig. 6P-1. What is its region of absolute convergence if the system is known to be causal but not stable? Noncausal but stable? Both noncausal and unstable?

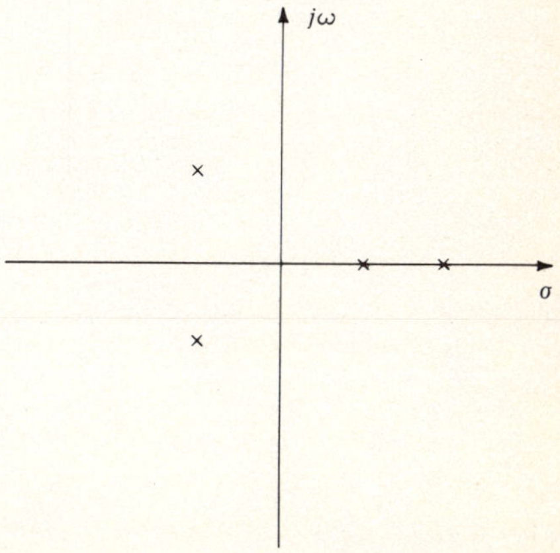

FIGURE 6P-1

6-5 Let $x(t)$ be a periodic signal of period 2.

$$x(t) = \begin{cases} \sin \pi t + (t+1) & -1 \leq t \leq 0 \\ \sin \pi t + (1-t) & 0 \leq t \leq 1 \end{cases}$$

Determine the exponential Fourier series representation of $x(t)$.

6-6 Determine the exponential Fourier series of the periodic signals shown in Fig. 6P-2, where

$$x(t) = \begin{cases} 1 & 0 < t < 1 \\ 0 & 1 < t < 2 \end{cases}$$

$$y(t) = \begin{cases} 2 & 0 < t < 1 \\ -1 & 1 < t < 2 \end{cases}$$

$$z(t) = \begin{cases} 2 + \frac{1}{2} \sin \pi t & 0 < t < 1 \\ -1 + \frac{1}{2} \sin \pi t & 1 < t < 2 \end{cases}$$

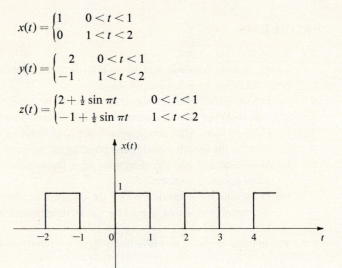

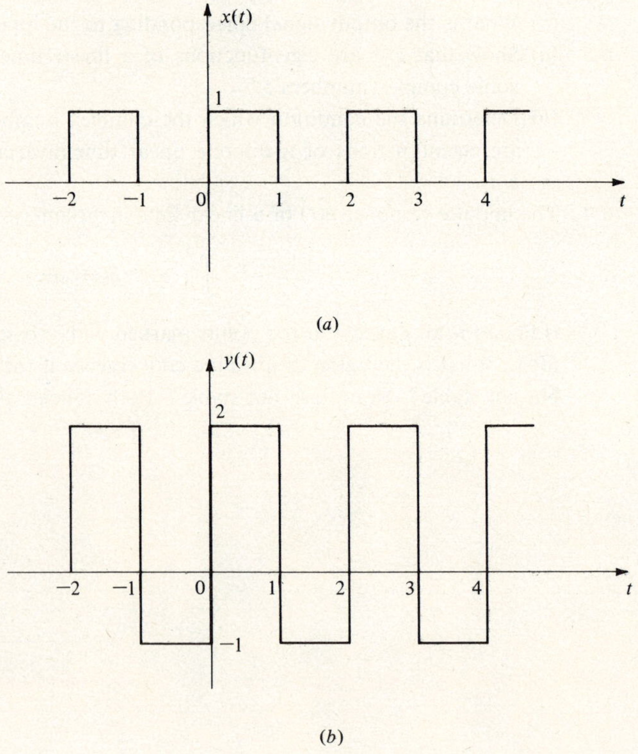

FIGURE 6P-2

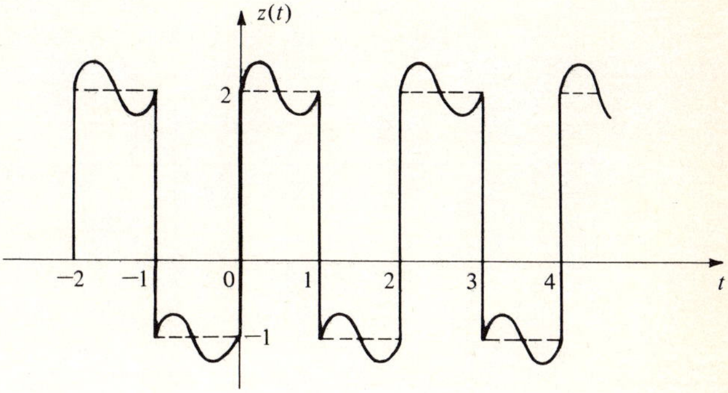

(c)

FIGURE 6P-2 (*continued*)

6-7 Let $x(t)$ and $y(t)$ be two periodic signals of period 4 as shown in Fig. 6P-3. Let $z(t)$ be another periodic signal of period 4 such that

$$\operatorname{Re}(z_n) = \operatorname{Re}(x_n)$$
$$\operatorname{Im}(z_n) = \operatorname{Im}(y_n)$$

for all n. Sketch $z(t)$. [Do not give the Fourier series representation of $z(t)$ as the answer.]

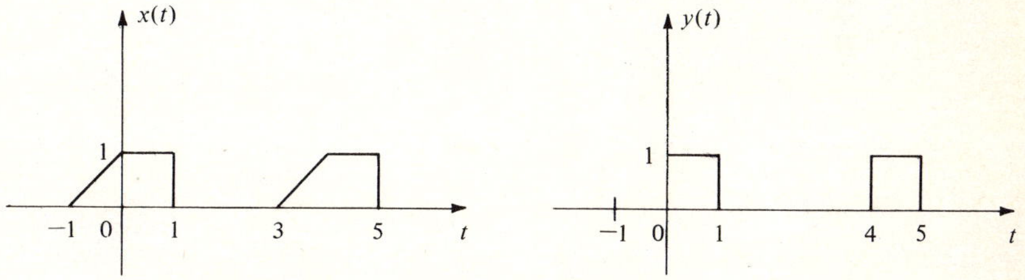

FIGURE 6P-3

6-8 Without evaluating their Fourier coefficients, state which of the time signals in Fig. 6P-4 have one or more of the following properties. Justify your answer.

1 It contains odd harmonics only.
2 Its Fourier coefficients are all real numbers.
3 Its Fourier coefficients are all imaginary numbers.

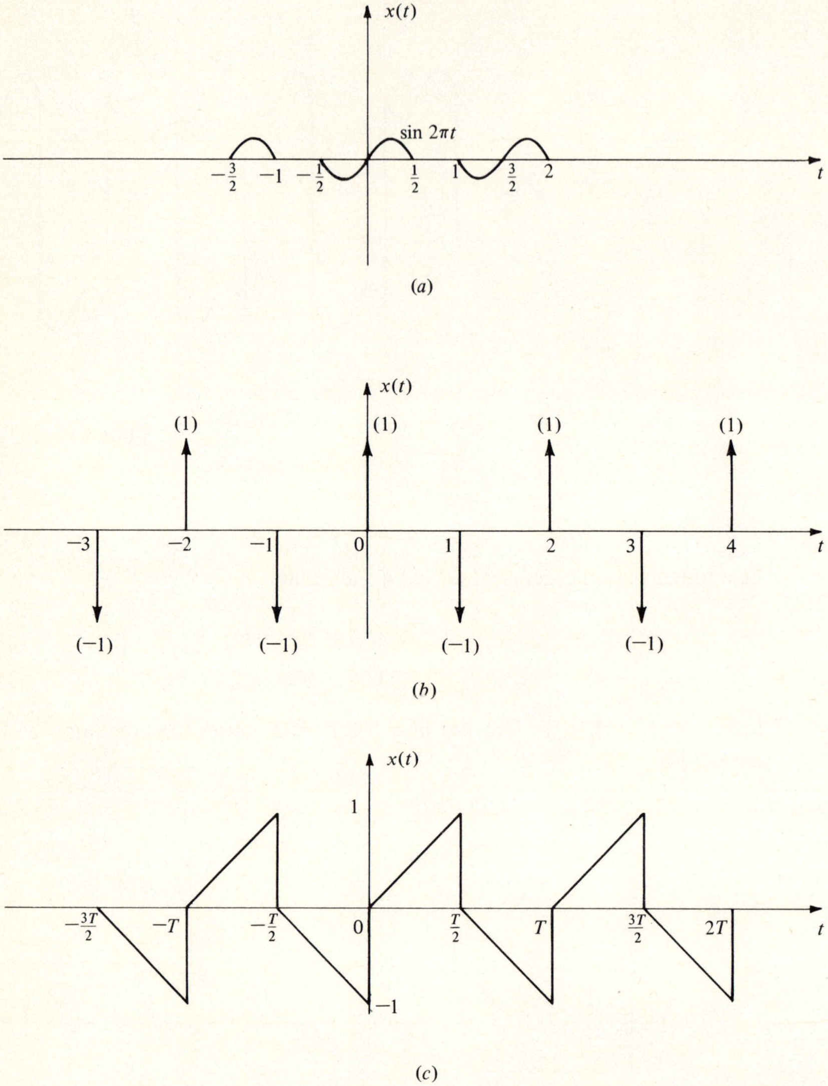

FIGURE 6P-4

6-9 (*a*) Show that the signals $x_{e1}(t)$, $x_{e2}(t)$, $x_{o1}(t)$, and $x_{o2}(t)$ defined in Sec. 6-4 are given by

$$x_{e2}(t) = \frac{1}{4}\left[x(t) + x\left(t + \frac{T}{2}\right) + x(-t) + x\left(-t + \frac{T}{2}\right)\right]$$

$$x_{e1}(t) = \frac{1}{4}\left[x(t) - x\left(t + \frac{T}{2}\right) + x(-t) - x\left(-t + \frac{T}{2}\right)\right]$$

$$x_{o2}(t) = \frac{1}{4}\left[x(t) + x\left(t + \frac{T}{2}\right) - x(-t) - x\left(-t + \frac{T}{2}\right)\right]$$

$$x_{o1}(t) = \frac{1}{4}\left[x(t) - x\left(t + \frac{T}{2}\right) - x(-t) + x\left(-t + \frac{T}{2}\right)\right]$$

(*b*) Determine $x_{e1}(t)$, $x_{e2}(t)$, $x_{o1}(t)$, and $x_{o2}(t)$ for the time signals shown in Fig. 6P-4*a* and *c*.

6-10 Find the Fourier series representation for each of the time signals in Fig. 6P-4. Express your answers in the exponential form as well as in the trigonometric form.

6-11 (*a*) Let

$$x_n = \begin{cases} 0 & n = 0 \\ \dfrac{1}{jn\pi} & n \neq 0 \end{cases}$$

be the Fourier coefficients for one of the two periodic signals shown in Fig. 6P-5*a*. Identify the time signal with the given set of Fourier coefficients. Justify your choice without evaluating the Fourier coefficients for each time signal.

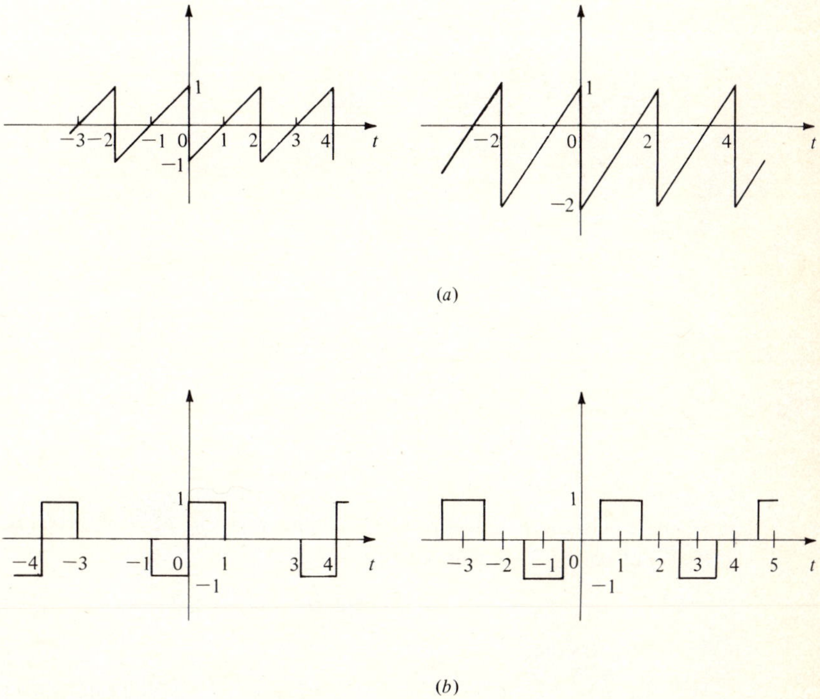

FIGURE 6P-5

(*b*) Let

$$x_n = \begin{cases} 0 & n \text{ even} \\ -\dfrac{j\sqrt{2}}{n\pi} & n \text{ odd} \end{cases}$$

Repeat part (*a*) for the signals shown in Fig. 6P-5*b*.

6-12 The periodic signal $x(t)$ in Fig. 6P-6 can be expressed as

$$x(t) = A_0 + \sum_{n=1}^{\infty} A_n \cos\frac{\pi t}{2} + \sum_{n=1}^{\infty} B_n \sin\frac{\pi t}{2}$$

(*a*) Sketch the signal

$$x_1(t) = A_0 + \sum_{n=1}^{\infty} \frac{A_n}{2} \cos\frac{\pi t}{2} + \sum_{n=1}^{\infty} 2B_n \sin\frac{\pi t}{2}$$

(*b*) Sketch the signal

$$x_2(t) = \sum_{n=1,3,5,\ldots} A_n \cos\frac{\pi t}{2} + \sum_{n=1,3,5,\ldots} B_n \sin\frac{\pi t}{2}$$

(*c*) Sketch the signal

$$x_3(t) = \sum_{n=1,3,5,\ldots} A_n \cos\frac{\pi t}{2} + \sum_{n=2,4,6,\ldots} B_n \sin\frac{\pi t}{2}$$

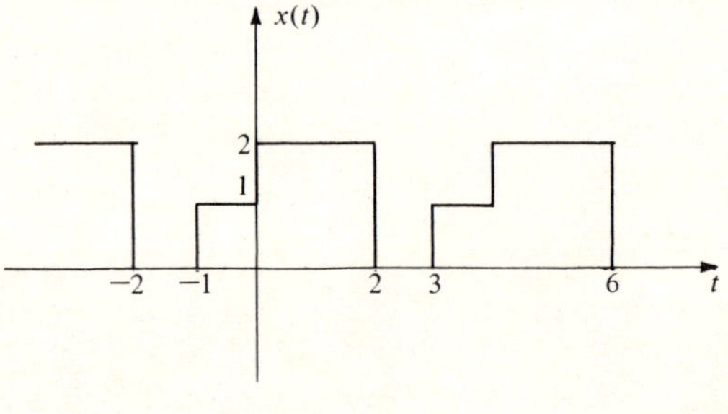

FIGURE 6P-6

6-13 Let $x(t)$ be a periodic signal of period 3 as shown in Fig. 6P-7. Let $y(t)$ be a periodic signal also of period 3. It is given that

$$y(t) = \sin \pi t \qquad 0 \le t \le 1$$

and in the expressions

$$x(t) = A_0 + \sum_{n=1}^{\infty} A_n \cos\frac{2\pi nt}{3} + \sum_{n=1}^{\infty} B_n \sin\frac{2\pi nt}{3}$$

$$y(t) = A'_0 + \sum_{n=1}^{\infty} A'_n \cos \frac{2\pi nt}{3} + \sum_{n=1}^{\infty} B'_n \sin \frac{2\pi nt}{3}$$

$$A_n = A'_n \qquad \text{and} \qquad B_n = B'_n$$

for all n not being a multiple of 3 ($n = 1, 2, 4, 5 \cdots$). Sketch $y(t)$.

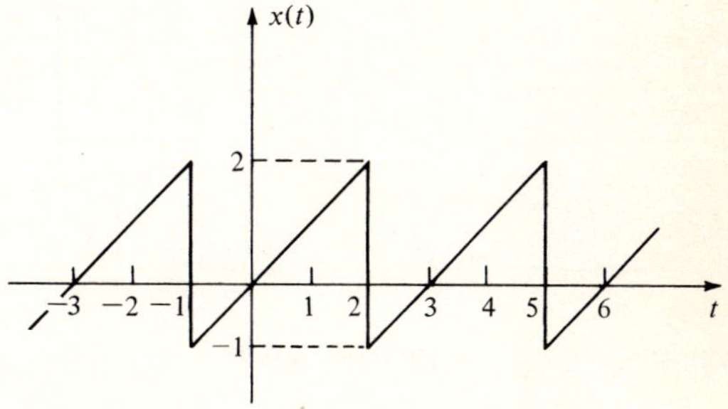

FIGURE 6P-7

6-14 (*a*) Let $x(t)$ be a periodic signal of period T, such that $x(t + T/3) = -x(t)$ for $0 \leq t \leq T/3$ and $x(t) = 0$ for $2T/3 \leq t \leq T$. Show that $x_n = 0$ if n is a multiple of 3.

(*b*) Let $x(t)$ be a periodic signal of period 3 as shown in Fig. 6P-8*a*. Show that $x_n = 0$ if n is a multiple of 3.

(*c*) Repeat part (*b*) for the signal $x(t)$ shown in Fig. 6P-8*b*.

(*d*) Give a general characterization of a periodic signal $x(t)$ whose Fourier coefficients x_n are zero for all n being multiples of 3.

(*e*) Give a general characterization of a periodic signal $x(t)$ whose Fourier coefficients x_n are zero for all n being multiples of k.

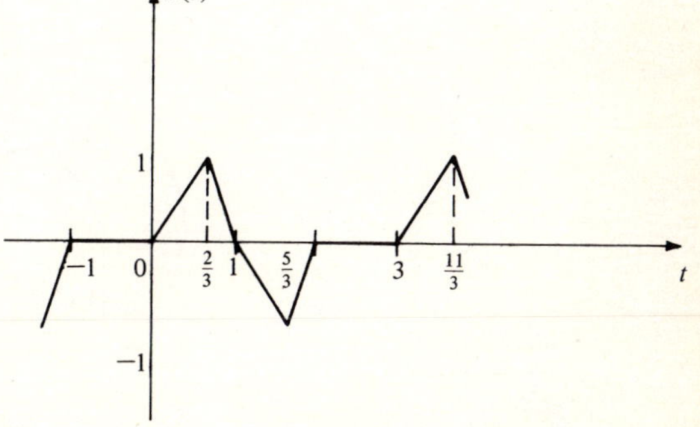

FIGURE 6P-8

(*a*)

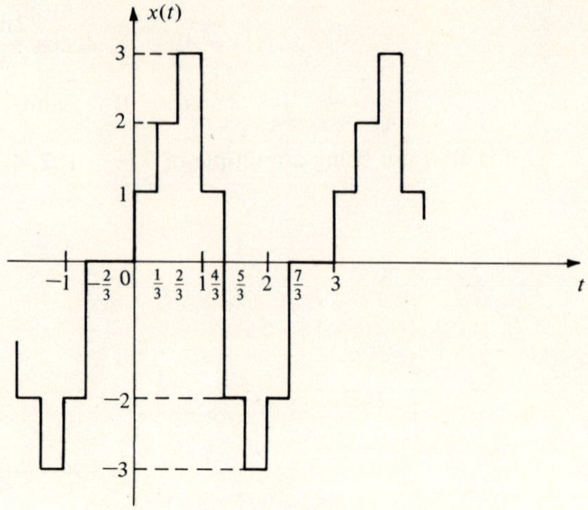

FIGURE 6P-8 (*b*)

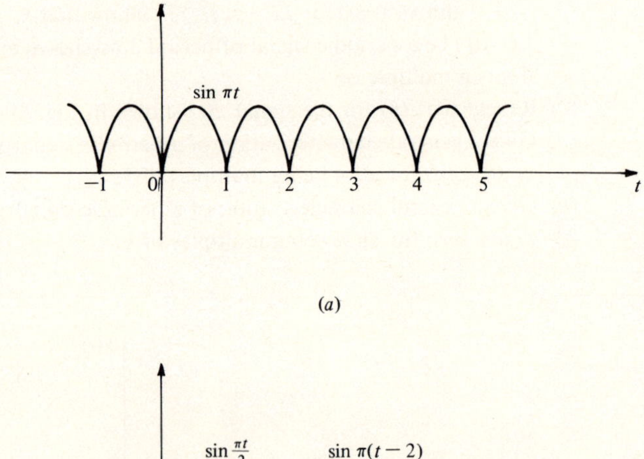

(*a*)

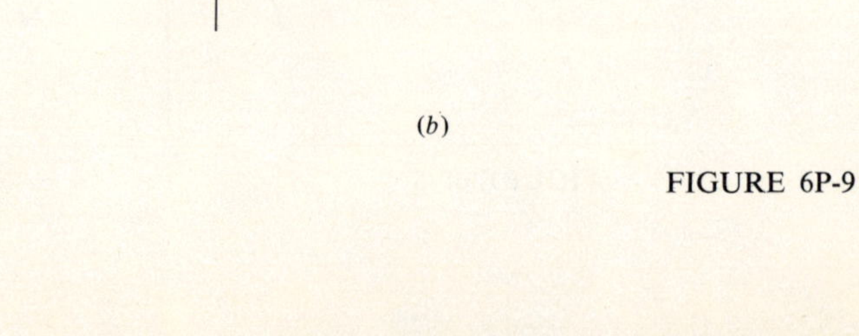

(*b*)

FIGURE 6P-9

6-15 (*a*) Let $x(t)$ be a periodic signal of period T. Let $y(t)$ be a periodic signal of period kT for some positive integer k and

$$y(t) = \begin{cases} x(t) & 0 \le t \le T \\ 0 & T \le t \le kT \end{cases}$$

Determine the Fourier coefficients of $x(t)$ in terms of those of $y(t)$. What if k is not an integer?

(*b*) Use the result in Eq. (6-10) to determine the Fourier coefficients of the periodic signal in Fig. 6P-9*a*.

(*c*) Discuss the case when k is not an integer. Determine the Fourier coefficients of the periodic signal in Fig. 6P-9*b*.

6-16 Find the exponential Fourier series of the periodic signals $x(t)$ and $y(t)$ shown in Fig. 6P-10. [Note that $x(t) + x(t+2) = \sin \pi t$ and $y(t) = x(t + \frac{1}{2}) - x(t - \frac{3}{2})$.]

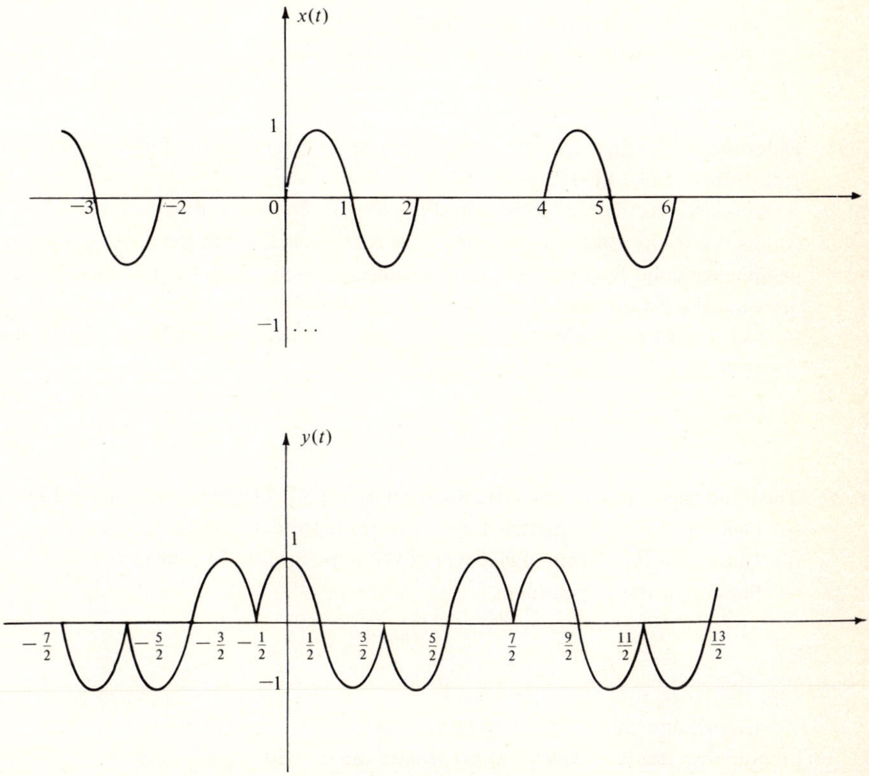

FIGURE 6P-10

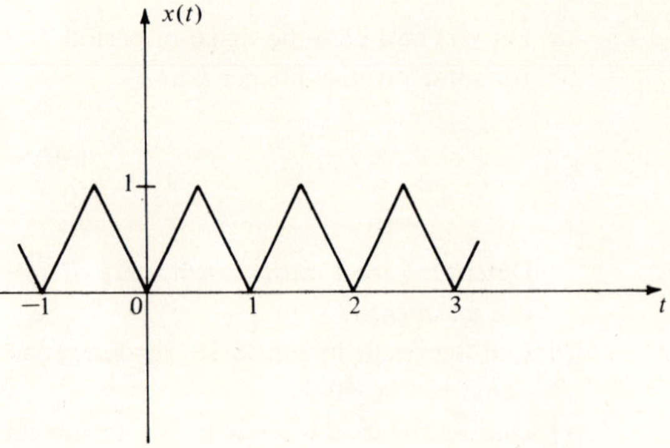

FIGURE 6P-11

6-17 (*a*) Without evaluating the Fourier coefficients x_n of the signal $x(t)$ shown in Fig. 6P-11, show that x_n decreases in magnitude for large values of n at least as rapidly as the ratio $1/n^2$.

(*b*) Let $y(t)$ be a periodic signal. Suppose that $y(t)$, $dy(t)/dt$, $d^2y(t)/dt^2, \ldots, d^ky(t)/dt^k$ are continuous everywhere, but $d^{k+1}y(t)/dt^{k+1}$ contains discontinuities. Show that the Fourier coefficients y_n of $y(t)$ decrease in magnitude for large values of n at least as rapidly as the ratio $1/n^{k+2}$.

6-18 Determine the trigonometric Fourier series representation of the signals $x(t)$ and $y(t)$ shown in Fig. 6P-12.

6-19 A half-wave rectifier is shown in Fig. 6P-13. In many practical cases, the diode conducts only for a very small portion of each cycle, so that the diode current $i(t)$ can be approximated by a periodic train of impulses with area A and period $2\pi/\omega$.

(*a*) Find the Fourier series for $e_3(t)$.

(*b*) Let $\bar{e}_3$ be the average value of $e_3(t)$ and E_3 be the rms value of the fundamental component of $e_3(t)$. Show that if $1/\omega C \ll \omega L$

$$\frac{E_3}{\bar{e}_3} \approx \frac{\sqrt{2}}{\omega^3 LC^2 R}$$

6-20 The input signal to the tank circuit shown in Fig. 6P-14*b*, $x(t)$, is shown in Fig. 6P-14*a*.

(*a*) Determine the exponential Fourier series representation of $x(t)$.

(*b*) Determine the output signal $y(t)$ of the tank circuit. Sketch $y(t)$.

6-21 (*a*) Show that the polynomials

$$p_1(t) = 1$$

$$p_2(t) = \sqrt{3}\,(1 - 2t)$$

are orthonormal in the time interval $0 \leq t \leq 1$.

(*b*) Suppose that we want to approximate the function

$$x(t) = e^t$$

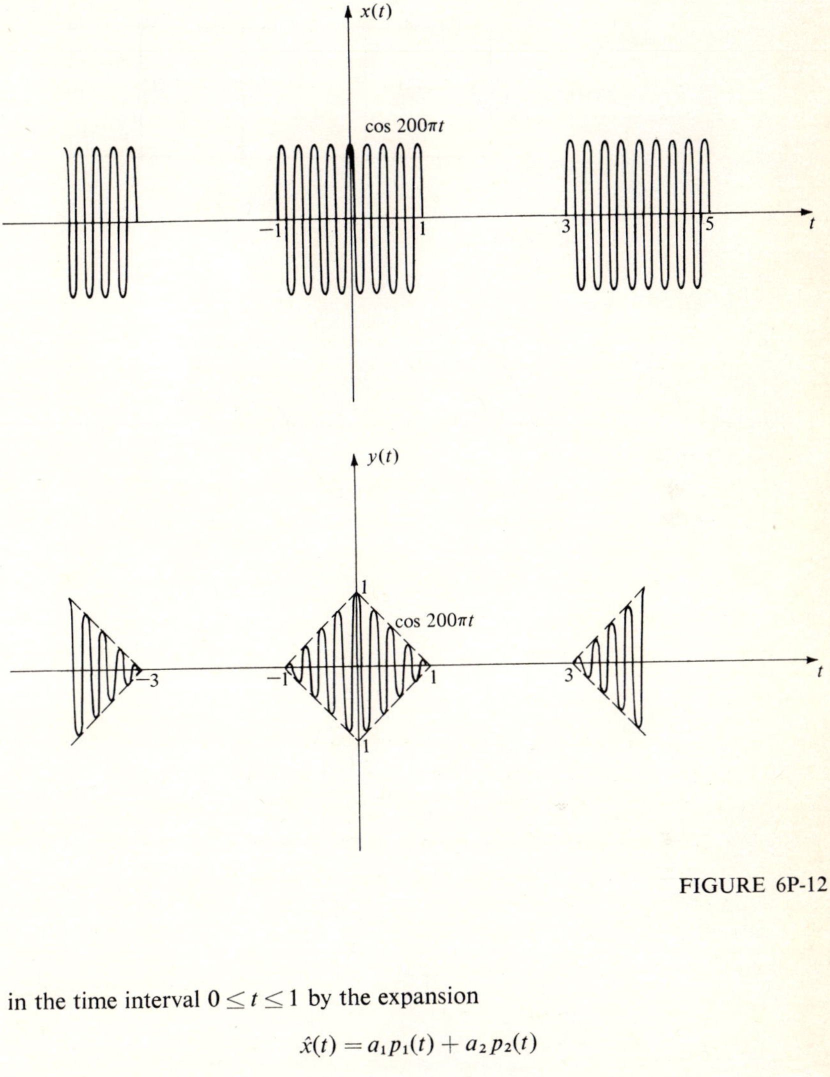

FIGURE 6P-12

in the time interval $0 \le t \le 1$ by the expansion

$$\hat{x}(t) = a_1 p_1(t) + a_2 p_2(t)$$

Determine the coefficients a_1 and a_2 so that the mean-square error

$$\int_0^1 |x(t) - \hat{x}(t)|^2 \, dt$$

is minimized.

6-22 (*a*) Show that the functions

$$\varphi_k(t) = \cos \frac{2\pi k t}{T} + \sin \frac{2\pi k t}{T} \qquad k = 0, \pm 1, \pm 2, \ldots$$

are orthogonal in the time interval $0 \le t \le T$.

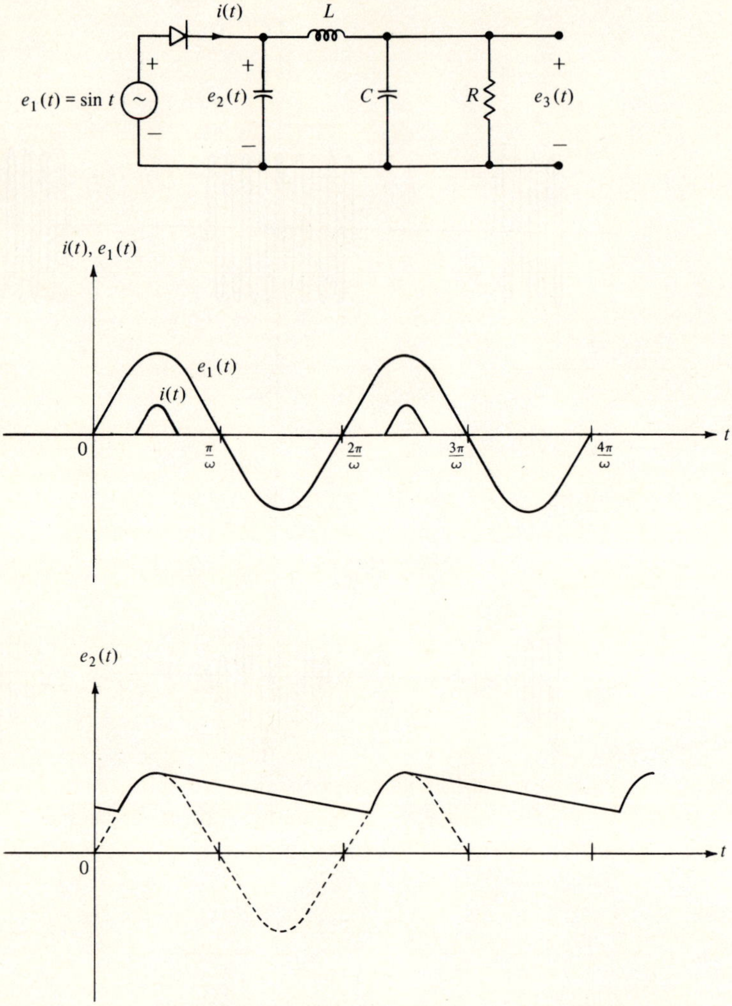

FIGURE 6P-13

(*b*) Let $x(t)$ be a real periodic function with Fourier coefficients x_n. Show that $x(t)$ can be written

$$x(t) = \sum_{k=-\infty}^{\infty} c_k \varphi_k(t)$$

by expressing the expansion coefficients c_k in terms of the Fourier coefficients of $x(t)$, x_n. Also show that

$$c_k = \frac{1}{T}\int_0^T x(t)\varphi_k(t)\, dt$$

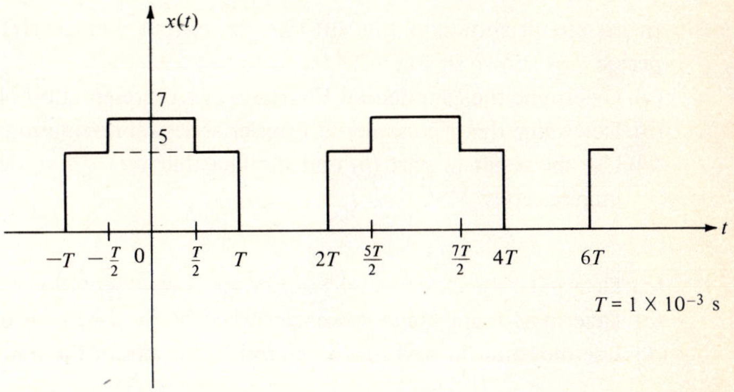

(a)

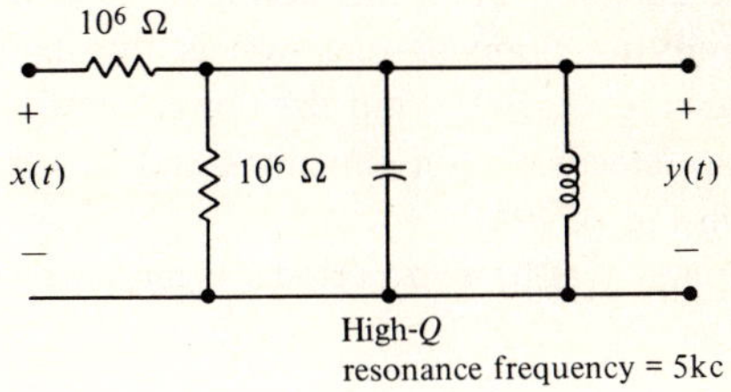

(b)

FIGURE 6P-14

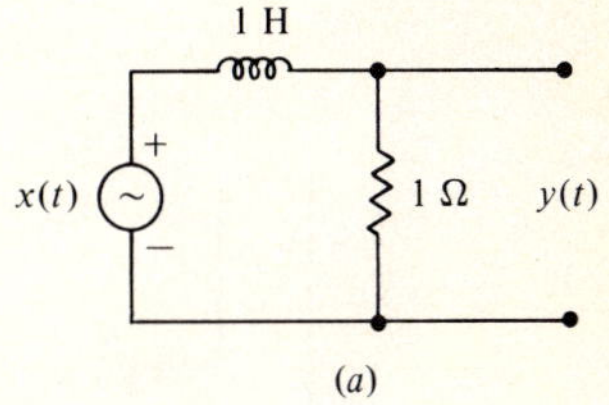

(a)

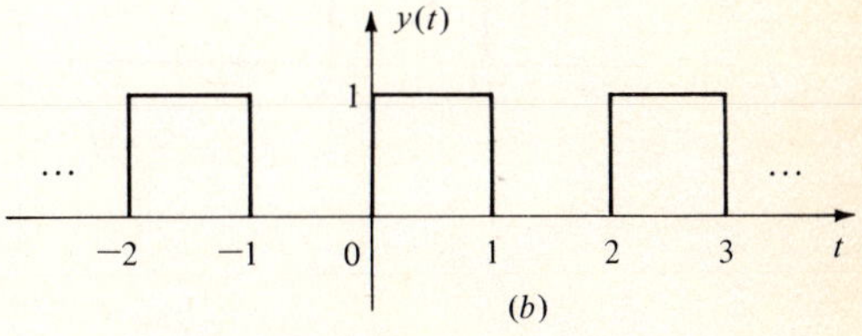

(b)

FIGURE 6P-15

6-23 In the circuit shown in Fig. 6P-15*a*, the output voltage $y(t)$ is a periodic signal of period 2 as shown in Fig. 6P-15*b*.

(*a*) Determine the exponential Fourier series representation of $y(t)$.

(*b*) Determine the trigonometric Fourier series representation of $y(t)$.

(*c*) Use the result in part (*b*) and the fact that $y(\frac{1}{2}) = 1$ to determine the sum of the infinite series

$$1 - \tfrac{1}{3} + \tfrac{1}{5} - \tfrac{1}{7} \cdots$$

(*d*) Determine the exponential Fourier series representation of $x(t)$.

(*e*) Determine the average power absorbed by the 1-Ω resistor.

(*f*) Use the result in part (*e*) to determine the sum of the infinite series

$$\frac{1}{1^2} + \frac{1}{3^2} + \frac{1}{5^2} + \frac{1}{7^2} + \cdots$$

6-24 Find the autocorrelation function and the power spectrum of the signal $x(t)$ shown in Fig. 6P-16. Suppose that $x(t)$ is the input signal of a system whose impulse response is $te^{-t}u_{-1}(t)$. Find the power spectrum of the output signal of the system.

6-25 Find the autocorrelation functions and power spectra of the signals $x(t)$ and $y(t)$ shown in Fig. 6P-10.

6-26 Let $x(t)$ and $y(t)$ be two periodic signals with period T. The periodic signal $r_{xy}(t)$ defined as

$$r_{xy}(t) = \frac{1}{T}\int_{-T/2}^{T/2} x(t)y(t+\tau)\,d\tau \qquad -\frac{T}{2} \le t \le \frac{T}{2}$$

is called the *cross-correlation function* of $x(t)$ and $y(t)$.

(*a*) Find the cross-correlation function of each pair of $x(t)$ and $y(t)$ shown in Fig. 6P-17.

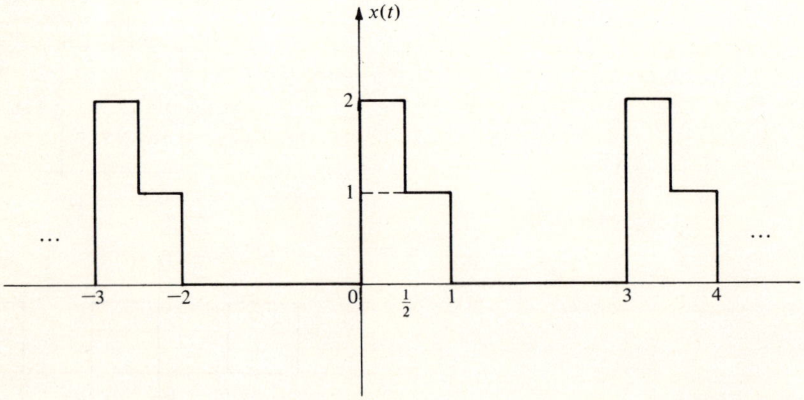

FIGURE 6P-16

(*b*) The Fourier coefficients of $r_{xy}(t)$ are called the *cross-power spectrum* of $x(t)$ and $y(t)$. Determine the cross-power spectrum of $x(t)$ and $y(t)$ in terms of the Fourier coefficients of $x(t)$ and $y(t)$.

(*c*) Find the cross-power spectrum for each pair of signals $x(t)$ and $y(t)$ in Fig. 6P-17.

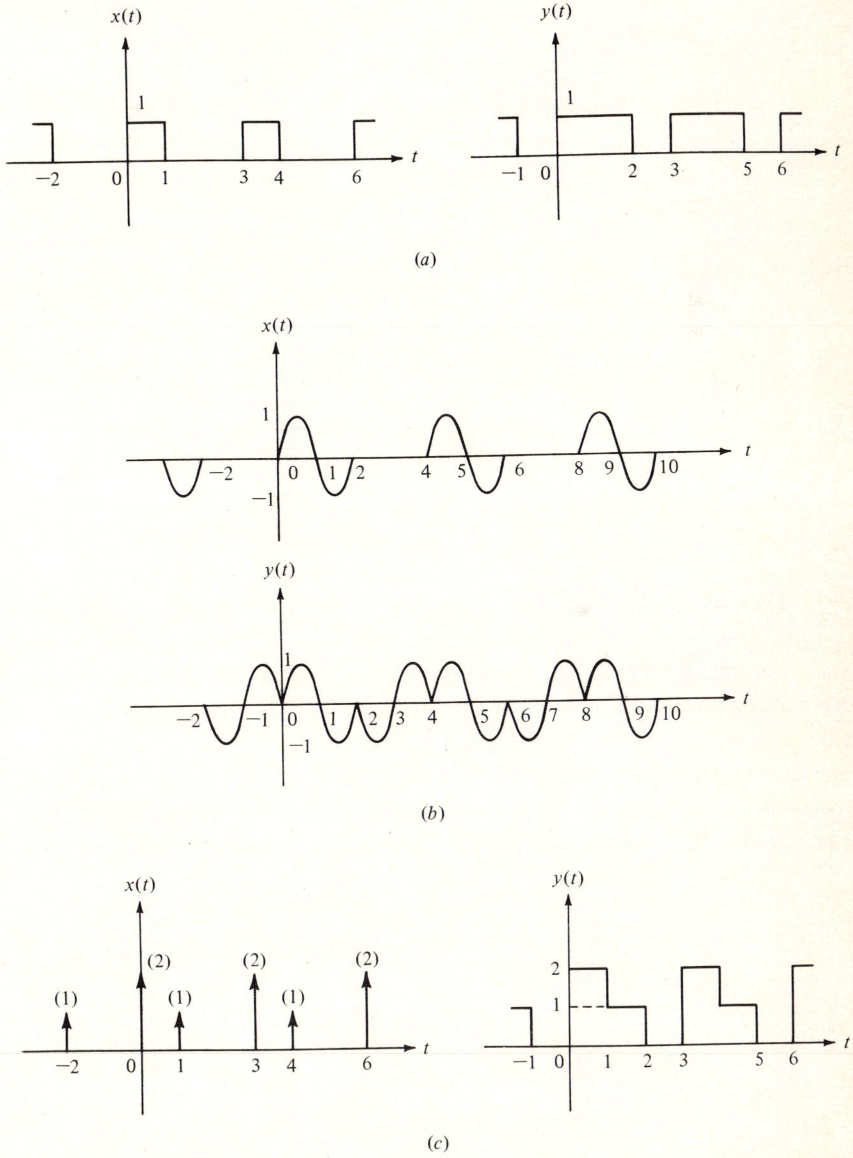

FIGURE 6P-17

7

FOURIER TRANSFORMATION

7-1 INTRODUCTION

As was pointed out in Chap. 6, an aperiodic signal cannot be expressed as a sum of complex exponential functions of the form $e^{jn\omega_0 t}$. In this chapter, we study a natural extension of the Fourier series representation of signals which leads to the possibility of expressing an aperiodic signal as an integral of complex exponential functions of the form $e^{j\omega t}$. Let $x(t)$ be a periodic signal of period T that satisfies the Dirichlet conditions. The Fourier series representation of $x(t)$ can be written

$$\begin{aligned} x(t) &= \sum_{n=-\infty}^{\infty} x_n e^{jn\omega_0 t} \\ &= \sum_{n=-\infty}^{\infty} \left[\frac{1}{T} \int_{-T/2}^{T/2} x(t) e^{-jn\omega_0 t}\, dt \right] e^{jn\omega_0 t} \\ &= \sum_{n=-\infty}^{\infty} \left[\frac{\omega_0}{2\pi} \int_{-T/2}^{T/2} x(t) e^{-jn\omega_0 t}\, dt \right] e^{jn\omega_0 t} \end{aligned} \tag{7-1}$$

When the period T becomes larger and larger, the angular velocity ω_0 becomes smaller and smaller. In the limit, as T approaches infinity, ω_0 becomes $d\omega$, $n\omega_0$ becomes ω, the infinite sum becomes an integral, and Eq. (7-1) becomes

$$x(t) = \int_{-\infty}^{\infty} \left[\frac{d\omega}{2\pi} \int_{-\infty}^{\infty} x(t)e^{-j\omega t}\, dt \right] e^{j\omega t} \tag{7-2}$$

Since, as T approaches infinity, $x(t)$ becomes an aperiodic signal, Eq. (7-2) suggests the possibility of expressing an aperiodic signal $x(t)$ as an integral of complex exponential functions:

$$x(t) = \frac{1}{2\pi} \int_{-\infty}^{\infty} X(\omega)e^{j\omega t}\, d\omega \tag{7-3}$$

where
$$X(\omega) = \int_{-\infty}^{\infty} x(t)e^{-j\omega t}\, dt \tag{7-4}$$

That is, if we let the "amplitude" of the complex exponential function $e^{j\omega t}$ be $(1/2\pi)X(\omega)\, d\omega$, then $x(t)$ is equal to an integral of these complex exponential functions.†

Such an argument is quite informal and is by no means a proof of the results. However, if the following conditions are satisfied:

1 $x(t)$ is absolutely integrable, that is, $\int_{-\infty}^{\infty} |x(t)|\, dt < \infty$;
2 $x(t)$ has a finite number of maxima and minima within any finite interval;
3 $x(t)$ has a finite number of discontinuities within any finite interval;

it can be shown that $x(t)$ can be expressed as an integral of complex exponential functions as in Eq. (7-3), where $X(\omega)$ is computed as in Eq. (7-4). The conditions 1 to 3 are called the *Dirichlet conditions.*

As was pointed out above, Eq. (7-3) can be viewed as a decomposition of the signal $x(t)$ into an integral of complex exponential functions of the form $e^{j\omega t}$, whereas Eq. (7-4) gives the amplitudes of these exponential functions. We can also take an alternative point of view. Mathematically, $x(t)$ and $X(\omega)$ are two functions related by Eqs. (7-3) and (7-4) so that $x(t)$ can be computed from $X(\omega)$ according to Eq. (7-4), and $X(\omega)$ can be computed from $x(t)$ according to Eq. (7-3). In this way we can view $X(\omega)$ as simply an alternative representation of $x(t)$, and $x(t)$ as an alternative representation of $X(\omega)$. (This indeed was the point of view we took when we discussed the z transformation of discrete signals in Chap. 5.) We call $X(\omega)$ the *Fourier transform* of $x(t)$, and $x(t)$ the *inverse Fourier transform* of $X(\omega)$. Also, $x(t)$ and $X(\omega)$ are called a *Fourier transform pair.*

† Following the terminologies used in Chap. 6, the terms $e^{j\omega t}$ are called the *frequency components* of the signal $x(t)$.

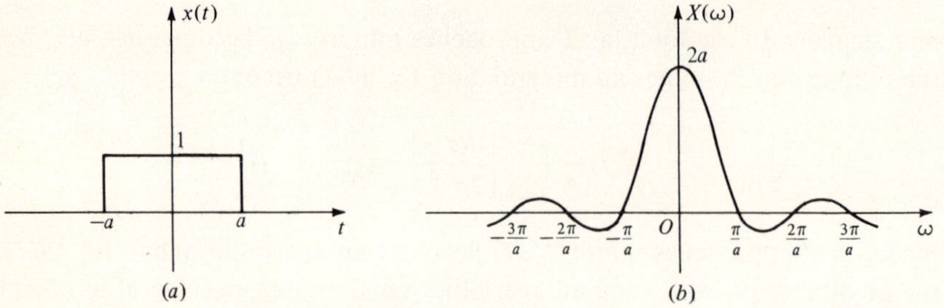

FIGURE 7-1

As an example, the Fourier transform of the signal $x(t)$ in Fig. 7-1a can be computed as

$$X(\omega) = \int_{-\infty}^{\infty} x(t)e^{-j\omega t}\,dt = \int_{-a}^{a} e^{-j\omega t}\,dt = \frac{1}{-j\omega}\,e^{-j\omega t}\Big|_{-a}^{a}$$

$$= \frac{1}{-j\omega}(e^{-j\omega a} - e^{j\omega a}) = \frac{2\sin a\omega}{\omega}$$

The function $X(\omega)$ is plotted in Fig. 7-1b.

As another example, the Fourier transform of the signal $e^{-\alpha t}u_{-1}(t)$ in Fig. 7-2a can be computed as

$$X(\omega) = \int_{0}^{\infty} e^{-\alpha t}e^{-j\omega t}\,dt = \frac{1}{\alpha + j\omega} \tag{7-5}$$

This example reminds us that $x(t)$ and $X(\omega)$ are, in general, complex functions of t and ω, respectively.

The complex function $X(\omega)$ is usually written in one of the two forms

$$X(\omega) = \text{Re}\,[X(\omega)] + j\,\text{Im}\,[X(\omega)]$$

$$X(\omega) = |X(\omega)|\,e^{j\angle X(\omega)}$$

where Re $[X(\omega)]$, Im $[X(\omega)]$, $|X(\omega)|$, and $\angle X(\omega)$ are all real functions of ω. Note that

$$|X(\omega)| = \sqrt{\text{Re}\,[X(\omega)]^2 + \text{Im}\,[X(\omega)]^2}$$

$$\angle X(\omega) = \tan^{-1}\frac{\text{Im}\,[X(\omega)]}{\text{Re}\,[X(\omega)]}$$

and

$$\text{Re}\,[X(\omega)] = |X(\omega)|\cos\,[\angle X(\omega)]$$

$$\text{Im}\,[X(\omega)] = |X(\omega)|\sin\,[\angle X(\omega)]$$

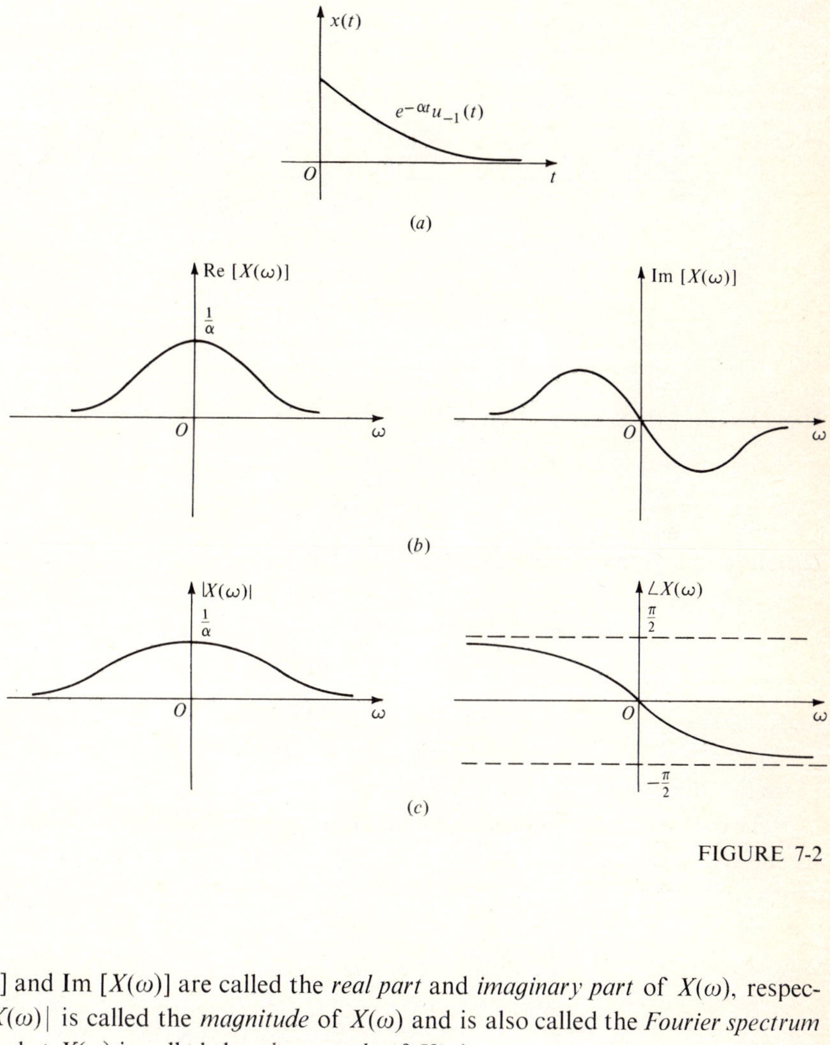

FIGURE 7-2

Re $[X(\omega)]$ and Im $[X(\omega)]$ are called the *real part* and *imaginary part* of $X(\omega)$, respectively; $|X(\omega)|$ is called the *magnitude* of $X(\omega)$ and is also called the *Fourier spectrum* of $x(t)$; and $\angle X(\omega)$ is called the *phase angle* of $X(\omega)$.

For example, the function $X(\omega)$ in Eq. (7-5) can be written

$$X(\omega) = \frac{\alpha}{\alpha^2 + \omega^2} - j\frac{\omega}{\alpha^2 + \omega^2}$$

Hence,

$$\text{Re}\,[X(\omega)] = \frac{\alpha}{\alpha^2 + \omega^2}$$

$$\text{Im}\,[X(\omega)] = \frac{-\omega}{\alpha^2 + \omega^2}$$

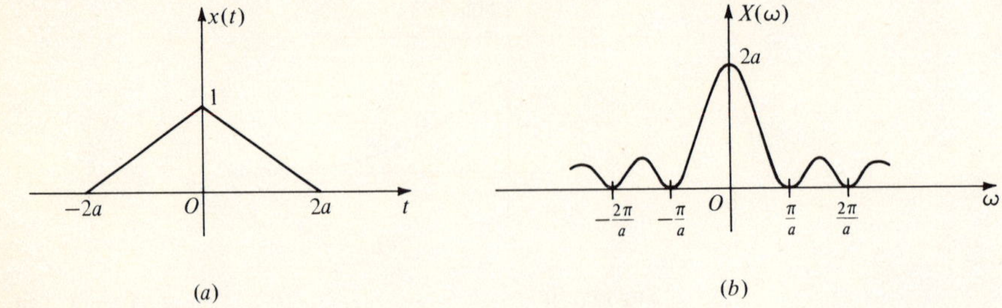

FIGURE 7-3

These two functions are plotted in Fig. 7-2*b*. Also, $X(\omega)$ can be written

$$X(\omega) = \frac{1}{\sqrt{\alpha^2 + \omega^2}} e^{-j\tan^{-1}(\omega/\alpha)}$$

Therefore, the Fourier spectrum of the signal $x(t)$ is

$$|X(\omega)| = \frac{1}{\sqrt{\alpha^2 + \omega^2}}$$

Also,
$$\angle X(\omega) = -\tan^{-1}\frac{\omega}{\alpha}$$

These two functions are plotted in Fig. 7-2*c*.

Two other frequently encountered time signals and their Fourier transforms are shown in Figs. 7-3 and 7-4. For the signal $x(t)$ in Fig. 7-3*a*, $X(\omega)$ can be computed as

$$\begin{aligned} X(\omega) &= \int_{-\infty}^{\infty} x(t)e^{-j\omega t}\,dt \\ &= \int_{-2a}^{0}\left(1 + \frac{t}{2a}\right)e^{-j\omega t}\,dt + \int_{0}^{2a}\left(1 - \frac{t}{2a}\right)e^{-j\omega t}\,dt \\ &= 2a\left(\frac{\sin a\omega}{a\omega}\right)^2 \end{aligned}$$

For the signal $x(t)$ in Fig. 7-4*a*, $X(\omega)$ can be computed as

$$\begin{aligned} X(\omega) &= \int_{-\infty}^{\infty} e^{-t^2/2b^2}e^{-j\omega t}\,dt \\ &= \left(\int_{-\infty}^{\infty} e^{-(1/2b^2)(t+j\omega b^2)^2}\,dt\right)e^{-b^2\omega^2/2} \\ &= \sqrt{2\pi}\,be^{-b^2\omega^2/2} \end{aligned}$$

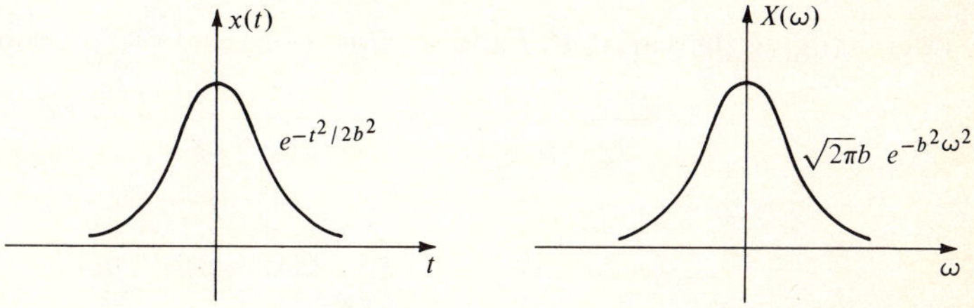

FIGURE 7-4

A reader might wonder why we have not yet given an example illustrating how $x(t)$ can be recovered from its Fourier transform $X(\omega)$. Conceptually, it is straightforward to compute $x(t)$ from $X(\omega)$ according to Eq. (7-3). However, in most cases, we shall encounter a complex integral that is quite difficult to evaluate. We shall return to the topic of evaluating complex integrals in Chap. 9 where a procedure similar to that presented in Chap. 5 for evaluating inverse z transformation will be presented. Here, we list several Fourier transform pairs in Table 7-1. Together with the pairs in Table 7-2 (Sec. 7-5) and making use of the many properties of Fourier transform pairs to be discussed in subsequent sections, we shall be able to compute $x(t)$ from $X(\omega)$ in most of the problems we shall encounter. We shall illustrate the computation of $x(t)$ from $X(\omega)$ by examples in subsequent sections.

If we replace t by $-t$ in the Fourier transform pair in Eqs. (7-3) and (7-4), we obtain

$$x(-t) = \frac{1}{2\pi}\int_{-\infty}^{\infty} X(\omega)e^{-j\omega t}\, d\omega \tag{7-6}$$

$$X(\omega) = \int_{-\infty}^{\infty} x(-t)e^{j\omega t}\, dt \tag{7-7}$$

Table 7-1

$x(t)$	$X(\omega)$
$u_{-1}(t+a) - u_{-1}(t-a)$	$2\,\dfrac{\sin a\omega}{\omega}$
$e^{-\alpha t}u_{-1}(t)$	$\dfrac{1}{\alpha + j\omega}$
$e^{\alpha t}u_{-1}(-t)$	$\dfrac{1}{\alpha - j\omega}$
$e^{-t^2/2b^2}$	$\sqrt{2\pi}\,be^{-b^2\omega^2/2}$
$\dfrac{1}{2\alpha}\,e^{-\alpha\lvert t\rvert}$	$\dfrac{1}{\alpha^2 + \omega^2}$

By interchanging the variables t and ω, Eqs. (7-6) and (7-7) can be written

$$2\pi x(-\omega) = \int_{-\infty}^{\infty} X(t)e^{-j\omega t}\,dt$$

$$X(t) = \frac{1}{2\pi}\int_{-\infty}^{\infty} 2\pi x(-\omega)e^{j\omega t}\,d\omega$$

We conclude that if $X(\omega)$ is the Fourier transform of the time signal $x(t)$, then $2\pi x(-\omega)$ is the Fourier transform of the time signal $X(t)$. This result is known as the *duality property* of Fourier transform pairs. Thus, from the Fourier transform pair in Fig. 7-1, we see that the Fourier transform of the time signal $(2 \sin at)/t$ is $2\pi[u_{-1}(\omega + a) - u_{-1}(\omega - a)]$. Also, from the Fourier transform pair in Fig. 7-2, we see that the Fourier transform of the complex time signal $1/(\alpha + jt)$ is $2\pi e^{\alpha\omega}u_{-1}(-\omega)$.

7-2 SYMMETRY PROPERTIES OF TIME SIGNALS

We study in this section how some of the symmetry properties of time signals are reflected in their Fourier transforms. Let $x(t)$ be an even time function. By a change of variable $t = -\tau$ in Eq. (7-4), we obtain

$$X(\omega) = \int_{-\infty}^{\infty} x(t)e^{-j\omega t}\,dt = \int_{-\infty}^{\infty} x(-\tau)e^{j\omega\tau}\,d\tau = \int_{-\infty}^{\infty} x(\tau)e^{j\omega\tau}\,d\tau$$
$$= X(-\omega)$$

That is, the Fourier transform of an even time signal is an even function of ω. For example, we see that the Fourier transform of the time signal in Fig. 7-1*a* is an even function of ω as shown in Fig. 7-1*b*. In a similar manner, we can show that if $x(t)$ is an odd time signal, $X(\omega)$ is an odd function of ω.

Let $x(t)$ be a real time signal. Since

$$X(\omega) = \int_{-\infty}^{\infty} x(t)e^{-j\omega t}\,dt$$

we have

$$X(-\omega) = \int_{-\infty}^{\infty} x(t)e^{j\omega t}\,dt$$

and
$$X^*(\omega) = \int_{-\infty}^{\infty} [x(t)e^{-j\omega t}]^*\,dt = \int_{-\infty}^{\infty} x(t)e^{j\omega t}\,dt$$

That is,

$$X(-\omega) = X^*(\omega)$$

or

$$\text{Re}\,[X(-\omega)] = \text{Re}\,[X(\omega)]$$

$$\text{Im}\,[X(-\omega)] = -\text{Im}\,[X(\omega)]$$

In other words, for a real time signal, the real part of its Fourier transform is an even function of ω and the imaginary part of its Fourier transform is an odd function of ω. As an example, we see in Fig. 7-2*b* that the real part and the imaginary part of the Fourier transform of the signal $e^{-\alpha t}u_{-1}(t)$ are even and odd functions of ω, respectively. In a similar manner, it can be shown that, for a real time signal, its Fourier spectrum is an even function of ω and the phase angle of its Fourier transform is an odd function of ω. We leave it to the reader to show that if $x(t)$ is an even and real time signal then its Fourier transform is an even and real function of ω. In other words, $\text{Re}\,[X(\omega)] = \text{Re}\,[X(-\omega)]$ and $\text{Im}\,[X(\omega)] = 0$. Also, if $x(t)$ is an odd and real time signal, its Fourier transform is an odd and imaginary function of ω. That is, $\text{Re}\,[X(\omega)] = 0$ and $\text{Im}\,[X(\omega)] = -\text{Im}\,[X(-\omega)]$.

It should also be noted that, for a real signal $x(t)$, Eq. (7-3) can be written

$$x(t) = \frac{1}{\pi}\int_0^\infty \text{Re}\,[X(\omega)] \cos \omega t \, d\omega - \frac{1}{\pi}\int_0^\infty \text{Im}\,[X(\omega)] \sin \omega t \, d\omega$$

In other words, $x(t)$ can be decomposed into an infinite number of sinusoids and cosinusoids, where the amplitude of $\sin \omega t$ is $-(1/\pi)\,\text{Im}\,[X(\omega)]\,d\omega$ and the amplitude of $\cos \omega t$ is $(1/\pi)\,\text{Re}\,[X(\omega)]\,d\omega$.

We show now that a real time signal $x(t)$ which is equal to zero for $t < 0$ is completely specified by the real part of its Fourier transform $X(\omega)$. Let us write $x(t)$ as

$$x(t) = x_e(t) + x_0(t)$$

where

$$x_e(t) = \tfrac{1}{2}[x(t) + x(-t)]$$

is an even time function, and

$$x_0(t) = \tfrac{1}{2}[x(t) - x(-t)]$$

is an odd time function. Since

$$x(t) = 0 \qquad \text{for } t < 0$$

we have

$$x_e(t) + x_0(t) = 0 \qquad \text{for } t < 0$$

or

$$x_0(t) = -x_e(t) \qquad \text{for } t < 0$$

It follows that

$$x_0(t) = x_e(t) \qquad \text{for } t > 0$$

or

$$x(t) = 2x_e(t)u_{-1}(t) \tag{7-8}$$

Let $X_e(\omega)$ and $X_0(\omega)$ denote the Fourier transforms of $x_e(t)$ and $x_0(t)$, respectively. As was discussed earlier, $X_e(\omega)$ is a real function of ω and $X_0(\omega)$ is an imaginary function of ω. Since†

$$X(\omega) = X_e(\omega) + X_0(\omega)$$

we conclude that

$$\begin{aligned} X_e(\omega) &= \text{Re}\,[X(\omega)] \\ X_0(\omega) &= j\,\text{Im}\,[X(\omega)] \end{aligned} \tag{7-9}$$

Therefore, according to Eqs. (7-9) we can determine $x_e(t)$ from Re $[X(\omega)]$, and according to Eq. (7-8), we can determine $x(t)$ from $x_e(t)$. Consequently, we can also determine Im $[X(\omega)]$ from Re $[X(\omega)]$. As a matter of fact, Re $[X(\omega)]$ and Im $[X(\omega)]$ are related by the equations

$$\text{Re}\,[X(\omega)] = \frac{1}{\pi}\int_{-\infty}^{\infty} \frac{\text{Im}\,[X(\sigma)]}{\omega - \sigma}\,d\sigma \tag{7-10}$$

$$\text{Im}\,[X(\omega)] = -\frac{1}{\pi}\int_{-\infty}^{\infty} \frac{\text{Re}\,[X(\sigma)]}{\omega - \sigma}\,d\sigma \tag{7-11}$$

The relations in Eqs. (7-10) and (7-11) will be proved in Sec. 7-5. This pair of integrals are known as *Hilbert transforms*.

As an example, for the time signal

$$x(t) = \begin{cases} e^{-\alpha t} & t > 0 \\ 0 & t < 0 \end{cases}$$

as shown in Fig. 7-5*a*, we have

$$x_e(t) = \begin{cases} \frac{1}{2}e^{-\alpha t} & t > 0 \\ \frac{1}{2}e^{\alpha t} & t < 0 \end{cases}$$

and

$$x_0(t) = \begin{cases} \frac{1}{2}e^{-\alpha t} & t > 0 \\ -\frac{1}{2}e^{\alpha t} & t < 0 \end{cases}$$

† We use an obvious result that the Fourier transform of the sum of two signals is equal to the sum of their Fourier transforms. See Sec. 7-3 for a proof.

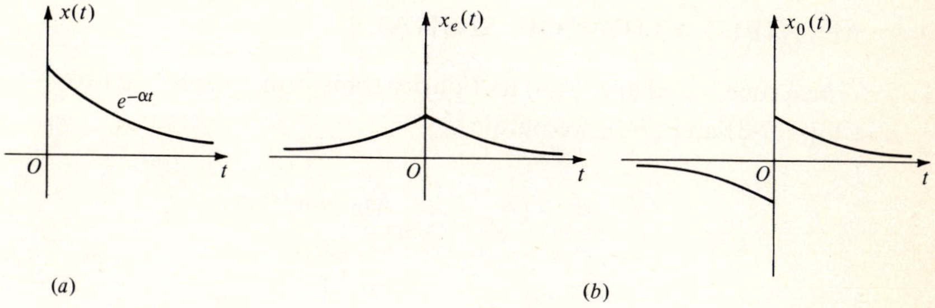

FIGURE 7-5

as shown in Fig. 7-5*b*. We found earlier that

$$X(\omega) = \frac{1}{\alpha + j\omega} = \frac{\alpha}{\alpha^2 + \omega^2} - j\frac{\omega}{\alpha^2 + \omega^2}$$

It follows that

$$X_e(\omega) = \frac{\alpha}{\alpha^2 + \omega^2}$$

$$X_0(\omega) = -\frac{\omega}{\alpha^2 + \omega^2}$$

We can also check the relation in Eq. (7-10). Indeed, we have

$$\begin{aligned}
\frac{1}{\pi}\int_{-\infty}^{\infty} \frac{\text{Im}\,[X(\sigma)]}{\omega - \sigma}\,d\sigma &= \frac{1}{\pi}\int_{-\infty}^{\infty} \frac{-\sigma}{(\alpha^2 + \sigma^2)(\omega - \sigma)}\,d\sigma \\
&= \frac{1}{\pi(\alpha^2 + \omega^2)}\int_{-\infty}^{\infty}\left(-\frac{\omega\sigma}{\alpha^2 + \sigma^2} + \frac{\alpha^2}{\alpha^2 + \sigma^2} - \frac{\omega}{\omega - \sigma}\right)d\sigma \\
&= \frac{1}{\pi(\alpha^2 + \omega^2)}\left[-\frac{\omega}{2}\ln(\alpha^2 + \sigma^2) + \alpha\tan^{-1}\frac{\sigma}{\alpha} - \omega\ln(\omega - \sigma)\right]\Bigg|_{-\infty}^{\infty} \\
&= \frac{\alpha}{\alpha^2 + \omega^2} \\
&= \text{Re}\,[X(\omega)]
\end{aligned}$$

We leave it to the reader to check the relation in Eq. (7-11) by showing that

$$-\frac{1}{\pi}\int_{-\infty}^{\infty} \frac{\alpha}{(\alpha^2 + \sigma^2)(\omega - \sigma)}\,d\sigma = \frac{\omega}{\alpha^2 + \omega^2}$$

7-3 MANIPULATION OF SIGNALS

Let $x(t)$ be a time signal and $X(\omega)$ its Fourier transform. Replacing t by $-t$ and ω by $-\omega$ in Eqs. (7-3) and (7-4), we obtain

$$x(-t) = \frac{1}{2\pi}\int_{-\infty}^{\infty} X(-\omega)e^{j\omega t}\,d\omega$$

$$X(-\omega) = \int_{-\infty}^{\infty} x(-t)e^{-j\omega t}\,dt$$

Thus, we conclude that the Fourier transform of the time signal $x(-t)$ is $X(-\omega)$. That is, transposing a time signal also transposes its Fourier transform.

Changing the variable t to at and the variable ω to ω/a for a positive constant a in Eqs. (7-3) and (7-4), we obtain

$$x(at) = \frac{1}{2\pi}\int_{-\infty}^{\infty} X\left(\frac{\omega}{a}\right)e^{j\omega t}\,\frac{d\omega}{a}$$

$$X\left(\frac{\omega}{a}\right) = \int_{-\infty}^{\infty} x(at)e^{-j\omega t}a\,dt$$

That is, the Fourier transform of the time signal $x(at)$ is $(1/a)X(\omega/a)$. We note from this result that contracting the time scale amounts to extending the frequency scale and extending the time scale amounts to contracting the frequency scale.

For example, since the Fourier transform of the signal $e^{-\alpha t}u_{-1}(t)$ is

$$\frac{1}{\alpha + j\omega}$$

the Fourier transform of $e^{\alpha t}u_{-1}(-t)$ is

$$\frac{1}{\alpha - j\omega}$$

Also, the Fourier transform of $e^{-t}u_{-1}(t)$, which can also be written $e^{-t}u_{-1}(t/\alpha)$, is

$$\alpha\,\frac{1}{\alpha + j\alpha\omega} = \frac{1}{1 + j\omega}$$

Changing the variable t to $t + t_0$ in Eqs. (7-3) and (7-4), we obtain

$$x(t + t_0) = \frac{1}{2\pi}\int_{-\infty}^{\infty} X(\omega)e^{j\omega t_0}e^{j\omega t}\,d\omega$$

$$X(\omega) = \int_{-\infty}^{\infty} x(t + t_0)e^{-j\omega t_0}e^{-j\omega t}\,dt$$

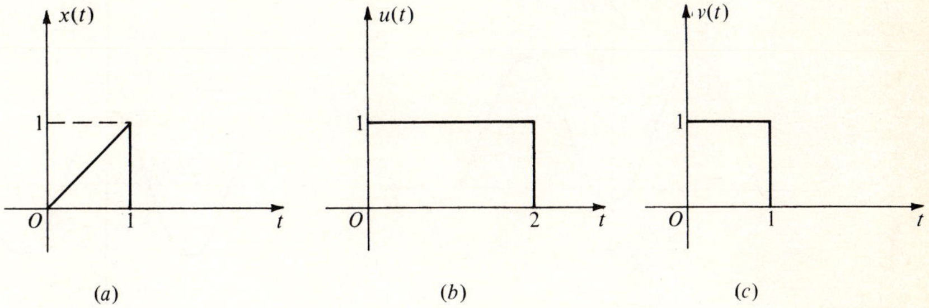

FIGURE 7-6

That is, the Fourier transform of $x(t + t_0)$ is $X(\omega)e^{j\omega t_0}$. Similarly, by changing the variable ω to $\omega + \omega_0$ in Eqs. (7-3) and (7-4), we can show that the Fourier transform of $e^{-j\omega_0 t}x(t)$ is $X(\omega + \omega_0)$.

Also, differentiating both sides of Eq. (7-3) with respect to t, we obtain

$$\frac{dx(t)}{dt} = \frac{1}{2\pi}\int_{-\infty}^{\infty} j\omega X(\omega)e^{j\omega t}\, d\omega$$

That is, the Fourier transform of $dx(t)/dt$ is $j\omega X(\omega)$ if the transform exists.† [Note that the Fourier transform of $dx(t)/dt$ might not exist even if the Fourier transform of $x(t)$ does.] Similarly, differentiating both sides of Eq. (7-4) with respect to ω, we show that the Fourier transform of $-jtx(t)$ is $dX(\omega)/d\omega$.

Let $X_1(\omega)$ be the Fourier transform of $x_1(t)$ and $X_2(\omega)$ be the Fourier transform of $x_2(t)$. Then, according to Eqs. (7-3) and (7-4),

$$a_1 x_1(t) + a_2 x_2(t) = \frac{1}{2\pi}\int_{-\infty}^{\infty} [a_1 X_1(\omega) + a_2 X_2(\omega)]e^{j\omega t}$$

$$a_1 X_1(\omega) + a_2 X_2(\omega) = \int_{-\infty}^{\infty} [a_1 x_1(t) + a_2 x_2(t)]e^{-j\omega t}\, dt$$

That is, the Fourier transform of $a_1 x_1(t) + a_2 x_2(t)$ is $a_1 X_1(\omega) + a_2 X_2(\omega)$.

As an illustrative example, we shall determine the Fourier transform of the signal $x(t)$ in Fig. 7-6*a*. According to the result in Fig. 7-1, the Fourier transform of $u(t)$ in Fig. 7-6*b* is

$$U(\omega) = \frac{2 \sin \omega}{\omega} e^{-j\omega}$$

† We caution the reader not to conclude immediately that the Fourier transform of $\int_{-\infty}^{t} x(\tau)\, d\tau$ is $X(\omega)/j\omega$. See Sec. 7-5 where the correct result $\pi X(0)u_0(\omega) + X(\omega)/j\omega$ is derived.

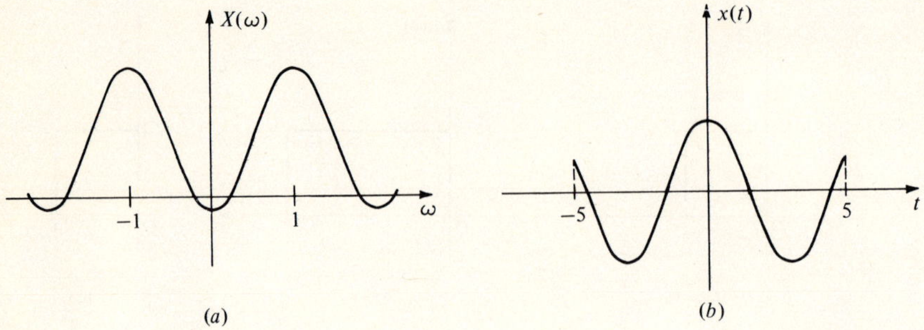

FIGURE 7-7

The Fourier transform of $v(t)$, which is equal to $u(2t)$, in Fig. 7-6c is

$$V(\omega) = \frac{2 \sin(\omega/2)}{\omega} e^{-j\omega/2}$$

Since

$$x(t) = tv(t)$$

we have

$$X(\omega) = j\frac{dV(\omega)}{d\omega} = \frac{j\cos(\omega/2)}{\omega} e^{-j\omega/2} + \frac{\sin(\omega/2)}{\omega} e^{-j\omega/2} - \frac{j2\sin(\omega/2)}{\omega^2} e^{-j\omega/2}$$

$$= \frac{je^{-j\omega/2}}{2\omega}(e^{j\omega/2} + e^{-j\omega/2} - e^{j\omega/2} + e^{-j\omega/2}) - \frac{j2\sin(\omega/2)}{\omega^2} e^{-j\omega/2}$$

$$= \frac{je^{-j\omega}}{\omega} - \frac{j2\sin(\omega/2)}{\omega^2} e^{-j\omega/2}$$

As another example, let us determine the time signal $x(t)$ whose Fourier transform is known to be

$$X(\omega) = \frac{\sin 5(\omega + 1)}{\omega + 1} + \frac{\sin 5(\omega - 1)}{\omega - 1}$$

as shown in Fig. 7-7a. Since the Fourier transform of $u_{-1}(t + 5) - u_{-1}(t - 5)$ is

$$\frac{2 \sin 5\omega}{\omega}$$

the Fourier transform of $e^{-jt}[u_{-1}(t + 5) - u_{-1}(t - 5)]$ is

$$\frac{2 \sin 5(\omega + 1)}{\omega + 1}$$

and the Fourier transform of $e^{jt}[u_{-1}(t+5) - u_{-1}(t-5)]$ is

$$\frac{2 \sin 5(\omega - 1)}{\omega - 1}$$

Consequently,

$$\begin{aligned} x(t) &= \tfrac{1}{2}(e^{jt} + e^{-jt})[u_{-1}(t+5) - u_{-1}(t-5)] \\ &= \cos t[u_{-1}(t+5) - u_{-1}(t-5)] \end{aligned}$$

which is plotted in Fig. 7-7*b*.

7-4 THE CONVOLUTION PROPERTY

Let $x(t)$ be the input signal to a linear time-invariant system whose impulse response is $h(t)$. Since

$$x(t) = \frac{1}{2\pi} \int_{-\infty}^{\infty} X(\omega) e^{j\omega t} \, d\omega$$

according to our discussion in Sec. 6-1, the output signal of the system is

$$y(t) = \frac{1}{2\pi} \int_{-\infty}^{\infty} \left[\int_{-\infty}^{\infty} h(\tau) e^{-j\omega\tau} \, d\tau \right] X(\omega) e^{j\omega t} \, d\omega \tag{7-12}$$

provided that $e^{j\omega t}$ are eigenfunctions of the system for all ω. Let us assume the existence of the Fourier transform of $h(t)$, which we shall denote $H(\omega)$. It follows from the Dirichlet condition that the region of absolute convergence of the system includes the $j\omega$ axis in the complex plane. Thus, Eq. (7-12) can be written

$$y(t) = \frac{1}{2\pi} \int_{-\infty}^{\infty} X(\omega) H(\omega) e^{j\omega t} \, d\omega$$

or

$$Y(\omega) = X(\omega) H(\omega) \tag{7-13}$$

That is, the Fourier transform of the output signal of a linear time-invariant system is equal to the product of the Fourier transform of the input signal and that of the impulse response of the system. The Fourier transform of the impulse response of a system is referred to as the *transfer function* or the *system function* of the system.

Indeed, the relation in Eq. (7-13) can be stated in slightly more general terms: Let $x(t)$ and $h(t)$ be two time signals. The Fourier transform of the convolution of two time signals $x(t) * h(t)$ is equal to the product of their Fourier transforms

$X(\omega)H(\omega)$.† We point out that many of the properties of the convolution algebra discussed in Sec. 4-4 can be derived easily from Eq. (7-13). For example, since

$$H(\omega)X(\omega) = X(\omega)H(\omega)$$

we have

$$h(t) * x(t) = x(t) * h(t)$$

We leave the details to the reader.

As an example, let $x(t) = 5e^{-2t}u_{-1}(t)$ be the input to a system whose impulse response is $h(t) = e^{-3t}u_{-1}(t)$. Since

$$X(\omega) = \frac{5}{2 + j\omega}$$

$$H(\omega) = \frac{1}{3 + j\omega}$$

the Fourier transform of the output is

$$Y(\omega) = \frac{5}{2 + j\omega}\,\frac{1}{3 + j\omega}$$

$$= \frac{5}{2 + j\omega} - \frac{5}{3 + j\omega}$$

It follows that

$$y(t) = (5e^{-2t} - 5e^{-3t})u_{-1}(t)$$

The reader can check the result by computing $x(t) * h(t)$.

The reader will undoubtedly realize that Eq. (7-13) suggests an indirect method of computing the convolution of two time signals. Instead of evaluating the convolution integral, we determine first the Fourier transform of the resultant signal and then carry out an inverse Fourier transformation. The following example illustrates that,

† Equation (7-13) can be derived directly without using the concept of eigenfunctions of systems: Since

$$y(t) = \int_{-\infty}^{\infty} x(\tau)h(t - \tau)\,d\tau$$

$$= \int_{-\infty}^{\infty} \left[\frac{1}{2\pi}\int_{-\infty}^{\infty} X(\omega)e^{j\omega\tau}\,d\omega\right] h(t - \tau)\,d\tau$$

$$= \frac{1}{2\pi}\int_{-\infty}^{\infty} X(\omega)\left[\int_{-\infty}^{\infty} h(t - \tau)e^{-j\omega(t-\tau)}\,d\tau\right] e^{j\omega t}\,d\omega$$

$$= \frac{1}{2\pi}\int_{-\infty}^{\infty} X(\omega)H(\omega)e^{j\omega t}\,d\omega$$

then $\qquad Y(\omega) = X(\omega)H(\omega)$

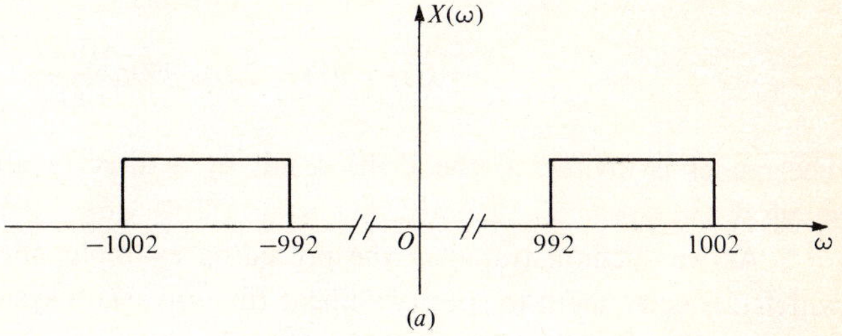

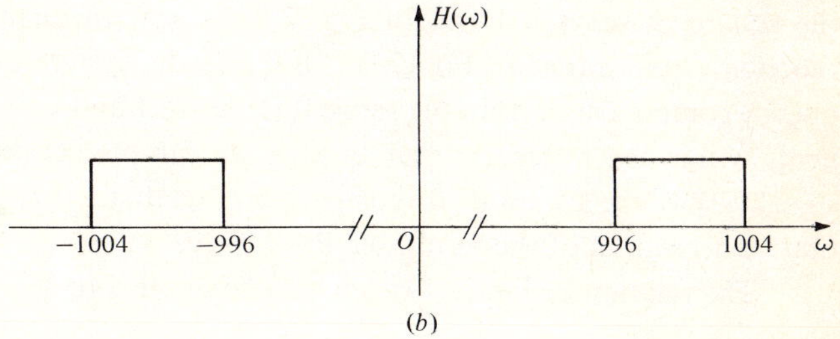

FIGURE 7-8

in many cases, the indirect computation might be easier than a direct evaluation of the convolution integral. Let

$$x(t) = 2 \cos 997t \frac{\sin 5t}{\pi t}$$

and

$$h(t) = 2 \cos 1000t \frac{\sin 4t}{\pi t}$$

We write

$$x(t) = (e^{j997t} + e^{-j997t}) \frac{\sin 5t}{\pi t}$$

Since the Fourier transform of $(\sin 5t)/\pi t$ is $[u_{-1}(\omega + 5) - u_{-1}(\omega - 5)]$, we obtain

$$X(\omega) = \begin{cases} 1 & 992 < |\omega| < 1002 \\ 0 & \text{otherwise} \end{cases}$$

as shown in Fig. 7-8*a*. Similarly, we obtain

$$H(\omega) = \begin{cases} 1 & 996 < |\omega| < 1004 \\ 0 & \text{otherwise} \end{cases}$$

as shown in Fig. 7-8*b*. It follows that

$$X(\omega)H(\omega) = \begin{cases} 1 & 996 < |\omega| < 1002 \\ 0 & \text{otherwise} \end{cases}$$

or
$$x(t) * h(t) = 2 \cos 999t \, \frac{\sin 3t}{\pi t}$$

The reader is invited to check the result by a direct evaluation of the convolution integral.

As was demonstrated in the preceding example, there are many occasions in which it is convenient to specify a linear time-invariant system by its transfer function $H(\omega)$ instead of its impulse response $h(t)$. To illustrate this point, let us suppose that the square pulse $x(t)$ shown in Fig. 7-9*a* is sent through a system whose transfer function $H(\omega)$ is that in Fig. 7-9*b*. Because the system attenuates the nonzero frequency components within the range $0 \le |\omega| \le 1$ and cuts off all frequency components within the range $1 < |\omega|$ in $x(t)$, we can predict qualitatively that the output signal $y(t)$ will be a smoothed version of $x(t)$, as that shown in Fig. 7-9*c*. For a quantitative derivation of the result, see Prob. 7-20.

The relation in Eq. (7-13) can also be applied to solve the integral equation

$$y(t) = \int_{-\infty}^{\infty} x(\tau)h(t - \tau)\, d\tau$$

for $h(t)$ when the signals $x(t)$ and $y(t)$ are given. As an example, let

$$x(t) = e^{-\alpha t}u_{-1}(t)$$

and
$$y(t) = te^{-\alpha t}u_{-1}(t)$$

be an input-output pair of a linear time-invariant system. We have

$$X(\omega) = \frac{1}{\alpha + j\omega}$$

To compute $Y(\omega)$, we recall that the Fourier transform of $-jte^{-\alpha t}u_{-1}(t)$ is equal to

$$\frac{d}{d\omega}\frac{1}{\alpha + j\omega} = \frac{-j}{(\alpha + j\omega)^2}$$

Thus,
$$Y(\omega) = \frac{1}{(\alpha + j\omega)^2}$$

It follows that

$$H(\omega) = \frac{Y(\omega)}{X(\omega)} = \frac{1}{\alpha + j\omega}$$

or
$$h(t) = e^{-\alpha t}u_{-1}(t)$$

Using the duality property, we note that if

$$y(t) = x(t)h(t)$$

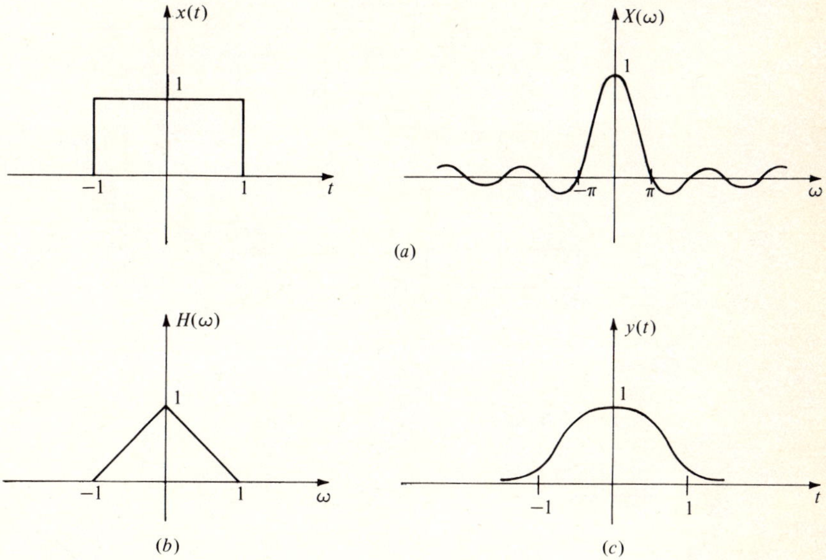

FIGURE 7-9

then $$Y(\omega) = \frac{1}{2\pi} X(\omega) * H(\omega)$$

In other words, the Fourier transform of the product of two time signals is equal to the convolution of their Fourier transforms.

As an example, let us determine the impulse response of the system shown in Fig. 7-10*a*, where

$$h_1(t) = \frac{\sin 2t}{\pi t}$$

$$h_2(t) = 2\pi \frac{\sin t}{\pi t} \frac{\sin 2t}{\pi t}$$

According to our discussion in Sec. 4-4, the impulse response of the overall system is equal to $h_1(t) * h_2(t)$. Thus, the transfer function of the overall system is equal to $H_1(\omega)H_2(\omega)$. We obtain

$$H_1(\omega) = u_{-1}(\omega + 2) - u_{-1}(\omega - 2)$$

as shown in Fig. 7-10*b* and

$$H_2(\omega) = [u_{-1}(\omega + 1) - u_{-1}(\omega - 1)] * [u_{-1}(\omega + 2) - u_{-1}(\omega - 2)]$$

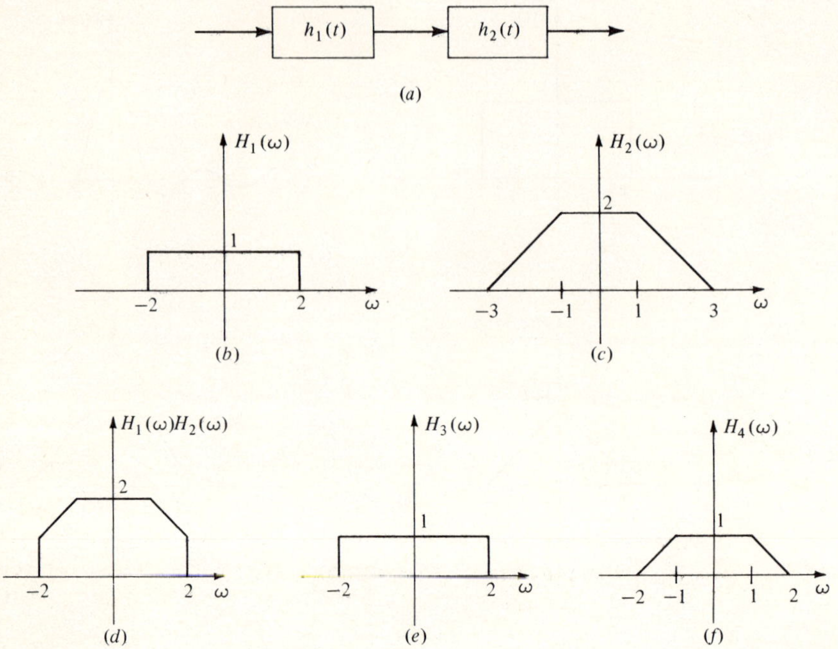

FIGURE 7-10

as shown in Fig. 7-10*c*. Consequently, $H_1(\omega)H_2(\omega)$ is that shown in Fig. 7-10*d*. $H_1(\omega)H_2(\omega)$ can be expressed as $H_3(\omega) + H_4(\omega)$, where $H_3(\omega)$ and $H_4(\omega)$ are shown in Fig. 7-10*e* and *f*. Since the inverse transform of $H_3(\omega)$ is

$$\frac{\sin 2t}{\pi t}$$

and the inverse transform of $H_4(\omega)$ is

$$2\pi \frac{\sin t/2}{\pi t} \frac{\sin 3t/2}{\pi t}$$

the impulse response of the overall system is

$$\frac{\sin 2t}{\pi t} + 2\pi \frac{\sin t/2}{\pi t} \frac{\sin 3t/2}{\pi t}$$

7-5 FOURIER TRANSFORMS OF SINGULARITY FUNCTIONS

Although time signals containing singularity functions do not satisfy the Dirichlet conditions, their Fourier transforms can be defined in terms of the theory of distributions. A rigorous treatment of the subject is beyond the scope of this book. Our discussion of the Fourier transformation of time signals containing singularity functions will be only an extension of the results in the preceding sections, making use of the properties of singularity functions defined earlier. The Fourier transform of the unit impulse $u_0(t)$ is

$$\int_{-\infty}^{\infty} u_0(t)e^{-j\omega t}\,dt = e^{-j\omega t}\Big|_{t=0} = 1$$

Also, the Fourier transform of the signal $u_0(t + t_0)$ is

$$\int_{-\infty}^{\infty} u_0(t+t_0)e^{-j\omega t}\,dt = e^{-j\omega t}\Big|_{t=-t_0} = e^{jt_0\omega} \tag{7-14}$$

Similarly, the Fourier transform of the unit n-tuplet, $u_n(t)$, $n > 0$ is

$$\int_{-\infty}^{\infty} u_n(t)e^{-j\omega t}\,dt = (-1)^n \frac{d^n}{dt^n} e^{-j\omega t}\Big|_{t=0} = (j\omega)^n$$

Since the Fourier transform of $u_0(t)$ is the constant 1, the duality property implies that the Fourier transform of the time signal $x(t) = 1$ is $2\pi u_0(\omega)$. Note that the time signal $x(t) = 1$ is not absolutely integrable and thus does not satisfy the Dirichlet conditions. However, the inclusion of singularity functions in our consideration enables us to define its Fourier transform. Indeed, once we have defined the Fourier transforms of singularity functions, making use of the duality property, we can also include singularity functions in the Fourier transforms of time signals. Thus, for example, from the result in Eq. (7-14), the Fourier transform of the time signal $\cos \omega_0 t$ will be

$$\pi[u_0(\omega + \omega_0) + u_0(\omega - \omega_0)]$$

We summarize some frequently used results in Table 7-2. The derivation of the not so obvious ones are left as exercises.

Some of the results in Table 7-2 give us further insight into the transformation representation of signals. That the Fourier transform of $u_0(t)$ is the constant 1 means that the time signal $u_0(t)$ contains all frequency components from $\omega = -\infty$ to $\omega = \infty$. Such an observation explains why the impulse response completely characterizes the input-output relationship of a linear time-invariant system. Since the impulse contains all frequency components, the impulse response of a system, $h(t)$, gives the response of the system to all frequency components. To be explicit, since

$$u_0(t) = \frac{1}{2\pi}\int_{-\infty}^{\infty} e^{j\omega t}\,d\omega$$

$$h(t) = \frac{1}{2\pi}\int_{-\infty}^{\infty} H(\omega)e^{j\omega t}\,d\omega$$

the response of the system to the stimulus $e^{j\omega t}$ is $H(\omega)e^{j\omega t}$. Consequently, the response of the system to any signal can be determined from $H(\omega)$ when the signal is expressed as an integral of complex exponential functions. Furthermore, we realize that the input-output relationship of a linear time-invariant system is completely characterized by the system response $y(t)$ to any input signal $x(t)$ whose frequency spectrum is nonzero in the whole frequency range $-\infty \leq \omega \leq \infty$.† That is, the response of the system to $X(\omega)e^{j\omega t}$ is $Y(\omega)e^{j\omega t}$. On the other hand, the reader can verify for himself that the response of a system to $x(t) = (\sin at)/\pi t$ does not completely characterize the input-output relationship of the system because $x(t)$ does not contain all frequency components.

That the Fourier transform of $x(t) = 1$ is $2\pi u_0(\omega)$ indicates that $x(t)$ has no frequency component other than a dc component ($\omega = 0$). Similarly, both $\sin \omega_0 t$ and $\cos \omega_0 t$ have only two frequency components corresponding to $e^{-j\omega_0 t}$ and $e^{j\omega_0 t}$. This can be seen either by noting that

$$\sin \omega_0 t = \frac{1}{2j} e^{j\omega_0 t} - \frac{1}{2j} e^{-j\omega_0 t}$$

Table 7-2

$x(t)$		$X(\omega)$
$u_0(t)$		1
1		$2\pi u_0(\omega)$
$u_n(t)$	$n > 0$	$(j\omega)^n$
$u_{-1}(t)$		$\pi u_0(\omega) + \dfrac{1}{j\omega}$
t^n		$2\pi j^n u_n(\omega)$
$\operatorname{sgn} t$		$\dfrac{2}{j\omega}$
$-\dfrac{1}{jt}$		$\pi \operatorname{sgn} \omega$
$\sin \omega_0 t$		$j\pi[u_0(\omega + \omega_0) - u_0(\omega - \omega_0)]$
$\cos \omega_0 t$		$\pi[u_0(\omega + \omega_0) + u_0(\omega - \omega_0)]$
$\displaystyle\sum_{n=-\infty}^{\infty} u_0(t - nT)$		$\dfrac{2\pi}{T} \displaystyle\sum_{n=-\infty}^{\infty} u_0\left(\omega - \dfrac{2n\pi}{T}\right)$

† This is an answer to a question raised in Chap. 4 whether any input-output pair of signals completely characterizes the input-output relationship of a linear time-invariant system.

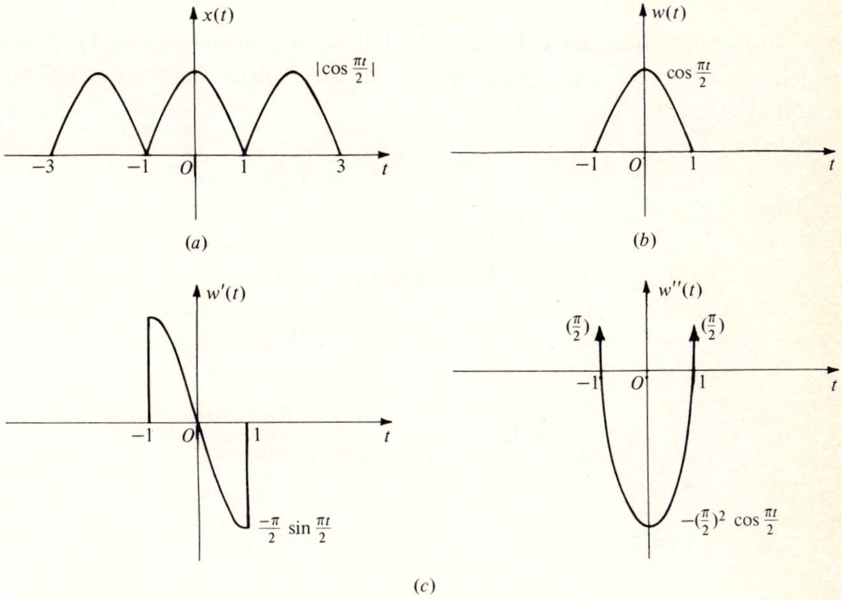

FIGURE 7-11

$$\cos \omega_0 t = \frac{1}{2} e^{j\omega_0 t} + \frac{1}{2} e^{-j\omega_0 t}$$

or by examining directly their Fourier transforms.

As an illustrative example, let us consider the problem of determining the Fourier transform of the signal $x(t)$ in Fig. 7-11*a*. We determine first the Fourier transform of the signal $w(t)$ in Fig. 7-11*b*. Differentiating $w(t)$ twice, we obtain

$$\left(\frac{\pi}{2}\right)^2 w(t) + w''(t) = \frac{\pi}{2} u_0(t+1) + \frac{\pi}{2} u_0(t-1)$$

where $w'(t)$ and $w''(t)$ are as shown in Fig. 7-11*c*. Thus, we have

$$\left(\frac{\pi}{2}\right)^2 W(\omega) + (j\omega)^2 W(\omega) = \frac{\pi}{2}(e^{j\omega} + e^{-j\omega})$$

or

$$W(\omega) = \frac{(\pi/2)(e^{j\omega} + e^{-j\omega})}{(\pi/2)^2 - \omega^2} = \frac{\pi}{(\pi/2)^2 - \omega^2} \cos \omega$$

It follows that

$$X(\omega) = \frac{\pi}{(\pi/2)^2 - \omega^2}(e^{j2\omega} + 1 + e^{-j2\omega}) \cos \omega$$

$$= \frac{\pi}{(\pi/2)^2 - \omega^2}(1 + 2\cos 2\omega) \cos \omega \tag{7-15}$$

It is indeed quite easy to prove the Hilbert transforms in Eqs. (7-10) and (7-11) which relate the real and imaginary parts of the Fourier transform $X(\omega)$ of a signal $x(t)$ that is equal to zero for $t < 0$. Since

$$x(t) = 2x_e(t)u_{-1}(t)$$

we have

$$\begin{aligned} X(\omega) &= \frac{1}{\pi} \operatorname{Re}[X(\omega)] * \left[\pi u_0(\omega) + \frac{1}{j\omega}\right] \\ &= \operatorname{Re}[X(\omega)] + \frac{1}{\pi}\int_{-\infty}^{\infty} \frac{\operatorname{Re}[X(\sigma)]}{j(\omega - \sigma)}\, d\sigma \\ &= \operatorname{Re}[X(\omega)] - j\frac{1}{\pi}\int_{-\infty}^{\infty} \frac{\operatorname{Re}[X(\sigma)]}{\omega - \sigma}\, d\sigma \end{aligned}$$

That is,

$$\operatorname{Im}[X(\omega)] = -\frac{1}{\pi}\int_{-\infty}^{\infty} \frac{\operatorname{Re}[X(\sigma)]}{\omega - \sigma}\, d\sigma$$

which is Eq. (7-11). In a similar manner, we can prove Eq. (7-10).

Also, let

$$y(t) = \int_{-\infty}^{t} w(\tau)\, d\tau \tag{7-16}$$

Equation (7-16) can be written

$$y(t) = w(t) * u_{-1}(t)$$

Thus, we obtain

$$\begin{aligned} Y(\omega) &= W(\omega)\left[\pi u_0(\omega) + \frac{1}{j\omega}\right] \\ &= \pi W(0)u_0(\omega) + \frac{W(\omega)}{j\omega} \end{aligned} \tag{7-17}$$

As an example, for

$$w(t) = e^{-\alpha t}u_{-1}(t)$$

we have

$$W(\omega) = \frac{1}{\alpha + j\omega}$$

Let

$$y(t) = \int_{-\infty}^{t} w(\tau)\, d\tau = \begin{cases} \dfrac{1}{\alpha}(1 - e^{-\alpha t}) & t \geq 0 \\ 0 & t \leq 0 \end{cases}$$

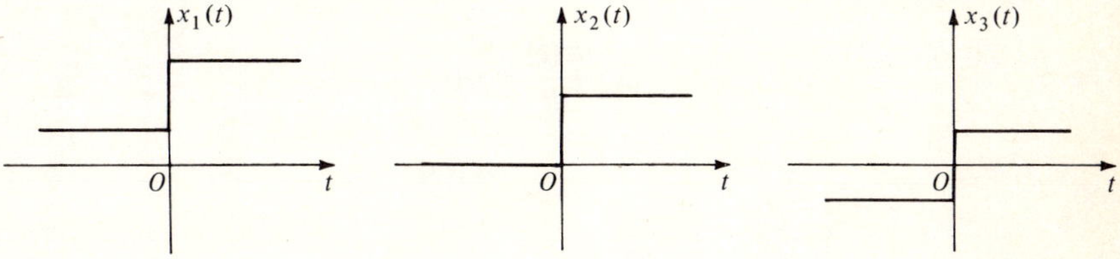

FIGURE 7-12

According to Eq. (7-17),

$$Y(\omega) = \pi \frac{1}{\alpha} u_0(\omega) + \frac{1}{j\omega(\alpha + j\omega)}$$

$$= \frac{\pi}{\alpha} u_0(\omega) + \frac{1}{-\omega^2 + j\alpha\omega}$$

We are now ready to clarify a point which was first brought up in Sec. 7-3, where we derived the result that the Fourier transform of $x'(t)$ can be computed as $j\omega$ times that of $x(t)$. One can easily be led to the false conclusion that the Fourier transform of $x(t)$ can then be computed as that of $x'(t)$ divided by $j\omega$. The falsity of this statement becomes obvious when we recall that we cannot recover completely a signal $x(t)$ from its derivative $x'(t)$. Consequently, we should not expect to determine uniquely the Fourier transform of $x(t)$ from that of $x'(t)$. For example, the three signals $x_1(t)$, $x_2(t)$, and $x_3(t)$ in Fig. 7-12 have the same derivative $u_0(t)$. It will be totally unreasonable if we expect to determine $X_1(\omega)$, $X_2(\omega)$, and $X_3(\omega)$ from the Fourier transform of $u_0(t)$. As a matter of fact, there is an infinite number of signals that have the same derivative, and they differ from each other by additive constants. Consequently, their Fourier transforms differ from each other by impulses at $\omega = 0$. For a given $w(t)$, there is a signal $y(t)$ such that

$$\frac{dy(t)}{dt} = w(t)$$

and

$$y(-\infty) = 0$$

In that case,

$$y(t) = \int_{-\infty}^{t} w(\tau)\, d\tau$$

and the Fourier transforms $Y(\omega)$ and $W(\omega)$ are related by Eq. (7-17). Also, there is a signal $z(t)$ such that

$$z(t) = y(t) - \tfrac{1}{2}W(0)$$

where $$\frac{dz(t)}{dt} = w(t)$$

and $$Z(\omega) = \frac{W(\omega)}{j\omega}$$

We now show an alternative method to compute the Fourier transform of the signal $x(t)$ in Fig. 7-11*a*. We write

$$x(t) = \cos\frac{\pi t}{2}\, v(t)$$

where $v(t)$ is the function shown in Fig. 7-13*a*. After computing $v'(t)$ as in Fig. 7-13*b*, we obtain

$$V(\omega) = \frac{1}{j\omega}(-e^{j3\omega} + 2e^{j\omega} - 2e^{-j\omega} + e^{-j3\omega})$$

$$= \frac{2}{\omega}(2 \sin \omega - \sin 3\omega)$$

Therefore, we have

$$X(\omega) = \frac{1}{2\pi}\pi\left[u_0\left(\omega + \frac{\pi}{2}\right) + u_0\left(\omega - \frac{\pi}{2}\right)\right] * V(\omega)$$

$$= \frac{1}{\omega + \pi/2}\left[2 \sin\left(\omega + \frac{\pi}{2}\right) - \sin 3\left(\omega + \frac{\pi}{2}\right)\right]$$

$$+ \frac{1}{\omega - \pi/2}\left[2 \sin\left(\omega - \frac{\pi}{2}\right) - \sin 3\left(\omega - \frac{\pi}{2}\right)\right]$$

$$= \frac{1}{\omega + \pi/2}(2 \cos \omega + \cos 3\omega)$$

$$+ \frac{1}{\omega - \pi/2}(-2 \cos \omega - \cos 3\omega)$$

$$= \frac{-\pi}{(\omega + \pi/2)(\omega - \pi/2)}(2 \cos \omega + \cos 3\omega)$$

So that we can check the result in (7-15), we rewrite $X(\omega)$:

$$X(\omega) = \frac{-\pi}{(\omega + \pi/2)(\omega - \pi/2)}(\cos \omega + \cos \omega + \cos 3\omega)$$

$$= \frac{-\pi}{\omega^2 - (\pi/2)^2}(\cos \omega + 2 \cos \omega \cos 2\omega)$$

$$= \frac{-\pi}{\omega^2 - (\pi/2)^2}(1 + 2 \cos 2\omega) \cos \omega$$

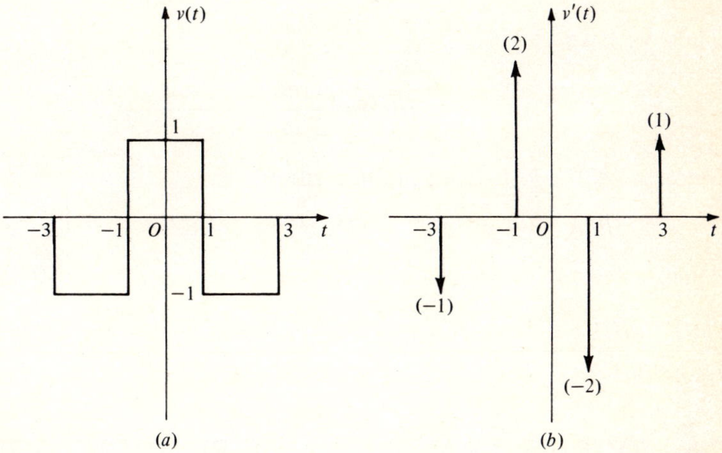

FIGURE 7-13

We now consider an example studied in Chap. 4. Let $x(t)$ and $y(t)$ in Fig. 7-14 be an input-output pair of a linear time-invariant system that is known to be causal. To determine the impulse response $h(t)$ of the system, we compute first $X(\omega)$ and $Y(\omega)$. Since

$$x(t) = e^{-t}u_{-1}(t) - e^{-1}e^{-(t-1)}u_{-1}(t-1)$$

we have

$$X(\omega) = \frac{1}{1+j\omega} - \frac{e^{-1}e^{-j\omega}}{1+j\omega} = \frac{1-e^{-(1+j\omega)}}{1+j\omega}$$

Also,
$$Y(\omega) = \frac{1}{1+j\omega}$$

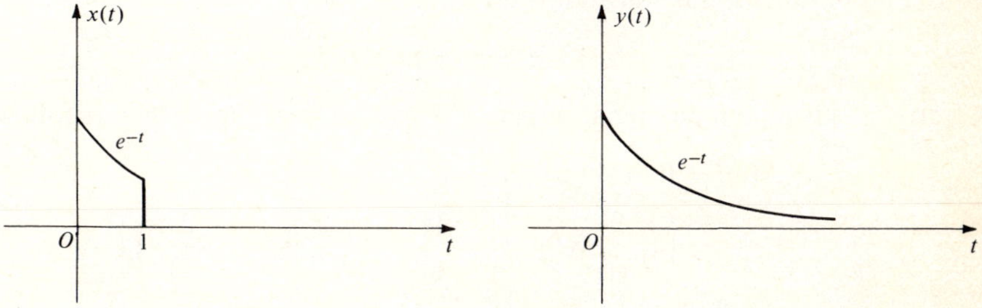

FIGURE 7-14

Thus, we have

$$H(\omega) = \frac{Y(\omega)}{X(\omega)} = \frac{1}{1 - e^{-(1+j\omega)}}$$

Since $|e^{-(1+j\omega)}| < 1$, $H(\omega)$ can be written

$$H(\omega) = 1 + e^{-1}e^{-j\omega} + e^{-2}e^{-j2\omega} + e^{-3}e^{-j3\omega} + \cdots + e^{-k}e^{-jk\omega} + \cdots$$

Thus,

$$h(t) = u_0(t) + e^{-1}u_0(t-1) + e^{-2}u_0(t-2) + e^{-3}u_0(t-3) + \cdots + e^{-k}u_0(t-k) + \cdots$$

7-6 FOURIER TRANSFORM OF PERIODIC SIGNALS

We discussed in Chap. 6 how a periodic signal can be represented as a Fourier series, and in this chapter how an aperiodic signal can be represented as a Fourier integral. However, there are many system analysis problems in which both periodic signals and aperiodic signals are encountered at the same time. For example, suppose that the impulse response $h(t)$ of a linear time-invariant system is that shown in Fig. 7-15*a*, and we want to determine the output signal of the system when the input signal $x(t)$ is the periodic square wave shown in Fig. 7-15*b*. In this case, we find that a direct computation of $x(t) * h(t)$ is rather lengthy, and yet the transformation techniques we have so far developed are not applicable because we do not know how the Fourier transform of periodic signals can be computed. Although periodic signals are not absolutely integrable, when singularity functions are included in our consideration, the Fourier transform of periodic signals that have a finite number of maxima, minima, and discontinuities in a period can be defined. Let $w(t)$ be a signal that is zero for $t < -T/2$ and for $t > T/2$, as shown in Fig. 7-16*a*. Let

$$v(t) = \sum_{-\infty}^{\infty} u_0(t - nT)$$

be a train of impulses as shown in Fig. 7-16*b*, and let

$$x(t) = w(t) * v(t)$$

Clearly, $x(t)$ is a periodic signal of period T as shown in Fig. 7-16*c*. It follows that

$$\begin{aligned} X(\omega) &= W(\omega)V(\omega) \\ &= \left[\int_{-T/2}^{T/2} w(t)e^{-j\omega t}\,dt\right]\left[\sum_{n=-\infty}^{\infty} \frac{2\pi}{T} u_0\left(\omega - \frac{2n\pi}{T}\right)\right]^{\dagger} \end{aligned}$$

† See Table 7-2.

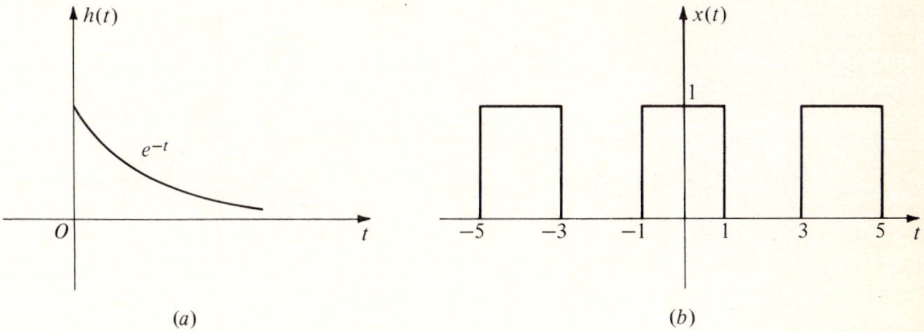

FIGURE 7-15

$$= \sum_{n=-\infty}^{\infty} \left[\frac{2\pi}{T} \int_{-T/2}^{T/2} w(t) e^{-j\omega t} \, dt \right] u_0\left(\omega - \frac{2n\pi}{T}\right)$$

$$= \sum_{n=-\infty}^{\infty} \frac{2\pi}{T} W\left(\frac{2n\pi}{T}\right) u_0\left(\omega - \frac{2n\pi}{T}\right)$$

That is, the Fourier transform of $X(\omega)$ is a train of impulses as shown in Fig. 7-17, where the envelope $(2\pi/T)W(\omega)$ means that the area of the impulse at $\omega = 2n\pi/T$ is equal to

$$\frac{2\pi}{T} \int_{-T/2}^{T/2} w(t) e^{-j2\pi nt/T} \, dt$$

We note immediately that the area of the impulse at $\omega = 2n\pi/T$ is equal to 2π times the value of the nth Fourier coefficient of the periodic signal $x(t)$. Indeed, for a periodic signal $x(t)$ the only difference between the Fourier transform of $x(t)$ and the Fourier series of $x(t)$ is that the former consists of a train of impulses and the latter consists of a discrete set of numbers.

Let us consider some simple examples. The time signal $x(t) = 1$ in Fig. 7-18*a* can be considered a periodic signal of period T for any arbitrary T. The Fourier coefficients of such a periodic function can be computed readily as

$$x_0 = 1$$

$$x_n = 0 \qquad n \neq 0$$

It follows that the Fourier transform of $x(t)$ is $2\pi u_0(\omega)$, as in Fig. 7-18*b*, a result stated in Sec. 7-5.

Also consider the time signal $x(t) = \cos \omega_0 t$ which is a periodic signal of period $2\pi/\omega_0$, as shown in Fig. 7-19*a*. The Fourier coefficients of $x(t)$ are

$$x_1 = x_{-1} = \tfrac{1}{2}$$

$$x_n = 0 \qquad n \neq \pm 1$$

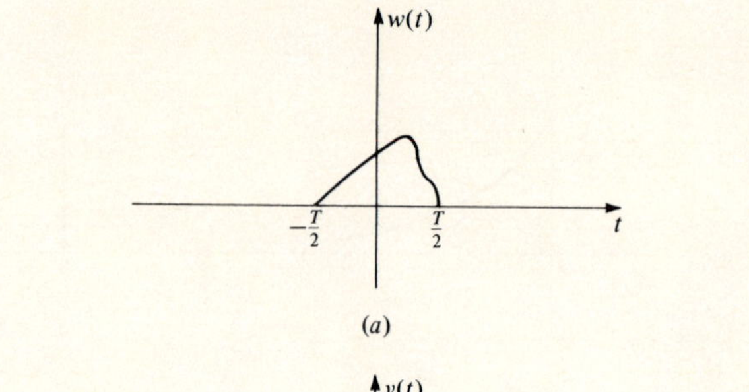

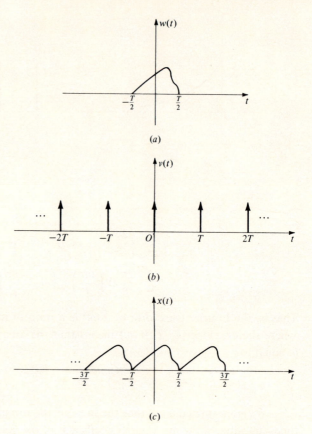

FIGURE 7-16

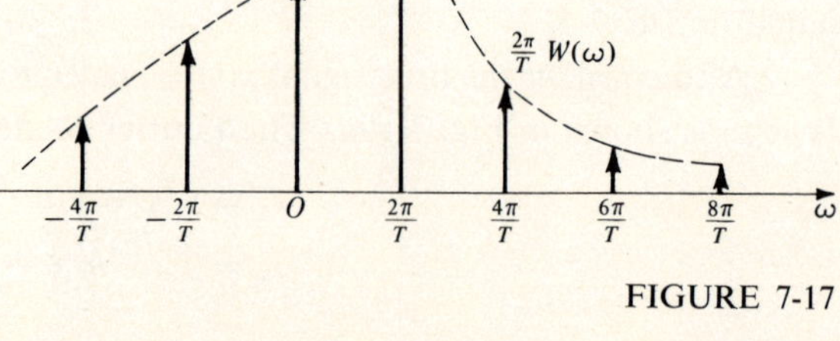

FIGURE 7-17

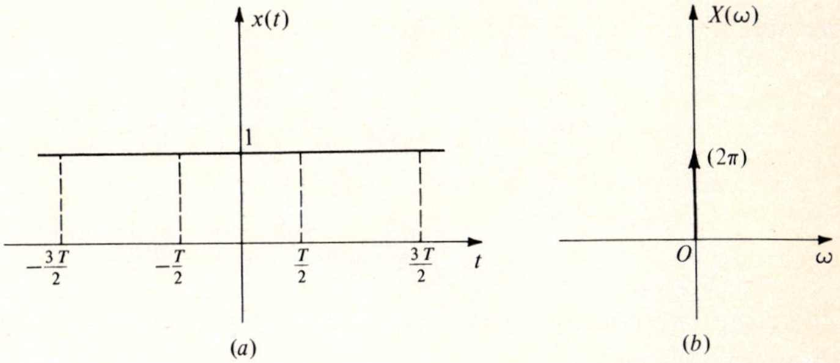

FIGURE 7-18

Therefore, the Fourier transform of $x(t)$ is

$$\pi u_0(\omega + \omega_0) + \pi u_0(\omega - \omega_0)$$

as shown in Fig. 7-19*b*.

The study of the Fourier transformation of periodic signals provides a unified approach to handling both periodic and aperiodic signals in the transform domain. As an illustrative example, let us compute the convolution of the signals $x(t)$ and $h(t)$ in Fig. 7-15*a* and *b*. Since

$$H(\omega) = \frac{1}{1 + j\omega}$$

$$X(\omega) = \sum_{n=-\infty}^{\infty} \frac{2}{n} \sin \frac{n\pi}{2} u_0\left(\omega - \frac{2n\pi}{4}\right)$$

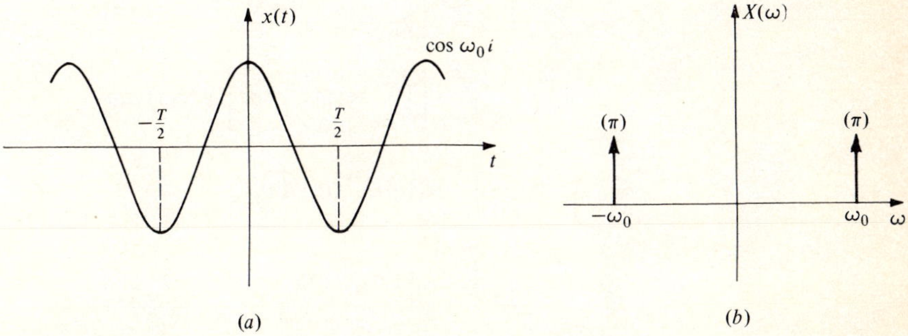

FIGURE 7-19

we obtain

$$X(\omega)H(\omega) = \sum_{n=-\infty}^{\infty} \frac{2}{n} \sin\frac{n\pi}{2} \frac{1}{1+j\omega} u_0\left(\omega - \frac{2n\pi}{4}\right)$$

$$= \sum_{n=-\infty}^{\infty} \frac{2}{n} \sin\frac{n\pi}{2} \frac{1}{1+jn\pi/2} u_0\left(\omega - \frac{n\pi}{2}\right)$$

Therefore, we conclude that $x(t) * h(t)$ is a periodic signal of period 4 whose nth Fourier coefficient is

$$\frac{1}{2\pi} \frac{2}{n} \sin\frac{n\pi}{2} \frac{1}{1+jn\pi/2}$$

That is,

$$x(t) * h(t) = \sum_{n=-\infty}^{\infty} \frac{1}{n\pi} \sin\frac{n\pi}{2} \frac{1}{1+jn\pi/2} e^{jn\pi t/2}$$

*7-7 ENERGY DISTRIBUTION IN APERIODIC SIGNALS

According to Eq. (6-22), the total energy in a signal $x(t)$ is defined to be

$$\int_{-\infty}^{\infty} |x(t)|^2 \, dt$$

which for an aperiodic signal can be expressed in terms of its Fourier transform:

$$\int_{-\infty}^{\infty} |x(t)|^2 \, dt = \int_{-\infty}^{\infty} x(t)x^*(t) \, dt$$

$$= \int_{-\infty}^{\infty} x(t)\left[\frac{1}{2\pi}\int_{-\infty}^{\infty} X^*(\omega)e^{-j\omega t} \, d\omega\right] dt$$

$$= \frac{1}{2\pi}\int_{-\infty}^{\infty} \left[\int_{-\infty}^{\infty} x(t)e^{-j\omega t} \, dt\right] X^*(\omega) \, d\omega$$

$$= \frac{1}{2\pi}\int_{-\infty}^{\infty} X(\omega)X^*(\omega) \, d\omega$$

$$= \frac{1}{2\pi}\int_{-\infty}^{\infty} |X(\omega)|^2 \, d\omega$$

This result is known as *Parseval's equality*.

As an example, for the signal

$$x(t) = e^{-\alpha t} u_{-1}(t)$$

we have

$$\int_{-\infty}^{\infty} |x(t)|^2 \, dt = \int_0^{\infty} e^{-2\alpha t} \, dt = \frac{1}{2\alpha}$$

and

$$\begin{aligned} \frac{1}{2\pi} \int_{-\infty}^{\infty} |X(\omega)|^2 \, d\omega &= \frac{1}{2\pi} \int_{-\infty}^{\infty} \left| \frac{1}{\alpha + j\omega} \right|^2 d\omega \\ &= \frac{1}{2\pi} \int_{-\infty}^{\infty} \frac{1}{\alpha^2 + \omega^2} \, d\omega \\ &= \frac{1}{2\pi} \frac{1}{\alpha} \tan^{-1} \frac{\omega}{\alpha} \bigg|_{-\infty}^{\infty} = \frac{1}{2\alpha} \end{aligned}$$

As another example, we recall that the Fourier transform of the signal

$$x(t) = 2 \cos 997t \, \frac{\sin 5\pi t}{\pi t} \tag{7-18}$$

is

$$X(\omega) = \begin{cases} 1 & 992 < |\omega| < 1002 \\ 0 & \text{otherwise} \end{cases}$$

Therefore, the total energy in $x(t)$ is

$$\begin{aligned} \frac{1}{2\pi} \int_{-\infty}^{\infty} |X(\omega)|^2 \, d\omega &= \frac{1}{2\pi} \left(\int_{-1002}^{-992} d\omega + \int_{992}^{1002} d\omega \right) \\ &= \frac{10}{\pi} \end{aligned}$$

The function $(1/2\pi)|X(\omega)|^2$ is called the *energy density spectrum* of $x(t)$ because $(1/2\pi)|X(\omega)|^2 \, d\omega$ gives the energy in the frequency components between ω and $\omega + d\omega$. Thus, for the signal $x(t)$ in Eq. (7-18), the energy of the signal is concentrated in the frequency components $e^{j\omega t}$ for $-1002 < \omega < -992$ and $992 < \omega < 1002$. Also, the energy in an impulse is distributed evenly among its frequency components, and the energy in a constant time signal is concentrated in the dc component only.

Similarly to the case of periodic signals, we define the autocorrelation function of an aperiodic signal to be

$$r(t) = \int_{-\infty}^{\infty} x^*(\tau) x(\tau + t) \, dt$$

It will be left to the reader to show that the Fourier transform of $r(t)$ is $(1/2\pi)|X(\omega)|^2$.

7-8 REMARKS AND REFERENCES

The Fourier transform representation of signals is a useful one. We now have two ways to describe signals: by its time domain representation and by its frequency domain representation. In many problems, one representation may be more convenient than another. For example, we recall that the convolution of two signals in the time domain corresponds to the multiplication of their Fourier transforms in the frequency domain, and in many cases, it is easier to compute the Fourier transforms than to evaluate directly the convolution integral. As a matter of fact, the application of the method of Fourier transformation is not limited to the study of systems and signals. It is probably one of the most powerful tools in engineering, physics, and mathematics, as the reader will undoubtedly discover later on.

By a change of variable $\omega = 2\pi f$, the definitions of Fourier transform and inverse Fourier transform in Eqs. (7-3) and (7-4) can be reduced to

$$x(t) = \int_{-\infty}^{\infty} X(f)e^{j2\pi ft}\, df$$

$$X(f) = \int_{-\infty}^{\infty} x(t)e^{-j2\pi ft}\, dt$$

In this way, we get rid of the multiplication factor $1/2\pi$ which, as the reader probably has discovered, is a small nuisance to carry along in all manipulation.

For further discussion of the Fourier transform, we recommend Bracewell [1] and Papoulis [5]. A rigorous treatment of the Fourier transformation of signals containing singularity functions can be found in Zemanian [7] and Arsac [1]. The notions of autocorrelation and energy density spectrum are very useful in communication theory; see, for example, Lee [4]. For a discussion of two-dimensional Fourier transformation and its application to optics, see Goodman [3] and Papoulis [6].

[1] ARSAC, J.: "Fourier Transforms and the Theory of Distributions," Prentice-Hall, Inc., Englewood Cliffs, N.J., 1966.

[2] BRACEWELL, R.: "The Fourier Transform and Its Applications," McGraw-Hill Book Company, New York, 1965.

[3] GOODMAN, J. W.: "Introduction to Fourier Optics," McGraw-Hill Book Company, New York, 1968.

[4] LEE, Y. W.: "Statistical Theory of Communication," John Wiley & Sons, Inc., New York, 1960.

[5] PAPOULIS, A.: "The Fourier Integral and Its Applications," McGraw-Hill Book Company, New York, 1962.

[6] PAPOULIS, A.: "Systems and Transforms with Applications in Optics," McGraw-Hill Book Company, New York, 1968.

[7] ZEMANIAN, A. H.: "Distribution Theory and Transform Analysis," McGraw-Hill Book Company, New York, 1965.

PROBLEMS

7-1 Determine the Fourier transform of each of the following time signals:

(*a*) $x(t) = \dfrac{\sin 2\pi(t-2)}{t-2} \qquad -\infty \leq t \leq \infty$

(*b*) $x(t) = \dfrac{2\alpha}{\alpha^2 + t^2} \qquad -\infty \leq t \leq \infty$

(*c*) $x(t) = \cos t^2 \qquad -\infty \leq t \leq \infty$

(*d*) $x(t) = \begin{cases} e^{-t} \cos \omega t & t \geq 0 \\ -e^{t} \cos \omega t & t \leq 0 \end{cases}$

(*e*) $x(t) = \sin 2\pi t[u_{-1}(t-2) - u_{-1}(t+2)]$

(*f*) $x(t) = t^2 e^{-2t} u_{-1}(t)$

7-2 Determine the Fourier transform of each of the time signals shown in Fig. 7P-1.

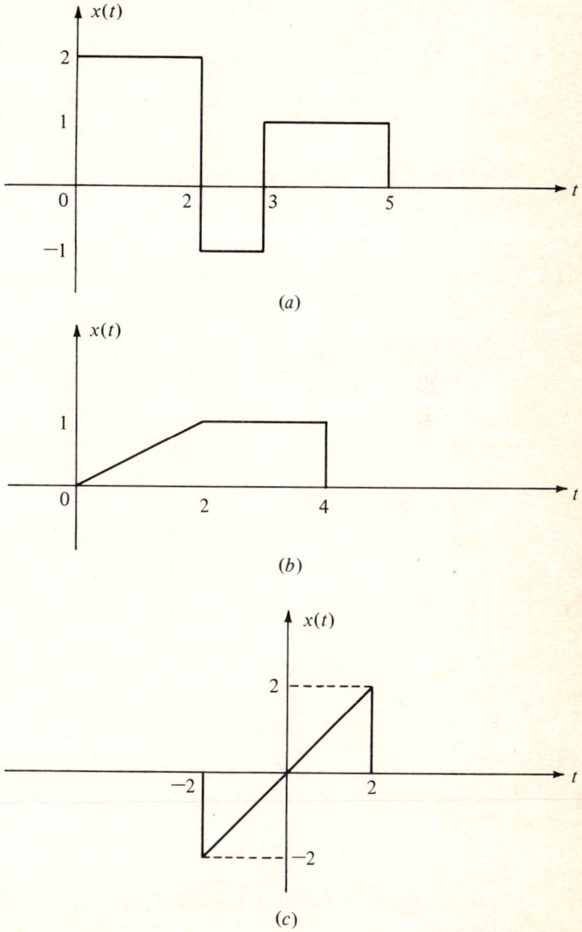

FIGURE 7P-1

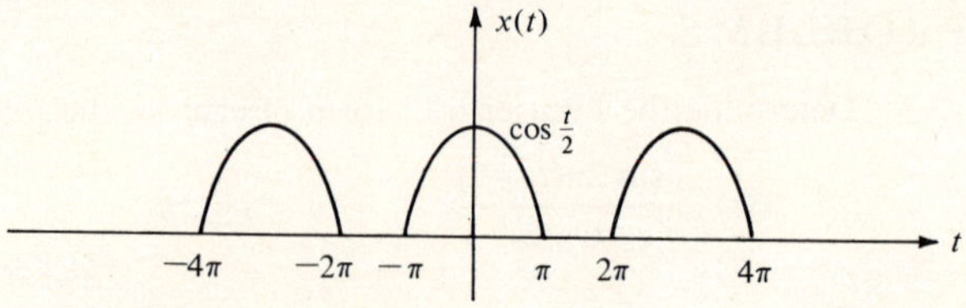

(d)

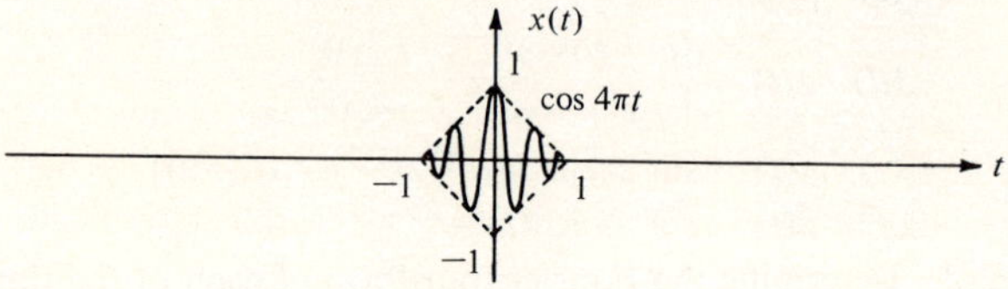

(e)

FIGURE 7P-1 (continued)

7-3 Determine the real part of the Fourier transform of the signal $x(t)$ shown in Fig. 7P-2.

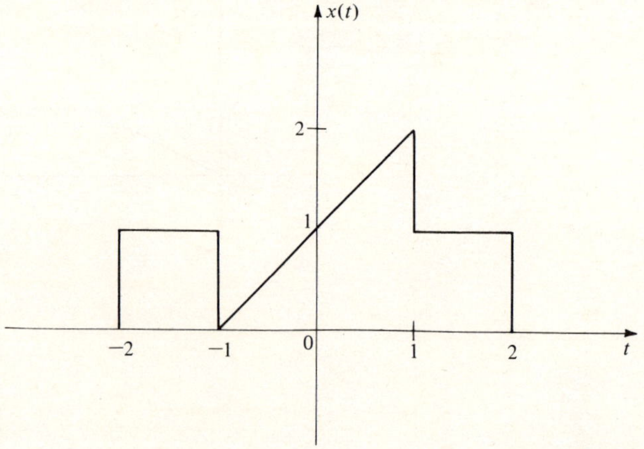

FIGURE 7P-2

7-4 For the signal $x(t)$ shown in Fig. 7P-3, determine the phase angle $\angle X(\omega)$ of its Fourier transform.

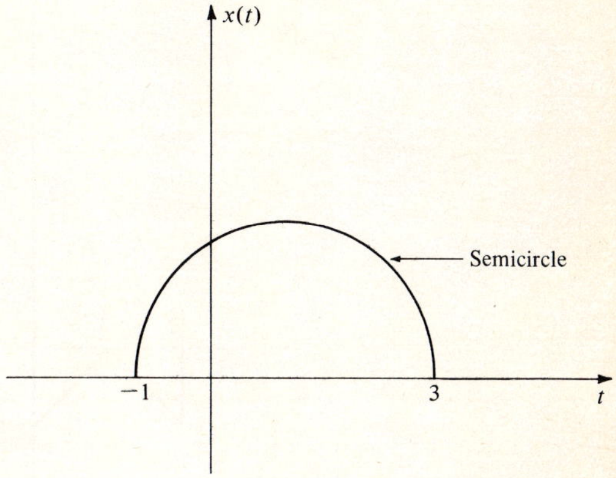

FIGURE 7P-3

7-5 The Fourier transform $X(\omega)$ of a time signal $x(t)$ is shown in Fig. 7P-4. Determine $x(t)$.

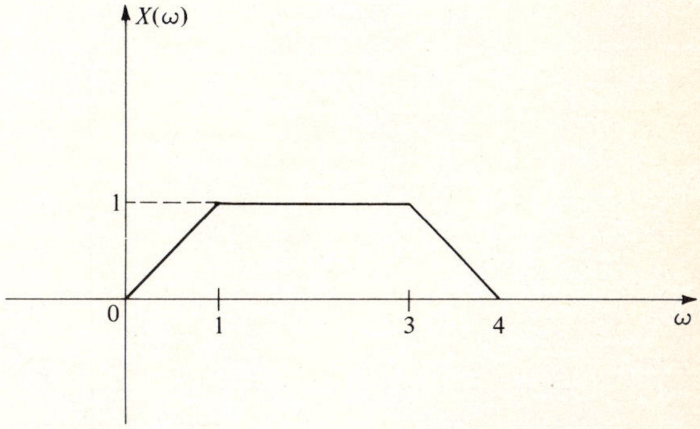

FIGURE 7P-4

7-6 (*a*) Show that if $x(t)$ is an even and real time signal, then its Fourier transform $X(\omega)$ is an even and real function of ω.

(*b*) Show that if $X(\omega)$ is an even and real function of ω, then $x(t)$ is an even and real time signal.

7-7 Determine the imaginary part of the Fourier transform of the signal $x(t)$ shown in Fig. 7P-5.

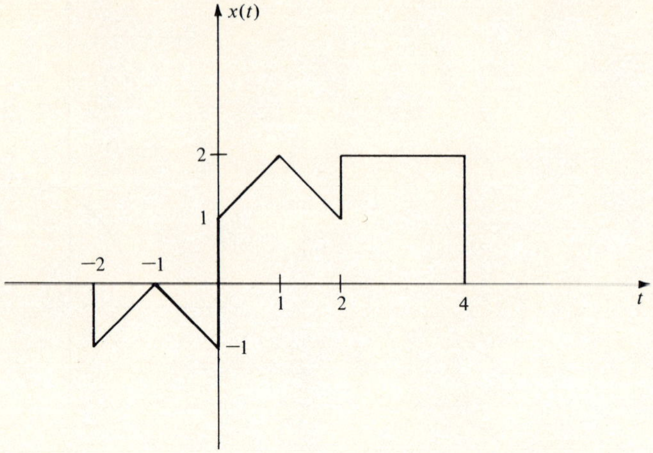

FIGURE 7P-5

7-8 Figure 7P-6 shows the portion of a signal $x(t)$ for $-\infty \leq t \leq 2$. Determine $x(t)$ for $2 \leq t \leq \infty$ if it is known that $\angle X(\omega) = j^{2\omega}$.

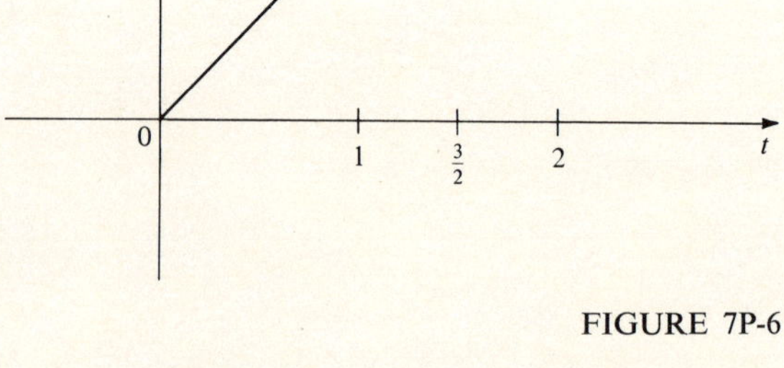

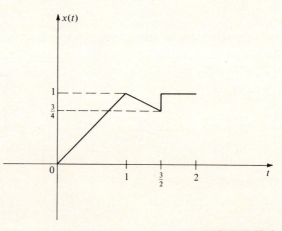

FIGURE 7P-6

7-9 For $\omega \geq 0$, the Fourier spectrum of a real time signal $x(t)$ is as shown in Fig. 7P-7*a*. The phase angle of its Fourier transform is shown in Fig. 7P-7*b*. Sketch the real and imaginary parts of $X(\omega)$.

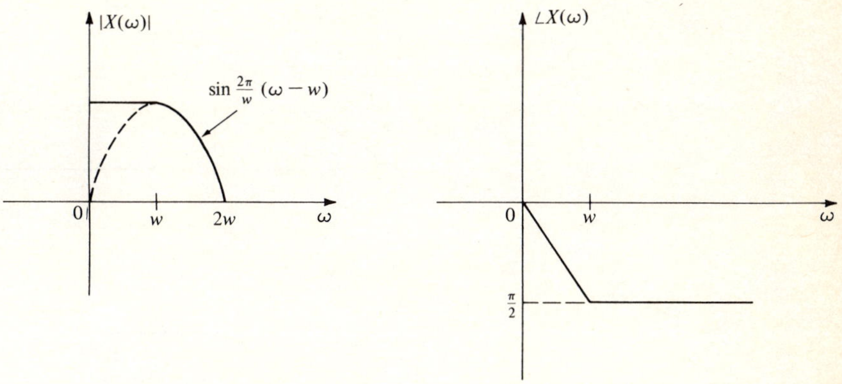

(a) (b)

FIGURE 7P-7

7-10 On many occasions, we wish to plot $|H(\omega)|$ versus ω and $\angle H(\omega)$ versus ω when an analytic expression of $H(\omega)$ is given. This, in general, is quite tedious because $|H(\omega)|$ and $\angle H(\omega)$ must be computed for many different values of ω. In this problem we introduce a way of sketching $|H(\omega)|$ and $\angle H(\omega)$ in an approximate manner. To be specific, we shall show how to obtain approximate plots for 20 log $|H(\omega)|$ versus log ω and $\angle H(\omega)$ versus log ω which are known as the Bode plots. For example, for $H(\omega) = 1 + j\omega$, 20 log $|H(\omega)|$ is approximated by

$$20 \log (\sqrt{1+\omega^2}) \doteq \begin{cases} 0 & \text{for } \omega \leq 1 \\ 20 \log \omega & \text{for } \omega \geq 1 \end{cases}$$

Hence 20 log $|H(\omega)|$ versus log ω is as shown in Fig. 7P-8*a*. The approximate plot consists of two line segments, one of slope 0, the other of slope 20, intersecting at the point $\omega = 1$. Also, we approximate $\angle H(\omega)$ as

$$\angle H(\omega) = \tan^{-1} \omega \doteq \begin{cases} 90^\circ & \omega \geq 10 \\ \text{varies linearly with } \log \omega & \text{for } 0.1 \leq \omega \leq 10 \\ 0^\circ & \text{for } \omega \leq 0.1 \end{cases}$$

Hence $\angle H(\omega)$ versus log ω is as shown in Fig. 7P-8*b*.

(*a*) Sketch the Bode plots for $H(\omega) = 1 + j10^{-1}\omega$.

(*b*) Sketch the Bode plots for $H(\omega) = 1/(1 + j\omega)$.

(*c*) Sketch the Bode plots for $H(\omega) = j\omega(1 + j\omega 10^{-2})^2/(1 + j\omega 10^{-3})^3$.

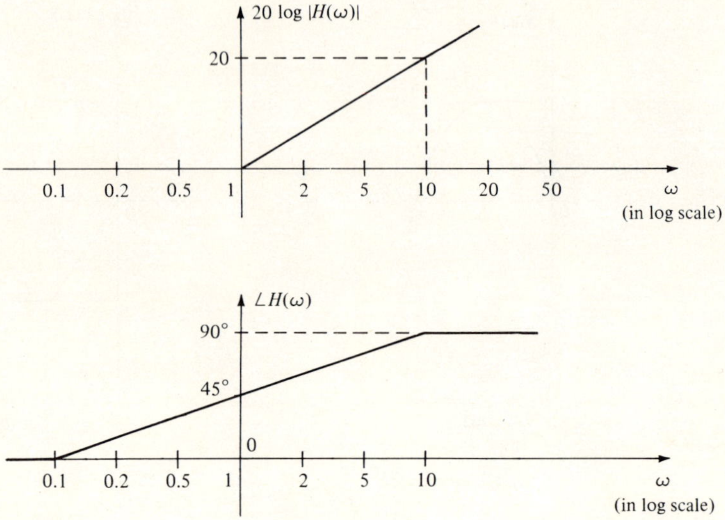

FIGURE 7P-8

7-11 Show that the values of both of the integrals

$$\int_{-\infty}^{\infty} \frac{\sin a\omega}{a\omega}\, d\omega$$

and

$$\int_{-\infty}^{\infty} \left(\frac{\sin a\omega}{a\omega}\right)^2 d\omega$$

are equal to π/a.

7-12 Find a time signal $x(t)$ whose Fourier transform is equal to $Kx(\omega)$, where K is a multiplying constant. Can you find many time signals with this property?

7-13 Let $x(t)$ be a signal whose Fourier transform $X(\omega)$ is equal to zero for $|\omega| > 2\pi$. Determine $x(t)$ if it is given that $x(t) + x(2t)$ is equal to

$$\frac{\sin 4\pi t}{\pi t} + 3\,\frac{\sin 2\pi t}{\pi t} + 2\,\frac{\sin \pi t}{\pi t}$$

7-14 Determine the Fourier transform of each of the time signals shown in Fig. 7P-9.

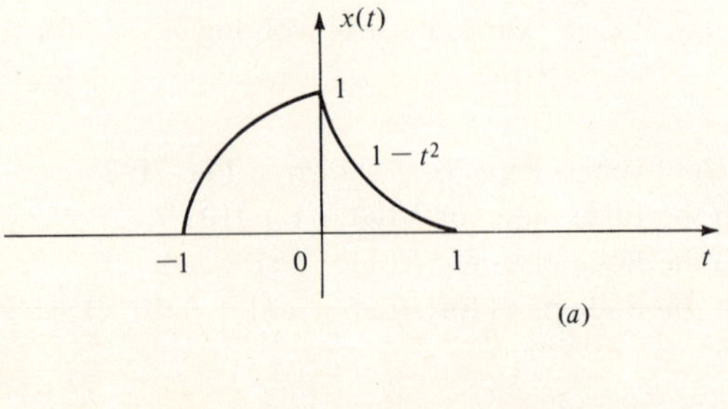

FIGURE 7-P9

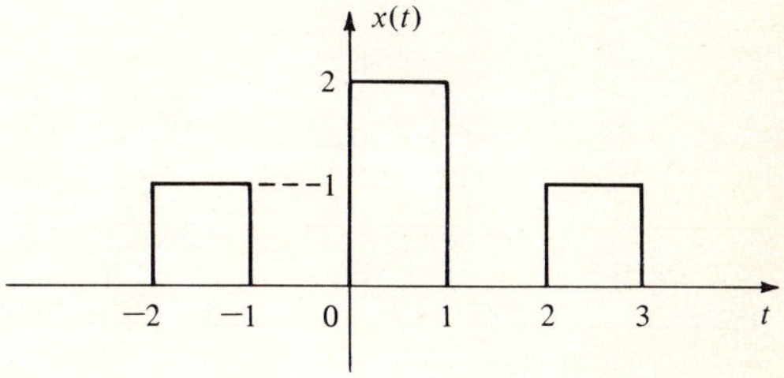

(b)

FIGURE 7P-9 (*continued*)

7-15 Let $x(t)$ be a continuous signal and

$$y(t) = \int_{-\infty}^{t} x(2(\tau - a))\, d\tau$$

where a is an arbitrary constant. Determine $Y(\omega)$ in terms of $X(\omega)$. It is known that $X(0) \neq 0$.

7-16 The real part of the transfer function of a causal system is given as

$$\frac{1}{\omega}(\sin \omega + \sin 2\omega)$$

Determine the imaginary part of the transfer function.

7-17 Suppose that the transfer function $H(\omega)$ of a linear time-invariant system is equal to 0 for all $|\omega| > w$.

(*a*) Show that for some $\omega_0 \gg w$

$$H\left(\frac{\omega^2 - \omega_0{}^2}{2\omega}\right) \approx H(\omega - \omega_0) + H(\omega + \omega_0)$$

and, hence, the time function

$$\frac{1}{2\pi}\int_{-\infty}^{\infty} H\left(\frac{\omega^2 - \omega_0{}^2}{2\omega}\right) e^{j\omega t}\, d\omega$$

is approximately equal to $h(t) \cos \omega_0 t$.

(*b*) Show that an *RLC* network with transfer function $H((\omega^2 - \omega_0{}^2)/2\omega)$ can be obtained from an *RLC* network with transfer function $H(\omega)$ by replacing each inductor L by the tune circuit in Fig. 7P-10*a* and each capacitor C by the tune circuit in Fig. 7P-10*b*.

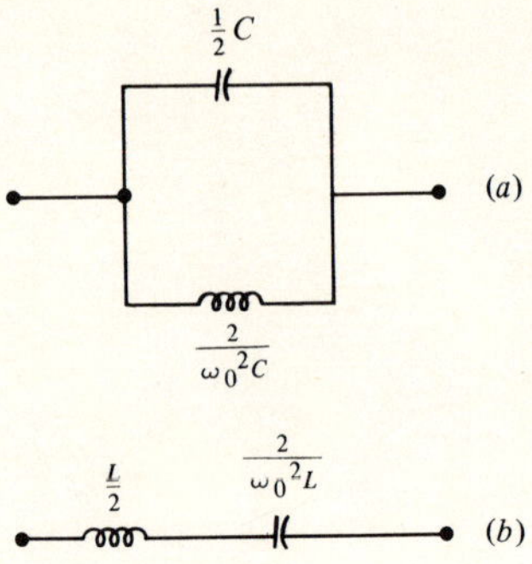

FIGURE 7P-10

7-18 Let $X(\omega)$ be the Fourier transform of a real time function. Show that the value of the integral

$$\int_{-\infty}^{\infty}\left[\int_{-\infty}^{\infty} X(\sigma)X(\omega-\sigma)\,d\sigma\right]e^{j\omega t_0}\,d\omega$$

is a nonnegative real number for any value of t_0.

7-19 Let $x(t) = \cos 2\pi t$ be the input signal and $h(t) = e^{-t}u_{-1}(t)$ be the impulse response of a linear time-invariant system. Determine the ouput signal $y(t)$ by the method of Fourier transform.

7-20 Let $x(t)$ shown in Fig. 7-9*a* be the input signal of a system whose transfer function $H(\omega)$ is that in Fig. 7-9*b*. Let $y(t)$ be its output signal. Sketch $y(t)$ carefully.

7-21 The transfer function of a linear time-invariant system is

$$H(\omega) = (1 - e^{-j2\omega})^2$$

Find the output signal when the input signal to this system is:

(*a*) $x(t) = \sin 3\pi t$

(*b*) $x(t) = \begin{cases} 0 & t \le 0 \\ \sin 3\pi t & t \ge 0 \end{cases}$

7-22 Find the output signal of the linear time-invariant system whose impulse response is

$$h(t) = \frac{\sin 4\pi(t-2)}{t-2} \qquad -\infty \le t \le \infty$$

when its input signal is

$$x(t) = \sin \frac{\pi}{2} t \qquad -\infty \le t \le \infty$$

7-23 Given that $x(t) * x'(t) = (1-t)e^{-t}u_{-1}(t)$, determine $x(t)$.

7-24 Determine the Fourier transform of each of the periodic signals shown in Fig. 7P-11.

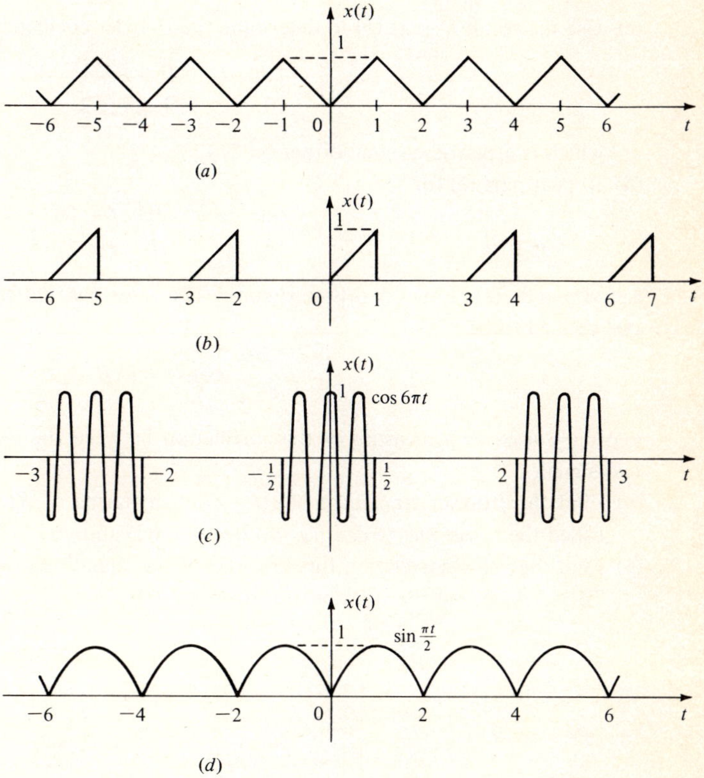

FIGURE 7P-11

7-25 The Fourier transform of a time signal $x(t)$ is

$$X(\omega) = e^{-\pi|\omega|}$$

Let $y(t)$ be a time signal which can be written

$$y(t) = \sum_{n=-\infty}^{n} x\left(t - \frac{n}{T}\right)$$

Find the Fourier transform $Y(\omega)$ of $y(t)$.

7-26 (*a*) Let $x(t)$ and $y(t)$ be two periodic signals of period T, and let $z(t)$ be their product, $x(t)y(t)$. Determine the Fourier coefficients of $z(t)$ in terms of those of $x(t)$ and $y(t)$.

(*b*) Determine the Fourier coefficients of the signal

$$x(t) = \begin{cases} 1 & 0 < t < 1 \\ 0 & 1 < t < 2 \end{cases}$$

which is a periodic signal of period 2.

(*c*) Use the result in part (*b*) to determine the Fourier coefficients of the signal

$$z(t) = \begin{cases} \sin \pi t & 0 \le t \le 1 \\ 0 & 1 \le t \le 2 \end{cases}$$

which is a periodic signal of period 2.

(*d*) Repeat part (*c*) for

$$z(t) = \begin{cases} \sin 2\pi t & 0 \le t \le 1 \\ 0 & 1 \le t \le 2 \end{cases}$$

7-27 Let $x(t)$ and $y(t)$ be two real time signals. The *cross-correlation function* of $x(t)$ and $y(t)$ defined to be

$$z(t) = \int_{-\infty}^{\infty} x(\tau)y(t+\tau)\, d\tau$$

is usually used as a measure of the correlation between the time variations in these two signals.

(*a*) Find the Fourier transform of $z(t)$, $Z(\omega)$, in terms of $X(\omega)$ and $Y(\omega)$. $Z(\omega)$ is called the *cross energy density spectrum* of $x(t)$ and $y(t)$.

(*b*) Find the cross-correlation function $z(t)$ of the signals $x(t)$ and $y(t)$ shown in Fig. 7P-12. Also find the Fourier transform of $z(t)$.

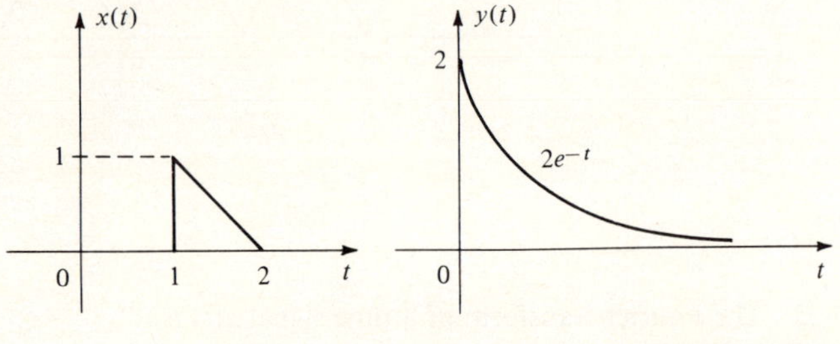

FIGURE 7P-12

7-28 Find the autocorrelation function and the energy density spectrum of each of the signals shown in Fig. 7P-13.

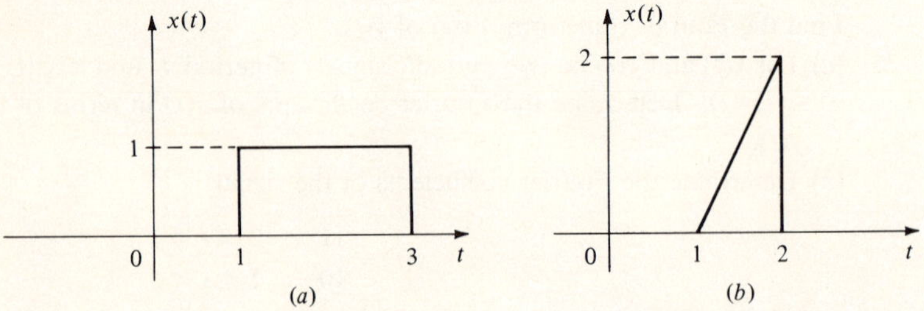

FIGURE 7P-13

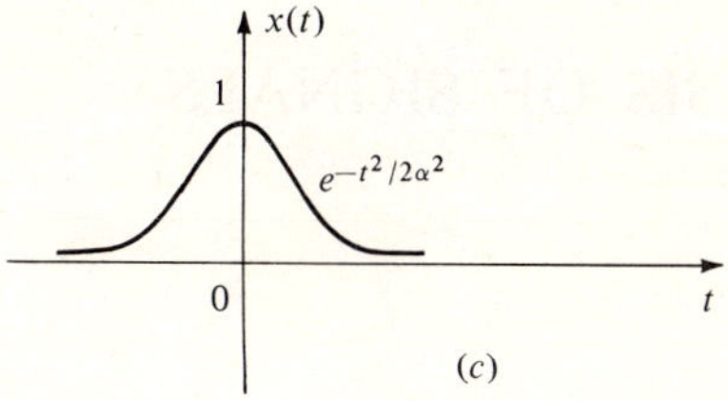

(c)

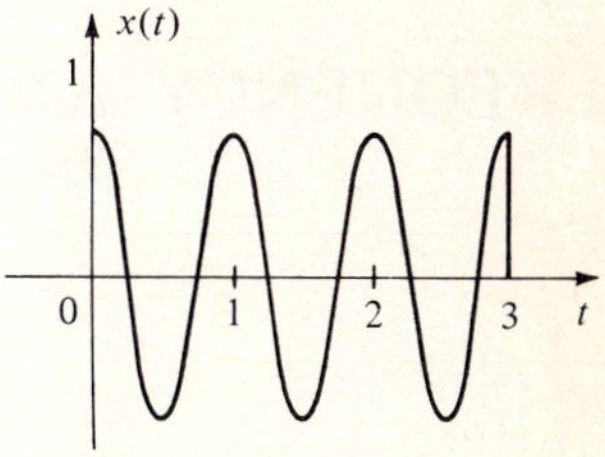

(d)

FIGURE 7P-13 *(continued)*

7-29 Let $x(t)$ and $y(t)$ be the input and output signals, respectively, of a linear time-invariant system whose impulse response is $h(t)$.

(*a*) Find the autocorrelation function of $y(t)$ in terms of $h(t)$ and the autocorrelation function of the input signal $x(t)$.

(*b*) Find the energy density spectrum of the output signal $y(t)$ in terms of the transfer function $H(\omega)$ of the system and the energy density spectrum of the input signal.

(*c*) Let $x(t)$ and $y(t)$ be the input and output signals of the circuit shown in Fig. 7P-14*a*. Find the total energy in the output signal when the input signal is as shown in Fig. 7P-14*b*.

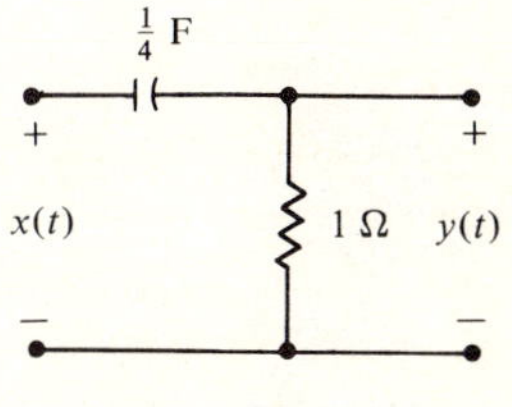

(*a*)

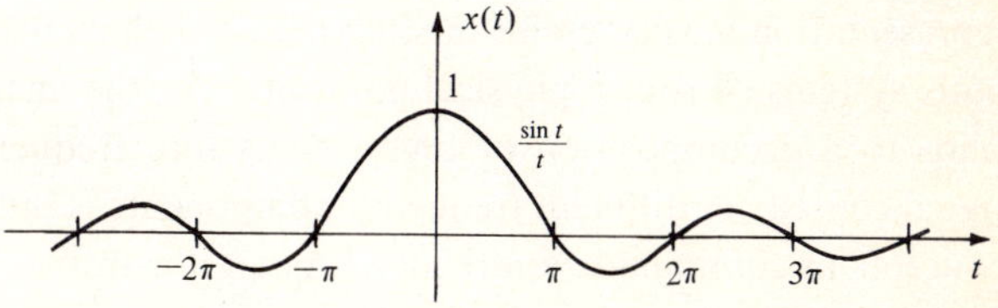

(*b*)

FIGURE 7P-14

8

FREQUENCY ANALYSIS OF SIGNALS

8-1 INTRODUCTION

From a mathematical point of view, the Fourier transform of a time signal is an alternative representation of the signal. As we have already observed in Chap. 7, such a representation is very useful in studying the behavior of linear time-invariant continuous systems. From a physical point of view, the method of Fourier transformation leads to a decomposition of time signals into frequency components. That a time signal consists of different frequency components is indeed one of the most important concepts in communication theory. As an illustration of the application of this concept, we shall study in this chapter modern communication systems such as radio and television systems. It is not our intention to give a detailed description of the principles of operations of these systems here. We wish only to point out that the mathematical analysis of decomposing signals into frequency components carried out in Chaps. 6 and 7 is indeed most useful in many practical problems.

8-2 FILTERING

Since a signal consists of different frequency components, we can talk about a "portion" of a signal, referring to some of its frequency components. For example, the human ears can detect only the audio frequency components of a signal, that is, those frequency components that are below 10,000 Hz.† Also, the intermediate frequency amplifier in a radio receiver is designed to amplify only the frequency components within a very narrow frequency range around 455 kHz in a signal.

To see how we can extract from a signal some of its frequency components, let us consider a linear time-invariant system whose transfer function is

$$H(\omega) = \begin{cases} 1 & -\omega_c \leq \omega \leq \omega_c \\ 0 & \text{otherwise} \end{cases}$$

as shown in Fig. 8-1*a*. For an input signal $x(t)$, the output signal of the system will contain only those frequency components of $x(t)$ whose frequencies are below $(1/2\pi)\omega_c$ Hz. A system is called a *filter* if it passes only some of the frequency components of the input signal and rejects the other frequency components. Thus, the linear system shown in Fig. 8-1*a* is a filter. In particular, it is known as a *low-pass filter* because it passes only the low-frequency components of the input signal. Similarly, a linear time-invariant system with a transfer function

$$H(\omega) = \begin{cases} 0 & -\omega_c \leq \omega \leq \omega_c \\ 1 & \text{otherwise} \end{cases}$$

as shown in Fig. 8-1*b* is another example of a filter, which is known as a *high-pass filter* because its output signal contains only the high-frequency components of the input signal. Also, a linear system with a transfer function

$$H(\omega) = \begin{cases} 1 & \omega_{c1} \leq |\omega| \leq \omega_{c2} \\ 0 & \text{otherwise} \end{cases}$$

as shown in Fig. 8-1*c* is known as a *bandpass filter* because it passes only the frequency components within the frequency range $(1/2\pi)\omega_{c1} \leq |f| \leq (1/2\pi)\omega_{c2}$ in the input signal. In general, a linear time-invariant system is a filter if the magnitude of its transfer function $|H(\omega)|$ differs substantially from zero for some frequency range, and it is essentially zero for the remaining frequency range. The frequency range within which $|H(\omega)|$ is substantially different from zero is called the *passband* of the filter. The frequency range within which $|H(\omega)|$ is essentially zero is called the *stopband* of

† A hertz is defined as one cycle per second. It is usually abbreviated Hz. Also, kilohertz and mepahertz are abbreviated kHz and MHz, respectively.

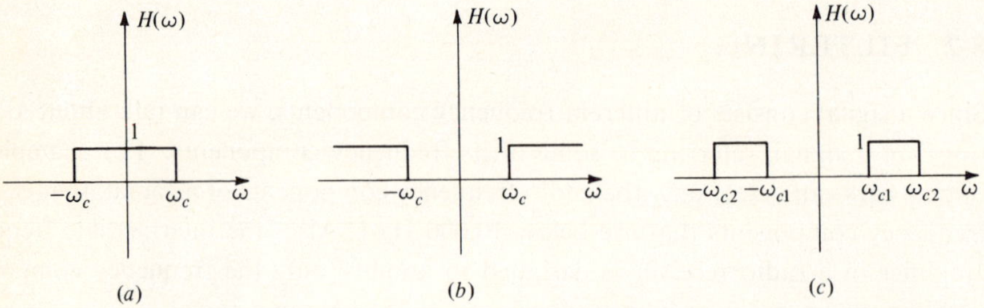

FIGURE 8-1

the filter. The width of the passband of a filter is called its *bandwidth*.† For example, the bandwidth of the filter in Fig. 8-1*a* is ω_c rad/s, the bandwidth of the filter in Fig. 8-1*b* is infinity, and the bandwidth of the filter in Fig. 8-1*c* is $\omega_{c2}-\omega_{c1}$ rad/s.

A filter is said to be *ideal* if the magnitude of its transfer function $|H(\omega)|$ is equal to a constant in its passband and is equal to 0 in its stopband. The filters shown in Fig. 8-1 are all ideal filters. Unfortunately, ideal filters are not physically realizable systems, as can be seen in the discussion in Prob. 8-4. (Intuitively, one can imagine the impossibility of building a physical system that will pass a frequency component whose frequency is $f_c - \epsilon$ but will reject a frequency component whose frequency is $f_c + \epsilon$ for arbitrarily small ϵ.) However, physically realizable filters with transfer functions closely approximating those of ideal filters can be designed and built. The electric circuit shown in Fig. 8-2*a* is an example of a low-pass filter, the magnitude of whose transfer function is shown in Fig. 8-2*b*. Further examples are discussed in Prob. 8-9.

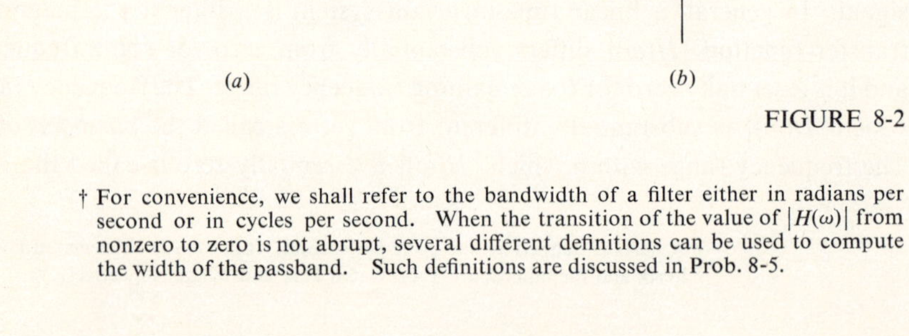

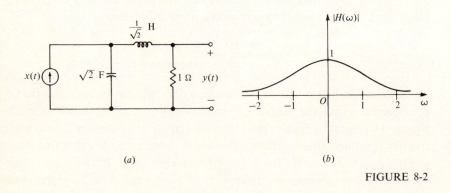

FIGURE 8-2

† For convenience, we shall refer to the bandwidth of a filter either in radians per second or in cycles per second. When the transition of the value of $|H(\omega)|$ from nonzero to zero is not abrupt, several different definitions can be used to compute the width of the passband. Such definitions are discussed in Prob. 8-5.

Extending these terminologies, we say that a continuous signal is *band-limited* if the magnitude of its Fourier transform is nonzero only within a finite frequency range. That is, a band-limited signal contains only frequency components that are within a finite frequency range. Specifically, a band-limited signal is said to be a *low-pass* signal if it contains only low-frequency components and is said to be a *band-pass* signal if it does not contain a zero frequency component. We also define the *bandwidth* of a signal to be the width of the frequency range within which its Fourier spectrum is nonzero.

8-3 THE SAMPLING THEOREM

When we studied continuous and discrete signals in Chap. 1, we pointed out the possibility of approximating a continuous signal by a discrete signal. For example, the continuous signal $x(t)$ shown in Fig. 8-3*a* can be approximated by the discrete signal $\hat{x}(n)$ shown in Fig. 8-3*b*, where $\hat{x}(n) = x(n\delta)$ for some fixed constant δ. We ask now whether it is possible to represent a continuous signal by a discrete signal *precisely* instead of just *approximately*. Clearly, not every continuous signal can be represented precisely by a discrete signal. However, the following rather trivial example indicates that such a possibility does exist for a class of signals. Let us consider a class of continuous signals $x(t)$ of the form

$$x(t) = k \qquad \text{for all } t$$

Clearly, such signals can be represented by discrete signals of the form

$$x(n) = k \qquad n = 0, \pm 1, \pm 2, \ldots$$

That is, given any such discrete signal we can unambiguously recover the corresponding continuous signal. The idea of representing a continuous signal by a discrete signal can be generalized in a nontrivial manner, as we shall discuss in this section.

Let $x(t)$ be a low-pass signal whose Fourier transform $X(\omega)$ is equal to zero for $|\omega| > 2\pi w$† as illustrated in Fig. 8-4*a*. Let $s(t)$ be a train of impulses:

$$s(t) = \sum_{n=-\infty}^{\infty} u_0\left(t - \frac{n}{2w}\right)$$

as shown in Fig. 8-4*b*. We note that the signal $x(t)s(t)$ is a sequence of impulses,

$$x(t)s(t) = \sum_{n=-\infty}^{\infty} x\left(\frac{n}{2w}\right) u_0\left(t - \frac{n}{2w}\right)$$

† Note that the dimension of w is cycles per second.

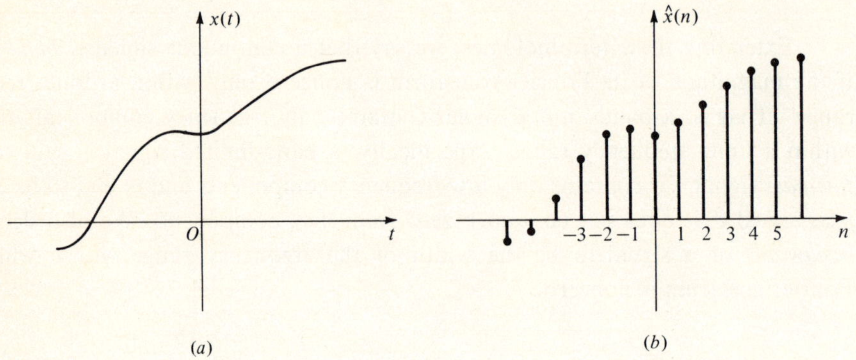

FIGURE 8-3

as shown in Fig. 8-5*a*. The Fourier transform of $x(t)s(t)$ is equal to $(1/2\pi)X(\omega) * S(\omega)$. Since $s(t)$ is a periodic signal of period $(1/2w)$, $S(\omega)$ can be computed as

$$S(\omega) = 4\pi w \sum_{n=-\infty}^{\infty} u_0(\omega - 4\pi nw)$$

as shown in Fig. 8-4*b*. It follows that $X(\omega) * S(\omega)$ consists of a series of scaled replicas of $X(\omega)$, as shown in Fig. 8-5*b*. Clearly, by passing $x(t)s(t)$ through an ideal low-pass filter whose transfer function is that shown in Fig. 8-5*c*, we can recover $x(t)$ from $x(t)s(t)$. The signal $x(t)s(t)$ is called a *sampled* signal of $x(t)$, and the impulses in $x(t)s(t)$ are called the *samples* of $x(t)$. The *sampling rate* is defined to be the number of

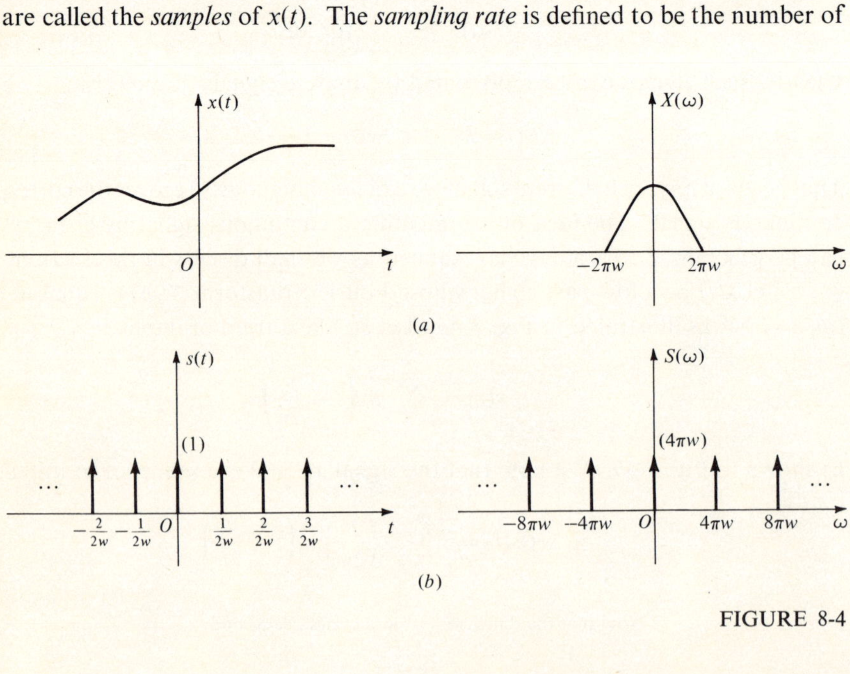

FIGURE 8-4

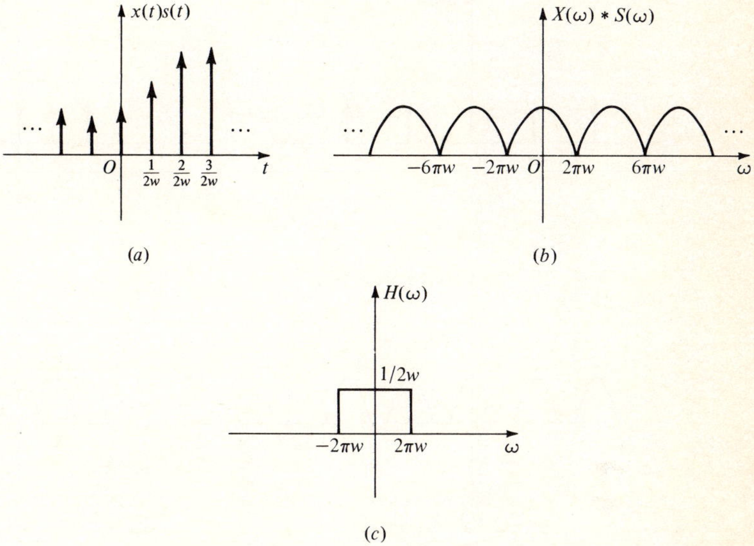

FIGURE 8-5

samples per second, which, in this case, is equal to $2w$. Since the signal $x(t)s(t)$ can be represented by a discrete signal $\hat{x}(n)$ such that

$$\hat{x}(n) = x\left(\frac{n}{2w}\right)$$

we conclude that the continuous signal $x(t)$ can be represented precisely by the discrete signal $\hat{x}(n)$.

Note that if we reduce the period of $s(t)$ to less than $(1/2w)$, $S(\omega)$ and $X(\omega) * S(\omega)$ will be that in Fig. 8-6*a* and *b*, respectively. That is, $X(\omega) * S(\omega)$ is a series of scaled replicas of $X(\omega)$ which are spaced farther apart along the frequency axis than those in Fig. 8-5*b*. Consequently, we can also recover $x(t)$ by passing $x(t)s(t)$ through a low-pass filter. On the other hand, if we increase the period of $s(t)$ to larger than $(1/2w)$, that is, if we sample $x(t)$ at a rate that is less than $2w$ samples per second, $S(\omega)$ will be that shown in Fig. 8-6*c*. In this case, $X(\omega) * S(\omega)$ is a sequence of overlapping scaled replicas of $X(\omega)$ as illustrated in Fig. 8-6*d*. Hence, it is generally impossible to recover $x(t)$ from the sampled signal $x(t)s(t)$.

Our discussion can be summarized as what is known as the *sampling theorem*. A low-pass signal of bandwidth w Hz is completely specified by the values of its samples at $t = 0, \pm(1/2a), \pm(2/2a), \pm(3/2a), \ldots$ for any $a \geq w$.† We should note that

† The sampling rate of $2w$ samples per second is known as the *Nyquist rate* of the signal.

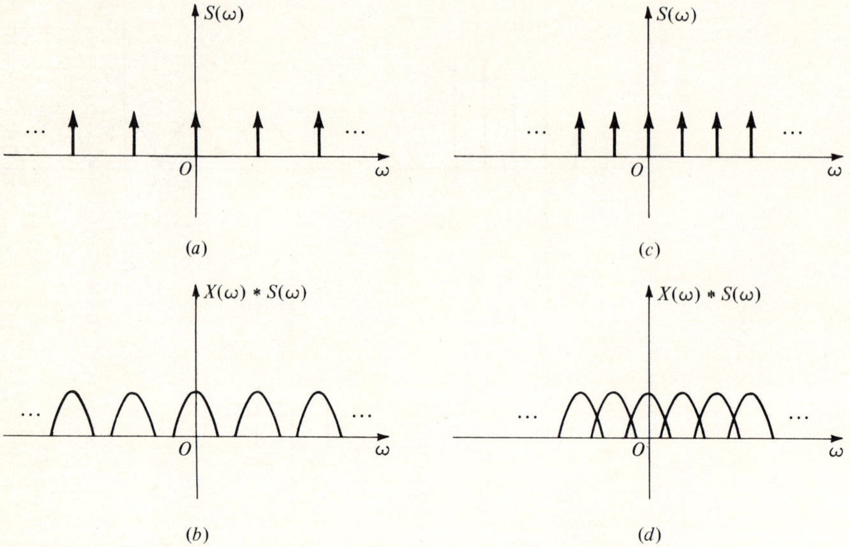

FIGURE 8-6

the sampling theorem can be explained qualitatively in a very intuitive way. A low-pass signal is a slow-varying signal because it does not contain any high-frequency components. Consequently, even if we know the values of the signal only at discrete time instants it is possible to recover the signal by interpolating between these discrete values. Moreover, the wider the bandwidth of a signal is, the faster the signal varies, and the higher the sampling rate should be.

The sampling theorem can be extended immediately to the case of bandpass signals: A bandpass signal of bandwidth w Hz is completely specified by the values of its samples at $t = 0$, $\pm(1/2a)$, $\pm(2/2a)$, $\pm(3/2a)$, ... for any $a \geq w$.† The interested reader is referred to Probs. 8-12 and 8-19.

It should also be noted that in practice the samples of a signal are not ideal impulses but rather are pulses of finite width. As a matter of fact, the physical process of sampling a signal is accomplished by connecting a switch in series with the signal source, as shown in Fig. 8-7*a*. If the switch closes for a fixed duration δ in every $1/2w$ seconds, the output signal from the switch will be pulses of width δ of the source signal as illustrated in Fig. 8-7*b*. The reader is asked to show that, in this case, the continuous signal can be recovered completely from these samples of finite duration in Prob. 8-13.

† Intuitively, although a bandpass signal might contain very-high-frequency components, its "envelope" is still a slow-varying signal.

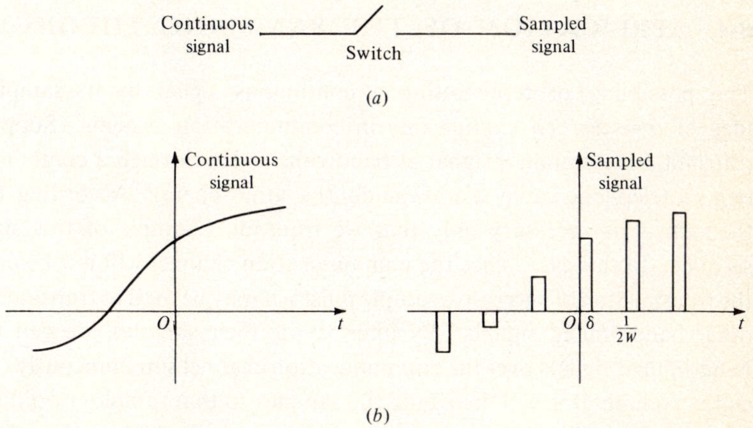

FIGURE 8-7

Many of the signals we encounter in practice either are band-limited signals or can be approximated as band-limited signals because they contain very little energy outside a finite frequency range. For example, the frequency spectrum of an audio signal is essentially zero outside the frequency range 10 kHz. Thus, if we sample an audio signal at a rate of 2×10^4 samples per second and pass the sampled signal through an ideal low-pass filter of bandwidth 10 kHz, the audio signal will be reproduced with very little distortion. In black-and-white television transmission, pictures are sampled and then transmitted at the rate of 30 picture frames per second. Each picture is divided into 525 horizontal lines, and each line is sampled at 650 points. Consequently, the number of samples transmitted is

$$30(525)(650) = 10{,}237{,}500$$

per second. According to the sampling theorem, if the bandwidth of a video signal w is such that

$$2w = 10{,}237{,}500$$

or

$$w = \frac{10{,}237{,}500}{2} \text{ Hz} \doteq 5 \text{ MHz}$$

the continuous video signal will be faithfully reproduced. Actually, the bandwidth of a video signal is larger than 5 MHz. However, the sampling rate 10,237,500 per second is high enough to yield pictures whose quality is quite acceptable to human eyes, although it does not allow a faithful reproduction of the video signal.

8-4 APPLICATION OF THE SAMPLING THEOREM

The possibility of representing a continuous signal by its samples leads to the idea of *time division multiplexing* in communication systems. Suppose we want to transmit a band-limited signal of bandwidth w Hz through a communication channel (e.g., a telephone cable, a waveguide, the atmosphere). According to the sampling theorem, it is necessary only that we transmit a sample of this signal every $1/2w$ seconds. In this case, since the communication channel will not be occupied between the transmission of successive sample pulses, it may be used to transmit sample pulses of other band-limited signals. By interleaving their samples, we can transmit several band-limited signals over the communication channel simultaneously. At the receiving end, a commutator will distribute the samples so that samples of different signals will be separated and individual signals can be recovered. Such a scheme for allowing two or more signals to share the use of the same communication channel with different signals occupying the channel at different time instants is known as a *time division multiplexing scheme*. For example, Fig. 8-8*b* shows how the samples of the two signals $x_1(t)$ and $x_2(t)$ in Fig. 8-8*a* can be interleaved, where the samples of $x_1(t)$ are shown as solid arrows and the samples of $x_2(t)$ are shown as dotted arrows.

Besides the concept of time division multiplexing, there is another important application of the sampling theorem. When a signal is transmitted through a communication channel, noise interference may distort the amplitude of the signal, as illustrated in Fig. 8-9. When the transmitted signal is band-limited, there are various schemes to transmit its samples so that the effect of noise interference will be reduced. For example, instead of transmitting the samples directly, which, by the way, is known as a *pulse amplitude modulation* (PAM) scheme, a sequence of pulses of constant amplitude but whose widths are proportional to the amplitudes of the samples can be transmitted. Such a scheme is known as a *pulse duration modulation* (PDM) scheme. For example, for the signal $x(t)$ in Fig. 8-10*a*, the pulses transmitted in a PAM scheme are shown in Fig. 8-10*b*, and the pulses transmitted in a PDM scheme are shown in Fig. 8-10*c*. Although noise interference might change the amplitudes of the transmitted pulses in a PDM system, it will not affect their widths significantly. Consequently, since the values of the samples are represented by the widths of the pulses, a PDM system offers higher quality of transmission than a PAM system does. Another scheme for transmitting the samples, known as a *pulse position modulation* (PPM) scheme, transmits a sequence of pulses of fixed amplitude and duration whose positions (with respect to fixed reference points) represent the values of the samples. For example, the pulses in a PPM system for the signal $x(t)$ in Fig. 8-10*a* is shown in Fig. 8-10*d*, where the dotted lines indicate the reference points. Another scheme of transmission, known as a *pulse code modulation* (PCM) scheme, is to encode the value of each sample as a sequence of binary digits. For example, a sample of value 9 can be represented as

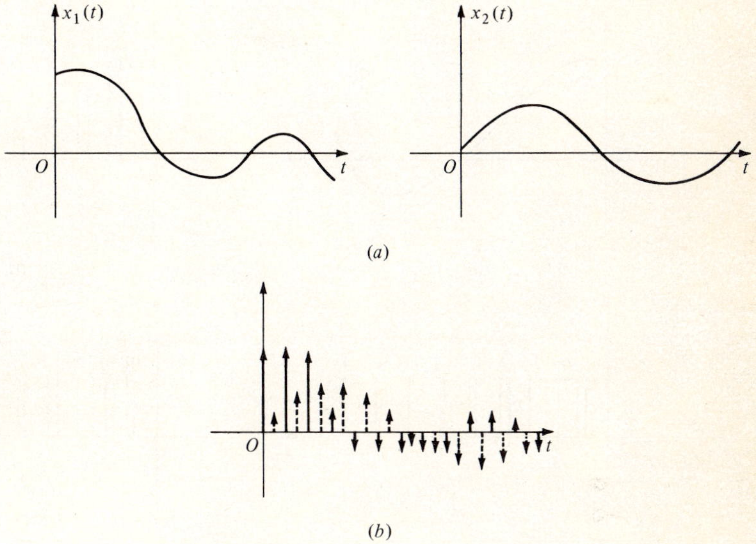

FIGURE 8-8

01001, where the leading 0 stands for a plus sign and 1001 is the binary number representation of 9. Also, a sample of value -3 can be represented as 10011, where the leading 1 stands for a minus sign and 0011 is the binary representation of 3. Thus, the samples of the signal $x(t)$ in Fig. 8-10*a* can be encoded as shown in Fig. 8-10*e*, where a 1 is represented by a positive pulse and a 0 is represented by a negative pulse. Although noise interference might affect the amplitude of these pulses, only under extreme circumstances will a positive pulse be recognized as a negative pulse or conversely at the receiving end.

We shall not analyze the performance of these transmission schemes here. It should be pointed out, however, when the PDM, PPM, or PCM schemes are employed,

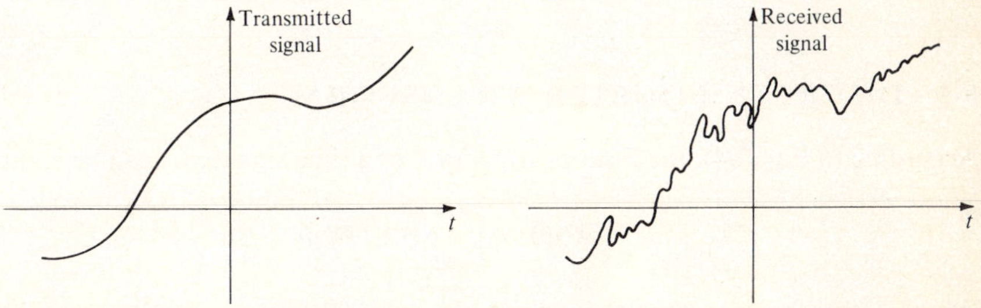

FIGURE 8-9

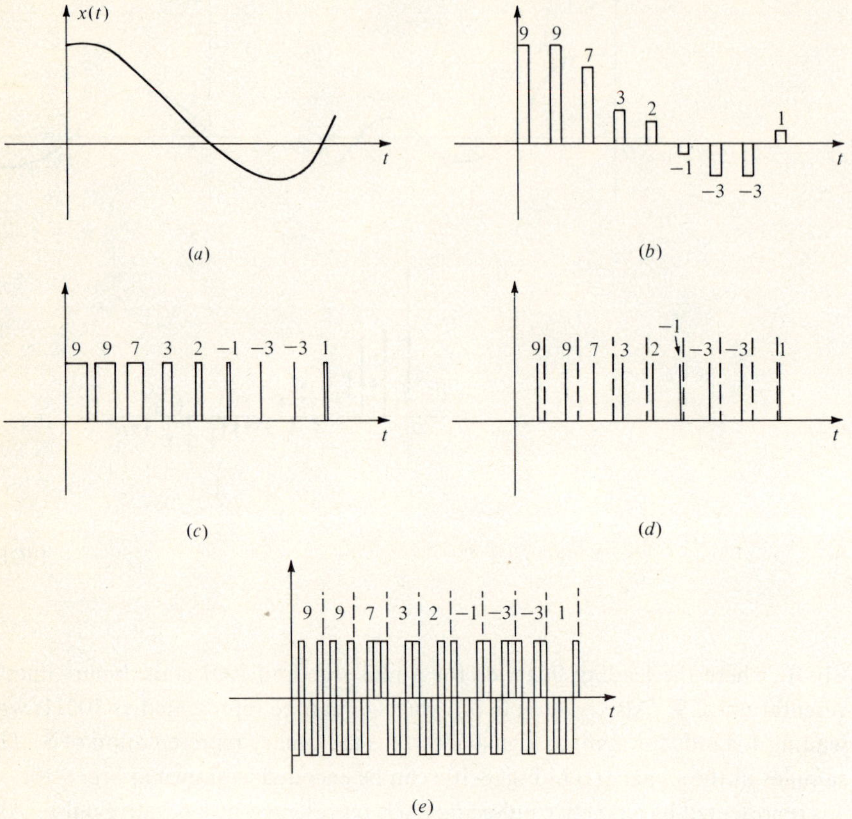

FIGURE 8-10

one is not getting something for nothing, which is always an impossibility in any engineering problem. Besides the cost of the equipment for encoding and decoding, the PDM, PPM, and PCM schemes require a larger bandwidth of transmission.

8-5 DISCRETE FOURIER TRANSFORM

According to Eq. (7-4), the Fourier transform of a time signal $x(t)$ can be computed as

$$X(\omega) = \int_{-\infty}^{\infty} x(t)e^{-j\omega t}\, dt \tag{8-1}$$

We have seen many examples in which this integral was evaluated to obtain closed-form expressions of $X(\omega)$. In many practical situations, however, the time signal $x(t)$

is determined by empirical measurement. Moreover, it is often the case that, instead of a continuous waveform, only sampled values of $x(t)$ are measured. For example, the output signal of an oscillator can either be displayed as a continuous waveform on an oscilloscope or be sampled at discrete times by using a voltmeter to measure the values of the samples. We can attempt to find an expression for $x(t)$ to approximate the measured waveform and then determine the corresponding expression of $X(\omega)$ from Eq. (8-1). Unfortunately, in many cases, even when a closed-form expression for $x(t)$ can be found as a good approximation of the measured waveform, the complex integral in Eq. (8-1) might be too complicated to be evaluated. Under such circumstances, it becomes necessary to determine the value of the integral in Eq. (8-1) by numerical methods.

To determine the Fourier transform of a time signal by numerical methods, we shall approximate the continuous signal $x(t)$ by a sampled version of $x(t)$, namely,

$$\sum_{k=-\infty}^{\infty} x(k\Delta)u_0(t-k\Delta) \tag{8-2}$$

for some sampling interval Δ. Let us, for simplicity of notation, denote $x(k\Delta)$ by x_k. The summation in (8-2) can be rewritten

$$\sum_{k=-\infty}^{\infty} x_k u_0(t-k\Delta) \tag{8-3}$$

Since in practice the value of $x(t)$ is measured only within a finite length of time, the infinite sum in (8-3) is further approximated by the finite sum†

$$\sum_{k=0}^{N-1} x_k u_0(t-k\Delta) \tag{8-4}$$

Here we assume the sample values of $x(t)$ are measured at $t = 0, \Delta, \ldots, (N-1)\Delta$. As a result of these approximations, the integral in Eq. (8-1) becomes

$$X(\omega) = \sum_{k=0}^{N-1} x_k e^{-jk\omega\Delta} \tag{8-5}$$

In practice, most of the time signals we encounter are band-limited. Hence, we can choose the sampling interval Δ to be small enough so that the sampling rate $1/\Delta$ is equal to or larger than twice the bandwidth of the signal, w.‡ In this case, the

† The summation in (8-4) can be obtained by multiplying the summation in (8-3) by the time function

$$u_{-1}(t+\tfrac{1}{2}\Delta) - u_{-1}(t-(N-\tfrac{1}{2})\Delta)$$

which is called the *data window*. Specifically, because of its shape, it is called a rectangular data window. For the effect of the shape and duration of the data window on the accuracy of the approximation in (8-4), see Prob. 8-22.

‡ If, because of other considerations, the sampling rate is chosen to be lower than the Nyquist rate of the time signal, error in the value of $X(\omega)$ will be introduced.

Fourier transform of the signal in (8-3) within the frequency range $|f| < w$ is equal to the Fourier transform of $x(t)$. In other words, $X(\omega)$ can be recovered from the Fourier transform of the signal in (8-3). However, we note that the Fourier transform $X(\omega)$ in Eq. (8-5), obtained from the finite summation in (8-4), is only an approximation of the $X(\omega)$ in Eq. (8-1). The accuracy of such an approximation depends on the properties of the time signal $x(t)$ as well as on the parameters Δ and N. The interested reader should study Probs. 8-23 and 8-24, where some of the pitfalls in numerical approximation are discussed.

Since $X(\omega)$ in Eq. (8-5) is the Fourier transform of a time signal of duration $N\Delta$, it follows from the sampling theorem that $X(\omega)$ is completely specified by its values at ω equals $\ldots -4\pi/N\Delta, -2\pi/N\Delta, 0, 2\pi/N\Delta, 4\pi/N\Delta, \ldots$. Hence, we shall determine only the values of $X(\omega)$ at these points. Let X_n denote $X(2\pi n/N\Delta)$ at $n = 0, \pm 1, \pm 2, \ldots$. Substituting $\omega = 2\pi n/N\Delta$ into Eq. (8-5), we obtain

$$X_n = \sum_{k=0}^{N-1} x_k e^{-j2\pi nk/N} \tag{8-6}$$

We recall that $X(\omega)$ is a periodic function of ω with period $1/\Delta$. We therefore need only compute X_n for $n = 0, 1, 2, \ldots, N-1$. X_n, as defined in Eq. (8-6), is known as a *discrete Fourier transform* (DFT) of the sampled signal x_k.

Similarly, to evaluate the inverse Fourier transform

$$x(t) = \frac{1}{2\pi}\int_{-\infty}^{\infty} X(\omega)e^{j\omega t}\, d\omega \tag{8-7}$$

by numerical methods, we shall approximate the continuous function $X(\omega)$ by a finite sampled version

$$\sum_{n=0}^{N-1} X\left(\frac{2\pi n}{N\Delta}\right)u_0\left(\omega - \frac{2\pi n}{N\Delta}\right) = \sum_{n=0}^{N-1} X_n u_0\left(\omega - \frac{2\pi n}{N\Delta}\right)$$

In other words, Eq. (8-7) is approximated as

$$x(t) = \frac{1}{2\pi}\sum_{n=0}^{N-1} X_n e^{j(2\pi n/N\Delta)t}$$

In particular, at time instants $0, \Delta, 2\Delta, \ldots, (N-1)\Delta$, we have

$$x_k = \frac{1}{2\pi}\sum_{n=0}^{N-1} X_n e^{j2\pi kn/N} \tag{8-8}$$

for $k = 0, 1, 2, \ldots N-1$. As defined in Eq. (8-8), x_k is known as an inverse discrete Fourier transform. We note from Eqs. (8-6) and (8-8) that the values of N sample points of $x(t)$ are needed to determine the values of N sample points of $X(\omega)$, and conversely.

Let us examine now the computational aspect of evaluating the discrete Fourier transform of a sampled signal. It is easy to see from Eq. (8-6) that to compute the value of X_n for any n $(0 \leq n \leq N-1)$ from the N sample values of $x(t)$ requires N multiplication and addition operations.† Hence, to determine the values of X_n at N points, N^2 multiplication and addition operations are needed. For $N = 2^m$ for some integer m,‡ the sums in Eqs. (8-8) and (8-9) can be evaluated by an algorithm that requires far fewer computation steps. This algorithm is known as the *fast Fourier transform* (FFT) algorithm which was discovered by Cooley and Tukey in 1965. Instead of presenting the algorithm in general terms, let us illustrate the steps by assuming $N = 4$.

For simplicity of notation, let

$$\alpha = e^{-j2\pi/4}$$

For $N = 4$, Eq. (8-6) can be rewritten

$$X_n = \sum_{k=0}^{3} x_k \alpha^{nk} \tag{8-9}$$

where $n = 0, 1, 2, 3$. Moreover, let us express the integers n and k as binary numbers. That is,

$$n = n_0 + 2n_1$$
$$k = k_0 + 2k_1$$

where n_1, n_0, k_1, k_0 are either 0 or 1. Let $X(n_1, n_0)$ and $x(k_1, k_0)$ denote X_n and x_k, respectively. We rewrite Eq. (8-9) as a double sum:

$$X(n_1, n_0) = \sum_{k_0=0}^{1} \sum_{k_1=0}^{1} x(k_1, k_0)\alpha^{(n_0+2n_1)(k_0+2k_1)}$$
$$= \sum_{k_0=0}^{1} \sum_{k_1=0}^{1} x(k_1, k_0)\alpha^{(n_0+2n_1)k_0}\alpha^{2(n_0+2n_1)k_1}$$

Since $\alpha^4 = 1$, this expression can be simplified to

$$X(n_1, n_0) = \sum_{k_0=0}^{1} \sum_{k_1=0}^{1} x(k_1, k_0)\alpha^{(n_0+2n_1)k_0}\alpha^{2n_0k_1} \tag{8-10}$$

Let

$$A(n_0, k_0) = \sum_{k_1=0}^{1} x(k_1, k_0)\alpha^{2n_0k_1} \tag{8-11}$$

† We assume that values of the complex numbers $e^{j2\pi nk/N}$ are stored in a table.

‡ When the samples of $x(t)$ are obtained empirically, it is possible that the number of measured samples M is not equal to a power of 2. In this case we can either use only $N = 2^m$ of the samples for some m such that $M \geq 2^m$ or append some zero-valued samples to the measured samples so that the total number of samples is a power of 2.

Then
$$X(n_1, n_0) = \sum_{k_0=0}^{1} A(n_0, k_0)\alpha^{(n_0+2n_1)k_0}\dagger \tag{8-12}$$

In other words, we have

$$A(0, 0) = x(0, 0)\alpha^0 + x(1, 0)\alpha^0$$
$$A(0, 1) = x(0, 1)\alpha^0 + x(1, 1)\alpha^0$$
$$A(1, 0) = x(0, 0)\alpha^0 + x(1, 0)\alpha^2$$
$$A(1, 1) = x(0, 1)\alpha^0 + x(1, 1)\alpha^2$$
$$X(0, 0) = A(0, 0)\alpha^0 + A(0, 1)\alpha^0$$
$$X(1, 0) = A(0, 0)\alpha^0 + A(0, 1)\alpha^2$$
$$X(0, 1) = A(1, 0)\alpha^0 + A(1, 1)\alpha^1$$
$$X(1, 1) = A(1, 0)\alpha^0 + A(1, 1)\alpha^3$$

Since
$$\alpha^0 = 1$$
$$\alpha^2 = -1$$
$$\alpha^1 = -\alpha^3$$

only 8 multiplication and addition operations are required to determine the value of any X_n instead of the 16 operations required when the summation in Eq. (8-9) is evaluated directly.

It can be shown (see Prob. 8-26) that for any $N = 2^m$ the summation in Eq. (8-9) can be rewritten as m subsums similar to those in Eqs. (8-11) and (8-12). The number

† Equations (8-10) to (8-12) can also be written as vector equations. Thus, Eq. (8-10) can be written

$$\begin{bmatrix} X(0,0) \\ X(0,1) \\ X(1,0) \\ X(1,1) \end{bmatrix} = \begin{bmatrix} \alpha^0 & \alpha^0 & \alpha^0 & \alpha^0 \\ \alpha^0 & \alpha^1 & \alpha^2 & \alpha^3 \\ \alpha^0 & \alpha^2 & \alpha^0 & \alpha^2 \\ \alpha^0 & \alpha^3 & \alpha^2 & \alpha^1 \end{bmatrix} \begin{bmatrix} x(0,0) \\ x(0,1) \\ x(1,0) \\ x(1,1) \end{bmatrix} \tag{8-13}$$

Equations (8-11) and (8-12) can be written

$$\begin{bmatrix} A(0,0) \\ A(0,1) \\ A(1,0) \\ A(1,1) \end{bmatrix} = \begin{bmatrix} \alpha^0 & 0 & \alpha^0 & 0 \\ 0 & \alpha^0 & 0 & \alpha^0 \\ \alpha^0 & 0 & \alpha^2 & 0 \\ 0 & \alpha^0 & 0 & \alpha^2 \end{bmatrix} \begin{bmatrix} x(0,0) \\ x(0,1) \\ x(1,0) \\ x(1,1) \end{bmatrix} \tag{8-14}$$

and

$$\begin{bmatrix} X(0,0) \\ X(1,0) \\ X(0,1) \\ X(1,1) \end{bmatrix} = \begin{bmatrix} \alpha^0 & \alpha^0 & 0 & 0 \\ \alpha^0 & \alpha^2 & 0 & 0 \\ 0 & 0 & \alpha^0 & \alpha^1 \\ 0 & 0 & \alpha^0 & \alpha^3 \end{bmatrix} \begin{bmatrix} A(0,0) \\ A(0,1) \\ A(1,0) \\ A(1,1) \end{bmatrix} \tag{8-15}$$

Hence the Cooley-Tukey algorithm can also be considered a method to factor the 4×4 matrix in Eq. (8-13) into the two 4×4 matrices in Eqs. (8-14) and (8-15). In general, for $N = 2^m$ the algorithm factors an $N \times N$ matrix into m $N \times N$ matrices.

of multiplication and addition operations required to compute the value of X_n is then equal to

$$\frac{N}{2}\log_2 N = m2^{m-1}$$

Therefore, for $N = 8$, the number of multiplications required is equal to 12 instead of $N^2 = 64$. For $N = 1024$, the number of operations required is equal to 5120 instead of $N^2 \approx 10^6$, representing a factor of 200 of reduction in computational steps when compared with determining the values of X_n directly.

With slight modification, the fast Fourier transform algorithm presented above can also be used to evaluate the inverse Fourier transform integral.

8-6 FREQUENCY DIVISION MULTIPLEXING

The idea of time division multiplexing suggests another possible way to allow two or more signals to share a transmission channel. Let us consider a simple illustrative example in which two signals with nonoverlapping frequency spectra are to be transmitted through the same communication channel. In this case, we can simply transmit the sum of the two signals and then separate them at the receiving end by filtering. The diagram in Fig. 8-11 shows how such a scheme works, where $x_1(t)$ is a low-pass signal and $x_2(t)$ is a bandpass signal. In general, the concept of using a communication channel to transmit simultaneously two or more signals that have nonoverlapping frequency spectra is known as *frequency division multiplexing*.

The diagrams in Fig. 8-12*a* and *b* illustrate how a communication channel is utilized in a time division multiplexing system and in a frequency division multiplexing system. In a time division multiplexing system, a transmitted signal occupies the communication channel only part of the time. However, during the transmission of any of the signals, the whole frequency range of the channel can be utilized by that signal. On the other hand, in a frequency division multiplexing system, every transmitted signal occupies the communication channel all the time. However, each signal may utilize only a portion of the frequency range of the channel.

It should be pointed out that no physical channel is capable of transmitting frequency components of all frequencies with reasonably high efficiency. Indeed, a channel can always be modeled as a bandpass filter which can be used to transmit only signals whose frequency spectra are within its passband. For example, a pair of telephone wires can be used to transmit frequency components from 0 to 2400 Hz, and the bandwidth of a coaxial cable is approximately 10^7 Hz. Consequently, there is a limitation on the number of signals a communication channel can transmit simultaneously.

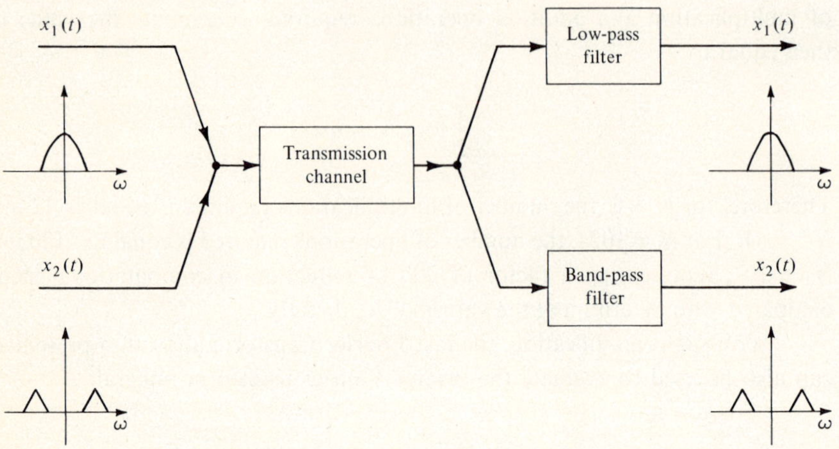

FIGURE 8-11

As a matter of fact, it can be shown quantitatively that the number of signals which can be transmitted in a time interval T over a channel of bandwidth W is of the order of $2TW$ (see Prob. 8-21).

In many cases, the signals to be transmitted have overlapping frequency spectra and therefore cannot be multiplexed directly. We shall illustrate how such cases can be handled when we discuss radio communication systems in the next section.

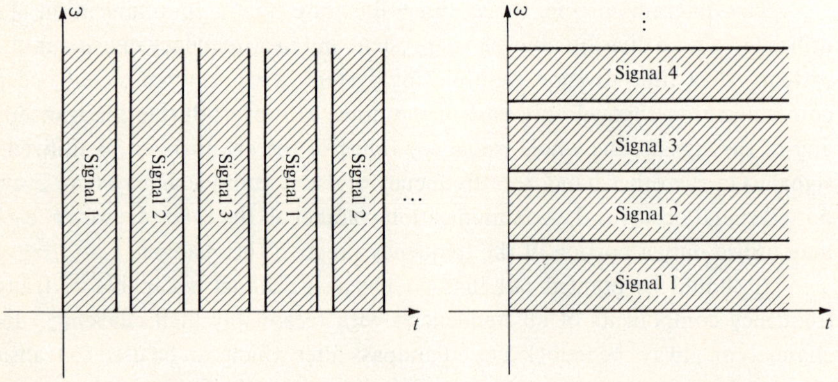

Time division multiplexing

(a)

Frequency division multiplexing

(b)

FIGURE 8-12

8-7 AMPLITUDE MODULATION

As was pointed out in the preceding section, two signals with overlapping frequency spectra cannot be multiplexed directly in a frequency division multiplexing system. For example, when two people in a room are talking at the same time the conversation becomes unintelligible because they are transmitting through the atmosphere two signals with overlapping frequency spectra. However, when the signals to be transmitted are band-limited signals, their frequency spectra can be "shifted" so that they no longer overlap. This indeed is the principle of radio communication where audio signals from different radio stations are transmitted through the atmosphere in a frequency division multiplexing arrangement. To illustrate how the frequency spectra of signals can be shifted, let us consider the transmission of two audio signals $x_1(t)$ and $x_2(t)$ whose frequency spectra $X_1(\omega)$ and $X_2(\omega)$ are shown in Fig. 8-13*a*. Let

$$y_1(t) = x_1(t) \cos 2\pi f_1 t$$

and

$$y_2(t) = x_2(t) \cos 2\pi f_2 t$$

The frequency spectra of $y_1(t)$ and $y_2(t)$ are

$$Y_1(\omega) = X_1(\omega) * \tfrac{1}{2}[u_0(\omega + 2\pi f_1) + u_0(\omega - 2\pi f_1)]$$

and

$$Y_2(\omega) = X_2(\omega) * \tfrac{1}{2}[u_0(\omega + 2\pi f_2) + u_0(\omega - 2\pi f_2)]$$

which are the frequency spectra of $x_1(t)$ and $x_2(t)$ shifted by f_1 and f_2 Hz, respectively. If the frequencies f_1 and f_2 are sufficiently separated, the frequency spectra $Y_1(\omega)$ and $Y_2(\omega)$ will not overlap, as illustrated in Fig. 8-13*b*. Consequently, simultaneous transmission of the two signals $y_1(t)$ and $y_2(t)$ over the same communication channel becomes possible. The signals $x_1(t)$ and $x_2(t)$ are known as the *modulating signals*, and the signals $y_1(t)$ and $y_2(t)$ are known as the *modulated signals*. The signals $\cos 2\pi f_1 t$ and $\cos 2\pi f_2 t$ are known as the *carrier signals*, and the frequencies f_1 and f_2 are known as the *carrier frequencies*. To shift the frequency spectrum of a signal by multiplying it with a carrier signal is known as *amplitude modulation* (AM), because the amplitude of the carrier signal is modulated by the modulating signal. In radio broadcast, the carrier frequencies f_1 and f_2 are assigned to different radio stations by the Federal Communications Commission so that stations in the same geographic area will have carrier frequencies sufficiently separated for minimal interference. Specifically, the frequency range 535 to 1605 kHz is used for AM broadcast, and each broadcast station is allocated a bandwidth of 10 kHz.

In radio communication, there is another compelling reason for shifting the frequency spectra of audio signals to a higher frequency range. In order to transmit an electromagnetic wave with high radiation efficiency, the transmitting antenna must be of a size at least one-tenth of the wavelength of the electromagnetic wave. To

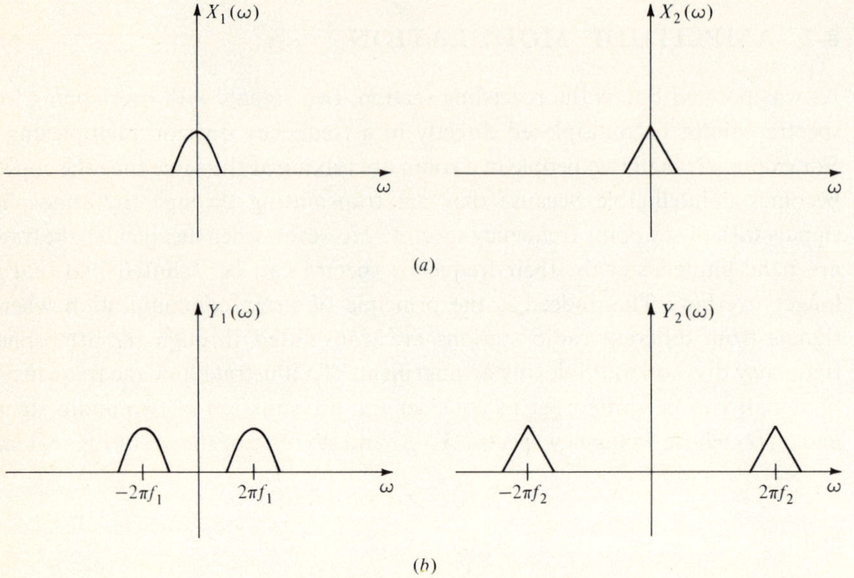

FIGURE 8-13

transmit an audio signal directly, an antenna several miles long would be needed.† By shifting the frequencies of the transmitted signals to the kilohertz and megahertz range, the sizes of the antennas become reasonable.

We now look into ways to recover the modulating signal from the modulated signal. Let $x(t)$ be an audio signal and

$$y(t) = x(t) \cos 2\pi f_0 t$$

be the corresponding transmitted signal in a radio communication system. At the receiving end, the signal $x(t)$ can be recovered from $y(t)$ by using the system in Fig. 8-14*a*. Since

$$z(t) = y(t) \cos 2\pi f_0 t = x(t) \cos^2 2\pi f_0 t = \tfrac{1}{2}x(t) + \tfrac{1}{2}x(t) \cos 4\pi f_0 t$$

the output of the low-pass filter is $\frac{1}{2}x(t)$, and the term $\frac{1}{2}x(t) \cos 4\pi f_0 t$ will be cut off. We can also deduce this result by observing the frequency spectrum $Z(\omega)$ in Fig. 8-14*b*. It follows immediately that the output of the low-pass filter is equal to $\frac{1}{2}x(t)$. Such a scheme for recovering the modulating signal from the modulated signal is known as *synchronous detection*, and the corresponding circuit is known as a *synchronous detector*.

† For example, the wavelength of a 10,000-Hz signal is $(3 \times 10^8)/10^4 = 30{,}000$ m = 98,400 ft.

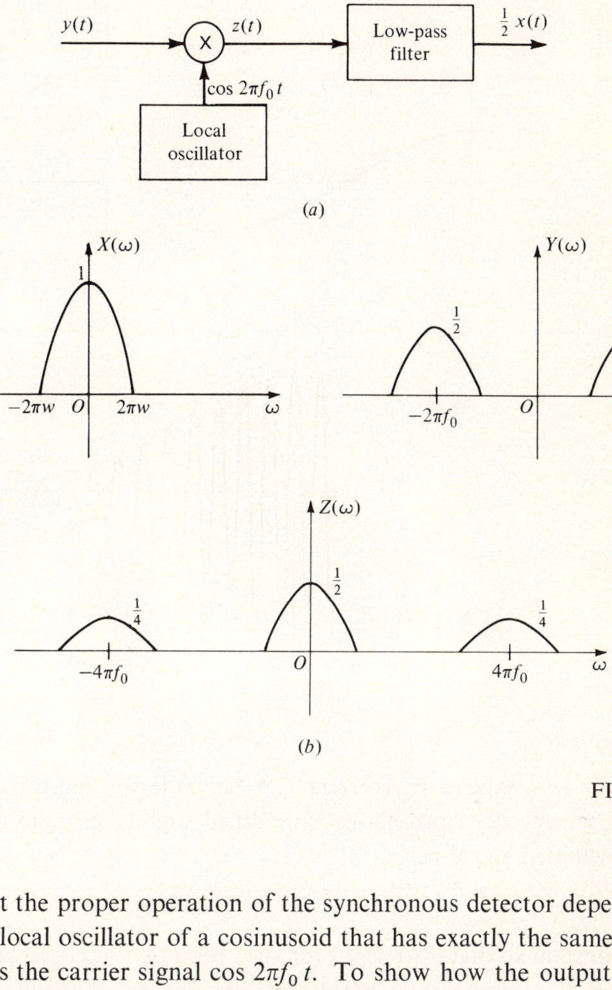

FIGURE 8-14

We note that the proper operation of the synchronous detector depends on the generation by the local oscillator of a cosinusoid that has exactly the same frequency and phase angle as the carrier signal $\cos 2\pi f_0 t$. To show how the output of a synchronous detector is affected when the local oscillator is out of synchronization with the carrier, let us suppose that the output of the local oscillator is $\cos(2\pi f_0 t + \theta)$. Thus, in Fig. 8-14*a*,

$$\begin{aligned} z(t) &= y(t)\cos(2\pi f_0 t + \theta) \\ &= x(t)\cos 2\pi f_0 t \cos(2\pi f_0 t + \theta) \\ &= \frac{x(t)}{2}\cos\theta + \frac{x(t)}{2}\cos(4\pi f_0 t + \theta) \end{aligned}$$

Consequently, the output of the low-pass filter is $[x(t)/2]\cos\theta$ which will be a distorted version of $x(t)$ when θ varies with time. In practice, expensive circuit elements are needed to build a local oscillator that attains the required precision.

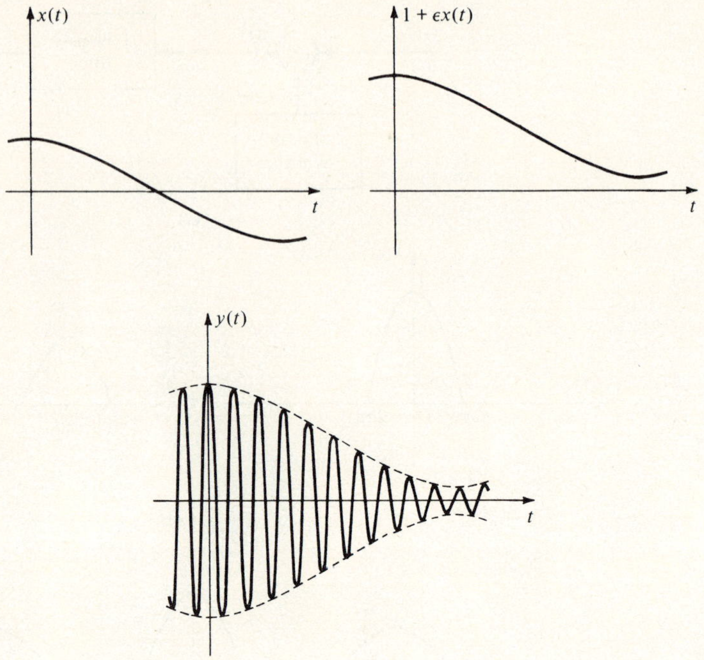

FIGURE 8-15

So that less expensive receivers can be built for mass communication, the modulation scheme described above is modified slightly in radio broadcast systems. Let the transmitted signal be

$$y(t) = [1 + \epsilon x(t)] \cos 2\pi f_0 t \tag{8-16}$$

where ϵ is constant so that $|\epsilon x(t)| \leq 1$ for all t. Because $1 + \epsilon x(t)$ is always nonnegative, the envelope of $y(t)$ is equal to $\pm[1 + \epsilon x(t)]$, as illustrated in Fig. 8-15. In this case, the simple circuit in Fig. 8-16*a* can be used to recover $x(t)$ from $y(t)$. The signals $y_2(t)$ and $y_3(t)$ in Fig. 8-16*a* are shown in Fig. 8-16*b*. Note that the signal $y_2(t)$ is approximately the envelope of $y_1(t)$, which is the positive portion of $y(t)$, because the capacitor C is charged when the value of $y_1(t)$ rises and will discharge only slowly when the value of $y_1(t)$ declines. The signal $y_3(t)$ is equal to $\epsilon x(t)$ because the series combination of C_1 and R_1 removes the dc component in $y_2(t)$. The circuit in Fig. 8-16*a* is called an *envelope detector*.

A good engineer would undoubtedly wonder what price must be paid in order to be able to use the envelope detector instead of the more expensive synchronous detector in the receivers. According to Eq. (8-16), the transmitted signal includes a copy of the carrier $\cos 2\pi f_0 t$. The carrier contains no useful information about the modulating

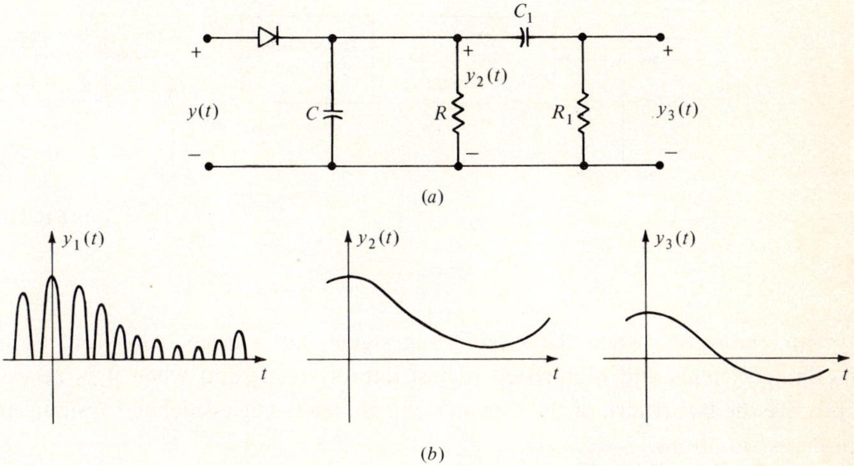

FIGURE 8-16

signal, and yet it consumes electric power at the transmitter. For an analysis of the power consumption in the carrier signal, see Prob. 8-27.

It should be pointed out that in an amplitude modulation system the bandwidth of the modulated signal is two times that of the modulating signal. As can be noted in Fig. 8-14*b*, the bandwidth of the audio signal $x(t)$ is w and that of the modulated signal $y(t)$ is $2w$. Examining the frequency spectrum of $y(t)$ in Fig. 8-14*b*, we realize that as far as recovering the signal $x(t)$ is concerned, it is redundant to transmit all the frequency components of $y(t)$. Rather, transmitting only the frequency components within the frequency range $f_0 - w \le |f| \le f_0$ or $f_0 \le |f| \le f_0 + w$ will be sufficient because there will be enough information to reconstruct the frequency spectrum of $x(t)$. The frequency components within the frequency range $f_0 - w \le |f| \le f_0$ are known as the *lower sideband*, and the frequency components within the frequency range $f_0 \le |f| \le f_0 + w$ are known as the *upper sideband* of the signal $y(t)$. A transmission system in which only the lower sideband or the upper sideband of the modulated signal is transmitted is known as a *single-sideband* (SSB) system.† Figure 8-17 shows the principle of operation of a single-sideband transmission and reception system, where the bandwidth of $x(t)$ is w Hz. We shall leave it to the reader to confirm that the receiver indeed reproduces the modulating signal $x(t)$. It is clear that equipment cost at both the transmitting and the receiving end will be higher in a single-sideband

† A system in which the transmitted signal is $[1 + \epsilon x(t)] \cos 2\pi f_0 t$ described above is known as a double-sideband (DSB) system. A system in which the transmitted signal is $x(t) \cos 2\pi f_0 t$ is called a double-sideband suppressed-carrier (DSB-SC) system.

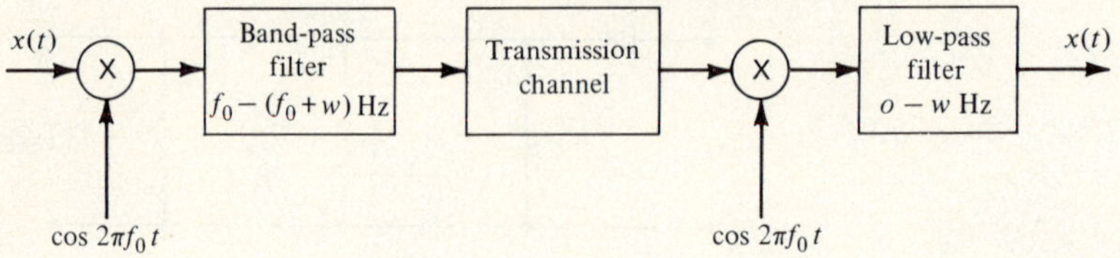

FIGURE 8-17

system. However, when the modulating signal has a large bandwidth such as in facsimile systems and high-speed digital data systems, and when it is necessary to conserve the bandwidth of the transmission channel, single-sideband systems are used in many situations.

In practice, it is difficult to build a bandpass filter such as that shown in Fig. 8-17 to cut off completely one of the sidebands of the modulated signal. A compromise is to transmit one of the sidebands together with a vestige of the other sideband of the modulated signal. Such a scheme is known as a *vestigial single-sideband* (VSB) system. We leave the details of such a transmission scheme to Prob. 8-28.

8-8 FREQUENCY MODULATION

A shortcoming of the amplitude modulation scheme is its relatively low quality of reception due to the fact that the transmitted signal is susceptible to noise interference. In this section, we shall study *frequency modulation* (FM) systems which, in terms of the quality of reception, are superior to amplitude modulation systems.

As was pointed out in the preceding section, the carrier signal in an amplitude modulation system is used to shift the frequency spectrum of a low-pass signal to a higher frequency range. We can also view the carrier as a signal that carries a low-pass signal to the destination of transmission. In particular, information concerning the low-pass signal is represented by the variation of the amplitude of the carrier. One might wonder whether information concerning the low-pass signal can be carried by the carrier in some other ways. In other words, the low-pass signal may modify some parameters of the carrier other than its amplitude. Since the carrier is specified by its amplitude and its frequency, it is just as natural that we investigate the possibility of representing the information concerning the low-pass signal by the variation in the frequency of the carrier. A transmission system in which the frequency of the carrier varies according to the amplitude of a low-pass signal is known as a frequency modulation system.

Let us define first the notion of the instantaneous frequency of signals of the form $\cos \theta(t)$. We recall that in the cosinusoid $\cos 2\pi f_0 t$, the quantities $2\pi f_0 t$, $2\pi f_0$, and f_0 are known as the angular displacement, the angular velocity, and the frequency, respectively, of the signal. Generalizing these notions, for a signal $\cos \theta(t)$, we define $\theta(t)$ to be the *angular displacement*, $d\theta(t)/dt$ to be the *instantaneous angular velocity*, and $(1/2\pi)[d\theta(t)/dt]$ to be the *instantaneous frequency* of the signal. Thus, for example, in the signal $\cos 2\pi f_0 t^2$, the angular displacement at time t is $2\pi f_0 t^2$, the instantaneous angular velocity is $4\pi f_0 t$, and the instantaneous frequency is $2f_0 t$.

Let $x(t)$ be a low-pass signal and $\cos 2\pi f_0 t$ be the carrier signal. Let us modify the carrier so that its instantaneous frequency is equal to

$$f(t) = f_0 + \epsilon x(t) \tag{8-17}$$

where ϵ is a multiplying constant known as the *frequency deviation constant*. It follows that the angular velocity is

$$\omega(t) = 2\pi f_0 + 2\pi\epsilon x(t)$$

and the total angular displacement is

$$\theta(t) = \int_0^t \omega(\tau)\, d\tau = 2\pi f_0 t + 2\pi\epsilon \int_0^t x(\tau)\, d\tau$$

Thus, the resultant signal is

$$y(t) = \cos \theta(t) = \cos \left[2\pi f_0 t + 2\pi\epsilon \int_0^t x(\tau)\, d\tau\right]$$

as illustrated in Fig. 8-18. Again, $x(t)$ is known as the modulating signal and $y(t)$ as the modulated signal. As a numerical example, let

$$x(t) = E \cos 2\pi 500t$$

and let f_0 be 10^6 Hz. Then,

$$\begin{aligned} y(t) &= \cos\left(2\pi 10^6 t + 2\pi\epsilon \int_0^t E \cos 2\pi 500\tau\, d\tau\right) \\ &= \cos\left(2\pi 10^6 t + \frac{\epsilon E}{500} \sin 2\pi 500t\right) \end{aligned}$$

Suppose that the frequency deviation constant ϵ is so chosen that

$$\frac{\epsilon E}{500} = 20\pi$$

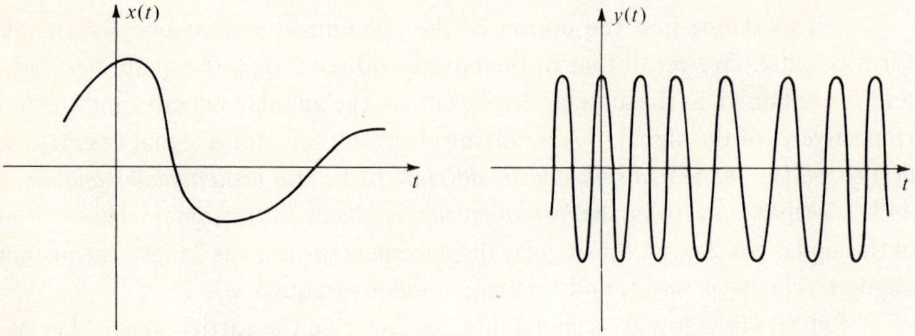

FIGURE 8-18

Then the instantaneous frequency of $y(t)$ will vary between 1,000,010 and 999,990 Hz. Moreover, the rate of such variation is 500 times per second. If we let

$$x(t) = 2E \cos 2\pi 500t$$

then
$$y(t) = \cos\left(2\pi 10^6 t + \frac{2\epsilon E}{500} \sin 2\pi 500t\right)$$

In this case, the instantaneous frequency of $y(t)$ will vary between 1,000,020 and 999,980 Hz at a rate of 500 times per second. However, if we let

$$x(t) = 2E \cos 2\pi 1000t$$

then
$$y(t) = \cos\left(2\pi 10^6 t + \frac{2\epsilon E}{1000} \sin 2\pi 1000t\right)$$

The instantaneous frequency of $y(t)$ will vary between 1,000,010 and 999,990 Hz, but at a rate of 1000 times per second.

We shall not discuss the problem of recovering $x(t)$ from the modulated signal $y(t)$ here. The interested reader may consult any book on communication engineering. We should point out, however, the reason that the quality of reception is higher in frequency modulation systems than in amplitude modulation systems. When a signal is transmitted through a communication channel, distortion caused by noise interference in the amplitude of the signal will be much more prominent than that in its instantaneous frequency. Since in a frequency modulation system the receiver will extract the modulating signal from the frequency variation of the modulated signal and will ignore distortions in the amplitude of the modulated signal, the effect of the noise will be much less than that in an amplitude modulation system.

In order to allocate frequency channels to different FM broadcast stations so that they can transmit over the atmosphere simultaneously, the bandwidth of a frequency-modulated signal must be determined. For many years, it was mistakenly believed that the bandwidth of a frequency-modulated signal can be made as small as desired. This misconception was not recognized until 1922, when Carson pointed it out, although the idea of frequency modulation was known before 1914. The false argument went as follows: According to Eq. (8-17), the instantaneous frequency of the modulated signal varies between $f_0 + \epsilon E_1$ and $f_0 + \epsilon E_2$, where E_1 and E_2 denote the largest and the smallest values of $x(t)$, respectively. Thus, the bandwidth of the modulated signal is equal to $\epsilon(E_1 - E_2)$. For the preceding example, since the instantaneous frequency of the modulated signal varies between 1,000,010 and 999,990 Hz, the bandwidth of the modulated signal is 20 Hz. If such an argument were indeed correct, we would have a nice way to conserve the use of the bandwidth of a communication channel: Since the frequency deviation of the modulated signal is independent of the frequency spectrum of the modulating signal and is dependent only on the amplitude of the modulating signal and the modulation index ϵ, the frequency deviation could be decreased by decreasing the value of ϵ. Consequently, the smaller the frequency deviation of the modulated signal is, the narrower its bandwidth will be. Unfortunately, there is a serious flaw in the argument, namely, the bandwidth of the modulated signal is *not* equal to its frequency deviation, as the following analysis shows.

It is rather complicated to derive a general expression for the frequency spectrum of $y(t)$ for an arbitrary low-pass signal $x(t)$. However, to illustrate the dependence of the frequency spectrum of the modulated signal $y(t)$ on both the amplitude and the frequency spectrum of the modulating signal $x(t)$, it is sufficient to consider the case in which

$$x(t) = E \cos 2\pi f_m t$$

It follows that

$$\begin{aligned} y(t) &= \cos\left(2\pi f_0 t + 2\pi\epsilon \int_0^t E \cos 2\pi f_m \tau \, d\tau\right) \\ &= \cos\left(2\pi f_0 t + \frac{\epsilon E}{f_m} \sin 2\pi f_m t\right) \\ &= \cos 2\pi f_0 t \cos\left(\frac{\epsilon E}{f_m} \sin 2\pi f_m t\right) - \sin 2\pi f_0 t \sin\left(\frac{\epsilon E}{f_m} \sin 2\pi f_m t\right) \end{aligned}$$

Since both $\cos [(\epsilon E/f_m) \sin 2\pi f_m t]$ and $\sin [(\epsilon E/f_m) \sin 2\pi f_m t]$ are periodic signals of period $1/f_m$, their Fourier transforms are sequences of impulses at $\omega = 2\pi n f_m$ for $n = 0, \pm 1, \pm 2, \ldots$. Consequently, the frequency spectrum of $y(t)$ consists of a sequence of impulses at $\omega = 2\pi(f_0 \pm n f_m)$, and we conclude that $y(t)$ is not a strictly band-limited signal at all. However, since the magnitudes of the Fourier coefficients

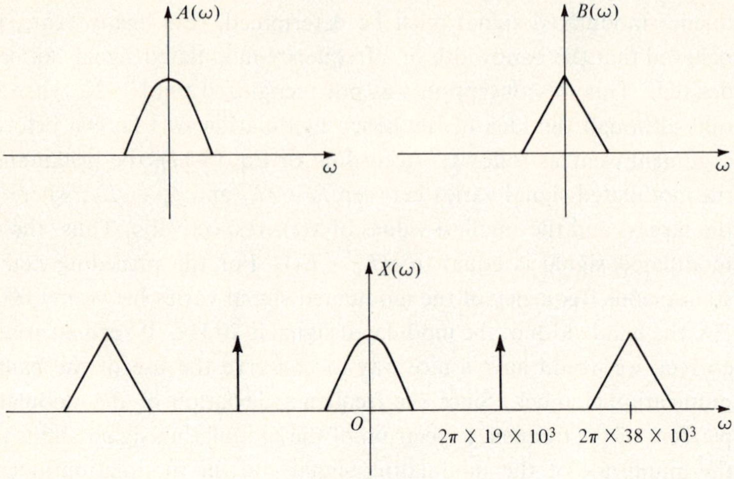

FIGURE 8-19

of cos $[(\epsilon E/f_m) \sin 2\pi f_m t]$ and sin $[(\epsilon E/f_m) \sin 2\pi f_m t]$ are negligibly small for $n > 5$, we can approximate $y(t)$ as a band-limited signal of bandwidth $10 f_m$ Hz. In FM broadcast, a bandwidth of 150 kHz is allocated to each station,† which is sufficient to accommodate frequency components up to 15 kHz in the audio signal. We recall that, in the case of AM broadcast, only frequency components up to 5 kHz can be accommodated in a bandwidth of 10 kHz allocated to each station.

We conclude this section by describing the principle of operation of stereophonic FM broadcast systems. In this case, two audio signals, the left-speaker signal $l(t)$ and the right-speaker signal $r(t)$, are to be broadcast simultaneously, using the bandwidth of 150 kHz allocated to an FM station. So that a monophonic receiver will also be able to receive a stereophonic broadcast, the two signals

$$a(t) = l(t) + r(t)$$
$$b(t) = l(t) - r(t)$$

are broadcast rather than $l(t)$ and $r(t)$ directly. As will be seen below, the transmission scheme is such that a monophonic receiver will produce the signal $a(t)$ as a monophonic output. On the other hand, a stereophonic receiver will receive both the signals $a(t)$ and $b(t)$ and will thus be able to recover from them the signals $l(t)$ and $r(t)$. In a stereophonic broadcast, the modulating signal is

$$x(t) = a(t) + b(t) \cos 2\pi \times 38 \times 10^3 t + \cos 2\pi \times 19 \times 10^3 t$$

† The frequency range 88 to 108 MHz is reserved for commercial FM broadcast.

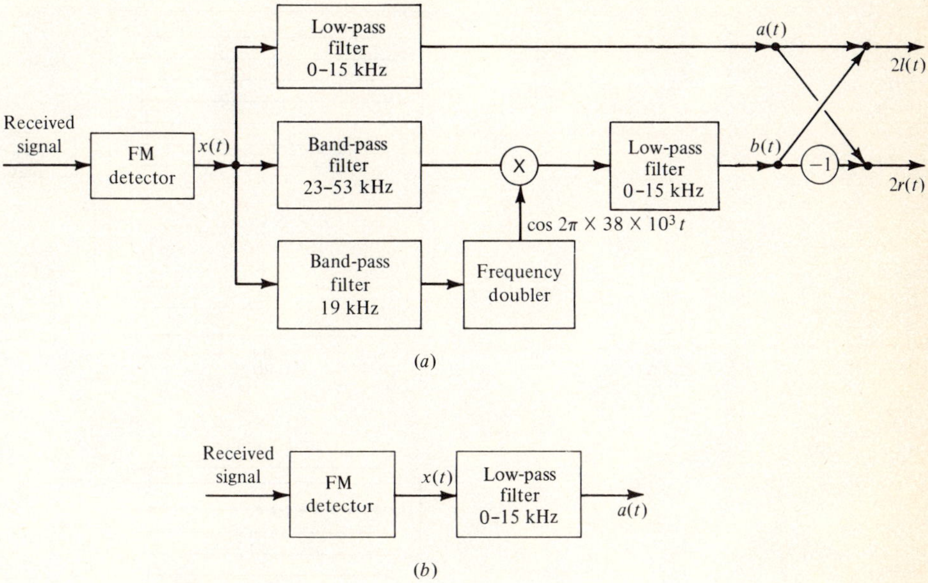

FIGURE 8-20

The frequency spectrum of $x(t)$, $X(\omega)$, is shown in Fig. 8-19, where $A(\omega)$ and $B(\omega)$ are the frequency spectra of the signals $a(t)$ and $b(t)$, respectively. In other words, the frequency spectrum $B(\omega)$ is shifted upward by 38 kHz in $X(\omega)$ so that it will not overlap with $A(\omega)$. At the receiving end, a stereophonic receiver will recover the signals $l(t)$ and $r(t)$, as the schematic diagram in Fig. 8-20*a* shows. Note that the signal $\cos 2\pi \times 19 \times 10^3 t$ in $x(t)$ is used by the receiver to generate a 38-kHz signal for detecting the signal $b(t)$. In most receivers, it is also used to drive a small indicator light to indicate the reception of a stereophonic broadcast. On the other hand, a monophonic receiver will reproduce only the signal $a(t)$, as illustrated in the schematic diagram in Fig. 8-20*b*.

8-9 TELEVISION SYSTEMS

As was mentioned in Sec. 8-3, in television transmission a frame of a picture is represented by a sequence of samples. Specifically, a picture is divided into 525 lines, as illustrated in Fig. 8-21*a*, and each line is further divided into 650 points, as illustrated in Fig. 8-21*b*. Consequently, a picture is represented by a sequence of

$$525(650) = 341{,}250$$

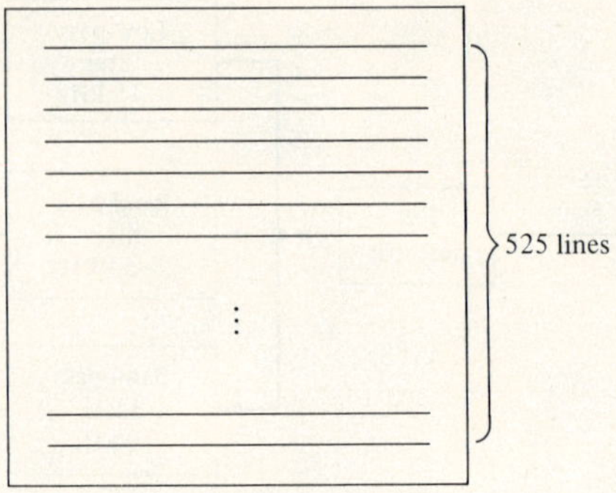

(a)

FIGURE 8-21 (b)

pulses whose amplitudes are proportional to the brightness of the picture at the sample points. At the receiver, the transmitted picture is recomposed by using these pulses to control the intensity of the light spots on the surface of the picture tube.

In practice, the pictures to be transmitted are scanned in a slightly different manner. First of all, a picture is not scanned horizontally but rather at a small angle, as illustrated in Fig. 8-22*a*. Such a way of scanning simplifies the timing problem in assembling the sample points to reproduce the picture at the receiving end. Second, a picture is scanned twice to improve the quality of the received picture. Specifically, instead of scanning the picture in 525 lines from the top of the picture down to the bottom of the picture, the scanner scans the odd-numbered lines from the top to the bottom of the picture and then scans the even-numbered lines, again from the top to the bottom of the picture. Such scanning is illustrated in Fig. 8-22*b*, where the solid lines indicate the lines in the first scanning and the dotted lines indicate the lines in the second scanning. In this way, there is less flickering in the received picture. One

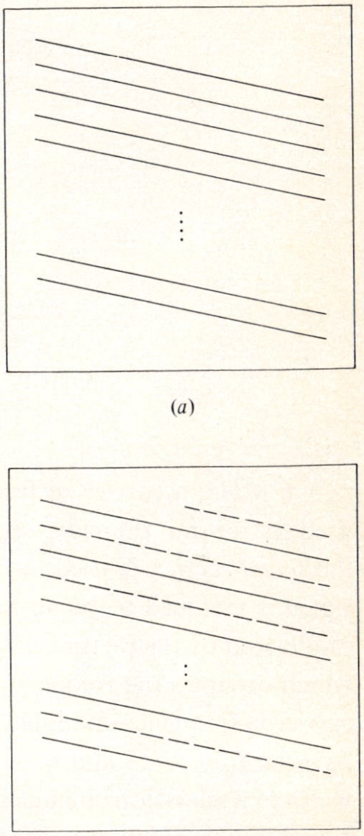

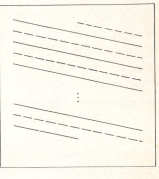

FIGURE 8-22 (a), (b)

notes immediately that in each scanning the picture is scanned in $262\frac{1}{2}$ lines instead of a whole number of lines. This again simplifies the timing problem because at the end of the first scanning the scanner can simply move vertically back to the top of the picture and start the second scanning from there.

We now consider the bandwidth requirement for television broadcast. The Federal Communications Commission allocates a bandwidth of 6 MHz to each television station.† For a television station occupying the frequency range f_0 MHz to

† Specifically, the assignments are as follows:

	Frequency range, MHz	Channel assignment
VHF	52–72	2, 3, and 4
	72–76	Radio astronomy and aeronautical radio navigation use
	76–88	5, 6
	88–108	FM broadcast
	174–216	7–13
UHF	470–890	14–83

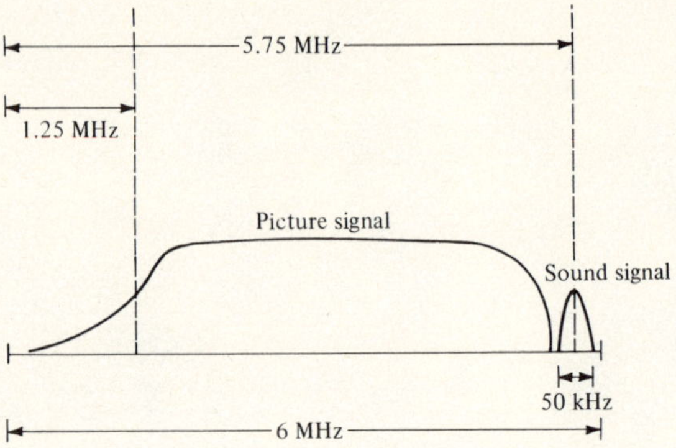

FIGURE 8-23

$f_0 + 6$ MHz, a carrier of frequency $f_0 + 1.25$ MHz, known as the *picture carrier*, is used to transmit the video signal and a carrier of frequency $f_0 + 5.75$ MHz, known as the *sound carrier*, is used to transmit the sound signal. The bandwidth of the picture signal is assumed to be no more than 4.25 MHz.† The picture carrier is amplitude-modulated by the picture signal and is transmitted as a vestigial single-sideband signal which occupies the frequency range between $f_0 + 0.5$ and $f_0 + 5.5$ MHz. The sound carrier is frequency-modulated by the sound signal and occupies the frequency range between $f_0 + 5.725$ and $f_0 + 5.775$ MHz.‡ Figure 8-23 shows how the 6-MHz bandwidth of a television channel is utilized. At the receiving end, the picture signal and the sound signal are recovered separately since the frequency spectra of their corresponding modulated signals do not overlap.

We turn now to the problem of transmitting color pictures. It is well known that a color picture is composed of three monochrome pictures, a red, a green, and a blue one. Since a monochrome picture can be transmitted in the same manner a black-and-white picture is transmitted, it seems that the problem is simply that of speeding up the rate of transmission by three times. In other words, instead of transmitting 30 frames of picture per second, 90 frames of picture would be transmitted per second. (The reader is reminded that a monochrome picture is represented by a sequence of pulses in the same way a black-and-white picture is. Only at the receiving end, the pulses representing a black-and-white picture are used to control the intensity of the light spot on the surface of the picture tube, and the pulses representing a monochrome picture are used to control the intensity of the corresponding color spot on the surface of the picture tube.) As it turns out, speeding up the rate of transmission by three times for color transmission is not the solution that was adopted. There are several

† In practice, the video signal is filtered to cut off frequency components above 4.25 MHz.

‡ Note that in FM broadcast a bandwidth of 150 kHz is allocated to each broadcast station. Here, only a bandwidth of 50 kHz is provided.

considerations to be borne in mind. First of all, according to our discussion above, to transmit 90 frames of pictures per second would require a bandwidth of 18 MHz and would, therefore, upset completely the existing assignment of channel frequencies. Second, the compatibility problem must be considered; that is, a black-and-white receiver should be able to receive, of course, a black-and-white picture in a color telecast. Third, there is the so-called reverse compatibility problem which means that a color receiver should be able to receive a black-and-white picture in a black-and-white telecast. We describe now the scheme used in the American television systems.

Let E_G, E_R, and E_B denote the three monochrome signals, green, red, and blue, respectively. Instead of transmitting these signals directly, we transmit three linear combinations of these signals, namely,

$$E_Y = 0.3E_R + 0.59E_G + 0.11E_B$$
$$E_I = 0.6E_R - 0.28E_G - 0.32E_B$$
$$E_Q = 0.21E_R - 0.52E_G + 0.31E_B$$

Notice that at the receiving end the signals E_G, E_R, E_B can be recovered from the signals E_Y, E_I, and E_Q. Specifically,

$$E_R = E_Y + 0.96E_I + 0.62E_Q$$
$$E_G = E_Y - 0.28E_I - 0.64E_Q$$
$$E_B = E_Y - 1.1E_I + 1.7E_Q$$

Moreover, according to physical measurement, E_Y is known as the *luminance signal* which determines the brightness of the picture. Consequently, E_Y is the only signal that is needed to reconstruct the black-and-white version of a color picture at the receiving end. E_I and E_Q are known as the *chrominance signals* which carry the color information of the picture.

We must now face the problem of having enough bandwidth to transmit the three signals E_Y, E_I, and E_Q. According to our discussion on the transmission of black-and-white pictures, the bandwidth of E_Y is slightly over 4 MHz. It seems then there will be no "room" for the other two signals E_I and E_Q if we are limited to a 6-MHz bandwidth. However, measurement of the frequency spectrum of E_Y reveals that the frequency components of E_Y cluster around frequencies which are multiples of 15,750 Hz, as illustrated in Fig. 8-24*a*. Such a result of measurement is not unexpected. We recall that a periodic signal of period 1/15,750 s contains only frequency components that are multiples of the fundamental frequency $f = 15{,}750$ Hz. Although the sequence of samples of a picture is not a periodic signal, it is quite close to a periodic one. Note that each of the lines in a picture is scanned in every $(1/30)(1/525) = 1/15{,}750$ s. Therefore, two points at the same vertical position but on successive lines, as shown in Fig. 8-24*b*, are scanned 1/15,750 s apart. Since variation in brightness is

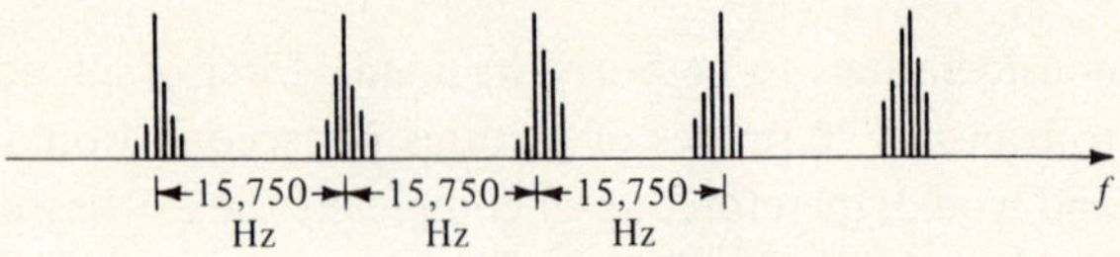

(a)

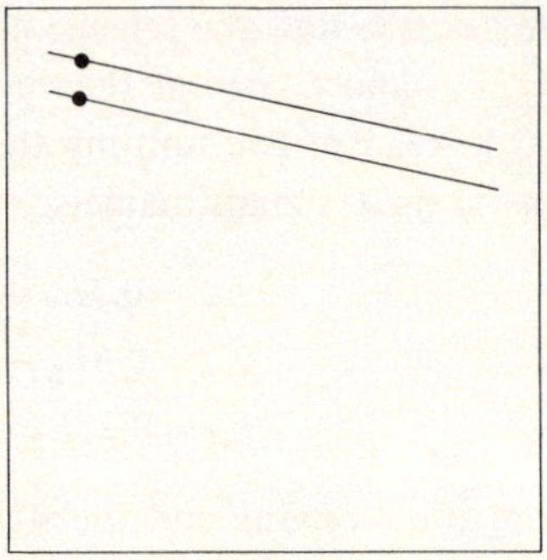

(b)

FIGURE 8-24

never very abrupt in most pictures, the amplitudes of the samples corresponding to every two such points are more or less the same. Consequently the sequence of samples is almost a periodic signal, which provides an intuitive explanation of the result of the physical measurement. Similarly, the frequency components of both E_Q and E_I also cluster around frequencies which are multiples of 15,750 Hz. We see now the possibility of shifting the frequency components of E_Q and E_I and interlacing them with that of E_Y. Physical measurement has shown that the bandwidth of E_I is about 1.5 MHz, and that of E_Q is about 0.5 MHz. Consequently, if we shift the frequency components of E_I by $(2m_1 + 1)(15{,}750)/2$ Hz and the frequency components of E_Q by $(2m_2 + 1)$ $(15{,}750)/2$ Hz for some appropriately chosen integers m_1 and m_2, they can be interlaced with that of E_Y, as illustrated in Fig. 8-25*a*. However, this is not exactly the way frequency components are interlaced in actual implementation. Rather, the signal

$$E_C = E_Q \cos 2\pi\hat{f}t + E_I \sin 2\pi\hat{f}t$$

where $\hat{f} = 3.579545 \times 10^6$ Hz, is generated and transmitted together with the luminance signal E_Y.† We shall refer to E_C as the *color signal* and $\cos 2\pi\hat{f}t$ as the *color subcarrier*.

† In actual implementation, the upper sideband of $E_I \sin 2\pi\hat{f}t$ is limited to about 0.5 MHz. See Fig. 8-29 and the discussion on bandwidth requirement later in this section.

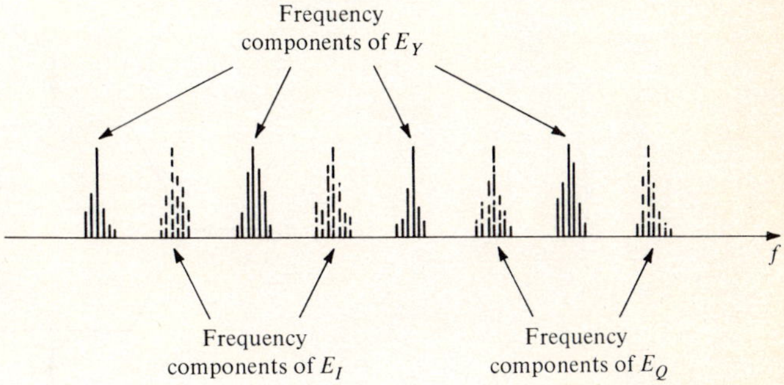

(a)

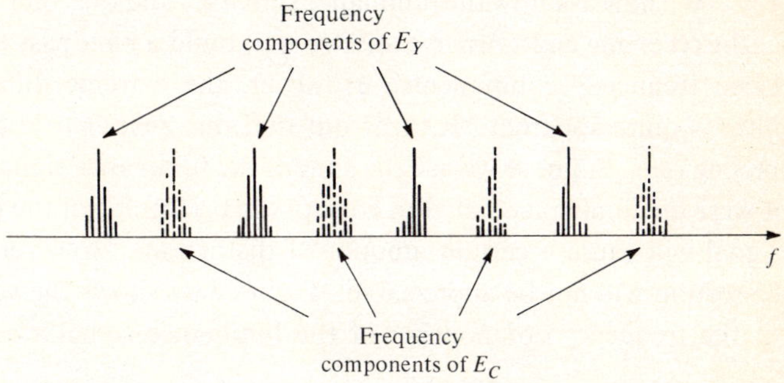

(b)

FIGURE 8-25

Note that

$$\hat{f} = 3.579545 \times 10^6 \text{ Hz} = \frac{2(227) + 1}{2} 15{,}750$$

Thus, in the resultant signal $E_Y + E_C$, the frequency components of E_Q and E_I are interlaced with that of E_Y, as illustrated in Fig. 8-25*b*. Let us point out that there is no difficulty in recovering E_Q and E_I from the color signal E_C at the receiving end. Since

$$\begin{aligned} E_C \cos 2\pi\hat{f}t &= (E_Q \cos 2\pi\hat{f}t + E_I \sin 2\pi\hat{f}t) \cos 2\pi\hat{f}t \\ &= \frac{E_Q}{2} + \frac{E_Q}{2} \cos 4\pi\hat{f}t + \frac{E_I}{2} \sin 4\pi\hat{f}t \end{aligned}$$

$$\begin{aligned} E_C \sin 2\pi\hat{f}t &= (E_Q \cos 2\pi\hat{f}t + E_I \sin 2\pi\hat{f}t) \sin 2\pi\hat{f}t \\ &= \frac{E_Q}{2} \sin 4\pi\hat{f}t - \frac{E_I}{2} \cos 4\pi\hat{f}t + \frac{E_I}{2} \end{aligned}$$

we can recover E_Q and E_I from E_C, as the schematic diagram in Fig. 8-26 shows.

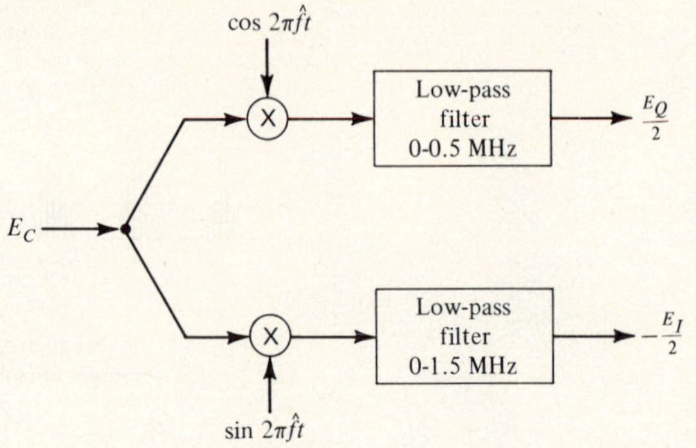

FIGURE 8-26

We now ask how the luminance signal E_Y and the color signal E_C can be separated at the receiving end. A naive answer is to build a bandpass filter to sieve out the interlaced frequency components. However, the extreme difficulty of building such a filter is quite apparent. It turns out that our answer to the question is a rather surprising one. Suppose we use the sum of the luminance signal and the color signal *as if* it were the luminance signal to control the brightness of the picture. Clearly, the color signal will cause a certain amount of distortion. However, we show now that such distortion will not be appreciable. Figure 8-27 shows the variation of the amplitudes of the frequency components of the luminance signal along successive lines of the

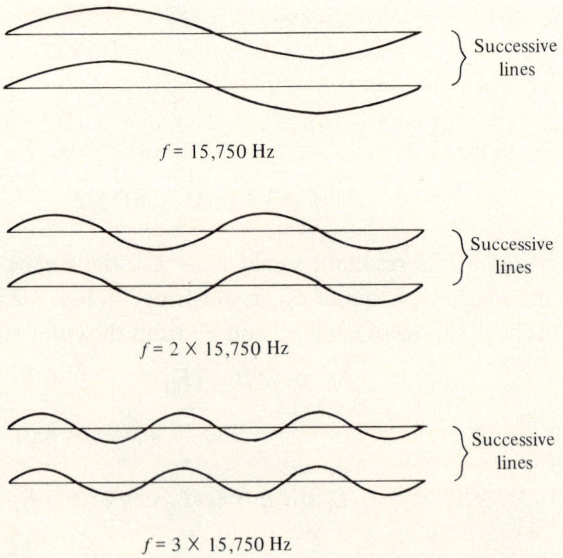

FIGURE 8-27

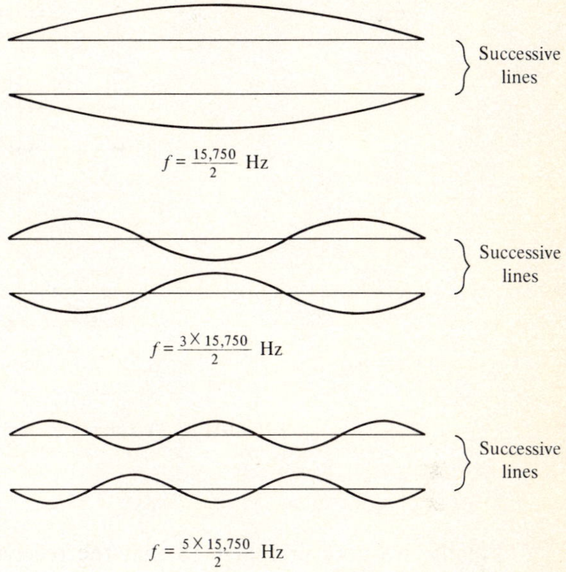

FIGURE 8-28

picture. Note that the frequencies of these components are essentially multiples of 15,750 Hz. Figure 8-28 shows the variation of the amplitudes of the frequency components of the color signal along successive lines of the picture. Note that the frequencies of these components are essentially $(2m + 1)(15{,}750)/2$ Hz. Thus, we see that distortion in the brightness of the picture due to the color signal has an opposite effect along successive lines of the picture. That is, if the color signal causes the received picture to become brighter than it should be along one line, it will cause the receiver picture to become dimmer than it should be along the next line. Since human eyes are not able to separate successive lines of the picture, the distortion in the brightness of the picture due to the color signal will be averaged out. In identical manner, a black-and-white receiver will use the sum of E_Y and E_C as the luminance signal. Thus, the compatibility problem is resolved. Similarly, the contamination of the color signal E_C by the frequency components of the luminance signal E_Y will also be insignificant, and we can recover the chrominance signals E_I and E_Q from $E_Y + E_C$. Figure 8-29 shows how the 6-MHz bandwidth assigned to a television station is occupied by the signals E_Y, E_Q, and E_I. Because the bandwidth of E_Q is only 0.5 MHz, both sidebands of $E_Q \cos 2\pi \hat{f} t$ are retained in E_C. On the other hand, because the bandwidth of E_I is 1.5 MHz, the upper sideband of $E_I \sin 2\pi \hat{f} t$ is limited to about 0.6 MHz. That is, the frequency components of frequencies above $\hat{f} + 0.6$ MHz in the upper sideband of $E_I \sin 2\pi \hat{f} t$ are filtered out prior to transmission.

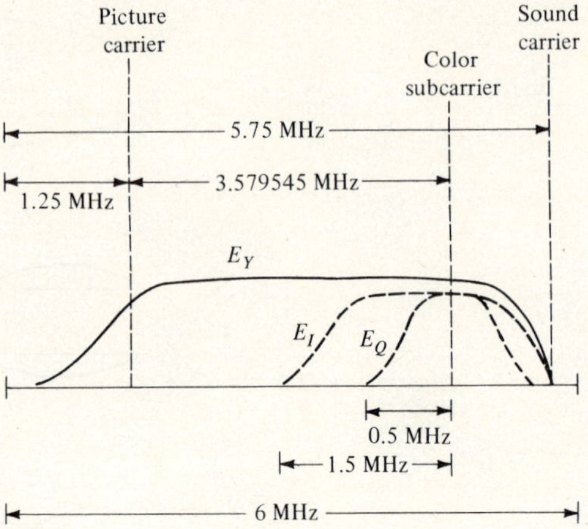

FIGURE 8-29

Finally, we note in Fig. 8-26 that the receiver uses the signals $\cos 2\pi\hat{f}t$ and $\sin 2\pi\hat{f}t$ to recover the signals E_I and E_Q from the color signal E_C. Therefore, the 3.579545-MHz color subcarrier is transmitted together with the picture so that the receiver can use the color subcarrier to synchronize its local oscillator that generates the signals $\cos 2\pi\hat{f}t$ and $\sin 2\pi\hat{f}t$. Moreover, the absence of the color subcarrier in the received signal indicates to a color receiver that it is receiving a black-and-white telecast. Consequently, a circuit, known as the *killer circuit*, cuts off the color section of the receiver, and a black-and-white picture is reproduced. Note that without such information a color receiver would interpret a black-and-white transmission as a color transmission. The result would be a low-quality color picture. Indeed, the transmission of the color subcarrier has provided a solution to the reverse compatibility problem.

8-10 REMARKS AND REFERENCES

Our discussion on communication systems in this chapter is a rather brief one. We hope, however, that the reader has seen a useful application of the method of Fourier transformation which leads to the concept of decomposing signals into frequency components. For further details on the circuitry implementation as well as on quantitative analysis of the performance of modern communication systems, the reader is referred to Carlson [1], Kiver [2], Lathi [3], Taub and Schilling [4], Terman [5], and Wozencraft and Jacobs [6]. Among these books, Carlson [1] and Lathi [3] are particularly useful at the introductory level.

[1] CARLSON, A. B.: "Communication Systems," McGraw-Hill Book Company, New York, 1968.
[2] KIVER, M. S.: "Color Television Fundamentals," 2d ed., McGraw-Hill Book Company, New York, 1964.
[3] LATHI, B. P.: "Communication Systems," John Wiley & Sons, Inc., New York, 1968.
[4] TAUB, H., and D. L. SCHILLING: "Principles of Communication Systems," McGraw-Hill Book Company, New York, 1971.
[5] TERMAN, F. E.: "Electronic and Radio Engineering," 4th ed., McGraw-Hill Book Company, New York, 1955.
[6] WOZENCRAFT, J. M., and I. M. JACOBS: "Principles of Communication Engineering," John Wiley & Sons, Inc., New York, 1965.

PROBLEMS

8-1 The transfer function of a bandpass filter is shown in Fig. 8P-1. Determine the output signal of the filter when its input signal is

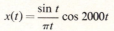

$$x(t) = \frac{\sin t}{\pi t} \cos 2000t$$

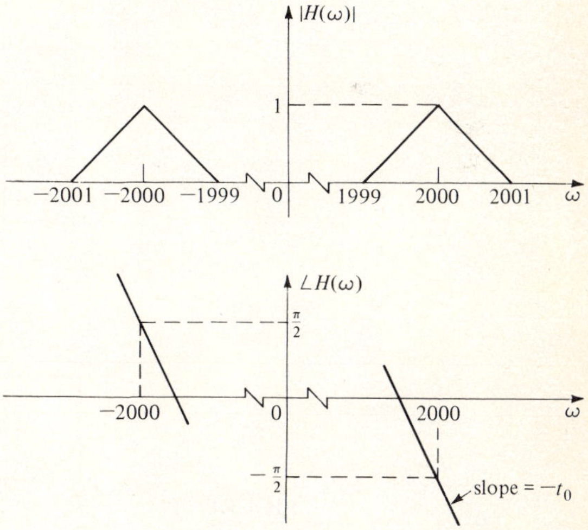

FIGURE 8P-1

8-2 Determine the output signal of the filter whose transfer function is shown in Fig. 8P-2*a* when its input is the periodic signal in Fig. 8P-2*b*.

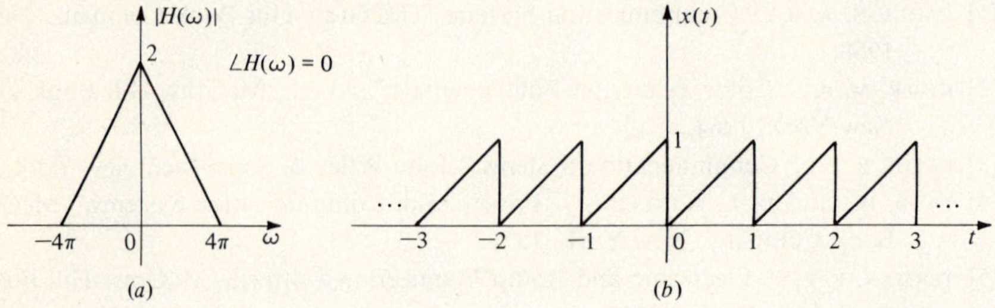

FIGURE 8P-2

8-3 Consider the circuit shown in Fig. 8P-3. Suppose that the input impedance of the bandpass filter is infinite. Find the exponential Fourier series representation of the signals $x(t)$ and $y(t)$. Evaluate the average power in the output signal $y(t)$.

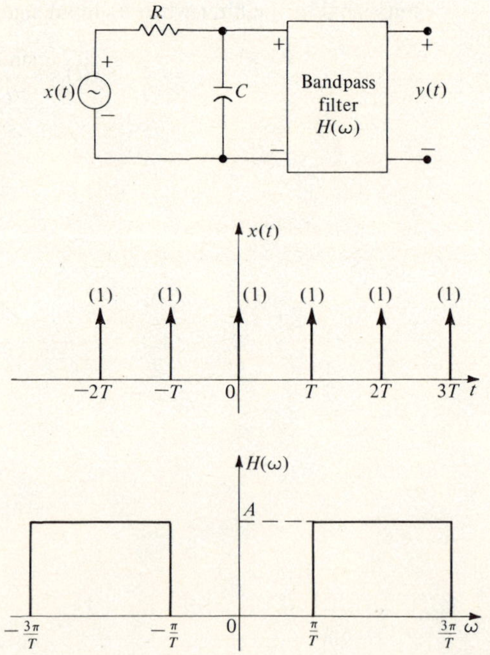

FIGURE 8P-3

8-4 (*a*) Find the impulse response of an ideal low-pass filter of bandwidth w and phase angle $\angle H(\omega) = -j\omega\tau_0$ and show that the filter is noncausal.

(*b*) Show that any low-pass filter whose Fourier spectrum $|H(\omega)|$ is equal to zero for $|\omega| > w$ for some constant w is noncausal. Similarly, show that any filter (bandpass or high-pass) whose Fourier spectrum is equal to zero in its stopband is noncausal.

(*c*) In general, a filter whose impulse response is square integrable (that is, $\int_{-\infty}^{\infty} [h(t)]^2\, dt < \infty$) is causal if and only if its transfer function $H(\omega)$ is such that

$$\int_{-\infty}^{\infty} \frac{|\ln H(\omega)|}{1+\omega^2}\, d\omega < \infty$$

This condition is known as the *Paley-Wiener theorem.* Discuss the implications of this theorem.

8-5 When the transition of the value of $|H(\omega)|$ between the passband and the stopband of a filter is not abrupt, the bandwidth of the filter cannot be defined as in Sec. 8-2. In this problem, we shall study other definitions of the bandwidth of filters.

(*a*) The half-power bandwidth Δw of a low-pass filter with transfer function $H(\omega)$ is equal to half the distance between the points at which $|H(\omega)| = (1/\sqrt{2})|H(0)|$. Find the half-power bandwidth Δw of the filters shown in Fig. 8P-4.

(*b*) Let

$$\Delta w_1 = \frac{1}{2}\frac{\int_{-\infty}^{\infty} H(\omega)\, d\omega}{|H(0)|}$$

be an alternative definition of the bandwidth of a filter. Find Δw_1 for the filters shown in Fig. 8P-4.

(*c*) Let

$$\Delta w_2 = \frac{1}{2}\frac{\int_{-\infty}^{\infty} |H(\omega)|^2\, d\omega}{|H(0)|^2}$$

be another definition of the bandwidth of a filter. Find Δw_2 for the filters shown in Fig. 8P-4. (Δw_2 is often called the equivalent noise bandwidth of a filter.)

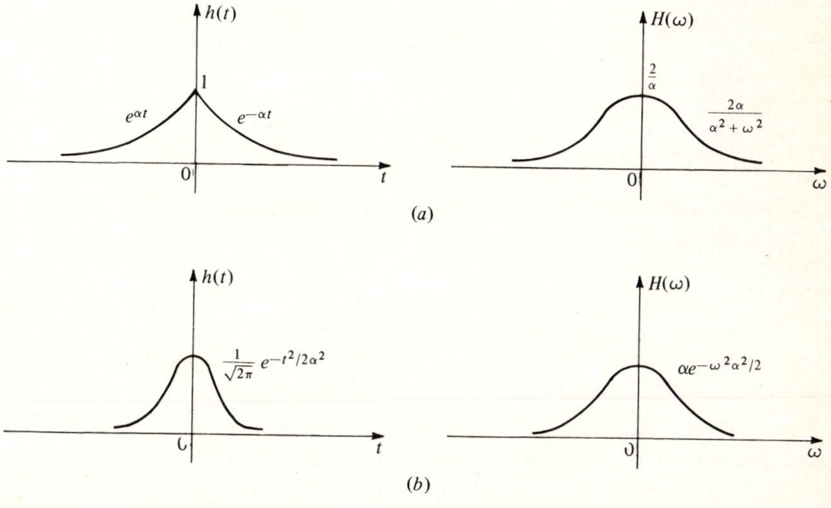

FIGURE 8P-4

8-6 This problem is concerned with more precise definitions of the duration of a time signal. Let

$$\Delta T_1 = \frac{\int_{-\infty}^{\infty} h(t)\, dt}{|h(0)|}$$

where $h(0) \neq 0$, and

$$\Delta T_2 = \frac{(\int_{-\infty}^{\infty} h(t)\, dt)^2}{\int_{-\infty}^{\infty} h^2(t)\, dt}$$

be two definitions of signal duration for the time signal $h(t)$. Find ΔT_1 and ΔT_2 for the filters shown in Fig. 8P-4.

8-7 In this problem, we shall demonstrate that signals of short duration must have large bandwidth and signals of narrow bandwidth must have long duration. In other words, the duration and bandwidth of any signal cannot be arbitrarily small simultaneously. This result is known as the *uncertainty principle*.

(*a*) Let us define the duration ΔT_1 of a real signal $h(t)$ to be

$$\Delta T_1 = \frac{\int_{-\infty}^{\infty} h(t)\, dt}{|h(0)|}$$

[assume that $h(0) \neq 0$] and its bandwidth Δw_1 to be

$$\Delta w_1 = \frac{1}{2} \frac{\int_{-\infty}^{\infty} |H(\omega)|\, d\omega}{|H(0)|}$$

For the signal shown in Fig. 8P-4*a*, find $\Delta T_1\, \Delta w_1$.

(*b*) Let us define the duration ΔT_2 of a real signal $h(t)$ to be

$$\Delta T_2 = \frac{(\int_{-\infty}^{\infty} h(t)\, dt)^2}{\int_{-\infty}^{\infty} h^2(t)\, dt}$$

and its bandwidth Δw_2 to be

$$\Delta w_2 = \frac{1}{2} \frac{\int_{-\infty}^{\infty} |H(\omega)|^2\, d\omega}{|H(0)|^2}$$

Find $\Delta T_2\, \Delta w_2$ for the signal shown in Fig. 8P-4*b*.

(*c*) Show that

$$\Delta T_1\, \Delta w_1 = \pi$$

for any real signal $h(t)$. Similarly, show that

$$\Delta T_2\, \Delta w_2 = \pi$$

for any real signal $h(t)$. Discuss the implications of these results.

8-8 The definitions of signal duration and bandwidth in Probs. 8-6 and 8-7 are of the "ratio of area to height" type. When the waveforms of a time signal and its Fourier transform have negative regions or widely separated multiple peaks, these definitions give poor measure of the time and frequency ranges over which the values of the time signal and its Fourier transform differ substantially from zero. In this case, a more useful measure of signal duration and bandwidth is given by the second moments of $h(t)$ and $H(\omega)$. Let $h(t)$ be a signal such that $\int_{-\infty}^{\infty} t[h(t)]^2\,dt = 0$. We define the duration of $h(t)$ to be

$$\Delta T_3 = \left(\frac{\int_{-\infty}^{\infty} t^2[h(t)]^2\,dt}{\int_{-\infty}^{\infty} [h(t)]^2\,dt} \right)^{1/2}$$

and its bandwidth to be

$$\Delta w_3 = \frac{1}{2} \left[\frac{\int_{-\infty}^{\infty} \omega^2 |H(\omega)|^2\,d\omega}{\int_{-\infty}^{\infty} |H(\omega)|^2\,d\omega} \right]^{1/2}$$

For the signal shown in Fig. 8P-4*b*, find $\Delta T_3\,\Delta w_3$. Show that for any real signal $h(t)$,

$$\Delta T_3\,\Delta w_3 \geq \tfrac{1}{2}$$

[*Hint:* Note that

$$\frac{1}{2\pi}\int_{-\infty}^{\infty} \omega^2 |H(\omega)|^2\,d\omega = \int_{-\infty}^{\infty} \left[\frac{dh(t)}{dt}\right]^2 dt$$

and

$$\int_{-\infty}^{\infty} g^2(t)\,dt \int_{-\infty}^{\infty} f^2(t)\,dt \geq \left(\int_{-\infty}^{\infty} g(t)f(t)\,dt \right)^2$$

where the equality sign holds if

$$g(t) = kf(t)$$

for some constant k.]

8-9 As was discussed in Sec. 8-2, the *RLC* circuit in Fig. 8-2*a* is a causal system which can be used to approximate an ideal low-pass filter. The Fourier spectrum of its impulse response is equal to $1/\sqrt{1+\omega^4}$. This filter is one of a sequence of filters whose transfer functions are equal to $1/\sqrt{1+\omega^{2n}}$ for integral values of n. These filters are known as Butterworth filters, and the value of n is referred to as the order of a filter of this type. Thus, the circuit in Fig. 8-2*a* is a second-order Butterworth filter.

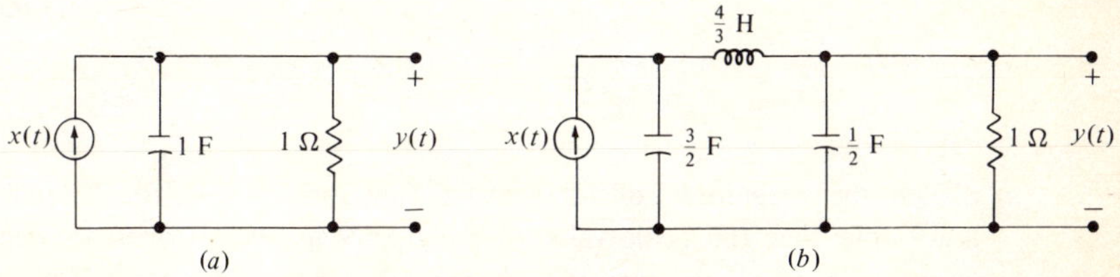

FIGURE 8P-5

(*a*) Find the transfer function of the filter shown in Fig. 8P-5*a*. Sketch its magnitude carefully. Compare it with the magnitude of the transfer function of the second-order Butterworth filter.

(*b*) Show that the circuit shown in Fig. 8P-5*b* is a third-order Butterworth filter. Sketch the magnitude of its transfer function.

(*c*) Sketch the function $1/\sqrt{1+\omega^{2n}}$ for $n = 4$ and 5. Note how a Butterworth filter can be used to approximate arbitrarily closely an ideal low-pass filter by choosing n to be arbitrarily large.

8-10 Consider the impulse response of a causal filter shown in Fig. 8P-6*a* obtained by truncating the impulse response of an ideal low-pass filter.

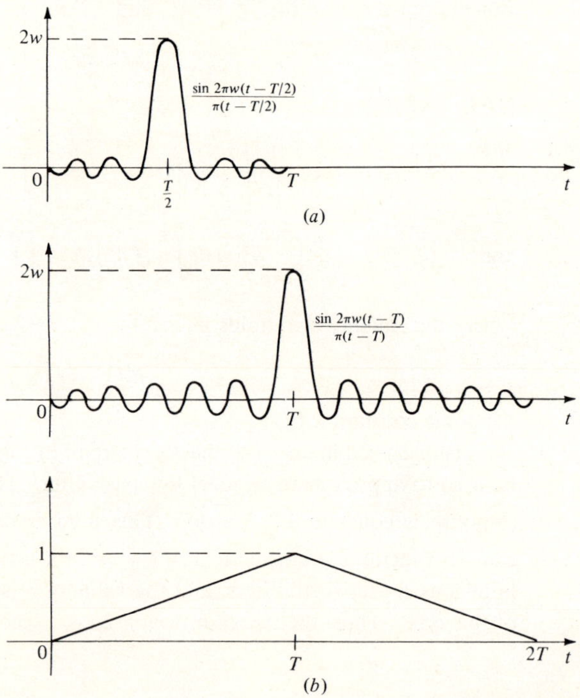

FIGURE 8P-6

(*a*) Sketch the magnitude of its transfer function $|H(\omega)|$ for T equals $4/w$ and $8/w$. Note that the peak value of the ripples in $|H(\omega)|$ does not decrease as T increases, but only the rate of variation in the ripples changes. This behavior is known as the Gibbs phenomenon.

(*b*) Consider the filter whose impulse response is equal to the product of the two time signals shown in Fig. 8P-6*b*. Sketch the magnitude of its transfer function for $T = 8/w$.

8-11 When an aircraft breaks the sound barrier, the waveform of the overpressure $p(t)$ exerted at human ears because of the sonic boom is approximately that shown in Fig. 8P-7*a*. The duration of the waveform, T, depends on the size of the aircraft and is usually of the order of tenths of a second. Since human ears can detect only the frequency components within the range of 30 Hz to 5 kHz, the loudness of the sound generated by this overpressure is a function of the energy in the output signal of the ideal bandpass filter shown in Fig. 8P-7*b*. Find the expression for the energy $E = \int_{-\infty}^{\infty} y^2(t)\, dt$ in terms of A and T.

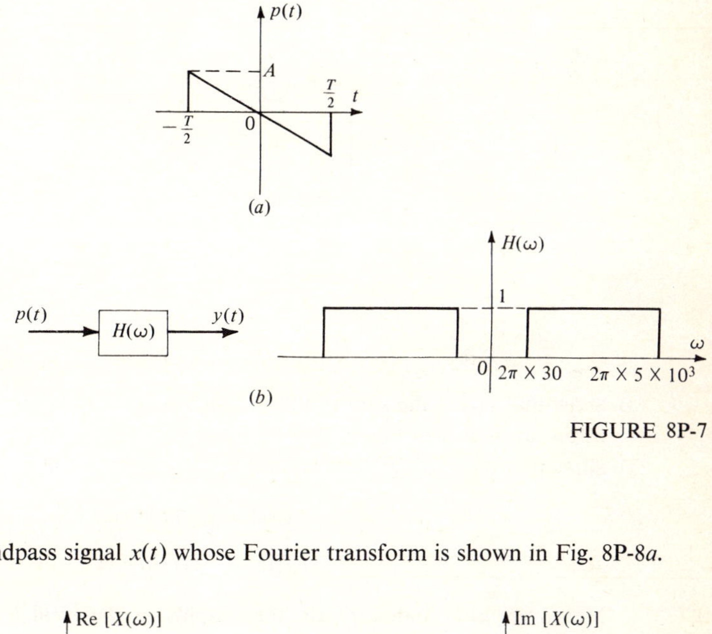

FIGURE 8P-7

8-12 Consider a bandpass signal $x(t)$ whose Fourier transform is shown in Fig. 8P-8*a*.

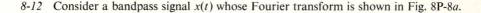

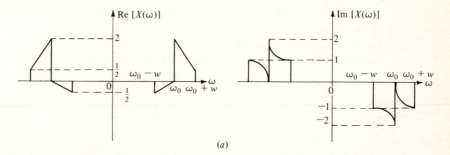

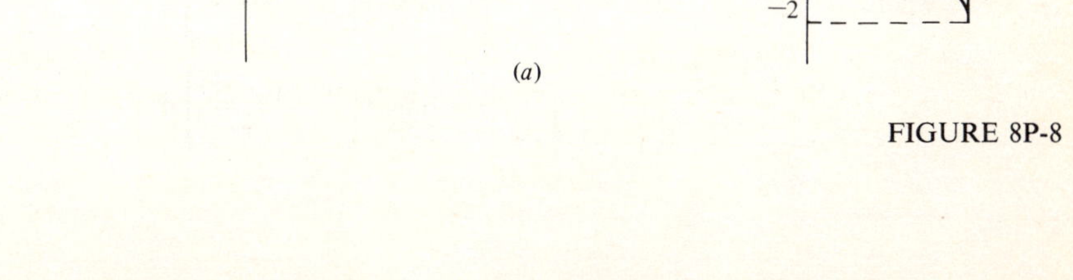

FIGURE 8P-8

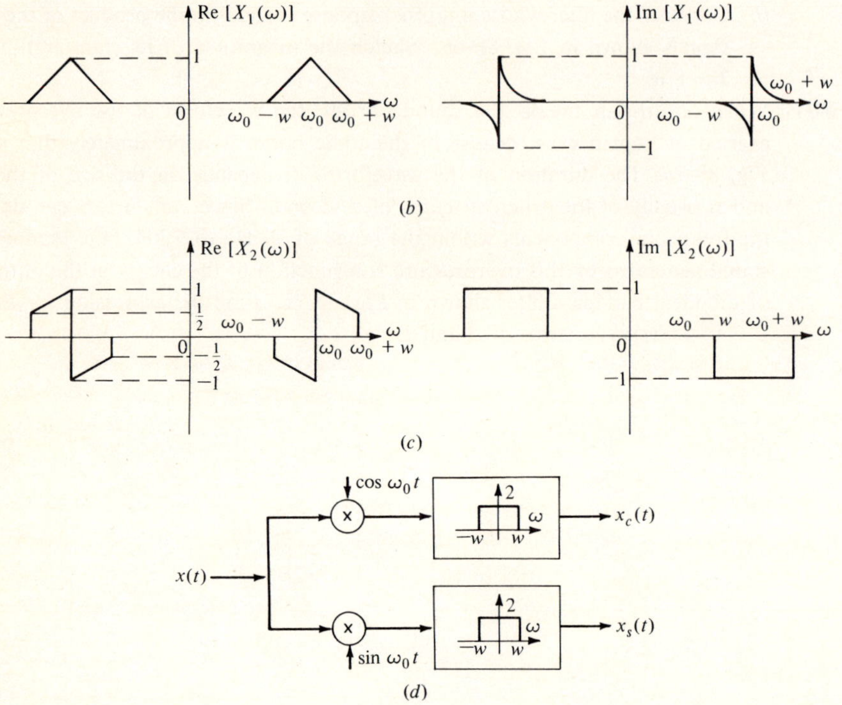

FIGURE 8P-8

(*a*) Show that $x(t)$ is the sum of the two signals $x_1(t)$ and $x_2(t)$ whose Fourier transforms are shown in Fig. 8P-8*b* and *c*, respectively.

(*b*) Show that the two signals $x_1(t)$ and $x_2(t)$ in part (*a*) can be written

$$x_1(t) = x_c(t) \cos \omega_0 t$$

and

$$x_2(t) = x_s(t) \sin \omega_0 t$$

Moreover, $x_c(t)$ and $x_s(t)$ are the outputs of the ideal low-pass filters in Fig. 8P-8*d*.

(*c*) Let $y(t)$ be any bandpass signal whose Fourier transform is nonzero only in the frequency range $w_a \leq |\omega| \leq w_b$. Show that $y(t)$ can be expressed as the sum

$$y(t) = y_c(t) \cos \omega_0 t + y_s(t) \sin \omega_0 t$$

where

$$y_c(t) = \frac{2}{\pi} \int_{-\infty}^{\infty} y(\tau) \cos \omega_0 t \left[\frac{\sin w(t-\tau)}{t-\tau} \right] d\tau$$

$$y_s(t) = \frac{2}{\pi} \int_{-\infty}^{\infty} y(\tau) \sin \omega_0 t \left[\frac{\sin w(t-\tau)}{t-\tau} \right] d\tau$$

and
$$\omega_0 = \frac{w_a + w_b}{2}$$

$$w = \frac{w_b - w_a}{2}$$

$y_c(t)$ is generally known as the *in-phase component* of $y(t)$, and $y_s(t)$ is known as the *quadrature component* of $y(t)$.

8-13 As was pointed out in Sec. 8-3, the samples of a signal are not ideal impulses in practice but rather are pulses of finite width. We show now that a continuous signal can be recovered from sample pulses of finite width such as those shown in Fig. 8-7*b*.

(*a*) Find the Fourier transform of the signal $s(t)$ in Fig. P8-9*a*.

(*b*) Show that the output signal of the ideal low-pass filter in Fig. P8-9*b* is $x(t)$ when its input signal is $x(t)s(t)$ for any low-pass signal $x(t)$ of bandwidth w or less.

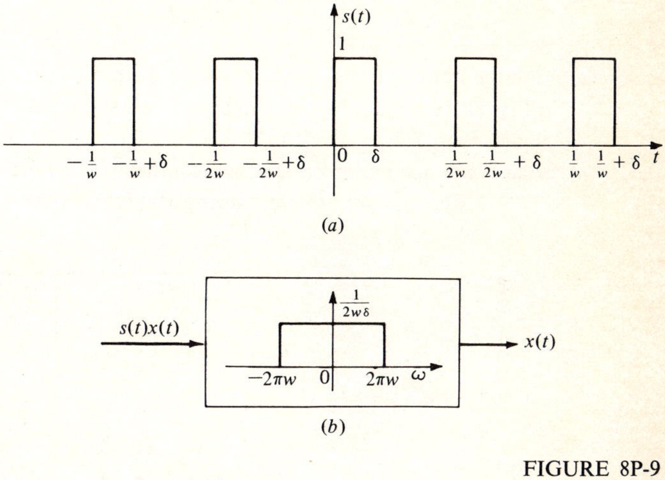

FIGURE 8P-9

8-14 Let $x(t)$ be a band-limited signal with bandwidth w Hz. What is the Nyquist rate for the signal $x(2t)$? What are the Nyquist rates for the signals $x^3(2t)$ and $x(t) * x(2t)$?

8-15 In the system shown in Fig. 8P-10, the input $x(t)$ is a low-pass signal of bandwidth w Hz.

(*a*) For the signal $s(t)$ and the transfer function $H(\omega)$ shown in Fig. P8-10, express the Fourier transform of $y(t)$ in terms of the Fourier transform of $x(t)$.

(*b*) Design a system for recovering $x(t)$ from $y(t)$.

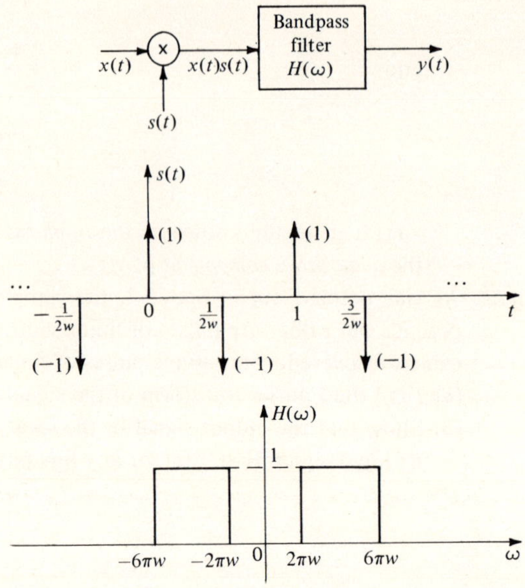

FIGURE 8P-10

8-16 (*a*) Find the Fourier transform of the output signal $y(t)$ of the sampling system shown in Fig. 8P-11, assuming that the bandwidth w (Hz) of the input signal $x(t)$ is much smaller than $1/T$.

(*b*) What is the maximum value of w such that $x(t)$ can be completely recovered from $y(t)$? Design a system for recovering $x(t)$ from $y(t)$.

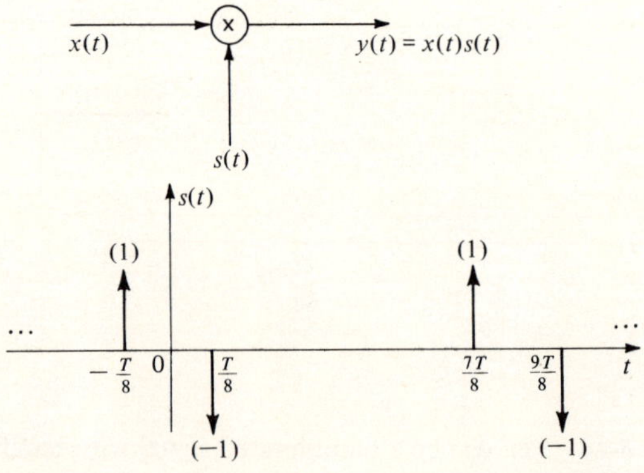

FIGURE 8P-11

8-17 Consider a low-pass signal $x(t)$ whose Fourier spectrum $|X(\omega)|$ is zero for $|\omega| \geq 2\pi w$.

(*a*) Consider a signal $y(t)$ whose Fourier transform $Y(\omega)$ is equal to $X(\omega)$ for $|\omega| < 2\pi w$ and is periodic in ω with a period $4\pi w$. Express the Fourier coefficients of $Y(\omega)$ in terms of the values of $x(t)$.

(*b*) Show that

$$x(t) = \sum_{-\infty}^{\infty} x(nT) \frac{\sin 2\pi w(t - nT)}{2\pi w(t - nT)}$$

where
$$T = \frac{1}{2w}$$

This is another way of presenting the sampling theorem in which an explicit expression of $x(t)$ in terms of the sample values is given.

8-18 Figure 8P-12*a* shows the block diagram of a sampling oscilloscope. The input $x(t)$ is a band-limited periodic signal. The multiplier and low-pass filter are designed so that the output signal $y(t)$ is equal to $x(at)$ for some constant $a < 1$. The signal $y(t)$ will be displayed on a cathode-ray tube as a time-stretched version of $x(t)$. Let the bandwidth and the period of $x(t)$ be w and T, respectively. The signal $g(t)$ is a train of unit impulses with period $(1 + \alpha)T$ for some constant α, $0 < \alpha < 1$. Hence, $x(t)g(t)$ is a sequence of samples of $x(t)$ separated in time by $(1 + \alpha)T$ as shown in Fig. 8P-12*b*.

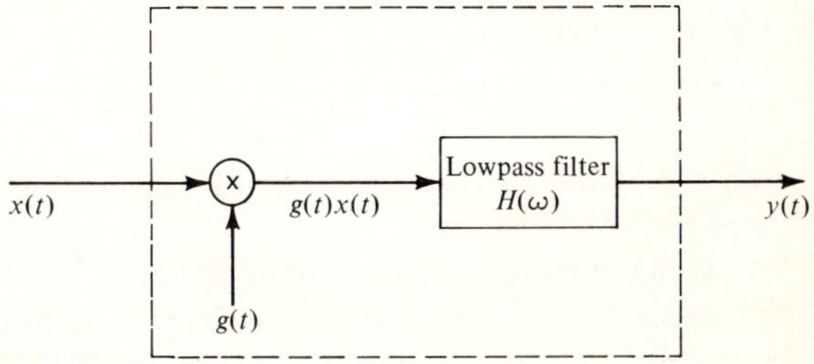

(*a*)

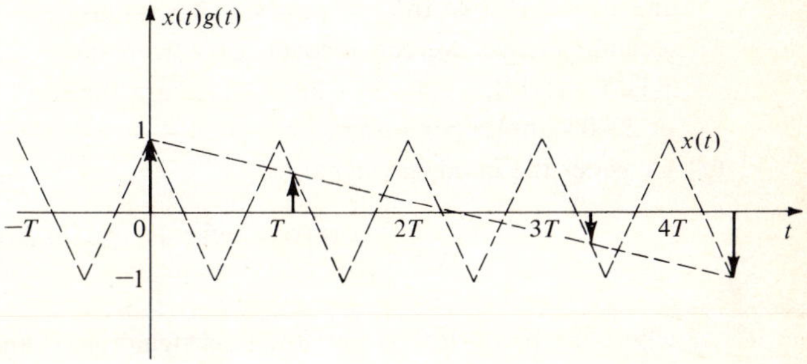

(*b*)

FIGURE 8P-12

(*a*) Suppose that the bandwidth of $x(t)$ is $4\pi/T$. Sketch a typical $X(\omega)$.

(*b*) Let the period of $g(t)$ be $(1+\frac{1}{4})T$. Sketch $G(\omega)$ and the Fourier transform of $x(t)g(t)$.

(*c*) If the transfer function of the low-pass filter is

$$H(\omega) = \begin{cases} 1 & |\omega| \leq \dfrac{\pi}{(1+\frac{1}{4})T} = \dfrac{4\pi}{5T} \\ 0 & \text{otherwise} \end{cases}$$

find $Y(\omega)$ and $y(t)$.

(*d*) Let $x(t)$ be a band-limited signal with bandwidth w, where $w < 1/2\alpha T$. Show that if the transfer function of the low-pass filter is

$$H(\omega) = \begin{cases} 1 & |\omega| \leq \dfrac{\pi}{(1+\alpha)T} \\ 0 & \text{otherwise} \end{cases}$$

its output signal $y(t)$ is proportional to $x(\alpha t/1+\alpha)$, that is, the input signal stretched in time by a factor of $(1+\alpha)/\alpha$.

8-19 Consider a signal

$$x(t) = x_c(t)\cos 2\pi 5000t + x_s(t)\sin 2\pi 5000t$$

where $x_c(t)$ and $x_s(t)$ are both low-pass signals whose Fourier transforms are equal to zero for $|f| \geq 1000$ Hz.

(*a*) Sketch the Fourier transform of $x(t)$ in terms of the Fourier transforms of $x_c(t)$ and $x_s(t)$. Also, express $x_c(t)$ and $x_s(t)$ in terms of $x(t)$ and the impulse response of an ideal low-pass filter with bandwidth 1000 Hz.

(*b*) Show that the signal $x(t)$ can be recovered from its samples if the samples were taken at a rate of 4000 samples per second.

(*c*) Suppose that, starting at $t=0$, $x(t)$ is sampled every 1/4000 s. Let $\hat{x}(0), \hat{x}(1), \ldots, \hat{x}(n)$ denote these samples. Show that $\hat{x}(0), (-1)\hat{x}(2), (-1)^2\hat{x}(4), \ldots, (-1)^n\hat{x}(2n), \ldots$ are the samples of the low-pass signal $x_c(t)$ taken at the rate of 2000 samples per second. Hence, we can recover $x_c(t)$ from these samples. Similarly, show that $\hat{x}(1), (-1)\hat{x}(3), \ldots, (-1)^n\hat{x}(2n+1), \ldots$ are the samples of $x_s(t)$ taken at the rate of 2000 samples per second.

(*d*) Consider the bandpass signal

$$y(t) = \begin{cases} y_c(t)\cos 2\pi f_0 t + y_s(t)\sin 2\pi f_0 t & t \geq 0 \\ 0 & t < 0 \end{cases}$$

where the bandwidth of the low-pass signals $y_c(t)$ and $y_s(t)$ is w Hz and f_0 equals $(2n+1)w$ for some even integer $n > 0$. Show that $y_c(t)$ can be recovered from the even samples, $\hat{y}(0), \hat{y}(2), \hat{y}(4), \ldots, \hat{y}(2k), \ldots$ of $y(t)$ taken at $t = 0, 2/4w, 4/4w, \ldots, 2k/4w, \ldots$ second. Similarly, $y_s(t)$ can be recovered from the odd samples

$\hat{y}(1), \hat{y}(3), \ldots, \hat{y}(2k+1), \ldots$ of $y(t)$ taken at $t = 1/4w,\ 3/4w, \ldots, (2k+1)/4w, \ldots$ second. Discuss the case in which f_0 is equal to $(2n+1)w$ for some odd integer $n \geq 1$.

(*e*) In parts (*c*) and (*d*), suppose that the samples were taken every $1/4w$ seconds but starting at some unknown time t_0. Show that $x(t)$ and $y(t)$ can be recovered similarly from their samples.

8-20 It is clear that the number of continuous signals of bandwidth w which can be transmitted simultaneously over a channel of bandwidth F is F/w in a frequency division multiplexing scheme. Show that by using a time division multiplexing scheme the number of signals of bandwidth w that can be transmitted simultaneously over a channel of bandwidth F is also F/w. (Note that the sampling pulses are of finite duration. Use any of the definitions of signal bandwidth given in Probs. 8-5 and 8-7.)

8-21 Let us consider a simple communication system in which one of four messages is transmitted over a communication channel each second. Suppose that the messages are represented by the time signals shown in Fig. 8P-13. That is, depending on what the message is, either $x_1(t)$, $x_2(t)$, $x_3(t)$, or $x_4(t)$ is transmitted.

(*a*) Find the half-power bandwidths of these time signals and hence the required bandwidth of the channel. (See Prob. 8-5 for a definition of half-power bandwidth.)

(*b*) Determine the required channel bandwidth if the messages are represented by the signals $x_1(2t)$, $x_2(2t)$, $x_3(2t)$, and $x_4(2t)$.

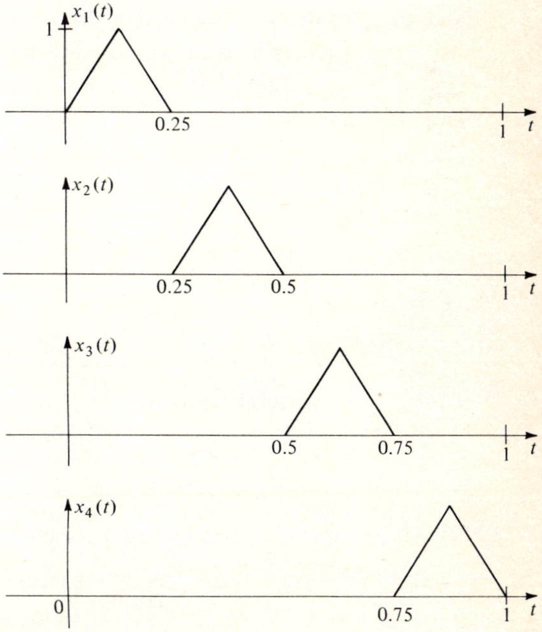

FIGURE 8P-13

8-22 To illustrate the effect of the shape of a data window (see the footnote on page 313) on the accuracy of using a discrete Fourier transform to approximate the Fourier transform of a continuous signal, let us consider the signal

$$x(t) = \cos 2\pi t$$

Suppose that the signal is sampled at $t = 0, \pm\frac{1}{4}, \pm\frac{1}{2}, \pm\frac{3}{4}, \ldots$. Let $x_k = x(k/4)$.

(*a*) Plot the Fourier transform of

$$\sum_{k=-\infty}^{\infty} x_k u_0\left(t - \frac{k}{4}\right)$$

Find the values of $X(2\pi n/4)$ which will be denoted X_n.

(*b*) Suppose that a rectangular data window $[u_{-1}(t+2) + u_{-1}(t-2)]$ is used. Plot the Fourier transform of

$$\sum_{k=-8}^{8} x_k u_0\left(t - \frac{k}{4}\right)$$

and find the corresponding values of X_n.

(*c*) Suppose that the triangular data window shown in Fig. 8P-14 is used. Plot the Fourier transform of

$$\sum_{k=-\infty}^{\infty} d(t)x(t)u_0\left(t - \frac{k}{4}\right)$$

and find the corresponding values of X_n.

Note that, for this time signal $x(t)$, the values of X_n in parts (*b*) and (*c*) differ from that in part (*a*) only by a known proportionality constant.

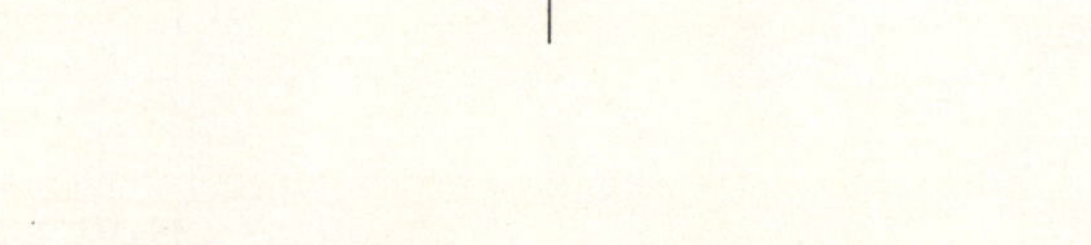

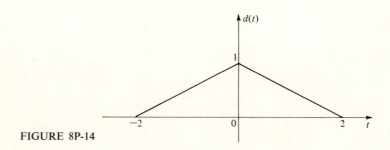

FIGURE 8P-14

8-23 Let $x(t)$ be a low-pass signal whose Fourier transform is

$$H(\omega) = \begin{cases} 1 & |\omega| \le \pi \\ 0 & \text{otherwise} \end{cases}$$

(a) Sketch the Fourier transform of the sampled signal $\sum_{k=-\infty}^{\infty} x(k\Delta)u_0(t-k\Delta)$ for $\Delta = \frac{1}{8}$ s.

(b) Sketch qualitatively the Fourier transform of the signal

$$\sum_{k=-\infty}^{\infty} x(k\Delta)u_0(t-k\Delta)[u_{-1}(t-2)-u_{-1}(t+2)]$$

(c) Repeat parts (a) and (b) for $\Delta = 2$ s.

(d) Repeat parts (a) and (b) for $\Delta = 1$ s.

8-24 (a) Suppose that the signal $x(t) = \cos 2\pi t$ is sampled at $t = 0, \pm\frac{1}{8}, \pm\frac{1}{4}, \pm\frac{3}{8}, \ldots$ s and is truncated, using the rectangular data window $[u_{-1}(t+\frac{5}{4}) - u_{-1}(t-\frac{5}{4})]$. Sketch the Fourier transform of $\sum_{k=-10}^{10} x_k u_0(t-k/8)$ and the corresponding frequency domain samples $X(4\pi n/5)u_0(\omega - 4\pi n/5)$, $n = 0, \pm 1, \pm 2, \ldots$.

The Fourier transform of the truncated signal no longer consists of two impulses but rather is a function with maxima centered at the impulses in $X(\omega)$ and a series of other peaks which are called side lobes. Hence, when the Fourier transform of the truncated time signal is sampled, we find the frequency domain sample values are nonzero at frequencies other than $\omega = \pm 2\pi$. This effect of time domain truncation is called *leakage*. Data windows other than the rectangular one have been designed to minimize the leakage effect.

(b) Repeat part (a) for the Hanning data window

$$d(t) = \begin{cases} \dfrac{1}{2} - \dfrac{1}{2}\cos\dfrac{4\pi}{5}\left(t + \dfrac{5}{4}\right) & -\dfrac{5}{4} \le t \le \dfrac{5}{4} \\ 0 & \text{otherwise} \end{cases}$$

(c) Repeat part (a) for the Parzen data window

$$d(t) = \begin{cases} \frac{4}{5}\left(\frac{5}{4} - |t|\right) & 0 \le |t| \le \frac{5}{4} \\ 0 & \text{otherwise} \end{cases}$$

8-25 (a) Sketch the Fourier transform of the signal

$$x(t) = \frac{\sin \pi t}{\pi t}$$

(b) Sketch the Fourier transform of the signal $y(t) = x(t)d(t)$ for the data window

$$d(t) = \begin{cases} \frac{1}{2}(2 - |t|) & 0 \le |t| \le 2 \\ 0 & \text{otherwise} \end{cases}$$

(c) For the truncated signal $y(t)$ in part (b), determine within 10 percent accuracy the sample values $Y(\pi n/2)$ of $Y(\omega)$ for $n = 0, \pm 1, \pm 2$.

(d) In most cases, we are interested in the estimated power spectral density $|X_n|^2$ of the signal $x(t)$. Find roughly the error in the estimated power spectral density of $x(t)$ for the data window in part (b).

(*e*) Repeat parts (*b*), (*c*), and (*d*) for the data window

$$d(t) = \begin{cases} \frac{1}{20}(20 - |t|) & 0 \leq |t| \leq 20 \\ 0 & \text{otherwise} \end{cases}$$

8-26 (*a*) Let

$$X(n_2, n_1, n_0) = X_n$$

and $$x(k_2, k_1, k_0) = x_k$$

where $$n = n_0 + 2n_1 + 4n_2$$

$$k = k_0 + 2k_1 + 4k_2$$

and n_0, n_1, n_2, k_0, k_1, and k_2 are either 0 or 1. Rewrite the summation

$$X_n = \sum_{k=0}^{7} x_k \alpha^{nk}$$

as three subsums similar to those in Eqs. (8-11) and (8-12).

(*b*) Show that only 12 multiplication and addition operations are required to determine the values of X_n, $n = 0, 1, 2, \ldots, 7$.

(*c*) Generalize part (*a*) for $n = 2^{10}$ and show that 10×2^9 multiplication operations are required to determine the values of X_n.

8-27 (*a*) The efficiency of transmission of an amplitude-modulated signal is defined to be the percentage ratio of the power in the sidebands of the signal to the total power carried by the signal. Show that the efficiency of transmission of the double-sideband amplitude-modulated signal

$$y(t) = (1 + \cos \omega t) \cos 2\pi f_0 t$$

is less than 33.3 percent.

(*b*) The peak factor P of a signal is defined to be the ratio of its peak value to its root-mean-square value. Let $x(t)$ be a periodic signal with zero average value and peak factor P. Show that the efficiency of transmission of the amplitude-modulated signal

$$y(t) = [1 + \epsilon x(t)] \cos 2\pi f_0 t$$

is less than $1/(1 + P^2)$ if $\epsilon |x(t)| \leq 1$.

8-28 As discussed in Sec. 8-7, it is difficult to build a bandpass filter with the sharp cutoff necessary to reject completely one of the sidebands of an amplitude-modulated signal. In this problem, we consider the vestigial sideband transmission system which offers a compromise between the SSB and DSB schemes. Let $y(t)$ denote the vestigial sideband amplitude-modulated signal generated by the system in Fig. 8P-15*a*.

(*a*) Given that the transfer function $H(\omega)$ of the bandpass filter is as shown in Fig. 8P-15*b*. Sketch the Fourier transform of the output signal $y(t)$ for a typical low-pass signal $x(t)$ of bandwidth w. Show that $x(t)$ can be recovered competely from $y(t)$ by using a synchronous detector (e.g., the system shown in Fig. 8-14*a*).

(*b*) Suppose that the transfer function $H(\omega)$ of the bandpass filter is as shown in Fig. 8P-15*c*. Sketch the Fourier transform of the output signal of a synchronous detector if its input signal is $y(t)$.

(*c*) Show that the modulating signal $x(t)$ can be recovered from the modulated signal $y(t)$ by using synchronous detection if the transfer function $H(\omega)$ of the bandpass filter is such that $H(\omega - \omega_0) + H(\omega + \omega_0)$ equals a constant for all $|\omega| \leq w$, where w is the bandwidth of $x(t)$. Discuss the physical implications of this condition.

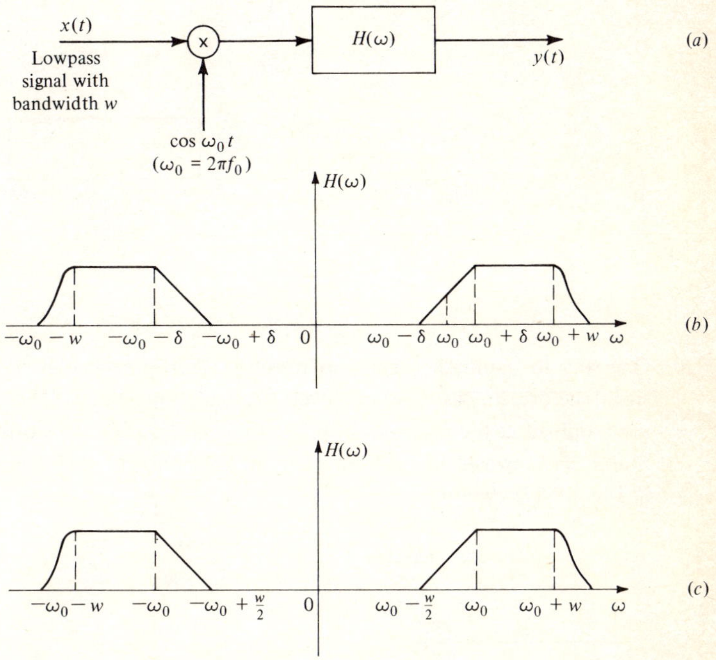

FIGURE 8P-15

8-29 Figure 8P-16 shows the block diagram of an amplitude modulation system. The input signal is a band-limited real signal of bandwidth w Hz. The signal $s(t)$ is a periodic impulse train of unit area and period $1/2w$.

(*a*) Show that the output signal $y(t)$ is equal to $Ax(t)\cos(4\pi wt + \theta)$.

(*b*) Determine the constants A and θ.

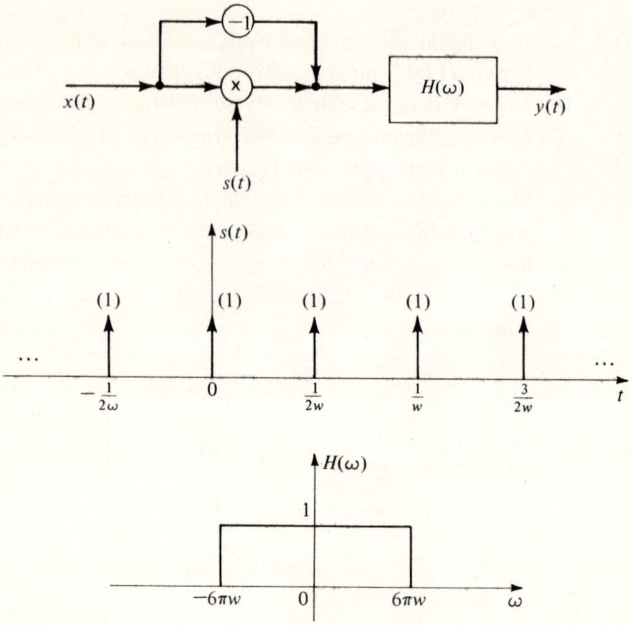

FIGURE 8P-16

8-30 One way to maintain privacy in telephone conversations is to scramble the speech signal before its transmission over the telephone lines. The transmitted signal is unscrambled at the receiver. Suppose that the Fourier transform of the speech signal $S(\omega)$ is as shown in Fig. 8P-17*a*. Design a scrambler which will change $S(\omega)$ to $C(\omega)$ in Fig. 8P-17*b*, where

$$C(\omega) = \begin{cases} S(6\pi \times 10^3 - \omega) & \omega > 0 \\ S(-6\pi \times 10^3 - \omega) & \omega < 0 \end{cases}$$

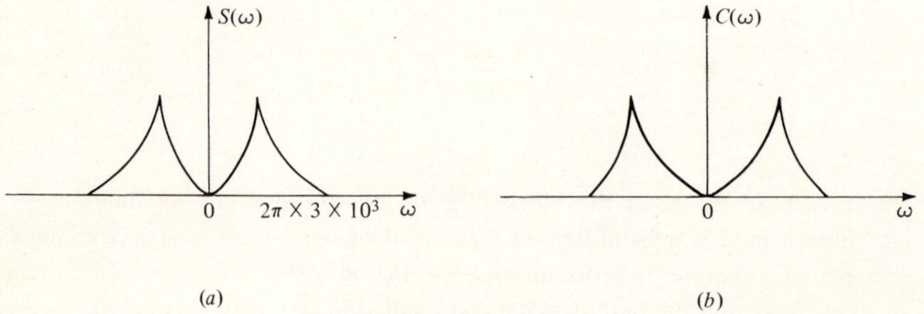

FIGURE 8P-17

8-31 Let the signal $s(t)$ in Fig. 8P-18*a* be the square wave shown in Fig. 8P-18*b*.

(*a*) Find the Fourier transform of the signal $z(t)$ if the impulse response $h_1(t)$ is $te^{-t}u_{-1}(t)$.

(*b*) If the input signal $x(t)$ is $\sin (3\pi/2)t$, find the corresponding output signal $y(t)$.

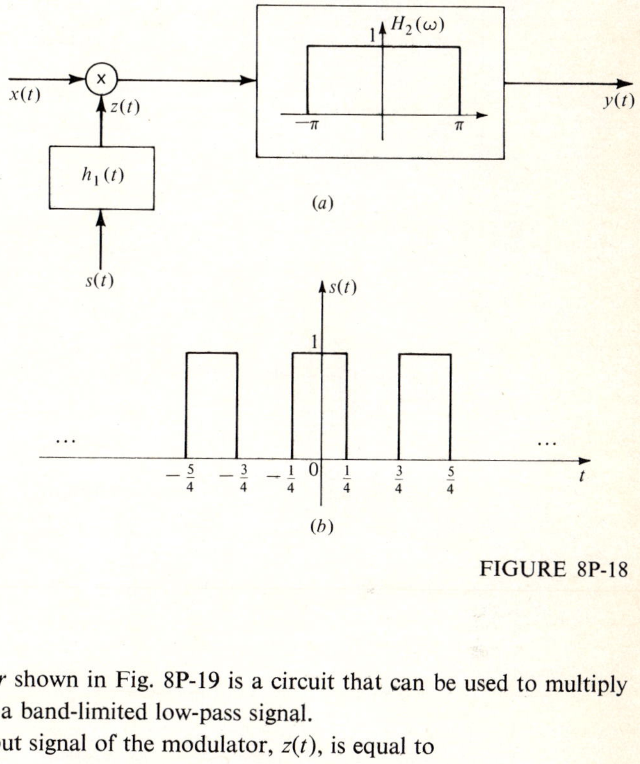

FIGURE 8P-18

8-32 The *balanced modulator* shown in Fig. 8P-19 is a circuit that can be used to multiply a sinusoidal carrier by a band-limited low-pass signal.

(*a*) Show that the output signal of the modulator, $z(t)$, is equal to

$$z(t) = \begin{cases} 2x(t) & \text{at time } t \text{ when } |x(t)| < y(t) \\ 0 & \text{at time } t \text{ when } |x(t)| \geq y(t) \end{cases}$$

(*b*) When the circuit is used to generate a modulated signal, the signal $y(t)$ is equal to $A \cos 2\pi f_0 t$, where $A \gg |x(t)|$ for all t. Hence, $z(t)$ can be written approximately as

$$z(t) = \begin{cases} 2x(t) & \text{if } \cos 2\pi f_0 t > 0 \\ 0 & \text{if } \cos 2\pi f_0 t \leq 0 \end{cases}$$

Sketch the Fourier transform of $z(t)$ for a typical low-pass signal $x(t)$ whose bandwidth w is small compared with f_0. Also show that the output signal of the bandpass filter can be written $Bx(t) \cos (2\pi f_0 t + \phi)$. Determine the constants B and ϕ.

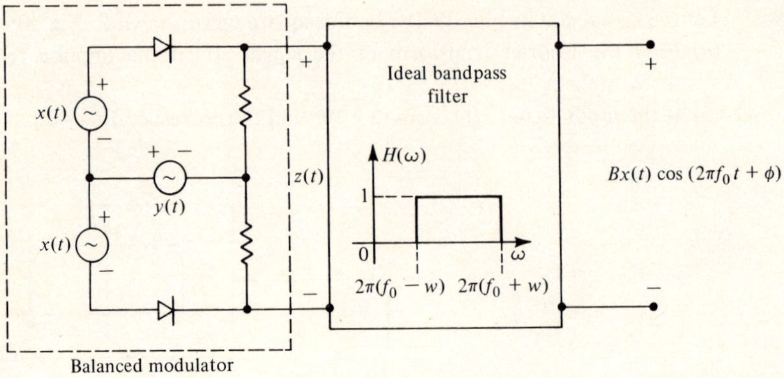

FIGURE 8P-19

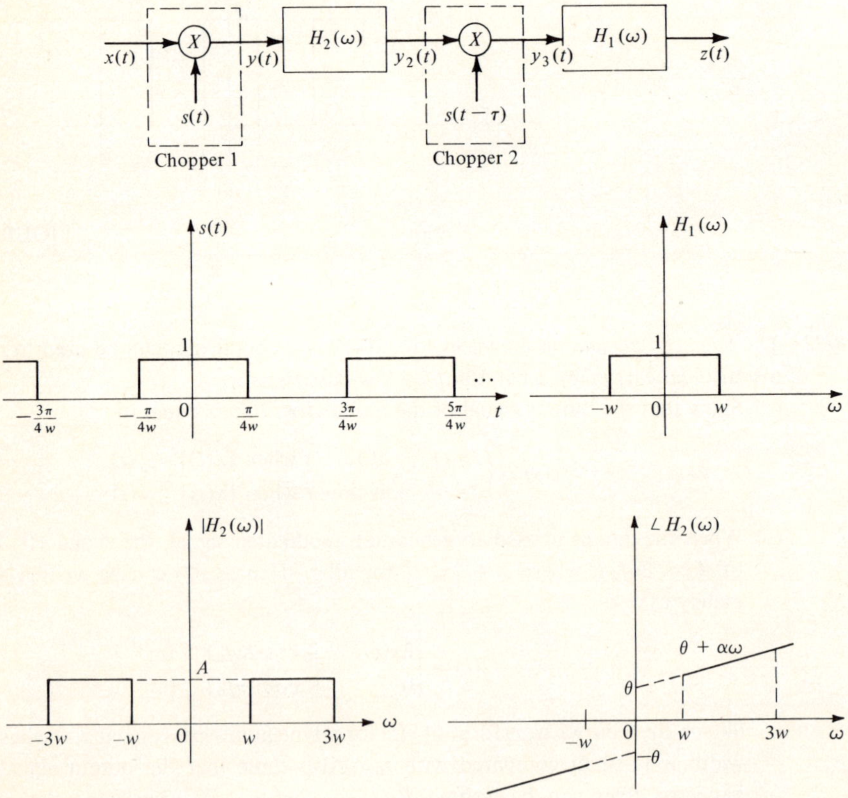

FIGURE 8P-20

8-33 In all practical circuits, the values of parameters of active elements drift slowly, because of the aging of the active elements, changes in ambient temperature, and fluctuations in power supply voltages, etc. In a dc amplifier, such slow changes in the characteristics of active elements will cause variations in the output signal that are indistinguishable from the variations due to changes in the input signal. Hence, when it is necessary to have relatively constant gain, the dc circuit shown in Fig. 8P-20 is often used. In this system, the input $x(t)$ is a low-pass signal of bandwidth w Hz. Chopper 1 is used to modulate the input signal onto a carrier so that an ac-coupled amplifier may be used to obtain the gain required. Chopper 2 is used as a synchronous detector to restore the signal to its original frequency range.

(*a*) Find the Fourier transforms for the periodic signals $s(t)$ and $s(t - \tau)$.

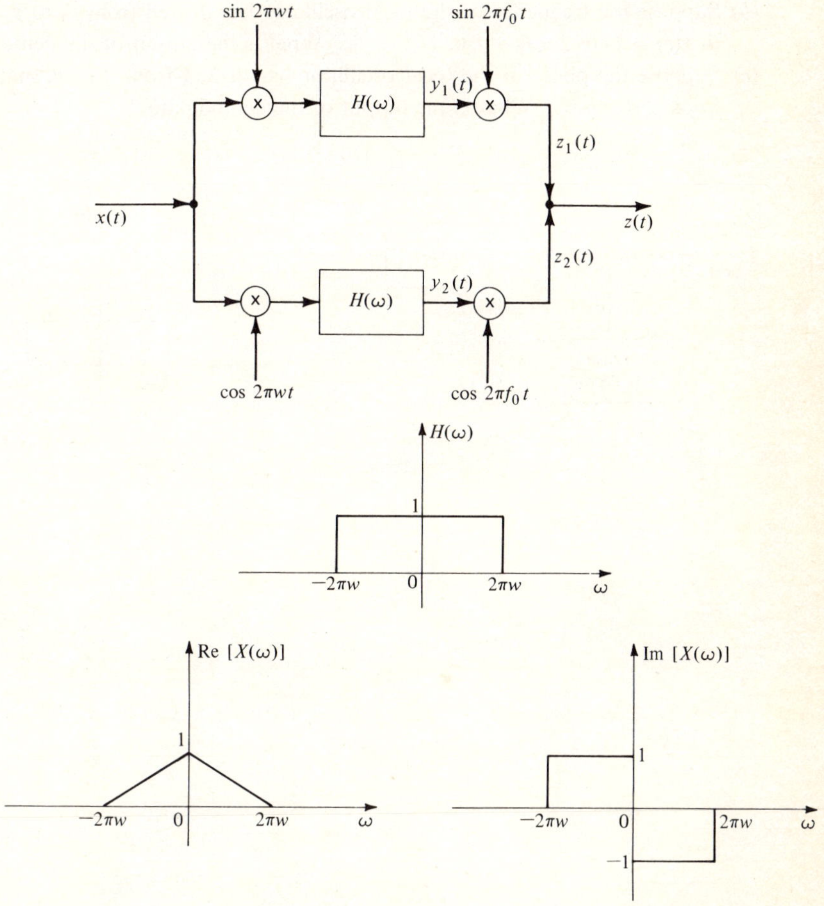

FIGURE 8P-21

(*b*) Sketch the Fourier transforms of $y_1(t)$, $y_2(t)$, and $y_3(t)$ for a typical low-pass input signal of bandwidth w Hz.

(*c*) Express $z(t)$ in terms of $x(t)$ and parameters A, θ, α, τ, and w.

8-34 The system shown in Fig. 8P-21 can be used to generate a single-sideband amplitude-modulated signal. (Note that the system contains neither Hilbert filters nor sharp cutoff bandpass filters.) Demonstrate how it works by sketching the real and imaginary parts of $Y_1(\omega)$, $Y_2(\omega)$, $Z_1(\omega)$, $Z_2(\omega)$, and $Z(\omega)$ for a typical low-pass signal $x(t)$ whose bandwidth w is such that $w \ll f_0$.

8-35 A carrier $\cos 2\pi f_0 t$ is modulated by an audio tone $x(t) = \cos 2\pi f_m t$, $f_m \ll f_0$, to produce a single-sideband signal $y(t) = \cos 2\pi(f_m - f_0)t$. The system in Fig. 8P-22 is used as a demodulator.

(*a*) Show that the demodulator indeed reproduces the audio tone $x(t)$, that is, $\hat{x}(t) = x(t)$.

(*b*) Suppose the frequency of the local oscillator has drifted from f_0 to $f_0 + f_a$, that is, $z(t) = 2 \cos 2\pi(f_0 + f_a)t$, $|f_a| \ll f_0$. What is the output of the demodulator?

(*c*) Suppose the phase of the local oscillator has drifted from 0 to θ, that is, $z(t) = 2 \cos (2\pi f_0 t + \theta)$. What is the output of the demodulator?

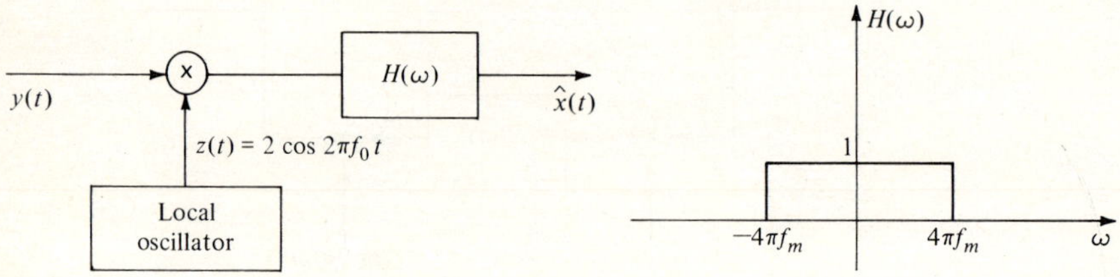

FIGURE 8P-22

9

BILATERAL LAPLACE TRANSFORMATION

9-1 INTRODUCTION

As was discussed in Chap. 5, when the region of absolute convergence of a linear time-invariant continuous system includes the $j\omega$ axis, complex exponential functions of the form $e^{j\omega t}$ are eigenfunctions of the system. In this case, the Fourier transform or Fourier series representation of signals is most useful in studying the input-output relationship of the system. In this chapter, we shall study a natural extension of the concepts developed in Chaps. 6 and 7: When the region of absolute convergence of a system does not include the $j\omega$ axis, we seek a representation of signals as integrals of complex exponential functions of the form e^{st}, where the complex numbers s are such that e^{st} are eigenfunctions of the system. (That is, the complex numbers s are in the region of absolute convergence of the system.) Moreover, as will be noted, many time signals that are not Fourier-transformable can be expressed as integrals of complex exponential functions of the form e^{st}. Consequently, such a representation adds further flexibility to our system analysis techniques. Also to be shown in this chapter is the relationship between the z transformation of discrete signals and the Laplace transformation of continuous signals, thus completing an overall picture of a transformation theory of both discrete and continuous signals.

9-2 BILATERAL LAPLACE TRANSFORMATION

We now study how a signal $x(t)$ can be expressed as an integral of complex exponential functions of the form e^{st} for the complex numbers s along a vertical line in the s plane (the complex plane). To this end, let us consider the signal $x(t)e^{-\sigma_0 t}$ for a real constant σ_0. If $x(t)$ has a finite number of maxima, minima, and discontinuities within any finite interval, and if

$$\int_{-\infty}^{\infty} |x(t)e^{-\sigma_0 t}|\, dt < \infty$$

the Fourier transform of the signal $x(t)e^{-\sigma_0 t}$ exists. Let us denote the Fourier transform of $x(t)e^{-\sigma_0 t}$ by $X(\omega)$. According to our discussion in Chap. 7, $X(\omega)$ is given by

$$X(\omega) = \int_{-\infty}^{\infty} x(t)e^{-\sigma_0 t}e^{-j\omega t}\, dt = \int_{-\infty}^{\infty} x(t)e^{-(\sigma_0 + j\omega)t}\, dt \tag{9-1}$$

Also, the inverse Fourier transformation formula yields

$$x(t)e^{-\sigma_0 t} = \frac{1}{2\pi}\int_{-\infty}^{\infty} X(\omega)e^{j\omega t}\, d\omega$$

or

$$x(t) = \frac{1}{2\pi}\int_{-\infty}^{\infty} X(\omega)e^{(\sigma_0 + j\omega)t}\, d\omega \tag{9-2}$$

Letting $s = \sigma_0 + j\omega$, we rewrite Eqs. (9-1) and (9-2) as

$$X(s) = \int_{-\infty}^{\infty} x(t)e^{-st}\, dt \tag{9-3}$$

$$x(t) = \frac{1}{2\pi j}\int_{\sigma_0 - j\infty}^{\sigma_0 + j\infty} X(s)e^{st}\, ds \tag{9-4}$$

As defined in Eq. (9-3), $X(s)$ is known as the *bilateral Laplace transform*† of the time signal $x(t)$, and Eq. (9-4) is the corresponding inverse transformation formula. Although our derivation of the formulas in Eqs. (9-3) and (9-4) is informal, it can be shown that if $x(t)$ is a time signal of bounded variation and if

$$\int_{-\infty}^{\infty} |x(t)e^{-st}|\, dt < \infty$$

for some s in the s plane, then the Laplace transform of $x(t)$ as defined in Eq. (9-3) exists. Moreover, $x(t)$ can be recovered from $X(s)$ by evaluating the integral in Eq. (9-4) along a vertical line from $s = \sigma_0 - j\infty$ to $s = \sigma_0 + j\infty$ in the s plane.

† From here on, the term "Laplace transform" will mean "bilateral Laplace transform."

The region in the s plane containing all s such that

$$\int_{-\infty}^{\infty} |x(t)e^{-st}| \, dt < \infty$$

is called the region of absolute convergence of the time signal $x(t)$. Of course, the notion of the region of absolute convergence of a time signal is not a totally new one. We discussed such a notion for discrete time signals when we studied the z transformation of discrete signals in Chap. 5. For convenience, the region of absolute convergence of a time signal $x(t)$ will also be referred to as the region of absolute convergence of its Laplace transform $X(s)$.

It should be noted that, according to the definition, the Laplace transform $X(s)$ of a time signal is a function defined on a vertical line in the region of absolute convergence of $x(t)$ [because for the complex exponential function e^{st} in Eqs. (9-3) and (9-4) the s are points along a vertical line in the s plane]. However, Eqs. (9-3) and (9-4) strongly suggest the extension of the definition so that $X(s)$ is defined as a function of the complex variable s within the region of absolute convergence of $x(t)$. Not only is such an extension possible, it can also be shown that $X(s)$ is an analytic function within its region of absolute convergence.† From now on, we shall consider $X(s)$ to be an analytic function in its region of absolute convergence.

As an example, let us determine the Laplace transform of the time signal

$$x(t) = e^{-\alpha t} u_{-1}(t)$$

in Fig. 9-1a. First of all, we note that

$$\int_{0}^{\infty} |e^{-\alpha t} e^{-st}| \, dt < \infty$$

for $-\text{Re}(s) - \alpha < 0$, that is, for $\text{Re}(s) > -\alpha$. Thus, the region of absolute convergence of $x(t)$ is the shaded area in Fig. 9-1b. The Laplace transform of $x(t)$ can be evaluated as

$$X(s) = \int_{0}^{\infty} e^{-\alpha t} e^{-st} \, dt = \frac{-1}{s+\alpha} e^{-(s+\alpha)t} \Big|_{0}^{\infty} = \frac{1}{s+\alpha}$$

Also, let us consider the signal

$$x(t) = \begin{cases} \sin \pi t & 0 \le t \le 2 \\ 0 & \text{otherwise} \end{cases}$$

† In fact, its derivative with respect to variable s in the region of absolute convergence is given by

$$\frac{dX(s)}{ds} = \int_{-\infty}^{\infty} (-t) x(t) e^{-st} \, dt$$

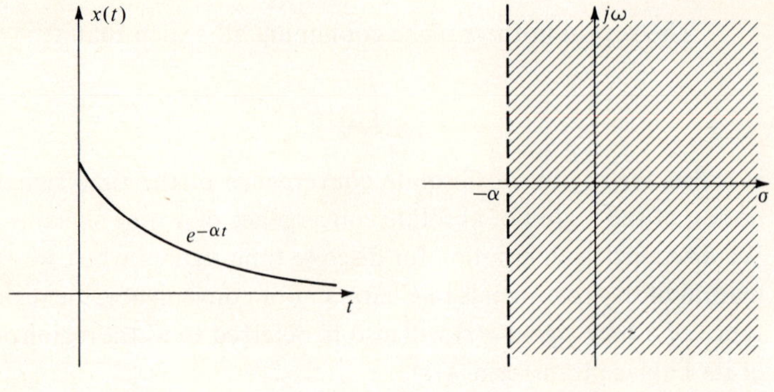

(a) (b)

FIGURE 9-1

Because

$$\int_{-\infty}^{\infty} |x(t)e^{-st}|\, dt = \int_{0}^{2} |e^{-st} \sin \pi t|\, dt < \infty$$

for all s, the region of absolute convergence of $x(t)$ is the entire s plane. The Laplace transform of $x(t)$ can be computed as

$$\begin{aligned} X(s) &= \int_0^2 e^{-st} \sin \pi t\, dt \\ &= \frac{1}{2j} \int_0^2 (e^{-(s-j\pi)t} - e^{-(s+j\pi)t})\, dt \\ &= \frac{\pi(1 - e^{-2s})}{s^2 + \pi^2} \end{aligned}$$

As another example, we note that the region of absolute convergence of the signal

$$x(t) = \frac{e^{-\alpha t}}{1 + t^2}$$

is the single vertical line at Re $(s) = -\alpha$, because the value of the integral

$$\int_{-\infty}^{\infty} \left| \frac{e^{-st} e^{-\alpha t}}{1 + t^2} \right| dt$$

is equal to π for Re $(s) = -\alpha$ and is equal to infinity for all other s. The Laplace transform of $x(t)$ is

$$X(s) = \int_{-\infty}^{\infty} \frac{e^{-(\alpha+s)t}}{1 + t^2}\, dt = \pi e^{-|\alpha+s|}$$

For the signal

$$x(t) = \begin{cases} t & t > 0 \\ e^{-t} & t < 0 \end{cases}$$

the Laplace transform does not exist. Observe that in

$$\int_{-\infty}^{\infty} |x(t)e^{-st}| \, dt = \int_{0}^{\infty} |te^{-st}| \, dt + \int_{-\infty}^{0} |e^{-(1+s)t}| \, dt$$

the value of the first integral is finite only when Re $(s) > 0$ and the value of the second integral is finite only when Re $(s) < -1$. Since these two regions do not overlap, $x(t)$ has no region of absolute convergence.

As in the case of z transformation of discrete signals, different time signals might have the same Laplace transform but different regions of absolute convergence. We have shown above that the Laplace transform of the signal $e^{-\alpha t}u_{-1}(t)$ is

$$\frac{1}{s + \alpha} \tag{9-5}$$

with its region of absolute convergence being Re $(s) > -\alpha$. Consider now the signal

$$x(t) = -e^{-\alpha t}u_{-1}(-t)$$

as shown in Fig. 9-2*a*. Since

$$\int_{-\infty}^{0} |-e^{-\alpha t}e^{-st}| \, dt < \infty$$

for Re $(s) < -\alpha$, the region of absolute convergence of $x(t)$ is the shaded area shown in Fig. 9-2*b*. The Laplace transform of $x(t)$ is

$$X(s) = \int_{-\infty}^{0} -e^{-\alpha t}e^{-st} \, dt = \frac{1}{s + \alpha} \, e^{-(s+\alpha)t} \Big|_{-\infty}^{0} = \frac{1}{s + \alpha}$$

which is the same expression as (9-5).

As another example, let us consider the signal

$$x(t) = \begin{cases} e^{-\alpha t} & t > 0 \\ e^{\alpha t} & t < 0 \end{cases}$$

shown in Fig. 9-3*a*. The region of absolute convergence of $x(t)$ contains all s so that

$$\int_{-\infty}^{0} |e^{\alpha t}e^{-st}| \, dt + \int_{0}^{\infty} |e^{-\alpha t}e^{-st}| \, dt < \infty$$

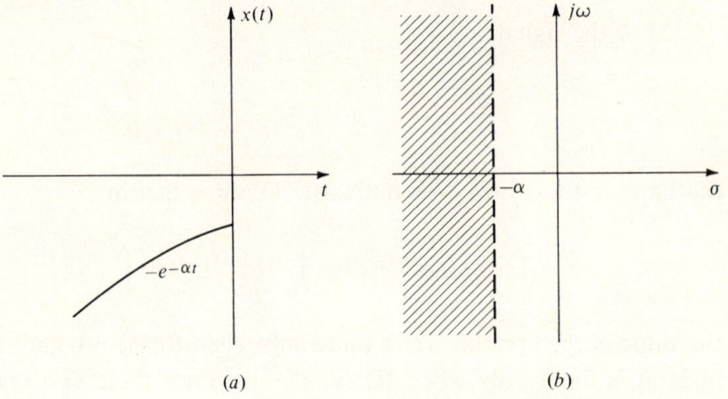

FIGURE 9-2

The first integral is finite for Re $(s) < \alpha$ and the second integral is finite for Re $(s) > -\alpha$. Thus, the region of absolute convergence is the vertical strip shown in Fig. 9-3*b*. The Laplace transform of $x(t)$ is

$$\int_{-\infty}^{0} e^{\alpha t}e^{-st}\,dt + \int_{0}^{\infty} e^{-\alpha t}e^{-st}\,dt = \frac{-1}{s-\alpha} + \frac{1}{s+\alpha} = \frac{-2\alpha}{s^2-\alpha^2}$$

Note that the Laplace transform of the signal

$$(e^{-\alpha t} - e^{\alpha t})u_{-1}(t)$$

shown in Fig. 9-3*c* is

$$\int_{0}^{\infty} (e^{-\alpha t} - e^{\alpha t})e^{-st}\,dt = \frac{1}{s+\alpha} - \frac{1}{s-\alpha} = \frac{-2\alpha}{s^2-\alpha^2}$$

with its region of absolute convergence being that shown in Fig. 9-3*d*, and the Laplace transform of the signal

$$(-e^{-\alpha t} + e^{\alpha t})u_{-1}(-t)$$

shown in Fig. 9-3*e* is

$$\int_{-\infty}^{0} (-e^{-\alpha t} + e^{\alpha t})e^{-st}\,dt = -\frac{1}{s+\alpha} + \frac{1}{s+\alpha} = \frac{-2\alpha}{s^2-\alpha^2}$$

with its region of absolute convergence being that shown in Fig. 9-3*f*.

These two examples illustrate that a signal $x(t)$ is specified by the expression $X(s)$ *together with* its region of absolute convergence. In other words, we cannot recover a signal from the expression of its Laplace transform alone. Rather, the corresponding region of absolute convergence must also be known. What remains

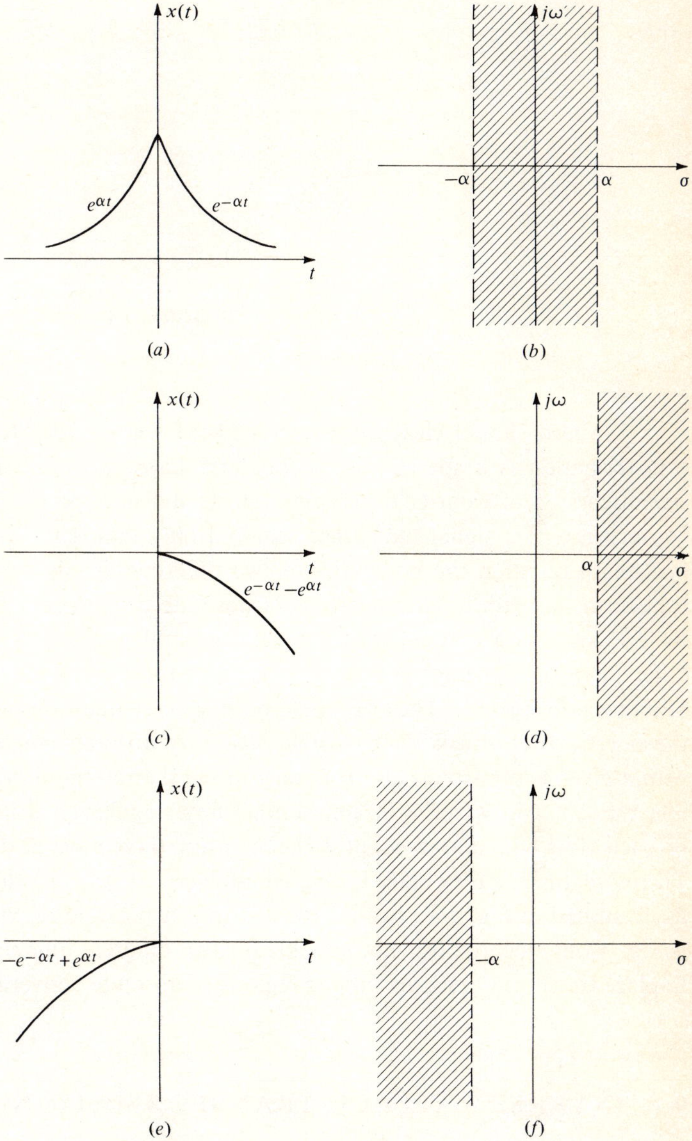

FIGURE 9-3

to be shown is that the Laplace transform $X(s)$ together with its region of absolute convergence *uniquely* determines $x(t)$. This will be proved in the next section where we shall show that the line integral in Eq. (9-4) can be evaluated to yield a unique result.

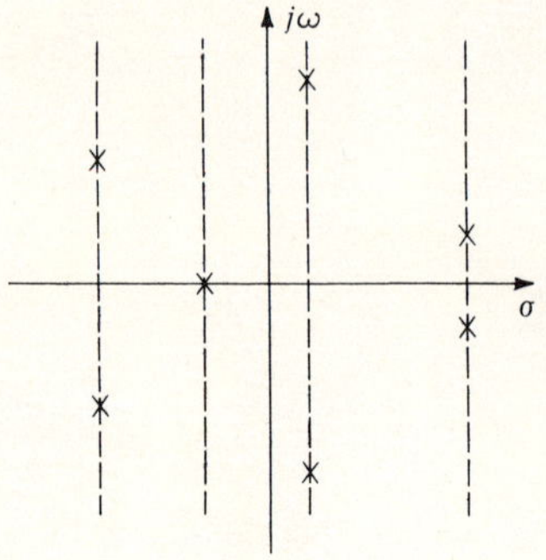

FIGURE 9-4

We also should clear up one point before going on. It is clear that Fourier transformation is a special case of bilateral Laplace transformation. However, in our discussion of Fourier transformation we did not specify the region of absolute convergence of a signal but rather required only that it include the $j\omega$ axis. Thus, one might question the possibility of two different signals having the same Fourier transform but with different regions of absolute convergence both of which include the $j\omega$ axis. As a matter of fact, a more general question is the possibility of two different signals having the same Laplace transform but with overlapping regions of absolute convergence. The answers to both of these questions are negative. To prove the assertion, let us ask what possible regions of absolute convergence are associated with a given expression $X(s)$. It is easy to see that a region of absolute convergence is a vertical strip which does not contain any singularity in its interior. Moreover, on each of its boundaries except at the infinities, there must be at least one singularity. Therefore, the complex plane is divided into vertical strips by the singularities of $X(s)$, as illustrated in Fig. 9-4. Each vertical strip is a possible region of absolute convergence. Consequently, it is not possible for two different time signals to have the same Laplace transform and overlapping regions of absolute convergence.

9-3 INVERSE LAPLACE TRANSFORMATION†

Given the Laplace transform $X(s)$, we can recover the time signal $x(t)$ by evaluating the integral in Eq. (9-4). We recall that the integral in Eq. (9-4) is a line integral along a vertical line in the region of absolute convergence. According to a result in the theory of functions of a complex variable, if $X(s) \to 0$ as $s \to \infty$, then for $t > 0$, the value of

† A reader who has not been exposed to the theory of functions of a complex variable may want to skip this section. In Sec. 9-4, we shall discuss performing the inverse transformation by consulting a table.

the line integral is equal to the sum of the residues of the function $X(s)e^{st}$ at the poles that are to the left of the vertical line $\text{Re}(s) = \sigma_0$; and for $t < 0$, the value of the line integral is equal to the minus of the sum of the residues of the function $X(s)e^{st}$ at the poles that are to the right of the vertical line $\text{Re}(s) = \sigma_0$, where the *residue* of the function $X(s)e^{st}$ at a simple pole s_0 is equal to

$$(s - s_0)X(s)e^{st}\Big|_{s=s_0}$$

and the residue of the function $X(s)e^{st}$ at a kth-order pole s_0 is equal to

$$\frac{1}{(k-1)!}\frac{d^{k-1}}{ds^{k-1}}[(s-s_0)^k X(s)e^{st}]\Big|_{s=s_0}$$

That we can evaluate the line integral in Eq. (9-4) by computing the residues of the function $X(s)e^{st}$ enables us to conclude that the value of the line integral in Eq. (9-4) is the same for any vertical line $\text{Re}(s) = \sigma_0$ in the region of absolute convergence along which the integration is carried out. In other words, a time signal can be determined uniquely from its Laplace transform and the region of absolute convergence.

As a simple example, consider the time signal

$$x(t) = e^{-3t}u_{-1}(t)$$

Its Laplace transform is

$$X(s) = \frac{1}{s+3}$$

with the region of absolute convergence being $\text{Re}(s) > -3$ as shown in Fig. 9-5. To recover $x(t)$ from $X(s)$, we evaluate the integral

$$\frac{1}{2\pi j}\int_{\sigma_0 - j\infty}^{\sigma_0 + j\infty} \frac{e^{st}}{s+3}\,ds$$

along a vertical line in the region $\text{Re}(s) > -3$. We obtain

$$x(t) = \begin{cases} \text{residue of } \dfrac{e^{st}}{s+3} \text{ at pole } s = -3 & t > 0 \\ 0 & t < 0 \end{cases}$$

that is,

$$x(t) = \begin{cases} e^{-3t} & t > 0 \\ 0 & t < 0 \end{cases}$$

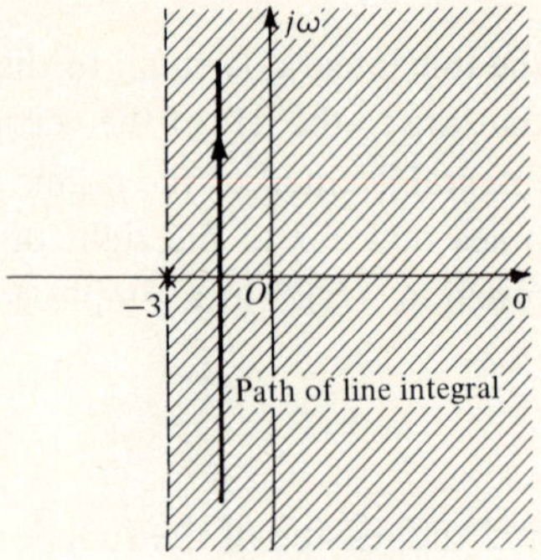

FIGURE 9-5

As another example, let

$$X(s) = \frac{2}{s(s+1)(s+2)}$$

with its region of absolute convergence being $-2 < \text{Re}\,(s) < -1$ as shown in Fig. 9-6. The inverse transform of $X(s)$ can be evaluated along any vertical line within the region of absolute convergence. We obtain, for $t > 0$,

$$x(t) = \text{residue of } \frac{2e^{st}}{s(s+1)(s+2)} \text{ at } s = -2$$

and for $t < 0$,

$$x(t) = -\left[\text{sum of residues of } \frac{2e^{st}}{s(s+1)(s+2)} \text{ at } s = 0 \text{ and } s = -1\right]$$

That is,

$$x(t) = \begin{cases} e^{-2t} & t > 0 \\ 2e^{-t} - 1 & t < 0 \end{cases}$$

As another example, let

$$X(s) = \frac{1}{s(s+1)^3}$$

with its region of absolute convergence being $\text{Re}\,(s) < -1$. Thus, for $t > 0$

$$x(t) = 0$$

and for $t < 0$

$$\begin{aligned} x(t) &= \frac{-1}{(s+1)^3} e^{st}\bigg|_{s=0} - \frac{1}{2!}\frac{d^2}{ds^2}\left(\frac{1}{s}e^{st}\right)\bigg|_{s=-1} \\ &= -1 + \tfrac{1}{2}(2 + 2t + t^2)e^{-t} \end{aligned}$$

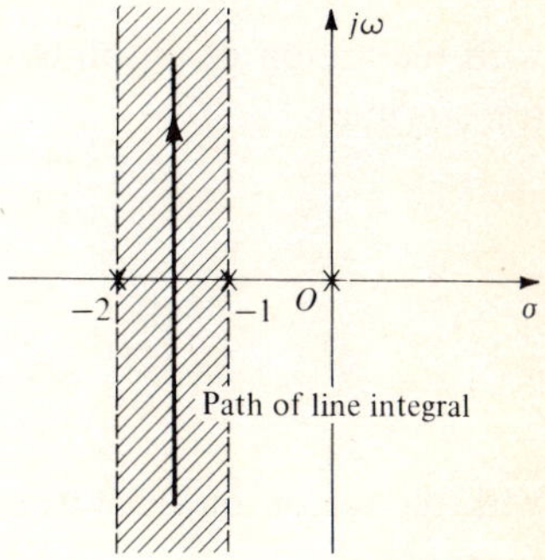

FIGURE 9-6

9-4 MANIPULATION OF SIGNALS

From the Laplace transform $X(s)$ of a time signal $x(t)$, we can determine the Laplace transforms of the transposed signal $x(-t)$, the translated signal $x(t + t_0)$, the derivative $dx(t)/dt$, and so on. Since Laplace transformation is merely a generalization of Fourier transformation, most of the properties of the Laplace transforms of signals can be derived in a manner similar to that in Chap. 7 for the Fourier transforms of signals. The only new question that arises is the region of absolute convergence associated with a Laplace transform.

Let the region of absolute convergence of $x(t)$ be $\sigma_1 < \text{Re}(s) < \sigma_2$. Since

$$\int_{-\infty}^{\infty} x(-t)e^{-st}\,dt = \int_{-\infty}^{\infty} x(\tau)e^{s\tau}\,d\tau = X(-s)$$

the Laplace transform of $x(-t)$ is $X(-s)$. Moreover, the region of absolute convergence of $x(-t)$ is $-\sigma_2 < \text{Re}(s) < -\sigma_1$.

Since, for $a > 0$,

$$\int_{-\infty}^{\infty} x(at)e^{-st}\,dt = \frac{1}{a}\int_{-\infty}^{\infty} x(\tau)e^{-s\tau/a}\,d\tau = \frac{1}{a}X\left(\frac{s}{a}\right)$$

the Laplace transform of $x(at)$ is $(1/a)X(s/a)$ for any positive a, and the corresponding region of absolute convergence is $a\sigma_1 < \text{Re}(s) < a\sigma_2$.

As an illustrative example, let us consider the signal

$$x(t) = \begin{cases} e^{-2t} & t > 0 \\ e^{3t} & t < 0 \end{cases}$$

We have

$$X(s) = \frac{1}{s+2} - \frac{1}{s-3} = \frac{-5}{(s+2)(s-3)}$$

with the region of absolute convergence being $-2 < \text{Re}(s) < 3$. Thus, the Laplace transform of

$$x(-t) = \begin{cases} e^{-3t} & t > 0 \\ e^{2t} & t < 0 \end{cases}$$

is

$$-\frac{1}{-s-3} + \frac{1}{-s+2} = \frac{-5}{(s+3)(s-2)}$$

with the region of absolute convergence being $-3 < \text{Re}(s) < 2$. The Laplace transform of

$$x(5t) = \begin{cases} e^{-10t} & t > 0 \\ e^{15t} & t < 0 \end{cases}$$

is

$$\frac{1}{5}\left(\frac{1}{s/5+2} - \frac{1}{s/5-3}\right) = \frac{1}{s+10} - \frac{1}{s-15} = \frac{-25}{(s+10)(s-15)}$$

with the region of absolute convergence being $-10 < \text{Re}(s) < 15$.

To show that the Laplace transform of $dx(t)/dt$ is $sX(s)$, we note that

$$\int_{-\infty}^{\infty} \frac{dx(t)}{dt} e^{-st}\, dt = x(t)e^{-st}\Big|_{-\infty}^{\infty} + s\int_{-\infty}^{\infty} x(t)e^{-st}\, dt$$

by carrying out the integration by parts. Because, within the region of absolute convergence of $x(t)$, $x(t)e^{-st} \to 0$ as $t \to \infty$ and $t \to -\infty$,† we obtain

$$\int_{-\infty}^{\infty} \frac{dx(t)}{dt} e^{-st}\, dt = s\int_{-\infty}^{\infty} x(t)e^{-st}\, dt$$

Thus, the Laplace transform of $dx(t)/dt$ is $sX(s)$ with its region of absolute convergence being (at least)‡ that of $x(t)$.

The Laplace transform of the signal $\int_{-\infty}^{t} x(\tau)\, d\tau$ is

$$\int_{-\infty}^{\infty} e^{-st}\, dt \int_{-\infty}^{t} x(\tau)\, d\tau$$

Interchanging the order of integration, we obtain

$$\int_{-\infty}^{\infty} x(\tau)\, d\tau \int_{\tau}^{\infty} e^{-st}\, dt \tag{9-6}$$

† Because, if this is not the case, the value of the integral $\int_{-\infty}^{\infty} |x(t)e^{-st}|\, dt$ will not be finite.

‡ If $X(s)$ has a simple pole at $s = 0$, $sX(s)$ will not have such a pole. In this case, the region of absolute convergence of $sX(s)$ might be larger than that of $X(s)$.

Since for Re $(s) > 0$

$$\int_{\tau}^{\infty} e^{-st}\,dt = -\frac{1}{s}e^{-st}\Big|_{\tau}^{\infty} = \frac{1}{s}e^{-s\tau}$$

the integral in (9-6) can be written

$$\frac{1}{s}\int_{-\infty}^{\infty} e^{-s\tau}x(\tau)\,d\tau$$

which, in turn, is equal to $(1/s)X(s)$ for all s in the region of absolute convergence of $x(t)$. Hence the Laplace transform of the signal $\int_{-\infty}^{t} x(\tau)\,d\tau$ is $(1/s)X(s)$ with its region of absolute convergence being the intersection of the region of absolute convergence of $x(t)$ and the right half plane Re $(s) > 0$. In a similar manner, we can show that the Laplace transform of $-\int_{t}^{\infty} x(\tau)\,d\tau$ is also $(1/s)X(s)$. However, the corresponding region of absolute convergence is the intersection of the region of absolute convergence of $x(t)$ and the left half plane Re $(s) < 0$.

Let us consider the signal

$$x(t) = \begin{cases} 1 & t > 0 \\ e^{t} & t < 0 \end{cases}$$

Its Laplace transform is

$$X(s) = \frac{1}{s} - \frac{1}{s-1} = -\frac{1}{s(s-1)}$$

with the region of absolute convergence being $0 < \text{Re}\,(s) < 1$. The Laplace transform of the signal

$$\frac{dx(t)}{dt} = \begin{cases} 0 & t > 0 \\ e^{t} & t < 0 \end{cases}$$

is

$$-\frac{1}{s-1}$$

which is indeed equal to $sX(s)$. The region of absolute convergence of $dx(t)/dt$ is Re $(s) < 1$ which contains the region of absolute convergence of $x(t)$. The Laplace transform of the signal

$$\int_{-\infty}^{t} x(\tau)\,d\tau = \begin{cases} 1+t & t > 0 \\ e^{t} & t < 0 \end{cases}$$

is

$$\frac{1}{s}X(s) = -\frac{1}{s^2(s-1)}$$

Its region of absolute convergence is also $0 < \text{Re}(s) < 1$. The reader may wish to check this result by computing the Laplace transform of $\int_{-\infty}^{t} x(\tau)\,d\tau$ directly.

We list several more Laplace transform pairs in Table 9-1. Their derivations are left to the reader.

Let us consider some illustrative examples. The Laplace transform of the signal

$$x(t) = e^{-\alpha t}u_{-1}(t)$$

is

$$X(s) = \frac{1}{s+\alpha}$$

and its region of absolute convergence is $\text{Re}(s) > -\alpha$. Since

$$e^{-\alpha t}\sin tu_{-1}(t) = \frac{1}{2j}e^{-\alpha t}(e^{jt} - e^{-jt})u_{-1}(t)$$

$$= \frac{1}{2j}[x(t)e^{jt} - x(t)e^{-jt}]$$

the Laplace transform of the signal $e^{-\alpha t}\sin tu_{-1}(t)$ is

$$\frac{1}{2j}[X(s-j) - X(s+j)] = \frac{1}{2j}\left[\frac{1}{(s-j)+\alpha} - \frac{1}{(s+j)+\alpha}\right]$$

$$= \frac{1}{(s+\alpha)^2 + 1}$$

TABLE 9-1

Signal	Laplace transform	Region of absolute convergence, given that the region of absolute convergence of $x(t)$ is $\sigma_1 < \text{Re}(s) < \sigma_2$
$x(-t)$	$X(-s)$	$-\sigma_2 < \text{Re}(s) < -\sigma_1$
$x(at)$	$\frac{1}{a}X\left(\frac{s}{a}\right)$	$a\sigma_1 < \text{Re}(s) < a\sigma_2,\ a > 0$
$x(t+t_0)$	$e^{st_0}X(s)$	$\sigma_1 < \text{Re}(s) < \sigma_2$
$e^{-s_0 t}x(t)$	$X(s+s_0)$	$\sigma_1 - \text{Re}(s_0) < \text{Re}(s) < \sigma_2 - \text{Re}(s_0)$
$\frac{dx(t)}{dt}$	$sX(s)$	At least $\sigma_1 < \text{Re}(s) < \sigma_2$
$tx(t)$	$-\frac{dX(s)}{ds}$	$\sigma_1 < \text{Re}(s) < \sigma_2$
$\sigma_1 x_1(t) + a_2 x_2(t)$	$a_1X_1(s) + a_2X_2(s)$	At least the intersection of the regions of absolute convergence of $x_1(t)$ and $x_2(t)$
$\int_{-\infty}^{t} x(\tau)d\tau$	$\frac{1}{s}X(s)$	At least the intersection of $\sigma_1 < \text{Re}(s) < \sigma_2$ and $\text{Re}(s) > 0$
$-\int_{t}^{\infty} x(\tau)\,d\tau$	$\frac{1}{s}X(s)$	At least the intersection of $\sigma_1 < \text{Re}(s) < \sigma_2$ and $\text{Re}(s) < 0$

Moreover, because the regions of absolute convergence of both $x(t)e^{jt}$ and $x(t)e^{-jt}$ are the region of absolute convergence of $x(t)$, the region of absolute convergence of $x(t) \sin t$ is also $\text{Re}(s) > -\alpha$.

The Laplace transform of the signal $te^{-\alpha t} \sin t u_{-1}(t)$ is

$$-\frac{d}{ds}\frac{1}{(s+\alpha)^2+1} = \frac{2(s+\alpha)}{[(s+\alpha)^2+1]^2}$$

with the region of absolute convergence being $\text{Re}(s) > -\alpha$.

In many cases, instead of evaluating the line integral in Eq. (9-4), we can compute the inverse Laplace transformation by using some known transformation pairs together with the results in Table 9-1. As an example, let

$$X(s) = \frac{2}{s(s+1)(s+2)}$$

with its region of absolute convergence being $-2 < \text{Re}(s) < -1$. Writing $X(s)$ as

$$X(s) = \frac{1}{s} - \frac{2}{s+1} + \frac{1}{s+2}$$

we obtain

$$x(t) = \begin{cases} e^{-2t} & t > 0 \\ 2e^{-t} - 1 & t < 0 \end{cases}$$

Note that, corresponding to the term $1/s$, we might have either the signal

$$x_1(t) = \begin{cases} 1 & t > 0 \\ 0 & t < 0 \end{cases}$$

with its region of absolute convergence being $\text{Re}(s) > 0$, or the signal

$$x_2(t) = \begin{cases} 0 & t > 0 \\ 1 & t < 0 \end{cases}$$

with its region of absolute convergence being $\text{Re}(s) < 0$. However, because only the region of absolute convergence of $x_2(t)$ overlaps with that of $x(t)$ the inverse transformation of the term $1/s$ yields the signal $x_2(t)$. Similarly, the inverse transformation of the term $-2/(s+1)$ yields the signal $2e^{-t}u_{-1}(-t)$ instead of the signal $-2e^{-t}u(t)$, and the inverse transformation of the term $1/(s+2)$ yields the signal $e^{-2t}u_{-1}(t)$ instead of the signal $-e^{-2t}u_{-1}(-t)$.

As another example, suppose that we want to find the inverse transform of

$$X(s) = \frac{e^{-2s}}{s^2 + \pi^2}$$

whose region of absolute convergence is Re $(s) > 0$. Because $X(s)$ does not approach 0 as s approaches ∞, the line integral in Eq. (9-4) cannot be evaluated by computing the residues of $X(s)e^{st}$. However, we note that for

$$Y(s) = \frac{1}{s^2 + \pi^2}$$

with the region of absolute convergence being Re $(s) > 0$, the inverse transformation formula yields

$$y(t) = \begin{cases} \dfrac{1}{\pi} \sin \pi t & t > 0 \\ 0 & t < 0 \end{cases}$$

$$= \frac{1}{\pi} \sin \pi t u_{-1}(t)$$

Hence, we have

$$x(t) = y(t-2)$$

$$= \frac{1}{\pi} \sin \pi(t-2) u_{-1}(t-2)$$

$$= \frac{1}{\pi} \sin \pi t u_{-1}(t-2)$$

Also, let

$$Z(s) = \frac{2 \sinh 2s}{s^2 + \pi^2}$$

with its region of absolute convergence being the entire complex s plane. Writing $Z(s)$ as

$$Z(s) = Y(s)(e^{2s} - e^{-2s})$$

we obtain

$$z(t) = \frac{1}{\pi} [\sin \pi(t+2) u_{-1}(t+2) - \sin \pi(t-2) u_{-1}(t-2)]$$

$$= \frac{1}{\pi} \sin \pi t [u_{-1}(t+2) - u_{-1}(t-2)]$$

Note that, although the regions of absolute convergence of $y(t+2)$ and of $y(t-2)$ are both Re $(s) > 0$, the region of absolute convergence of $y(t+2) - y(t-2)$ is the whole s plane.

As a further example, let

$$Y(s) = -\tan^{-1} s$$

with its region of absolute convergence Re $(s) > 0$. We note that

$$-\frac{dY(s)}{ds} = \frac{1}{s^2 + 1}$$

It follows that

$$ty(t) = \begin{cases} \sin t & t > 0 \\ 0 & t < 0 \end{cases}$$

$$= \sin t u_{-1}(t)$$

or

$$y(t) = \frac{\sin t}{t} u_{-1}(t)$$

9-5 LAPLACE TRANSFORMATION OF SINGULARITY FUNCTIONS

Following the development in Chap. 7, we want to enlarge our scope of discussion to include singularity functions in our consideration of time signals.

We shall define the Laplace transforms of singularity functions operationally. A rigorous mathematical discussion on their derivations is beyond the scope of this book. According to Eq. (1-9), the Laplace transform of the unit impulse is

$$\int_{-\infty}^{\infty} u_0(t)e^{-st}\, dt = 1$$

Since

$$\int_{-\infty}^{\infty} |u_0(t)e^{-st}|\, dt = \int_{-\infty}^{\infty} u_0(t)e^{-\sigma_0 t}\, dt = 1$$

for all σ_0, the region of absolute convergence of $u_0(t)$ is the entire s plane. It follows that† the Laplace transform of $u_n(t)$, $n > 0$, is s^n, with the entire s plane being the region of absolute convergence.

The Laplace transform of the unit step is

$$\int_{-\infty}^{\infty} u_{-1}(t)e^{-st}\, dt = \frac{1}{s}$$

† See Table 9-1.

with its region of absolute convergence being Re $(s) > 0$. In general, the Laplace transform of $u_{-n}(t)$, $n > 0$, is $1/s^n$ with the region of absolute convergence being Re $(s) > 0$.

9-6 RELATIONSHIP BETWEEN z TRANSFORMATION AND LAPLACE TRANSFORMATION

The theory of transformation of signals can be better understood by establishing a connection between the z transformation of discrete signals and the Laplace transformation of continuous signals. For a given discrete signal $x(n)$, let us define a *continuous* signal $x(t)$ such that

$$x(t) = \sum_{n=-\infty}^{\infty} x(n)u_0(t-n)$$

Let $X(s)$ denote the bilateral Laplace transform of $x(t)$. That is,

$$\begin{aligned} X(s) &= \int_{-\infty}^{\infty} x(t)e^{-st}\,dt \\ &= \sum_{n=-\infty}^{\infty} x(n)e^{-sn} \end{aligned} \tag{9-7}$$

Also, consider the following line integral which is similar to the inverse Laplace transformation integral:

$$\begin{aligned} \frac{1}{2\pi j}\int_{\sigma_0-j\pi}^{\sigma_0+j\pi} X(s)e^{sn}\,ds &= \frac{1}{2\pi j}\int_{\sigma_0-j\pi}^{\sigma_0+j\pi}\left[\sum_{k=-\infty}^{\infty} x(k)e^{-sk}\right]e^{sn}\,ds \\ &= \sum_{k=-\infty}^{\infty}\frac{1}{2\pi j}\int_{\sigma_0-j\pi}^{\sigma_0+j\pi} x(k)e^{s(n-k)}\,ds \end{aligned} \tag{9-8}$$

Because

$$\int_{\sigma_0-j\pi}^{\sigma_0-j\pi} e^{s(n-k)}\,ds = \begin{cases} 0 & n \neq k \\ 2\pi j & n = k \end{cases}$$

Eq. (9-8) can be simplified to

$$\frac{1}{2\pi j}\int_{\sigma_0-j\pi}^{\sigma_0+j\pi} X(s)e^{sn}\,ds = x(n) \tag{9-9}$$

Thus we can view $X(s)$ as a transformation of the signal $x(n)$, and Eq. (9-9) is the corresponding inverse transformation formula.

Let

$$s_0 = \sigma_0 + j\omega_0$$

and
$$s_1 = s_0 + j2\pi = \sigma_0 + j(\omega_0 + 2\pi)$$

We note that

$$X(s_1) = \sum_{n=-\infty}^{\infty} x(n)e^{[\sigma_0 + j(\omega_0 + 2\pi)]n} = \sum_{n=-\infty}^{\infty} x(n)e^{\sigma_0 + j\omega_0} = X(s_0)$$

Indeed, if we divide the s plane into horizontal strips of width 2π as illustrated in Fig. 9-7*a*, the behavior of $X(s)$ in every strip $(2k-1)\pi \le \omega \le (2k+1)\pi$, $k = \pm 1, \pm 2, \ldots$, is identical to that of $X(s)$ in the strip $-\pi \le \omega \le \pi$. Such an observation suggests a change of variables $z = e^s$ which maps each of the horizontal strips into the z plane. Thus, Eq. (9-7) becomes

$$X(z) = \sum_{n=-\infty}^{\infty} x(n)z^{-n}$$

which is the definition of the z transform of $x(n)$. Since the change of variable $z = e^s$ maps the line segment AB in the s plane into a circle of radius e^{σ_0} in the z plane, as shown in Fig. 9-7*b*, the integral in Eq. (9-9) becomes

$$x(n) = \frac{1}{2\pi j}\oint_C X(z)z^{n-1}\,dz$$

which is the inverse z transformation formula in Eq. (5-7).

Since the vertical line Re $(s) = \sigma_0$ in the s plane is also mapped into the circle of radius e^{σ_0} in the z plane, the half plane to the left of the vertical line Re $(s) = \sigma_0$ is mapped into the interior of the circle of radius e^{σ_0}, and the half plane to the right of the vertical line Re $(s) = \sigma_0$ is mapped into the exterior of the circle of radius e^{σ_0}. Consequently, a vertical strip in the s plane is mapped into an annulus in the z plane, which is another way to explain why the region of absolute convergence of a discrete signal is, in general, an annulus in the z plane. Note that, specifically, the half plane to the left of the $j\omega$ axis in the s plane is mapped into the interior of the unit circle, and the half plane to the right of the $j\omega$ axis is mapped into the exterior of the unit circle in the z plane.

What we wish to point out is the conceptual resemblance between the Laplace transformation of continuous signals and the z transformation of discrete signals. Indeed, many of the results concerning Laplace transformation are also valid for z transformation, and conversely. From this point on, many of our discussions will be either in terms of Laplace transformation or in terms of z transformation. A parallel development can easily be supplied by the reader.

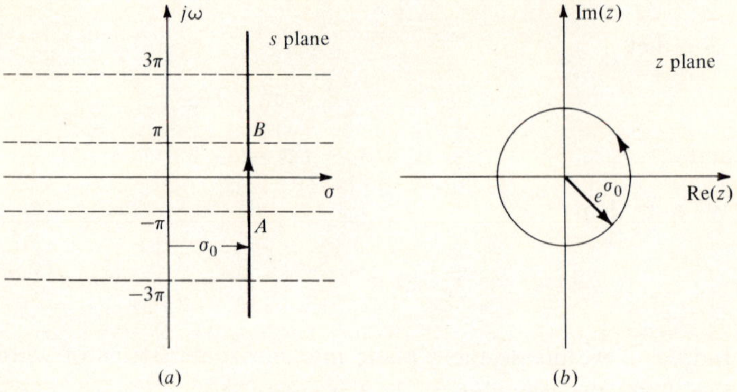

FIGURE 9-7

9-7 THE CONVOLUTION PROPERTY

Let $h(t)$ be the impulse response of a linear time-invariant system. Let $x(t)$ be an input signal and $y(t)$ its corresponding output signal. According to Eq. (9-4), $x(t)$ can be decomposed as an integral of complex exponential functions $e^{(\sigma_0+j\omega)t}$ for some σ_0 within the region of absolute convergence of $x(t)$. That is,

$$x(t) = \frac{1}{2\pi j}\int_{\sigma_0 - j\infty}^{\sigma_0 + j\infty} X(s)e^{st}\, ds$$

If σ_0 is also within the region of absolute convergence of $h(t)$, the complex exponential functions $e^{(\sigma_0+j\omega)t}$ are eigenfunctions of the system. Therefore, we have

$$y(t) = \frac{1}{2\pi j}\int_{\sigma_0 - j\infty}^{\sigma_0 + j\infty}\left[\int_{-\infty}^{\infty} e^{-st}h(t)\, dt\right] X(s)e^{st}\, ds \tag{9-10}$$

It follows from Eq. (9-10) that

$$Y(s) = X(s)H(s) \tag{9-11}$$

Moreover, the region of absolute convergence of $y(t)$ is at least the overlapping area of the regions of absolute convergence of $x(t)$ and $h(t)$. Again, $H(s)$ is called the *transfer function* or the *system function* of the system.†

Let us consider a linear time-invariant system whose impulse response is

$$h(t) = e^{-t}u_{-1}(t)$$

† The notion of eigenfunctions of discrete systems and the derivation of the result $Y(z) = X(z)H(z)$ for discrete systems, using the concept of eigenfunctions, are discussed in Prob. 9-8.

For the input signal

$$x_1(t) = \begin{cases} e^{-2t} & t > 0 \\ e^{3t} & t < 0 \end{cases}$$

we want to determine its corresponding output signal $y_1(t)$. Since

$$H(s) = \frac{1}{s+1}$$

with its region of absolute convergence being Re $(s) > -1$, and

$$X_1(s) = \frac{-5}{(s+2)(s-3)}$$

with its region of absolute convergence being $-2 <$ Re $(s) < 3$, we obtain

$$Y_1(s) = \frac{-5}{(s+1)(s+2)(s-3)}$$

with its region of absolute convergence being $-1 <$ Re $(s) < 3$. It follows that

$$y_1(t) = \begin{cases} \frac{5}{4}e^{-t} - e^{-2t} & t > 0 \\ \frac{1}{4}e^{3t} & t < 0 \end{cases}$$

One may ask whether the method of Fourier transformation can also be used to solve this problem. The answer is affirmative because the regions of absolute convergence of $h(t)$, $x_1(t)$, and $y_1(t)$ all include the $j\omega$ axis. Indeed, the reader can immediately check that the Fourier transforms of $h(t)$, $x(t)$, and $y(t)$ are

$$H(\omega) = \frac{1}{j\omega + 1}$$

$$X_1(\omega) = \frac{-5}{(j\omega + 2)(j\omega + 3)}$$

$$Y_1(\omega) = H(\omega)X(\omega) = \frac{-5}{(j\omega + 1)(j\omega + 2)(j\omega - 3)}$$

Suppose, however, the input signal is

$$x_2(t) = \begin{cases} e^{2t} & t > 0 \\ e^{3t} & t < 0 \end{cases}$$

we have

$$X_2(s) = \frac{-1}{(s-2)(s-3)}$$

with its region of absolute convergence being $2 < \text{Re}(s) < 3$. Hence, the Laplace transform of the output signal $y_2(t)$ is

$$Y_2(s) = \frac{-1}{(s+1)(s-2)(s-3)}$$

with its region of absolute convergence being $2 < \text{Re}(s) < 3$. It follows that

$$y_2(t) = \begin{cases} -\frac{1}{12}e^{-t} + \frac{1}{3}e^{2t} & t > 0 \\ \frac{1}{4}e^{3t} & t < 0 \end{cases}$$

Note that in this case, because the Fourier transform of $x_2(t)$ does not exist, the method of Fourier transformation is not applicable. Suppose the input signal is

$$x_3(t) = \begin{cases} 0 & t > 0 \\ -e^{-2t} + e^{3t} & t < 0 \end{cases}$$

Then
$$X_3(s) = \frac{-5}{(s+2)(s-3)}$$

with its region of absolute convergence being $\text{Re}(s) < -2$. Note that although the expressions $X_1(s)$ and $X_3(s)$ are identical, because the regions of absolute convergence of $H(s)$ and $X_3(s)$ do not overlap, the corresponding output signal $y_3(t)$ is not defined. Indeed, by convolving $x_3(t)$ with $h(t)$ we can show that $y_3(t)$ is equal to infinity at all t.

We recall our discussion in Sec. 4-5: Given a pair of input and output signals of a linear time-invariant system, $x(t)$ and $y(t)$, we can determine the impulse response of the system $h(t)$ by solving the integral equation:

$$y(t) = \int_{-\infty}^{\infty} x(\tau)h(t-\tau)\,d\tau \tag{9-12}$$

Although some examples illustrating how Eq. (9-12) can be solved for $h(t)$ were given in Sec. 4-5, it was also pointed out that the method of solution used in those examples is limited only to simple cases. Using the Fourier transform representation of signals, we have shown in Sec. 7-4 that Eq. (9-12) can be solved by computing the Fourier transform of $h(t)$:

$$H(\omega) = \frac{Y(\omega)}{X(\omega)} \tag{9-13}$$

According to Eq. (9-11), we can also solve the integral equation (9-12) by computing the Laplace transform of $h(t)$:

$$H(s) = \frac{Y(s)}{X(s)} \tag{9-14}$$

One might wonder in computing the impulse response $h(t)$ whether there is any difference between using the Fourier transform representation of signals as in Eq. (9-13) and using the Laplace transform representation of signals as in Eq. (9-14). In general, the solution of the integral equation (9-12) is not unique. Computing $H(\omega)$ according to Eq. (9-13) will yield only one of the solutions, whereas computing $H(s)$ according to Eq. (9-14) will yield all possible solutions. Let us consider the following illustrative example.

Let

$$x(t) = u_0(t) - 4e^{-2t}u_{-1}(t)$$

and

$$y(t) = e^{-t}u_{-1}(t)$$

be the input and output signals of a linear time-invariant system. We have

$$X(s) = 1 - \frac{4}{s+2} = \frac{s-2}{s+2}$$

with its region of absolute convergence being $-2 < \mathrm{Re}\,(s)$. It follows that the transfer function of the system is

$$H(s) = \frac{Y(s)}{X(s)} = \frac{s+2}{(s+1)(s-2)}$$

To determine $h(t)$ from $H(s)$, we must know its region of absolute convergence. Because $H(s)$ has a pole at $s = -1$ and another pole at $s = 2$, the region of absolute convergence of $h(t)$ might be any one of the three shown in Fig. 9-8*a*, *b*, and *c*. However, since the region of absolute convergence of $y(t)$ is at least equal to the overlapping area between that of $x(t)$ and $h(t)$, the region of absolute convergence in Fig. 9-8*a* is not a possible one. Corresponding to the region of absolute convergence in Fig. 9-8*b*, we have

$$h(t) = \begin{cases} -\frac{1}{3}e^{-t} & t > 0 \\ -\frac{4}{3}e^{2t} & t < 0 \end{cases} \tag{9-15}$$

Corresponding to the region of absolute convergence in Fig. 9-8*c*, we have

$$h(t) = \begin{cases} -\frac{1}{3}e^{-t} + \frac{4}{3}e^{2t} & t > 0 \\ 0 & t < 0 \end{cases} \tag{9-16}$$

Indeed, this example shows that the solution of the integral equation (9-12) is not unique. We leave it to the reader to check that $y(t) = x(t) * h(t)$ for both $h(t)$ found above. However, if the linear time-invariant system is known to be stable, its impulse response is that in Eq. (9-15). Similarly, if the system is known to be causal, its impulse response is that in Eq. (9-16).

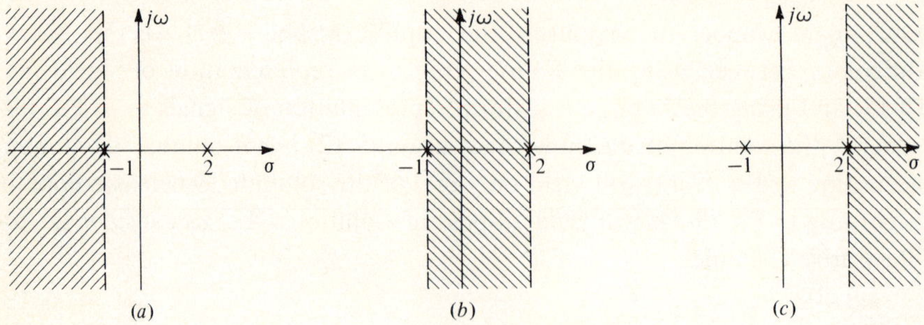

FIGURE 9-8

If we solve the same problem by the method of Fourier transformation, we have

$$X(\omega) = 1 - \frac{4}{2+j\omega} = \frac{-2+j\omega}{2+j\omega}$$

$$Y(\omega) = \frac{1}{1+j\omega}$$

Thus, we obtain

$$H(\omega) = \frac{2+j\omega}{(1+j\omega)(-2+j\omega)}$$

and

$$h(t) = \begin{cases} -\frac{1}{3}e^{-t} & t > 0 \\ -\frac{4}{3}e^{2t} & t < 0 \end{cases}$$

which is the result in Eq. (9-15). The reason that we obtain only this result but not the result in Eq. (9-16) is quite clear. When we represent signals by their Fourier transforms, we have tacitly assumed that the regions of absolute convergence of these signals include the $j\omega$ axis. Consequently, we miss the other solution of $h(t)$ in Eq. (9-16) whose region of absolute convergence does not include the $j\omega$ axis.

From this example as well as our discussion in Sec. 4-5, we observe that the uniqueness of $h(t)$ depends on the uniqueness of the region of absolute convergence of $H(s)$. Note that the poles of $Y(s)$ and the zeros of $X(s)$ are the poles of $H(s)$. In other words, the poles of $Y(s)$ and the zeros of $X(s)$ divide the complex plane into vertical strips which are potentially regions of absolute convergence of $H(s)$. Since the region of absolute convergence of $Y(s)$ is at least the intersection of that of $X(s)$ and $H(s)$, a vertical strip is a region of absolute convergence of $H(s)$ if it is within both the region of absolute convergence of $Y(s)$ and that of $X(s)$. Consequently, $H(s)$ can have two or more possible regions of absolute convergence only if one or more of the zeros of $X(s)$ are within the region of absolute convergence of $X(s)$.† In the preceding example, because of the zero $s = 2$ in $X(s)$, $H(s)$ has two possible regions of absolute convergence, as shown in Fig. 9-8*b* and *c*. Therefore, the impulse response $h(t)$ is not unique.

There is an entirely parallel discussion for the determination of the unit response of a discrete system from a given pair of input and output signals of the system. It suffices that we limit ourselves to an illustrative example. Let

$$x(n) = \begin{cases} 1 & n = 0 \\ -1 & n = 1 \\ 0 & \text{otherwise} \end{cases}$$

and

$$y(n) = \begin{cases} 1 & n = 0 \\ 0 & \text{otherwise} \end{cases}$$

be a pair of input and output signals of a linear time-invariant discrete system. To determine the unit response of the system, we note that

$$X(z) = 1 - z^{-1}$$
$$Y(z) = 1$$

with both of their regions of absolute convergence being the entire z plane. Thus,

$$H(z) = \frac{Y(z)}{X(z)} = \frac{1}{1 - z^{-1}}$$

Since $H(z)$ has a pole at $z = 1$, its region of absolute convergence is either $|z| > 1$ or $|z| < 1$. In this case, both of these regions are possible regions of absolute convergence because the region of absolute convergence of $Y(z)$ includes the intersection of that of $X(z)$ and either one of the two regions $|z| > 1$ and $|z| < 1$. Therefore, the unit response of the system is not unique. Corresponding to the region of absolute convergence $|z| > 1$,

$$h(n) = \begin{cases} 1 & n \geq 0 \\ 0 & n < 0 \end{cases}$$

Corresponding to the region of absolute convergence $|z| < 1$,

$$h(n) = \begin{cases} 0 & n \geq 0 \\ -1 & n < 0 \end{cases}$$

† We can imagine that these zeros divide the region of absolute convergence of $X(s)$ into vertical strips. Each of these strips that intersects with the region of convergence of $Y(s)$ will be a possible region of absolute convergence of $H(s)$.

9-8 TRANSFER FUNCTIONS OF SYSTEMS DESCRIBED BY DIFFERENTIAL EQUATIONS

Consider a system whose input-output relationship is described by a linear differential equation with constant coefficients. According to our discussion in Sec. 4-7, the transfer function of the system can be determined by solving the differential equation for the impulse response of the system. However, a more direct computation is possible, as we shall show below. Let

$$C_0 \frac{d^r y(t)}{dt^r} + C_1 \frac{d^{r-1} y(t)}{dt^{r-1}} + \cdots + C_{r-1} \frac{dy(t)}{dt} + C_r y(t)$$

$$= E_0 \frac{d^m x(t)}{dt^m} + E_1 \frac{d^{m-1} x(t)}{dt^{m-1}} + \cdots + E_{m-1} \frac{dx(t)}{dt} + E_m x(t) \qquad (9\text{-}17)$$

be a differential equation describing the input-output relationship of a continuous system. Recalling that the Laplace transform of $d^k x(t)/dt^k$ is $s^k X(s)$, we obtain from Eq. (9-17)

$$(C_0 s^r + C_1 s^{r-1} + \cdots + C_{r-1} s + C_r) Y(s) = (E_0 s^m + E_1 s^{m-1} + \cdots + E_{m-1} s + E_m) X(s)$$

or

$$H(s) = \frac{Y(s)}{X(s)} = \frac{E_0 s^m + E_1 s^{m-1} + \cdots + E_{m-1} s + E_m}{C_0 s^r + C_1 s^{r-1} + \cdots + C_{r-1} s + C_r} \qquad (9\text{-}18)$$

Moreover the region of absolute convergence of $H(s)$ is a right half plane.† For example, given that

$$\frac{d^2 y(t)}{dt^2} + 3 \frac{dy(t)}{dt} + 2y(t) = 2 \frac{dx(t)}{dt} + x(t) \qquad (9\text{-}19)$$

we obtain

$$H(s) = \frac{2s + 1}{s^2 + 3s + 2} \qquad (9\text{-}20)$$

with its region of absolute convergence being Re $(s) > -1$. Note that the transfer function of a system described by a linear differential equation with constant coeffi-

† We recall that the impulse response of a system is the response of the system to an impulse when the system is initially at rest. According to our discussion in Sec, 2-5, a system whose input-output relationship can be described by a linear differential equation with constant coefficients is causal if it is initially at rest.

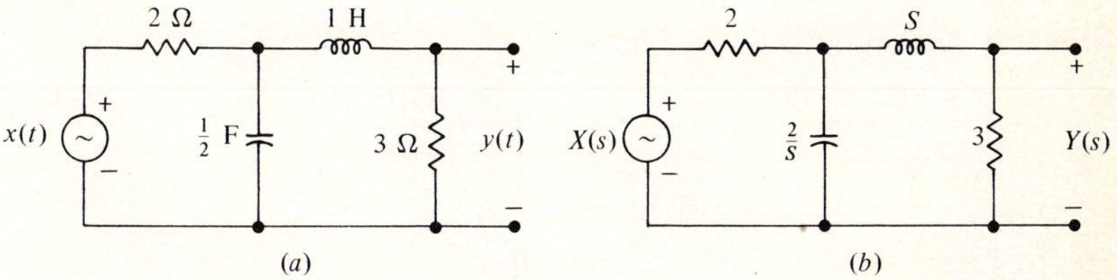

FIGURE 9-9

cients is always a rational function of s. Moreover, the poles of $H(s)$ are the characteristic roots of the differential equation.

As an example, let us consider the electric circuit shown in Fig. 9-9a. The input-output relationship of the circuit is described by the differential equation

$$\frac{d^2y}{dt^2} + 4\frac{dy}{dt} + 5y(t) = 3x(t)$$

Hence the transfer function of the system is

$$H(s) = \frac{Y(s)}{X(s)} = \frac{3}{s^2 + 4s + 5}$$

$$= \frac{3}{(s+2-j)(s+2+j)} \qquad \text{(9-21)}$$

with its region of absolute convergence being Re $(s) > -2$. The reader may recall from a course in circuit theory that the transfer function of the system can also be obtained directly from the circuit shown in Fig. 9-9b, where the impedance of an inductor with inductance L is equal to sL and the impedance of a capacitor with capacitance C is equal to $1/sC$. That is, we have

$$2\left[\frac{1}{3}\,Y(s) + \frac{1}{3}\,Y(s)(s+3)\,\frac{s}{2}\right] + \frac{1}{3}\,Y(s)(s+3) = 3X(s) \qquad \text{(9-22)}$$

which simplifies to Eq. (9-21). Such a computational procedure can be justified easily in terms of Laplace transformation of signals. Note that the current in a capacitor, $i(t)$, and the voltage across the capacitor, $v(t)$, are related by

$$i(t) = C\frac{dv(t)}{dt}$$

that is,

$$\frac{V(s)}{I(s)} = \frac{1}{sC}$$

Similarly, for an inductor, we have

$$v(t) = L\frac{di(t)}{dt}$$

or

$$\frac{V(s)}{I(s)} = sL$$

Consequently, the computation in Eq. (9-22) follows.

Equation (9-18) immediately suggests the possibility of solving linear differential equations with constant coefficients by the method of Laplace transformation. Specifically, writing Eq. (9-18) as

$$Y(s) = \frac{E_0 s^m + E_1 s^{m-1} + \cdots + E_{m-1}s + E_m}{C_0 s^r + C_1 s^{r-1} + \cdots + C_{r-1}s + C_r} X(s)$$

we see how $y(t)$ can be determined for a given $x(t)$. It should be pointed out, however, such a method of solution is applicable only when $x(t)$ is specified for $-\infty \leq t \leq \infty$ and when the system is initially at rest. In the case that $x(t)$ is specified only for $t \geq t_0$, the method of unilateral Laplace transformation is useful. We shall study such a method of solution in Chap. 11.

Our discussion also suggests an alternative method for determining the differential equation description of a system from the impulse response of the system. If the Laplace transform of the impulse response is a rational function of s such as that in Eq. (9-18), a differential equation description of the input-output relationship such as that in Eq. (9-17) can be obtained directly from Eq. (9-18). For example, let

$$h(t) = (e^{-t} - e^{-6t})u_{-1}(t)$$

be the impulse response of a linear time-invariant system. Since

$$H(s) = \frac{5}{s^2 + 6s + 7}$$

we obtain

$$\frac{d^2y(t)}{dt^2} + 6\frac{dy(t)}{dt} + 7y(t) = 5x(t)$$

as a differential equation description of the system.

We are now ready to complete a discussion in Sec. 4-7 where we pointed out that the differential equation description of a system is not uniquely determined by its impulse response. Consider the transfer function in Eq. (9-20) which can be written, in an obviously redundant manner,

$$H(s) = \frac{(2s + 1)(s - 1)}{(s^2 + 3s + 2)(s - 1)} = \frac{2s^2 - s - 1}{s^3 + 2s^2 - s - 2}$$

Therefore, we conclude that the impulse response of the differential equation

$$\frac{d^3y(t)}{dt^3} + 2\frac{d^2y(t)}{dt^2} - \frac{dy(t)}{dt} - 2y(t) = 2\frac{d^2x(t)}{dt^2} - \frac{dx(t)}{dt} - x(t)$$

is the same as that of Eq. (9-19). Indeed, for any transfer function $H(s)$ that is a rational function of s,

$$H(s) = \frac{N(s)}{D(s)}$$

where $N(s)$ and $D(s)$ are polynomials of s, we can write

$$H(s) = \frac{N(s)P(s)}{D(s)P(s)} \tag{9-23}$$

for an arbitrary polynomial $P(s)$. From Eq. (9-23) a differential equation description of a system whose transfer function is equal to $H(s)$ can be determined immediately.

Similarly, let

$$\begin{aligned} C_0 y(n) + C_1 y(n-1) + \cdots + C_{r-1}y(n-r+1) + C_r y(n-r) \\ = E_0 x(n) + E_1 x(n-1) + \cdots + E_{m-1}x(n-m+1) + E_m x(n-m) \end{aligned}$$

be a difference equation describing the input-output relationship of a discrete system. It follows that the transfer function of the system is

$$\begin{aligned} H(z) &= \frac{E_0 + E_1 z^{-1} + \cdots + E_{m-1}z^{-m+1} + E_m z^{-m}}{C_0 + C_1 z^{-1} + \cdots + C_{r-1}z^{-r+1} + C_r z^{-r}} \\ &= \frac{z^{r-m}(E_0 z^m + E_1 z^{m-1} + \cdots + E_{m-1}z + E_m)}{C_0 z^r + C_1 z^{r-1} + \cdots + C_{r-1}z + C_r} \end{aligned}$$

with its region of absolute convergence being the exterior of a circle of finite radius in the z plane.†

As an example, let

$$y(n) + 3y(n-1) + 2y(n-2) = 2x(n) + 3x(n-1)$$

be a difference equation relating the input and output signals of a discrete system that is initially at rest. The transfer function of the system is

$$H(z) = \frac{2 + 3z^{-1}}{1 + 3z^{-1} + 2z^{-2}} = \frac{1}{1 + 2z^{-1}} + \frac{1}{1 + z^{-1}}$$

† Again, because the system is causal if it is initially at rest.

Because $H(z)$ has two poles at $z = -2$ and $z = -1$, the region of absolute convergence of $H(z)$ is $|z| > 2$. It follows that the unit response of the system is

$$h(n) = \begin{cases} (-2)^n + (-1)^n & n \geq 0 \\ 0 & n < 0 \end{cases}$$

Suppose we are given that

$$x(n) = \begin{cases} 3^n & n \geq 0 \\ 0 & n < 0 \end{cases}$$

Since

$$X(z) = \frac{1}{1 - 3z^{-1}}$$

with its region of absolute convergence being $|z| > 3$, we obtain

$$Y(z) = X(z)H(z) = \frac{2 + 3z^{-1}}{(1 - 3z^{-1})(1 + 2z^{-1})(1 + z^{-1})}$$

with its region of absolute convergence being $|z| > 3$. Writing $Y(z)$ as

$$Y(x) = \frac{\frac{27}{20}}{1 - 3z^{-1}} + \frac{\frac{2}{5}}{1 + 2z^{-1}} + \frac{\frac{1}{4}}{1 + z^{-1}}$$

we obtain

$$y(n) = \begin{cases} \frac{27}{20}(3)^n + \frac{2}{5}(-2)^n + \frac{1}{4}(-1)^n & n \geq 0 \\ 0 & n < 0 \end{cases}$$

9-9 REMARKS AND REFERENCES

We have now completed our development of a transformation theory for continuous signals. Furthermore, we also observed a close relationship between the method of z transformation for discrete signals and the method of Laplace transformation for continuous signals. Indeed, as we have seen in Secs. 9-7 and 9-8, many of the results for continuous signals and systems in terms of Laplace transformation can be derived parallelly for discrete signals and systems in terms of z transformation, and conversely. We shall continue to see such parallel development in the next two chapters. Besides the general references cited in Chap. 1, we also recommend Doetsch [1]. Widder [3] is an important reference on Laplace transformation. It is, however, too advanced for our level of presentation.

The method of Laplace transformation is a useful tool not only in system analysis problems but also in system synthesis problems. Very often an engineering specification of a system is given in terms of its transfer function instead of its impulse repsonse mainly because the transfer function exhibits clearly the frequency response characteristic of the system. Techniques for designing systems from their transfer functions have been studied extensively. See, for example, Guillemin [2].

[1] DOETSCH, G.: "Guide to the Application of the Laplace and z-transforms," 2d ed., Van Nostrand Reinhold Company, New York, 1971.

[2] GUILLEMIN, E. A.: "Synthesis of Passive Networks," John Wiley & Sons, Inc., New York, 1957.

[3] WIDDER, D. V.: "Laplace Transform," Princeton University Press, Princeton, N.J., 1946.

PROBLEMS

9-1 Determine the Laplace transform and the region of absolute convergence for each of the following time signals.

(*a*) $x(t) = |\sin \alpha t| u_{-1}(t)$

(*b*) $x(t) = e^{-\alpha|t|} \cos \alpha t$

(*c*) $x(t) = t \cos \alpha t u_{-1}(t)$

(*d*) $x(t) = e^{-\alpha t^2}$

(*e*) $x(t) = \sinh \alpha t u_{-1}(t)$

(*f*) $x(t) = \sin \alpha t \sinh \alpha t u_{-1}(t)$

(*g*) $x(t) = t^2 e^{-|t|}$

(*h*) $x(t) = \dfrac{\sin \alpha(t - \pi)}{(t - \pi)} u_{-1}(t - \pi)$

(*i*) $x(t) = (4t^2 - \sin t) u_{-1}(t)$

(*j*) $x(t) = \sqrt{t}\, u_{-1}(t)$

(*k*) $x(t) = \begin{cases} e^{-2t} & t \geq 0 \\ e^{t} & t \leq 0 \end{cases}$

9-2 Determine the Laplace transform and the region of absolute convergence for each of the time signals shown in Fig. 9P-1.

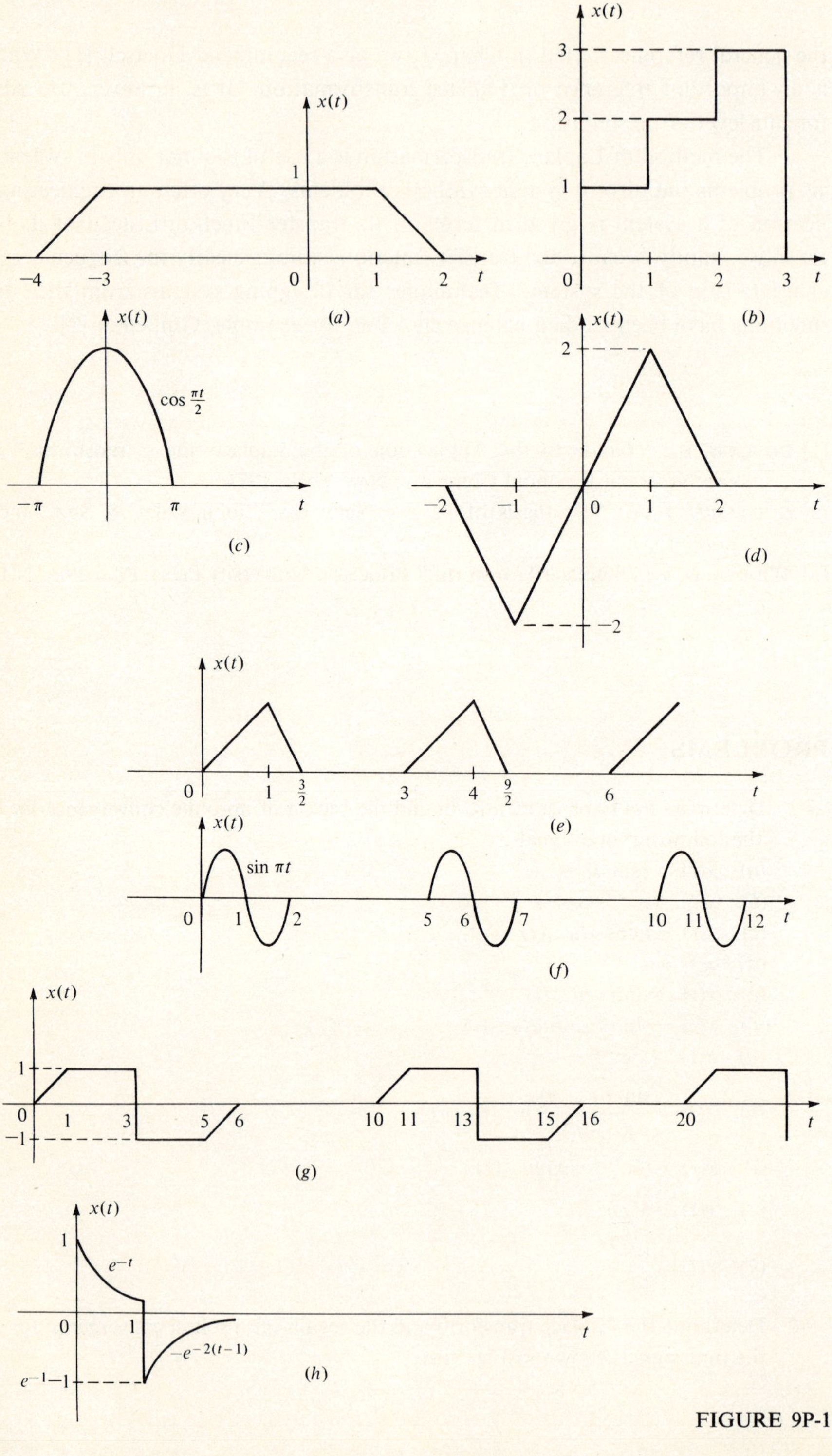
x(t)
1
−4
−3
0
1
2
t
(a)
x(t)
3
2
1
0
1
2
3
t
(b)
x(t)
cos πt/2
−π
π
t
(c)
x(t)
2
−2
−1
0
1
2
t
−2
(d)
x(t)
0
1
3/2
3
4
9/2
6
t
(e)
x(t)
sin πt
0
1
2
5
6
7
10
11
12
t
(f)
x(t)
1
0
1
3
5
6
10
11
13
15
16
20
t
−1
(g)
x(t)
1
e^{−t}
0
1
t
−e^{−2(t−1)}
e^{−1}−1
(h)

FIGURE 9P-1

9-3 Determine the time signals corresponding to the following Laplace transforms:

(*a*) $X(s) = \dfrac{6s - 1}{(s+1)(s+2)(s-3)} \qquad -1 < \text{Re}(s) < 3$

(*b*) $X(s) = \dfrac{s^2 - 3s - 1}{(s+1)^3(s^2+4)} \qquad -1 < \text{Re}(s) < 0$

(*c*) $X(s) = \dfrac{2s}{s^2 + 4s + 5} \qquad -2 < \text{Re}(s)$

(*d*) $X(s) = \dfrac{s}{1 - e^s} \qquad \text{all } s$

(*e*) $X(s) = \dfrac{1 - e^{-(s+1)}}{(s+1)(1 - e^{-2s})} \qquad 0 < \text{Re}(s)$

(*f*) $X(s) = \log \dfrac{1 - s}{1 + s} \qquad -1 < \text{Re}(s) < 1$

(*g*) $X(s) = \dfrac{1}{(s+1)(s^2 + s + 1)} \qquad \text{Re}(s) < -1$

(*h*) $X(s) = \dfrac{1}{s \cosh s} \qquad 0 < \text{Re}(s)$

(*i*) $X(s) = \dfrac{e^{-3s}}{(s+1)(s+2)(s-3)^3} \qquad -1 < \text{Re}(s) < 3$

9-4 Let

$$X(s) = \frac{2}{s^2 - 4}$$

be the Laplace transform of a signal $x(t)$. It is known that

$$\int_{-\infty}^{t} x(\tau)\, d\tau = -\tfrac{1}{4}e^{2t} \qquad \text{for } t \leq 0$$

Determine $x(t)$.

9-5 Let $x(t)$ and $y(t)$ be two time signals whose Laplace transforms are $X(s)$ and $Y(s)$, respectively.

(*a*) Find the Laplace transform of the time signal

$$z(t) = x(t)y(t)$$

Express it in terms of $X(s)$ and $Y(s)$. Suppose that the regions of absolute convergence of $x(t)$ and $y(t)$ are $\sigma_1 < \text{Re}(s) < \sigma_2$ and $\delta_1 < \text{Re}(s) < \delta_2$, respectively. What is the region of absolute convergence of $z(t)$?

(*b*) Use the result in part (*a*) to find the Laplace transform of the time signal

$$z(t) = (1 + \sin t) \cos(\alpha t - \theta) u_{-1}(t)$$

9-6 Let the Laplace transform of a time signal $x(t)$ be

$$X(s) = \frac{2s}{s^2 + 1}$$

and its region of absolute convergence be Re $(s) > 0$. Find the Laplace transform and region of absolute convergence of the signal

$$y(t) = \int_{-\infty}^{t} x(4 - 5\tau)e^{-\tau}\, d\tau$$

9-7 Let $X(s)$ be the Laplace transform of a time signal $x(t)$ which is equal to zero for $t < 0$.

(*a*) Let

$$\bar{t} = \frac{\int_0^\infty t x(t)\, dt}{\int_0^\infty x(t)\, dt}$$

$$\delta^2 = \frac{\int_0^\infty (t - \bar{t})^2 x(t)\, dt}{\int_0^\infty x(t)\, dt}$$

Show that

$$\bar{t} = -\frac{1}{X(0)} \frac{dX(s)}{ds}\bigg|_{s=0}$$

$$\delta^2 = \frac{1}{X(0)} \frac{d^2 X(s)}{ds^2}\bigg|_{s=0} - \bar{t}^2$$

(*b*) Let $x(t)$ and $y(t)$ be two signals which equal zero for $t < 0$, and let $z(t) = x(t) * y(t)$. Find

$$\frac{\int_0^\infty t z(t)\, dt}{\int_0^\infty z(t)\, dt}$$

and

$$\frac{\int_0^\infty (t - \bar{t})^2 z(t)\, dt}{\int_0^\infty z(t)\, dt}$$

Express them in terms of $X(s)$ and $Y(s)$, the Laplace transforms of $x(t)$ and $y(t)$ respectively.

9-8 Let $x(n)$ and $y(n)$ be the input and output signals, respectively, of a discrete linear time-invariant system with unit response $h(n)$. Let $X(z)$, $Y(z)$, and $H(z)$ be the z transforms of $x(n)$, $y(n)$, and $h(n)$, respectively. Show that

$$Y(z) = X(z)H(z)$$

9-9 Let

$$x(t) = \begin{cases} e^{-2t} & t > 0 \\ 4e^{t} & t < 0 \end{cases}$$

$$h(t) = \begin{cases} e^{-3t} & t > 0 \\ -e^{-t} & t < 0 \end{cases}$$

Determine $x(t) * h(t)$ by the method of Laplace transformation.

9-10 Find the output voltage $y(t)$ of the circuit shown in Fig. 9P-2, when the input voltage is

$$x(t) = e^t u_{-1}(-t)$$

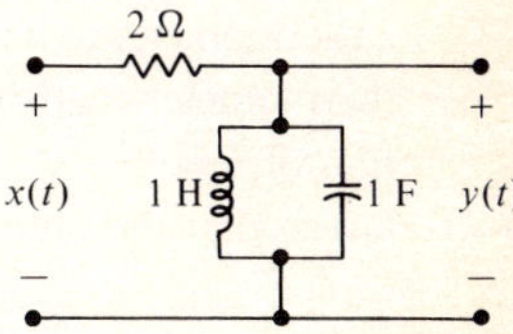

FIGURE 9P-2

9-11 Find the output voltage $y(t)$ of the circuit shown in Fig. 9P-3*a* when the input voltage $x(t)$ is as shown in Fig. 9P-3*b*.

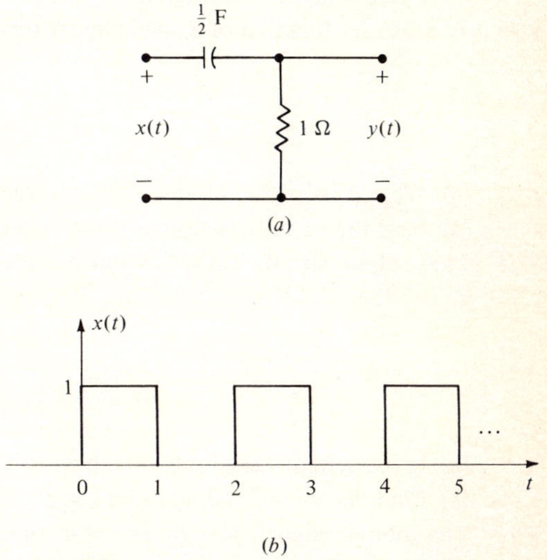

FIGURE 9P-3

9-12 The system function $H(s)$ of a linear time-invariant causal system is given by

$$H(S) = \frac{s+1}{s^2 + 2s + 2}$$

(*a*) Is the system stable?
(*b*) Determine the impulse response of the system.
(*c*) Find and sketch the output signal $y(t)$ when the input signal $x(t)$ is given by

$$x(t) = e^{-|t|} \qquad -\infty \leq t \leq \infty$$

9-13 An altimeter measures the altitude of a balloon every second. Suppose that the meter reads $n(1-e^{-n})\ u_{-1}(n)$ when the actual balloon altitude is $nu_{-1}(n)$. Find the actual balloon altitude as a function of time if the meter reads $[2-4(3^{-n})-n2^{-n+1}]u_{-1}(n)$.

9-14 The input signal of a thermometer is the temperature. Suppose that the response of a thermometer to a unit step is $(1-e^{-2t}-te^{-2t})u_{-1}(t)$. If, during a measurement, the thermometer reads $(1-2e^{-2t}+e^{-3t})u_{-1}(t)$, what is the true temperature as a function of time?

9-15 Given that the output signal $y(t)$ of a causal system with transfer function

$$H(s)=\frac{s(s-1)}{(s+2)(s+1)}$$

is

$$y(t)=e^{-2t}\sin tu_{-1}(t)$$

(*a*) Find all possible input signals $x(t)$ of the system.
(*b*) Determine the input signal $x(t)$, if it is known that $x(t)=0$ for $t<0$.

9-16 The transfer function of a stable linear time-invariant system is

$$H(s)=\frac{1}{s^4+1}$$

(*a*) What is its region of absolute convergence?
(*b*) Find the impulse response of the system.

9-17 The output signal $y(n)$ of a linear time-invariant system with transfer function $z/(1-2z)(1-e^{-1}z)$ is

$$y(n)=\begin{cases}1 & n>0\\ 2^{-n} & n\le 0\end{cases}$$

(*a*) Is the system causal? Is it stable?
(*b*) Find the corresponding input signal.

9-18 The output signal $y(t)$ of a linear time-invariant system with transfer function $1/(s^2-1)$ is

$$y(t)=\begin{cases}e^{-t} & t>0\\ 1 & t<0\end{cases}$$

(*a*) Is the system causal? Is it stable? Explain.
(*b*) Find the corresponding input signal.

9-19 Let $h(t)$ denote the impulse response of a causal linear time-invariant system whose input-output relationship can be described by a second-order differential equation. Let $y(t)=x(t)*h(t)$. Given that the denominator of $Y(s)$ is equal to $(s^2-1)(s^2-4)$ and $y(t)$ is an even time signal, determine $x(t)$ and $h(t)$.

9-20 Let $h(t)$ denote the impulse response of a causal and stable system. Let $x(t)$ and $y(t)$ be a pair of its input-output signals. It is known that

$$Y(s) = \frac{2s}{s^2 - 1}$$

(*a*) Determine $h(t)$, $x(t)$, and $y(t)$, given that $x(t) = 0$ for $t < 0$.

(*b*) Repeat part (*a*) if given that $x(t) = 0$ for $t > 0$.

9-21 (*a*) Find the differential equation relating the output signal $y(t)$ to the input signal $x(t)$ for the circuit shown in Fig. P9-4.

(*b*) Find the transfer function $H(s)$ of the circuit and identify its region of absolute convergence.

(*c*) Find the output signal $y(t)$ for the input signal $x(t) = e^{-2|t|}$.

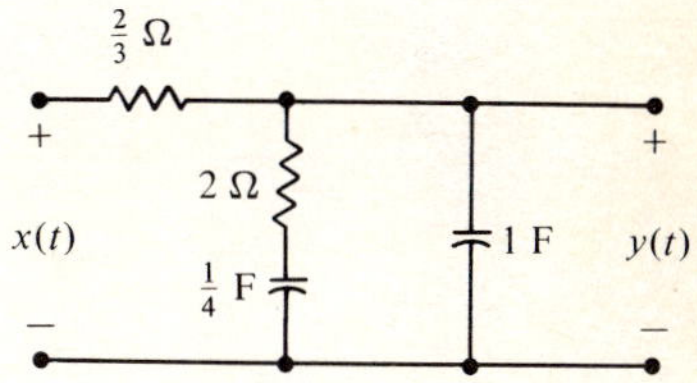

FIGURE 9P-4

9-22 The circuit shown in Fig. 9P-5 is a third-order Butterworth filter.

(*a*) Find the differential equation relating the input and output signals $x(t)$ and $y(t)$.

(*b*) Find the impulse response of the circuit by solving the differential equation in part (*a*).

(*c*) Find the transfer function and the impulse response of the circuit.

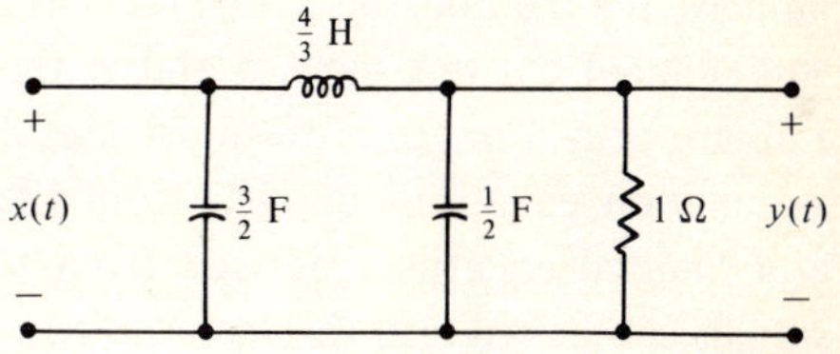

FIGURE 9P-5

10

INTERCONNECTION OF SYSTEMS

10-1 INTRODUCTION

In Chaps. 1 and 4, we discussed briefly how complex systems can be built up by interconnecting less complex systems. We are now ready to give the subject a more thorough treatment as we have developed the proper mathematical tools, namely, the methods of z transformation and bilateral Laplace transformation. As was pointed out in Chaps. 5 and 9, the methods of z transformation and bilateral Laplace transformation are most useful in many system analysis problems because summation equations derived from the convolution sum of discrete time signals and integral equations derived from the convolution integral of continuous time signals can be reduced to algebraic equations relating their z transforms or Laplace transforms. Such a reduction also makes it much easier to describe the behavior of complex interconnected systems, as we shall see in this chapter.

Undoubtedly, the reader has understood quite well by now the conceptual resemblance between the z transformation representation of discrete signals and the Laplace transformation representation of continuous signals. Our presentation will be mostly in terms of continuous systems; a straightforward parallel development for discrete systems will be left to the reader.

10-2 CASCADE AND PARALLEL CONNECTION OF SYSTEMS

We discussed in Sec. 1-8 the two basic ways to interconnect systems, namely, cascade connection and parallel connection. Furthermore, it was shown in Sec. 4-4 how the impulse responses of the overall systems for such interconnections can be obtained from those of the subsystems. We now extend our discussion to show how the transfer functions of the overall systems can be obtained from those of the subsystems.

For a cascade connection of two continuous systems as shown in Fig. 10-1*a*, since

$$Y_1(s) = X(s)H_1(s)$$

and

$$Y(s) = Y_1(s)H_2(s) = X(s)H_1(s)H_2(s)$$

the transfer function of the overall system, $H(s)$, is equal to $H_1(s)H_2(s)$. According to our discussion in Sec. 9-7, the region of absolute convergence of the overall system is at least the overlapping area of the regions of absolute convergence of the two subsystems. As an example, suppose that

$$H_1(s) = \frac{1}{s+1}$$

with its region of absolute convergence being $\text{Re}\,(s) < -1$ and

$$H_2(s) = \frac{1}{s+2}$$

with its region of absolute convergence being $\text{Re}\,(s) > -2$. The transfer function of the cascade connection of these two systems is

$$H(s) = \frac{1}{(s+1)(s+2)}$$

with its region of absolute convergence being $-2 < \text{Re}\,(s) < -1$.

Clearly, for a cascade connection of two discrete systems as shown in Fig. 10-1*b*, the transfer function of the overall system, $H(z)$, is equal to $H_1(z)H_2(z)$ with its region of absolute convergence being at least the overlapping area of that of $H_1(z)$ and $H_2(z)$.

According to the relation $H(s) = H_1(s)H_2(s)$, the poles of $H(s)$ are those of $H_1(s)$ and $H_2(s)$. However, if $H_1(s)$ has a zero which is also a pole of $H_2(s)$, this pole of $H_2(s)$ will be canceled and will not be a pole of $H(s)$. Consequently, the region of absolute convergence of $H(s)$ may include an area which is not in the over-

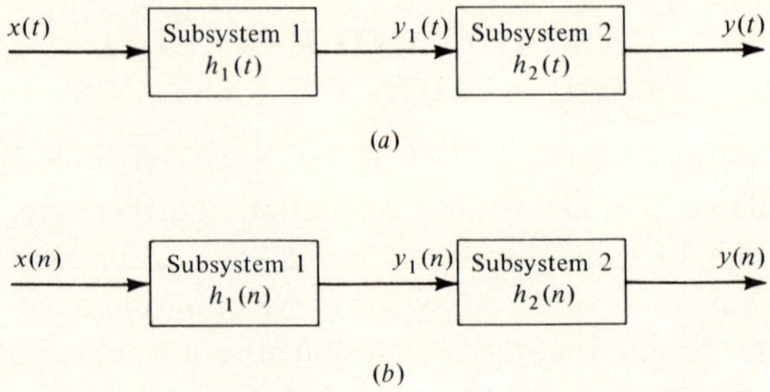

FIGURE 10-1

lapping area of the regions of absolute convergence of $H_1(s)$ and $H_2(s)$. For example, suppose that

$$H_1(s) = \frac{s-2}{s+2}$$

with its region of absolute convergence being Re $(s) > -2$ and

$$H_2(s) = \frac{1}{s-2}$$

with its region of absolute convergence being Re $(s) > 2$. The transfer function of the cascade system is

$$H(s) = \frac{1}{s+2}$$

The region of absolute convergence of the overall system is Re $(s) > -2$ which contains a strip $-2 <$ Re $(s) < 2$ that is not in the overlapping area of the regions of absolute convergence of $H_1(s)$ and $H_2(s)$.

A system is said to be an *inverse system* of another system if the cascade connection of the two systems yields an overall system whose output signal is always a replica of its input signal. For the case of linear time-invariant systems, a system is said to be an inverse system of another system if the convolution of their impulse responses is a unit impulse. That is, let $h_1(t)$ and $h_2(t)$ denote the impulse responses of the two systems; we have

$$h_1(t) * h_2(t) = u_0(t)\dagger \qquad (10\text{-}1)$$

† Since $h_1(t) * h_2(t) = h_2(t) * h_1(t)$, the first system is also an inverse system of the second system.

We can give a more physical interpretation of the notion of inverse systems: Suppose a signal $x(t)$ is to be transmitted from one place to another place. However, the signal $x(t)$ was sent through a system prior to transmission. That is, the transmitted signal is actually the output signal of this system. To recover the signal $x(t)$ from the transmitted signal, we need only to pass the transmitted signal through an inverse system of the first system. For example, a radio receiver is a (nonlinear) inverse system of the radio transmitter because it reproduces the sound signal from the modulated signal.

Let $H_1(s)$ and $H_2(s)$ denote the Laplace transforms of $h_1(t)$ and $h_2(t)$, respectively. According to Eq. (10-1),

$$H_1(s)H_2(s) = 1$$

That is,

$$H_2(s) = \frac{1}{H_1(s)}$$

As was pointed out in Sec. 9-7, the region of absolute convergence of $H_2(s)$ is not unique if one or more of the zeros of $H_1(s)$ lie within the region of absolute convergence of $H_1(s)$. That is, for a given $h_1(t)$ the solution of the integral equation (10-1) for $h_2(t)$ might not be unique. In other words, it is possible that a system might have more than one inverse system.

As an example, let

$$h_1(t) = u_0(t) - 4e^{-2t}u_{-1}(t)$$

be the impulse response of a linear time-invariant system. Thus,

$$H_1(s) = 1 - \frac{4}{s+2} = \frac{s-2}{s+2}$$

with its region of absolute convergence being Re $(s) > -2$. To determine the impulse response of its inverse system, we note that

$$H_2(s) = \frac{1}{H_1(s)} = \frac{s+2}{s-2} = 1 + \frac{4}{s-2}$$

The two possible regions of absolute convergence of $H_2(s)$ are Re $(s) > 2$ and Re $(s) < 2$ because both of them overlap with the region of absolute convergence of $H_1(s)$. The corresponding impulse responses are

$$h_2(t) = u_0(t) + 4e^{2t}u_{-1}(t)$$

and

$$h_2(t) = u_0(t) - 4e^{2t}u_{-1}(-t)$$

We leave it to the reader to check that, for both of these $h_2(t)$,

$$h_1(t) * h_2(t) = u_0(t)$$

As another example, let

$$h_1(n) = \begin{cases} (\frac{1}{2})^n & n \geq 0 \\ 0 & n < 0 \end{cases}$$

be the unit response of a linear time-invariant system. We have

$$H_1(z) = \frac{1}{1 - \frac{1}{2}z^{-1}} = \frac{2z}{2z - 1}$$

with its region of absolute convergence being $|z| > \frac{1}{2}$. The transfer function of its inverse system is

$$H_2(z) = \frac{1}{H_1(z)} = \frac{2z - 1}{2z} = 1 - \frac{1}{2}z^{-1}$$

Since the only possible region of absolute convergence of $H_2(z)$ is $|z| > 0$, the inverse system is unique with its unit response being

$$h_2(n) = \begin{cases} 1 & n = 0 \\ -\frac{1}{2} & n = 1 \\ 0 & \text{otherwise} \end{cases}$$

One can immediately check that

$$h_1(n) * h_2(n) = \begin{cases} 1 & n = 0 \\ 0 & \text{otherwise} \end{cases}$$

For a parallel connection of two continuous systems as shown in Fig. 10-2*a*, since

$$\begin{aligned} Y(s) &= Y_1(s) + Y_2(s) \\ &= X(s)H_1(s) + X(s)H_2(s) \\ &= X(s)[H_1(s) + H_2(s)] \end{aligned}$$

the transfer function of the overall system, $H(s)$, is equal to $H_1(s) + H_2(s)$. Moreover, the region of absolute convergence of the overall system is at least the intersection of that of subsystems 1 and 2. For example, let

$$H_1(s) = \frac{1}{s + 1}$$

with its region of absolute convergence being $\text{Re}(s) < -1$ and

$$H_2(s) = \frac{1}{s + 2}$$

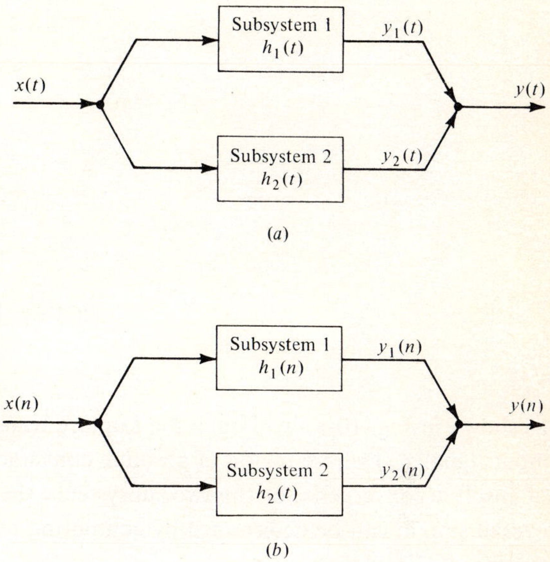

FIGURE 10-2

with its region of absolute convergence being Re $(s) > -2$. The transfer function of the parallel connection of these two systems is

$$H(s) = \frac{1}{s+1} + \frac{1}{s+2} = \frac{2s+3}{(s+1)(s+2)}$$

with its region of absolute convergence being $-2 < \text{Re}\,(s) < -1$.

Similarly, for a parallel connection of two discrete systems as shown in Fig. 10-2*b*, the transfer function of the overall system, $H(z)$, is equal to $H_1(z) + H_2(z)$ with its region of absolute convergence being at least the overlapping area of that of $H_1(z)$ and $H_2(z)$.

It should be pointed out, however, that in a parallel connection of two systems, even when the regions of absolute convergence of these systems do not overlap, the overall system might still have a well-defined impulse response. For example, let

$$h_1(t) = e^t u_{-1}(t)$$
$$h_2(t) = u_{-1}(-t)$$

The impulse response $h(t)$ of the parallel combination of these two systems is

$$h(t) = h_1(t) + h_2(t) = \begin{cases} e^t & t > 0 \\ 1 & t < 0 \end{cases}$$

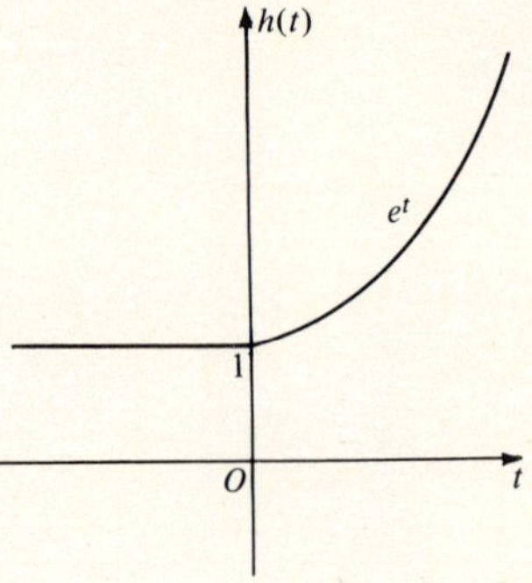

FIGURE 10-3

as shown in Fig. 10-3. Although the Laplace transform of $h(t)$ does not exist, for an input signal $x(t)$ whose region of absolute convergence overlaps with both the regions of absolute convergence of the two subsystems, the corresponding output signal of the overall system can be determined by computing the output signals of the subsystems separately. Thus, for

$$x(t) = \begin{cases} -\frac{1}{4}e^{-2t} & t > 0 \\ -\frac{1}{4}e^{2t} & t < 0 \end{cases}$$

we have

$$X(s) = \frac{1}{s^2 - 4}$$

with its region of absolute convergence being $-2 < \text{Re}(s) < 2$. The Laplace transform of the output signal of subsystem 1 is

$$Y_1(s) = X(s)H_1(s) = \frac{1}{s^2 - 4}\frac{1}{s - 1}$$

with its region of absolute convergence being $1 < \text{Re}(s) < 2$. It follows that

$$y_1(t) = \begin{cases} -\frac{1}{3}e^{t} + \frac{1}{12}e^{-2t} & t > 0 \\ -\frac{1}{4}e^{2t} & t < 0 \end{cases}$$

Similarly, the Laplace transform of the output signal of subsystem 2 is

$$Y_2(s) = X(s)H_2(s) = \frac{1}{s^2 - 4}\left(-\frac{1}{s}\right)$$

with its region of absolute convergence being $-2 < \text{Re}(s) < 0$. It follows that

$$y_2(t) = \begin{cases} -\frac{1}{8}e^{-2t} & t > 0 \\ -\frac{1}{4} + \frac{1}{8}e^{2t} & t < 0 \end{cases}$$

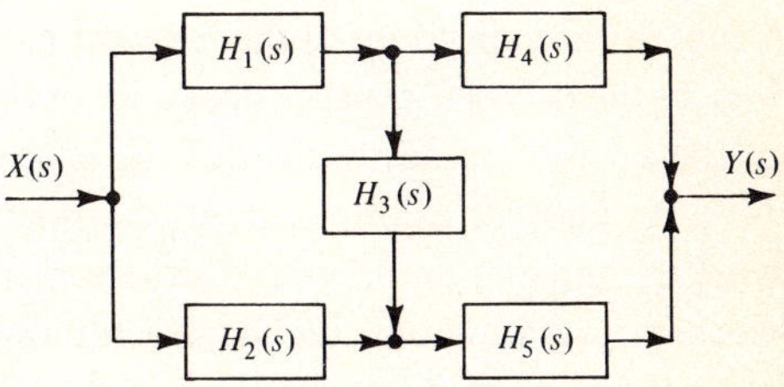

FIGURE 10-4

Consequently, the output signal of the overall system is

$$y(t) = y_1(t) + y_2(t) = \begin{cases} -\frac{1}{3}e^t - \frac{1}{24}e^{-2t} & t > 0 \\ -\frac{1}{4} - \frac{1}{8}e^{2t} & t < 0 \end{cases}$$

Although many complex systems can be built up by cascade and parallel connection of simple subsystems, it should be pointed out that there are systems made up of subsystems that are not connected in a cascade or parallel manner. Figure 10-4 shows one such connection. The reader can check immediately that the transfer function of the overall system is

$$H(s) = H_1(s)H_4(s) + [H_2(s) + H_1(s)H_3(s)]H_5(s)$$

10-3 FEEDBACK CONNECTION OF SYSTEMS

There is another important and interesting way to interconnect two systems; it is known as a *feedback connection* and is illustrated in Fig. 10-5*a* and *b*, where $x(t)$ and $y(t)$, $x(n)$ and $y(n)$, are the input and output signals of the overall system, respectively. Note that the output signal of the overall system $y(t)$ or $y(n)$ is fed back, through subsystem 2, to be part of the input signal applied to subsystem 1. Subsystem 1 is said to be in the *forward path*, and subsystem 2 is said to be in the *feedback path* of the overall system.

There are many examples of physical systems in which the output signal of a system is fed back as part of the input signal to the subsystem in the forward path. Consider how the speed of a car is adjusted by the driver of the car. When the driver steps on the gas pedal, the car attains a certain speed. We can imagine a predetermined position of the gas pedal as the input and the speed of the car as the output of the overall system. The engine of the car is the subsystem in the forward path which transduces the position of the gas pedal into the speed of the car. That the driver reads the speedometer and adjusts the position of his foot on the gas pedal accordingly amounts to feeding the output signal of the system back as part of the input signal to the subsystem in the forward path. Specifically, the eyes, the brain, and the foot of the driver form the subsystem in the feedback path which transduces the speed of the car into an adjustment of the position of the gas pedal.

As another example, let us consider the relationship between the migration of workers and the employment situation in a metropolitan area. Let the number of workers migrating into the area each month be the input and the monthly unemployment index be the output of a system describing the relationship between these two figures. We note that the system is a feedback connection of two subsystems. In the forward path, the migration of workers affects directly the employment situation. Specifically, when more workers move into the area the unemployment index will go up, and, conversely, when fewer workers move into the area the unemployment index will go down. In the feedback path of the system, information concerning the employment situation is fed back and will influence the migration of workers. A poor employment situation will discourage migration into the area, and a good employment situation will cause the number of workers in the area to increase.

To determine the impulse response of the overall system in Fig. 10-5*a*, let us derive its transfer function first. Since

$$z(t) = x(t) + y(t) * h_2(t) \tag{10-2}$$

and

$$y(t)\, z(t) = * h_1(t) \tag{10-3}$$

we have

$$Z(s) = X(s) + Y(s)H_2(s)$$

and

$$Y(s) = Z(s)H_1(s)$$

which can be combined to yield

$$Y(s) = X(s)H_1(s) + Y(s)H_1(s)H_2(s) \tag{10-4}$$

Equation (10-4) is simplified to

$$Y(s)[1 - H_1(s)H_2(s)] = X(s)H_1(s)$$

or

$$Y(s) = X(s)\frac{H_1(s)}{1 - H_1(s)H_2(s)}$$

In other words, the transfer function $H(s)$ of the overall system is

$$H(s) = \frac{H_1(s)}{1 - H_1(s)H_2(s)} \tag{10-5}$$

To determine the region of absolute convergence of $H(s)$, we note that the overall system can be viewed as a cascade connection of two systems whose transfer functions are $H_1(s)$ and $1/[1 - H_1(s)H_2(s)]$. Consequently, the region of absolute convergence of $H(s)$ must include the intersection of that of $H_1(s)$ and

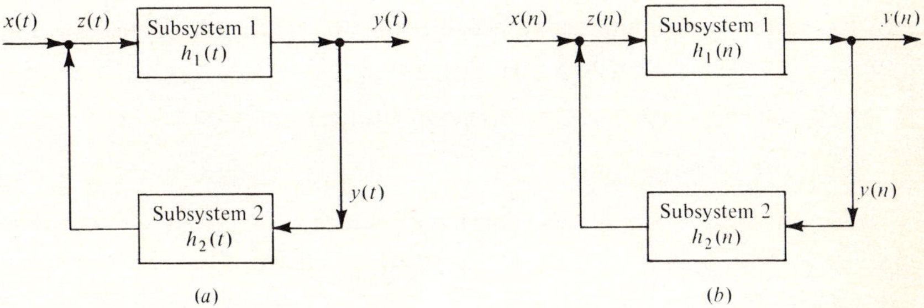

FIGURE 10-5

$1/[1 - H_1(s)H_2(s)]$. The system whose transfer function is $1/[1 - H_1(s)H_2(s)]$ is an inverse system of a system whose transfer function is $1 - H_1(s)H_2(s)$. According to our discussion in Sec. 10-2, if $1 - H_1(s)H_2(s)$ has one or more zeros within its region of absolute convergence, the region of absolute convergence of $1/[1 - H_1(s)H_2(s)]$ is not unique. Therefore, the region of absolute convergence of $H(s)$ and thus the impulse response of the overall system might not be unique either.

As an example, for the feedback connection in Fig. 10-5*a*, let

$$h_1(t) = e^{-t}u_{-1}(t)$$

and

$$h_2(t) = -u_0(t)$$

It follows that

$$H_1(s) = \frac{1}{s+1}$$

with its region of absolute convergence being Re $(s) > -1$, and

$$H_2(s) = -1$$

with its region of absolute convergence being the entire s plane. Since

$$1 - H_1(s)H_2(s) = 1 + \frac{1}{s+1} = \frac{s+2}{s+1}$$

with its region of absolute convergence being Re $(s) > -1$, the region of absolute convergence of

$$\frac{1}{1 - H_1(s)H_2(s)} = \frac{s+1}{s+2}$$

is Re $(s) > -2$. Consequently, the transfer function of the overall system is

$$H(s) = \frac{H_1(s)}{1 - H_1(s)H_2(s)} = \frac{1}{s+2}$$

with its region of absolute convergence being Re $(s) > -2$. Thus, the impulse response of the overall system is

$$h(t) = e^{-2t}u_{-1}(t)$$

Suppose we have instead

$$h_1(t) = e^{-t}u_{-1}(t)$$

and

$$h_2(t) = 2u_0(t)$$

We then have

$$1 - H_1(s)H_2(s) = 1 - \frac{2}{s+1} = \frac{s-1}{s+1}$$

with its region of absolute convergence being Re $(s) > -1$. It follows that

$$\frac{1}{1 - H_1(s)H_2(s)} = \frac{s+1}{s-1}$$

which has two possible regions of absolute convergence, namely, Re $(s) > 1$ and Re $(s) < 1$. Consequently, the transfer function of the overall system is

$$H(s) = \frac{1}{s-1}$$

with its region of absolute convergence being either Re $(s) > 1$ or Re $(s) < 1$. Thus, the impulse response of the overall system is either

$$h(t) = e^{t}u_{-1}(t) \tag{10-6}$$

or

$$h(t) = -e^{t}u_{-1}(-t) \tag{10-7}$$

Note that in this case, although both subsystems 1 and 2 are stable in the bounded-input–bounded-output sense, the overall system in Eq. (10-6) is not. Also note that, although both subsystems 1 and 2 are causal, the overall system in Eq. (10-7) is non-causal. Indeed, there are many examples in which feedback connection of subsystems enables us to build up systems whose transfer functions cannot be realized by employing only cascade and parallel connections of physical systems.†

At this point, one might wonder why a feedback connection of two well-defined systems would not yield a unique overall system. Mathematically, it is not difficult to see the reason. According to Eqs. (10-2) and (10-3),

$$y(t) = [x(t) + y(t) * h_2(t)] * h_1(t)$$

† For example, the transfer function of any series and parallel connection of resistors, inductors, capacitors, and amplifiers does not have poles in the right half of the *s* plane. However, it is possible to realize a transfer function with poles in the right half of the *s* plane by a feedback connection of *RLC* elements and amplifiers.

That is,

$$y(t) * [u_0(t) - h_2(t) * h_1(t)] = x(t) * h_1(t) \tag{10-8}$$

In other words, for a given input signal $x(t)$, the corresponding output signal $y(t)$ must be such that the integral equation (10-8) is satisfied. Since, in general, the solution for $y(t)$ may not be unique, the transfer function of the overall system may not be unique either. However, since all physical systems are causal systems, an interconnection of physical systems will always yield an overall system that is also a causal system. Therefore, although the mathematical solution for $y(t)$ in Eq. (10-8) might not be unique, only the solution corresponding to a causal system is the response of a physical system.

Similarly, for a feedback connection of discrete systems as shown in Fig. 10-5*b*, one can derive the transfer function of the overall system to be

$$H(z) = \frac{H_1(z)}{1 - H_1(z)H_2(z)}$$

following exactly the steps in the derivation of Eq. (10-5). Furthermore, the discussion on the region of absolute convergence of the overall system for the continuous case presented above also applies to the discrete case. As an example, let us consider the problem of the relationship between the migration of workers and the employment situation in a metropolitan area discussed earlier in this section. For the block diagram in Fig. 10-5*b*, we let $x(n)$ denote the number of workers who might want to move into the area, $z(n)$ denote the number of workers who actually move into the area, and $y(n)$ denote the number of workers who cannot find employment opportunity after moving into the area, all on a monthly basis. Let us assume that

$$x(n) = \begin{cases} 840 & n \geq 0 \\ 0 & n < 0 \end{cases}$$

Furthermore, let us assume that

$$h_1(n) = \begin{cases} 0.05 & n = 0 \\ 0 & \text{otherwise} \end{cases}$$

meaning that 5 percent of the workers moving into the area in each month become unemployed, and that

$$h_2(n) = \begin{cases} -1 & n = 1 \\ 0 & \text{otherwise} \end{cases}$$

meaning that the number of unemployed workers in each month will affect the number of workers who will move into the area in the following month. Since

$$H_1(z) = 0.05$$

and

$$H_2(z) = -z^{-1}$$

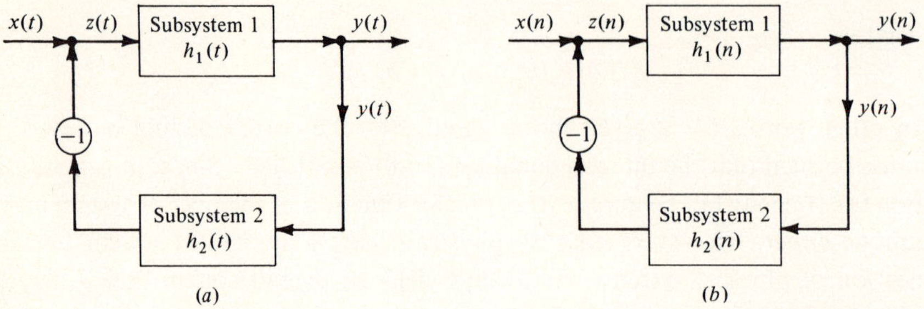

FIGURE 10-6

the transfer function of the overall system is

$$H(z) = \frac{0.05}{1 + 0.05z^{-1}}$$

with the region of absolute convergence being $|z| > 0.05$. It follows that the z transform of $y(n)$ is

$$\begin{aligned} Y(z) &= X(z)H(z) \\ &= \frac{840}{1 - z^{-1}} \frac{0.05}{1 + 0.05z^{-1}} \\ &= \frac{40}{1 - z^{-1}} + \frac{2}{1 + 0.05z^{-1}} \end{aligned}$$

Thus,

$$y(n) = \begin{cases} 40 + 2(-0.05)^n & n \geq 0 \\ 0 & n < 0 \end{cases}$$

Note that the value of $y(n)$ fluctuates and approaches 40 as a limit.

A variation of the feedback connection in Fig. 10-5*a* and *b* is shown in Fig. 10-6*a* and *b*. Because the output signal is fed back and added to the input signal in the connection in Fig. 10-5*a* and *b*, it is called a *positive feedback connection*, and because the output signal is fed back and subtracted from the input signal in the connection in Fig. 10-6*a* and *b*, it is called a *negative feedback connection*. As an example of negative feedback systems, let us consider the case of a feedback amplifier. Suppose we want to build an amplifier with a gain equal to 10. We can use a simple transistor amplifier whose equivalent circuit is shown in Fig. 10-7*a*. Note that

$$y(t) = gx(t)r = 10x(t)$$

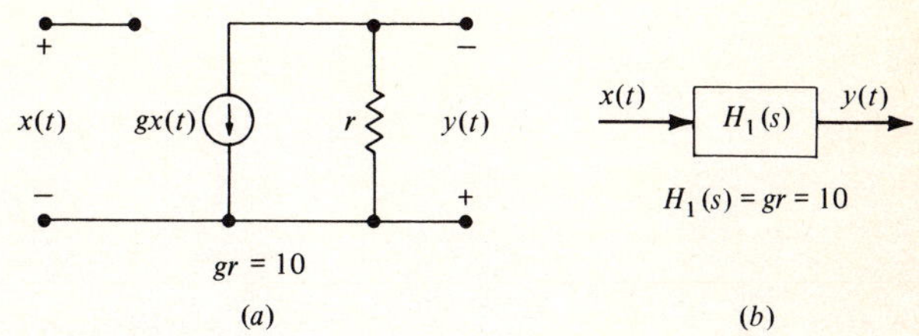

FIGURE 10-7

A block diagram corresponding to this circuit is shown in Fig. 10-7*b*. Since it is not uncommon for the value of the current gain g of a transistor to vary by as much as 50 percent when the transistor ages and the ambient temperature and the voltage of the power supply fluctuate, the value of the gain of the amplifier may vary from 5 to 15. In applications where such variation in the gain of the amplifier cannot be tolerated, an amplifier whose equivalent circuit is shown in Fig. 10-8*a* can be used. Note that in Fig. 10-8*a*

$$gz(t) = \frac{z(t) + y(t)}{r_1} + \frac{y(t)}{r} \tag{10-9}$$

$$x(t) - z(t) = [x(t) + y(t)]\frac{r_2}{r_1 + r_2} \tag{10-10}$$

Since $r_1 \gg r$, Eq. (10-9) can be approximated as

$$grz(t) = y(t) \tag{10-11}$$

and since $r_2/(r_1 + r_2) \ll 1$, Eq. (10-10) can be approximated as

$$x(t) - z(t) = \frac{r_2}{r_1 + r_2}\,y(t) \tag{10-12}$$

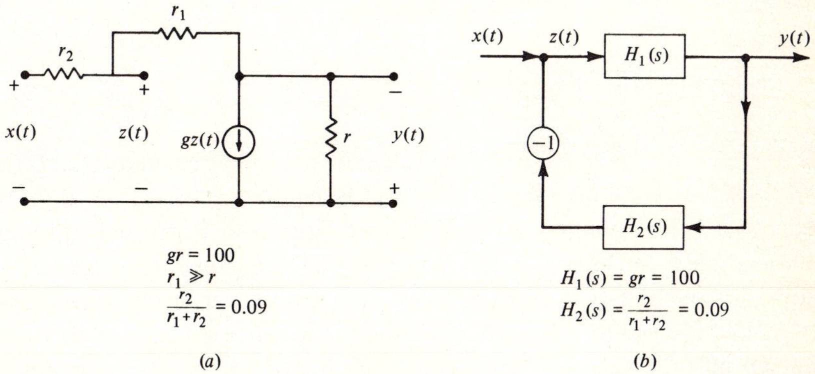

FIGURE 10-8

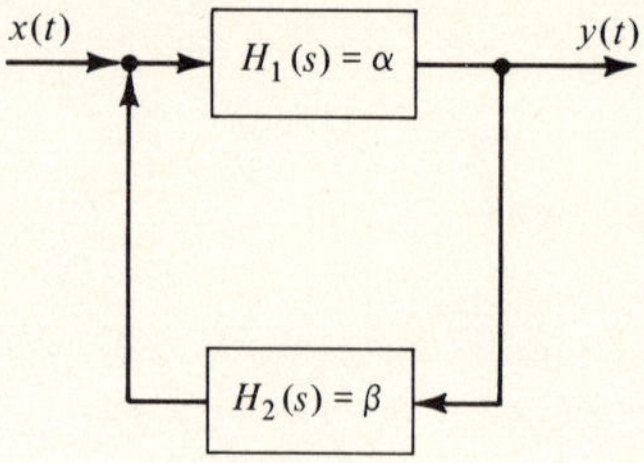

FIGURE 10-9

A block diagram corresponding to the relations in Eqs. (10-11) and (10-12) is shown in Fig. 10-8*b*. Thus, the gain of the overall system is

$$\frac{100}{1 + 0.09(100)} = 10$$

For a 50 percent variation in the value of the current gain g, the value of gr will vary between 50 and 150. Consequently, the value of the gain of the overall system will vary between

$$\frac{50}{1 + 0.09(50)} = 9.09$$

and

$$\frac{150}{1 + 0.09(150)} = 10.34$$

which is only a 10 percent variation from the designed value 10.

On the other hand, consider the positive feedback system shown in Fig. 10-9. The gain of the overall system is equal to

$$H(s) = \frac{\alpha}{1 - \alpha\beta}$$

If the value of $\alpha\beta$ is equal to 0.9, the value of $H(s)$ will be equal to 10α. If the value of $\alpha\beta$ is equal to 0.99, the value of $H(s)$ will be equal to 100α. If we let the value of $\alpha\beta$ approach 1, the value of $H(s)$ will become larger and larger and will approach ∞. Indeed, this is the basic concept of building an oscillator. If the gain of a system is equal to ∞, the system will produce a nonzero output in the absence of an input (the input is equal to zero), which is exactly the definition of an oscillator. We shall not go into the details of the design and analysis of oscillator circuits here. The interested reader can consult a book on electronic circuits.

*10-4 A GENERAL FORMULA

Our discussion in the preceding sections can be extended to the analysis of more complex systems. Since our results will be valid for both continuous systems and discrete systems, throughout this section we shall use capital letters to denote the z transforms of discrete signals and the Laplace transforms of continuous signals, with the arguments z and s omitted. That is, we shall write X, Y, H, H_1, H_2, ... instead of $X(z)$, $Y(z)$, $H(z)$, $H_1(z)$, $H_2(z)$, ... or $X(s)$, $Y(s)$, $H(s)$, $H_1(s)$, $H_2(s)$, ...

As an example of the analysis of complex systems, let us consider the system in Fig. 10-10a. According to Eq. (10.5), the system in Fig. 10-10a can be reduced to that in Fig. 10-10b. Applying the relation in Eq. (10-5) again, we obtain

$$H = \frac{H_1}{1 - H_1 H_2 [H_3/(1 - H_3 H_4)]}$$

$$= \frac{H_1(1 - H_3 H_4)}{1 - H_3 H_4 - H_1 H_2 H_3}$$

as the transfer function of the overall system.

As another example, let us consider the system in Fig. 10-11, where X_1 and X_2 denote the output signals of distributing devices labeled b and c, respectively. Since

$$X_1 = X H_1$$

$$X_2 = XH_2 + X_1 H_3 = XH_2 + XH_1 H_3$$

we obtain

$$Y = X_1 H_4 + X_2 H_5 = XH_1 H_4 + XH_2 H_5 + XH_1 H_3 H_5$$

or

$$H = \frac{Y}{X} = H_1 H_4 + H_2 H_5 + H_1 H_3 H_5$$

These two examples illustrate quite clearly how a step-by-step reduction can be carried out to determine the transfer functions of complex systems. However, for systems containing a large number of subsystems, to be able to organize the reduction steps in a more systematic manner is highly desirable. Let us consider again the system in Fig. 10-10a which is repeated in Fig. 10-12, where X_1, X_2, X_3, Y are the output signals of the distributing devices b, c, d, e, respectively. If, for each of the distributing devices, we write an equation describing its input-output relationship, we obtain

$$\begin{aligned} X_1 &= X + X_2 \\ X_2 &= X_3 H_3 \\ X_3 &= X_2 H_4 + Y H_3 \\ Y &= X_1 H_1 \end{aligned} \qquad (10\text{-}13)$$

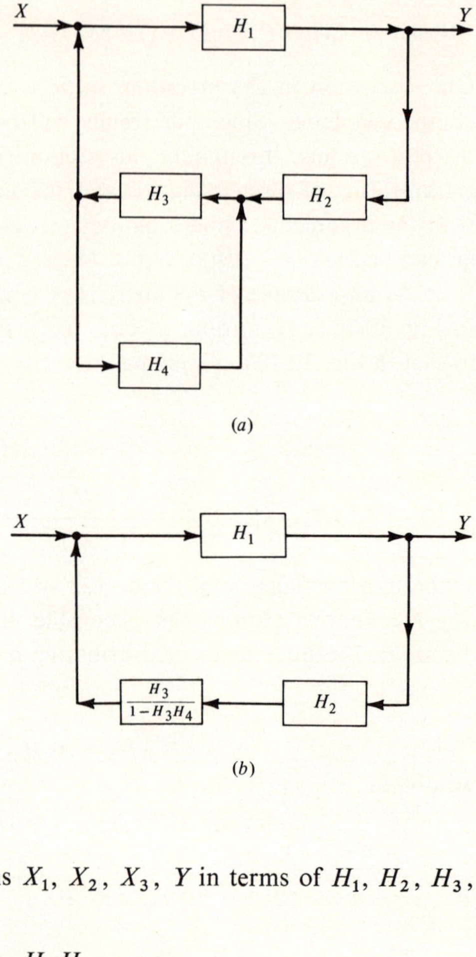

FIGURE 10-10 (b)

Solving these equations for the unknowns X_1, X_2, X_3, Y in terms of H_1, H_2, H_3, H_4, X, we obtain

$$X_1 = \frac{1 - H_3 H_4}{1 - H_3 H_4 - H_1 H_2 H_3} X$$

$$X_2 = \frac{H_1 H_2 H_3}{1 - H_3 H_4 - H_1 H_2 H_3} X$$

$$X_3 = \frac{H_1 H_2}{1 - H_3 H_4 - H_1 H_2 H_3} X$$

$$Y = \frac{H_1(1 - H_3 H_4)}{1 - H_3 H_4 - H_1 H_2 H_3} X$$

Our discussion suggests immediately a general procedure for determining the transfer functions of complex systems. Suppose that a system contains n distributing devices, and the output of one of these devices is the output of the overall system Y. Let

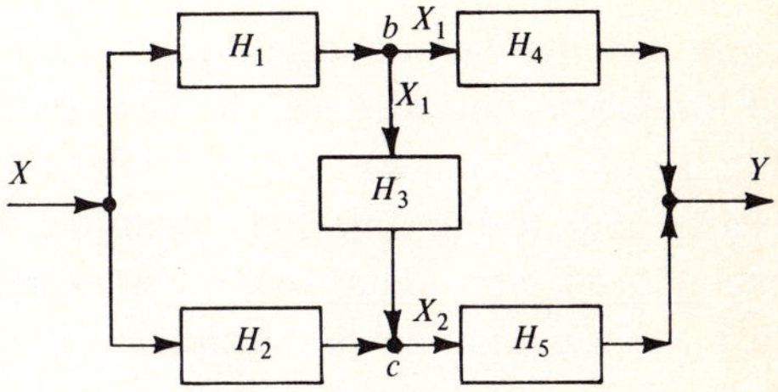

FIGURE 10-11

$X_1, X_2, \ldots, X_{n-1}$ denote the outputs of the other $n-1$ distributing devices. For each of the n distributing devices, we can write an equation expressing the output of the device as a sum of the inputs to the device. We thus obtain a set of n equations in the n unknowns $X_1, X_2, \ldots, X_{n-1}, Y$. We can then solve these n equations to express Y in terms of X and the transfer functions of the subsystems.

A direct consequence of this general procedure is a rather surprising formula which can be used to determine the transfer function of the overall system by inspection. We shall give this formula here but not its derivation. It suffices to say that the formula is derived from a systematic procedure for solving the simultaneous equations relating $X_1, X_2, \ldots, Y$. This point will be illustrated in a later example.

We shall represent the interconnection of subsystems by a graph, called a *flow graph*, in which there is a node corresponding to each distributing device. Moreover, for clarity, we always assume that there is a distributing device at the input terminal and a distributing device at the output terminal of the overall system. The node corresponding to the distributing device at the input terminal is called the *source node*, and the node corresponding to the distributing device at the output terminal is called the *sink node*. Between two nodes in the graph there will be directed edges which are labeled with the transfer functions of the subsystems connected between the two

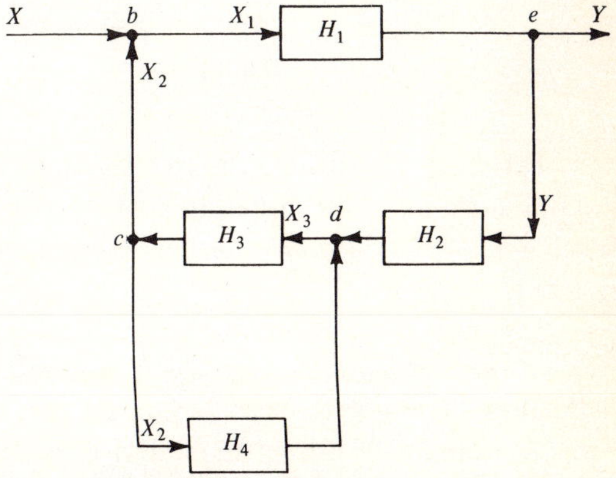

FIGURE 10-12

corresponding distributing devices. Such labels are referred to as the *transmissions* of the edges. For example, the flow graph for the system in Fig. 10-12 is shown in Fig. 10-13. A (simple) *path* in a flow graph is a sequence of edges leading from the source node to the sink node along the direction of the edges such that no node is encountered more than once. For example, the sequence of edges $\{(a, b), (b, e), (e, z)\}$† is a path in the flow graph in Fig. 10-13. The *transmission of a path* is defined to be the product of the transmissions of the edges in the path. A *loop* is a simple path leading from a node back to itself. For example, $\{(b, e)\,(e, d)\,(d, c)\,(c, b)\}$ is a loop in the flow graph in Fig. 10-13, and so is $\{(d, c), (c, d)\}$. The *transmission of a loop* is defined to be the product of the transmissions of the edges in the loop. Two paths or loops are said to be *disjoint* if they do not have any node in common. A collection of paths or loops is said to be disjoint if every two of them are disjoint. The *determinant* of a flow graph is defined to be

$$\begin{aligned}\Delta = 1 &- \text{(sum of transmissions of all loops)}\\ &+ \text{(sum of products of transmissions of all pairs of disjoint loops)}\\ &- \text{(sum of products of transmissions of all triples of disjoint loops)}\\ &+ \cdots\end{aligned}$$

For a path in the flow graph, we define the *cofactor* of the flow graph with respect to this path to be

$$\begin{aligned}\Delta_i = 1 &- \text{(sum of transmissions of all loops that are disjoint from path)}\\ &+ \text{(sum of products of transmissions of all pairs of disjoint loops that are disjoint from path)}\\ &- \text{(sum of products of transmissions of all triples of disjoint loops that are disjoint from path)}\\ &+ \cdots\end{aligned}$$

Let $g_1, g_2, \ldots$ denote the transmissions of the paths from the source node to the sink node in a flow graph, and let $\Delta_1, \Delta_2, \ldots$ be their corresponding cofactors. What is known as *Mason's formula* states that the transfer function of the overall system is

$$H = \frac{1}{\Delta} \sum_i g_i \Delta_i$$

† We use (a, b) to denote the directed edge from node a to node b, and use braces to enclose the sequence of edges in a path or a loop.

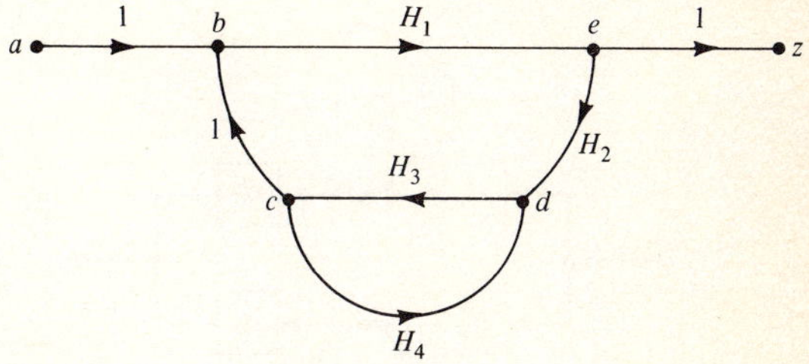

FIGURE 10-13

As an example, for the flow graph in Fig. 10-13, there are two loops: $\{(b, e)(e, d)(d, c)(c, b)\}$, and $\{(d, c)(c, d)\}$. Moreover, these two loops are not disjoint. Thus, the determinant of the flow graph is

$$\Delta = 1 - (H_1 H_2 H_3 + H_3 H_4)$$

The only path from the source node to the sink node is $\{(a, b), (b, e), (e, z)\}$. Its transmission is

$$g_1 = H_1$$

Since $\{(d, c), (c, d)\}$ is the only loop that is disjoint from the path $\{(a, b), (b, e), (e, z)\}$, the cofactor of the flow graph with respect to this path is

$$\Delta_1 = 1 - H_3 H_4$$

Consequently, the transfer function of the overall system is

$$H = \frac{H_1(1 - H_3 H_4)}{1 - (H_1 H_2 H_3 + H_3 H_4)} \tag{10-14}$$

To remove some of the mystery of Mason's formula, let us check the result in Eq. (10-14) by solving the simultaneous equations (10-13) for Y by Cramer's rule. Rewriting Eqs. (10-13) as

$$\begin{aligned} X_1 - X_2 \qquad\qquad\qquad &= X \\ X_2 - X_3 H_3 \qquad\quad &= 0 \\ - X_2 H_4 + X_3 - YH_2 &= 0 \\ - X_1 H_1 \qquad\qquad\qquad + Y \quad &= 0 \end{aligned}$$

we obtain

$$Y = \frac{\Lambda_1}{\Delta}$$

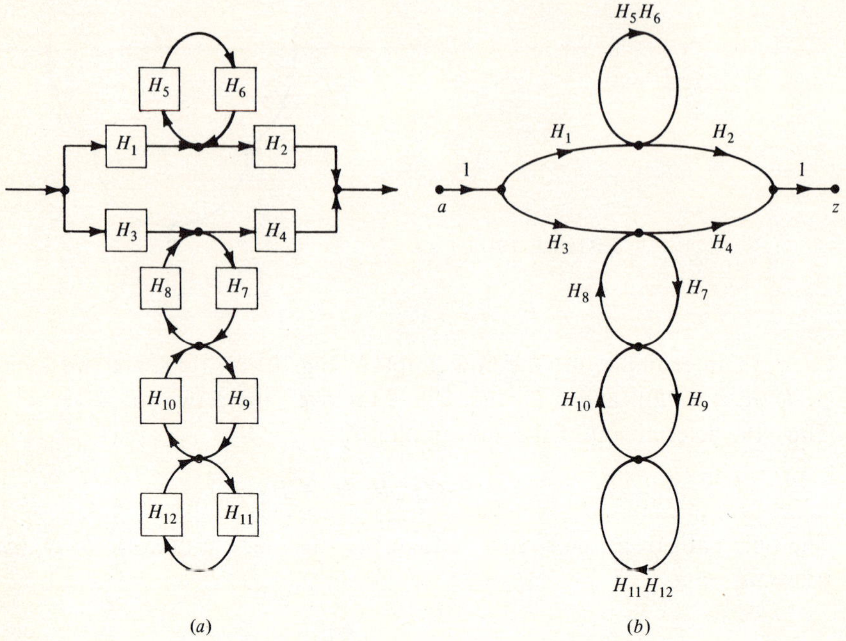

FIGURE 10-14

where

$$\Delta = \begin{vmatrix} 1 & -1 & 0 & 0 \\ 0 & 1 & -H_3 & 0 \\ 0 & H_4 & 1 & -H_2 \\ -H_1 & 0 & 0 & 1 \end{vmatrix} \tag{10-15}$$

and

$$\Lambda_1 = \begin{vmatrix} 1 & -1 & 0 & X \\ 0 & 1 & -H_3 & 0 \\ 0 & -H_4 & 1 & 0 \\ -H_1 & 0 & 0 & 0 \end{vmatrix} \tag{10-16}$$

Note that Δ is indeed the determinant of the flow graph and Λ_1 is the product of the transmission of the path $\{(a, b), (b, e), (e, z)\}$ and the cofactor of the flow graph with respect to the path. Moreover, the Laplace expansions of the determinants in Eqs. (10-15) and (10-16) contain terms which are products of the transmissions of disjoint loops as was specified in Mason's formula.

As another example, let us consider the system in Fig. 10-14*a*. The corresponding flow graph is shown in Fig. 10-14*b*. We obtain

$$H = \frac{1}{\Delta}\{H_1H_2[1 - (H_7H_8 + H_9H_{10} + H_{11}H_{12}) - H_7H_8H_{11}H_{12}]$$
$$+ H_3H_4[1 - (H_5H_6 + H_9H_{10} + H_{11}H_{12}) + (H_5H_6H_9H_{10} + H_5H_6H_{11}H_{12})]\}$$

where

$$\Delta = 1 - (H_5H_6 + H_7H_8 + H_9H_{10} + H_{11}H_{12}) + (H_5H_6H_7H_8 + H_5H_6H_9H_{10}$$
$$+ H_5H_6H_{11}H_{12} + H_7H_8H_{11}H_{12}) - (H_5H_6H_7H_8H_{11}H_{12})$$

Finally, we should point out that the region of absolute convergence of the transfer function of a complex system can be determined only if we reduce the system in a step-by-step manner as illustrated in the beginning of this section. However, since any interconnection of physical systems yields a causal system, the region of absolute convergence of the overall system must be a right half plane in any physical problem. Consequently, we can apply Mason's formula to determine the transfer function of a system and then determine the region of absolute convergence by examining the positions of the poles of the transfer function.

10-5 REMARKS AND REFERENCES

The most important concept in modern control theory is that of a feedback connection of systems. In order to generate a desired output, we use the actual output of the overall system to regulate the input to the subsystem in the forward path. A great deal more can be said on the analysis and synthesis of feedback systems. The interested reader may consult a book on control theory such as Melsa and Schultz [8] (introductory) and Athans and Falb [1] (advanced).

The concept of flow graphs was first published in Refs. [5] and [6]. See also Mason and Zimmermann [7, chap. 4]. Other useful references on the subject are Chow and Cassignol [2] and Lorens [4]. Besides application to system analysis, the method of flow graphs is also useful in circuit theory problems and in probability theory problems concerning Markov processes. See Mason and Zimmermann [7, chap. 5] and Howard [3].

[1] ATHANS, M., and P. L. FALB: "Optimal Control," McGraw-Hill Book Company, New York, 1966.

[2] CHOW, Y., and E. CASSIGNOL: "Linear Signal-flow Graphs and Applications," John Wiley & Sons, Inc., New York, 1962.

[3] HOWARD, R. A.: "Dynamic Probabilistic Systems," vol. I, Markov Models, John Wiley & Sons, Inc., New York, 1971.
[4] LORENS, C. S.: "Flowgraphs," McGraw-Hill Book Company, New York, 1964.
[5] MASON, S. J.: Some Properties of Signal Flow Graphs, *Proc. IRE*, vol. 41, pp. 1144–1156, Sept. 1953.
[6] MASON, S. J.: Feedback Theory—Further Properties of Signal Flow Graphs, *Proc. IRE*, vol. 44, pp. 920–926, July 1956.
[7] MASON, S. J., and H. J. ZIMMERMANN: "Electronics Circuits, Signals, and Systems," John Wiley & Sons, Inc., New York, 1960.
[8] MELSA, J. L., and D. G. SCHULTZ: "Linear Control Systems," McGraw-Hill Book Company, New York, 1969.

PROBLEMS

10-1 Let $y(t)$ be the output of the *RC* oscillator shown in Fig. 10P-1*a*. (The input signal of the oscillator is always equal to zero.) Draw the equivalent block diagram of the oscillator as an interconnection of the subsystems shown in Fig. 10P-1*b*.

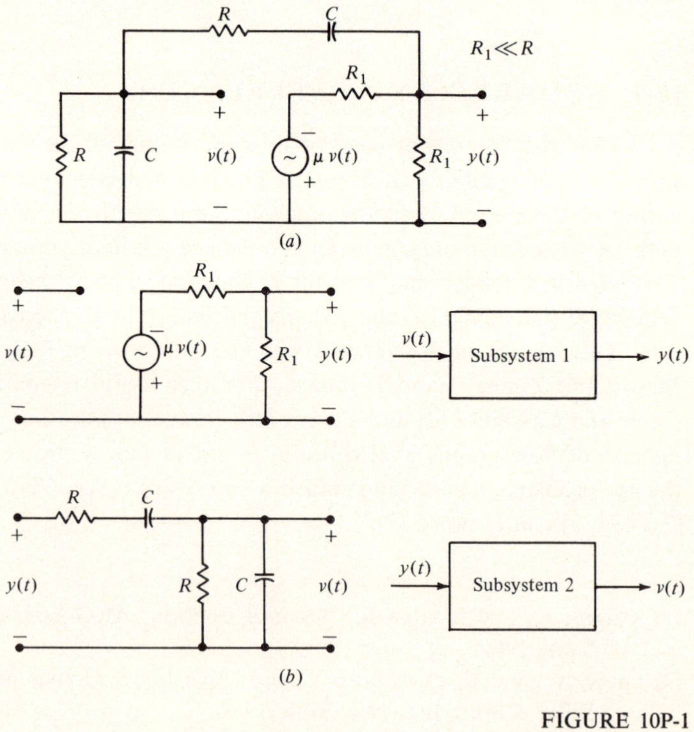

FIGURE 10P-1

10-2 An electronically controlled system for driving machine tools at variable speeds is shown in Fig. 10P-2*a*. Let $x(t)$ and $\omega(t)$ be the input and output signals of the system, respectively. Draw the block diagram of this system as an interconnection of subsystems shown in Fig. 10P-2*b*. Determine the transfer function of the system.

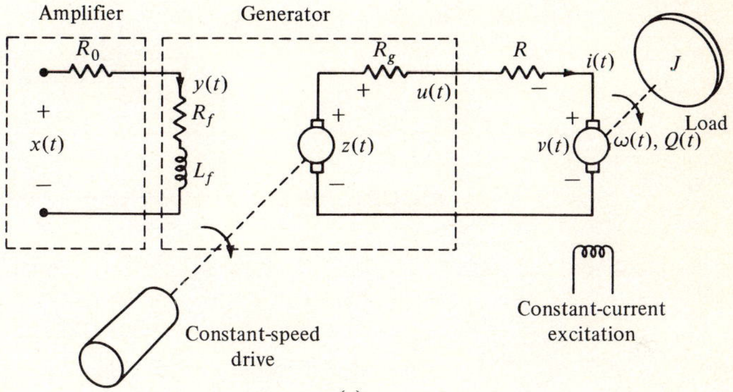

(*a*)

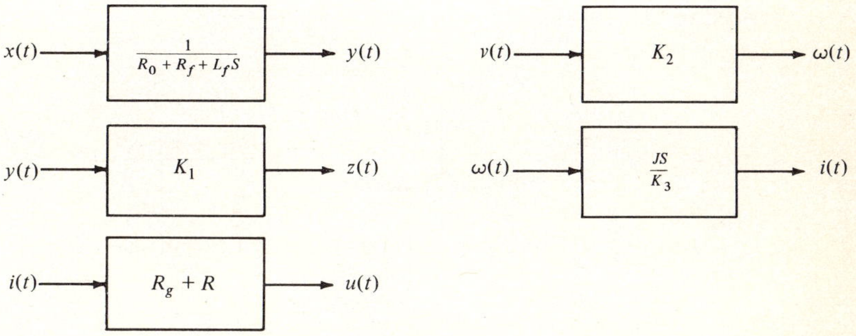

(*b*)

FIGURE 10P-2

10-3 In the system shown in Fig. 10P-3, both the support S and the mass M are constrained to move vertically only. Let the vertical position, $x(t)$, of the support S be the input signal of the system and the vertical position, $y(t)$, of the mass M be its output signal. Suppose that the inertia and mass of the spring and the mass and resilience of the damper are all negligible. Hence, the force on the mass M due to the spring, f_k, and that due to the damper, f_B, are equal to $k[x(t) - y(t)]$ and $B(d/dt)[x(t) - y(t)]$, respectively. Draw an equivalent block diagram of the system as an interconnection of amplifiers, integrators, and differentiators.

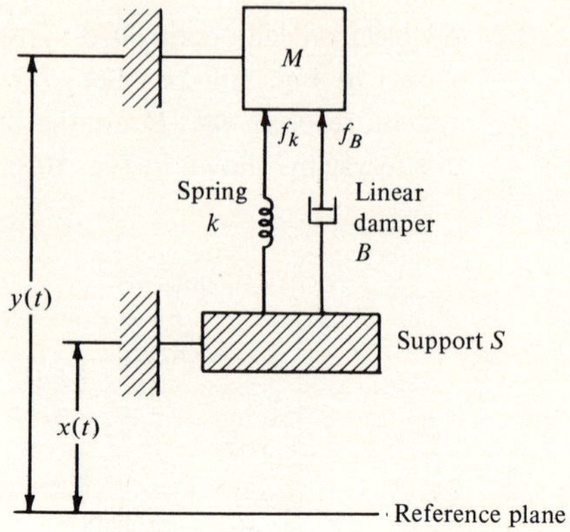

FIGURE 10P-3

10-4 In the systems shown in Fig. 10P-4*b* and *c*, the values of the input and output signals at any time instant are equal to either 1 or 0. (Such signals are called binary signals.) The input-output relationship of the modulo-2 adder in the block diagrams is described in Fig. 10P-4*a*.

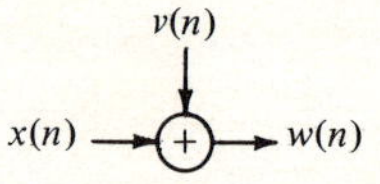

$v(n)$	$x(n)$	$w(n)$
0	0	0
0	1	1
1	0	1
1	1	0

(*a*)

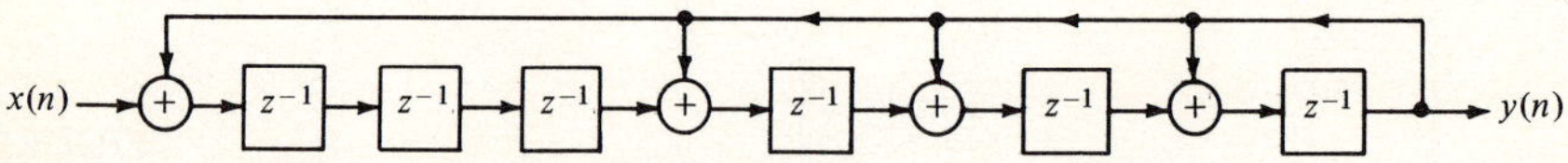

(*b*)

(*c*)

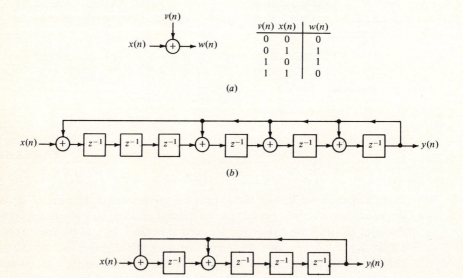

FIGURE 10P-4

(*a*) Given that the input signal of the system shown in Fig. 10P-4*b* is

$$x(n) = \begin{cases} 1 & 0 \le n \le 4 \\ 0 & \text{otherwise} \end{cases}$$

find its output signal $y(n)$ for $n \ge 0$.

(*b*) Given that the input signal of the system shown in Fig. 10P-4*c* is

$$x(n) = \begin{cases} 1 & n = 0 \\ 0 & \text{otherwise} \end{cases}$$

find the corresponding output signal $y(n)$ for $n \ge 0$.

10-5 Determine the transfer functions for the systems shown in Fig. 10P-5.

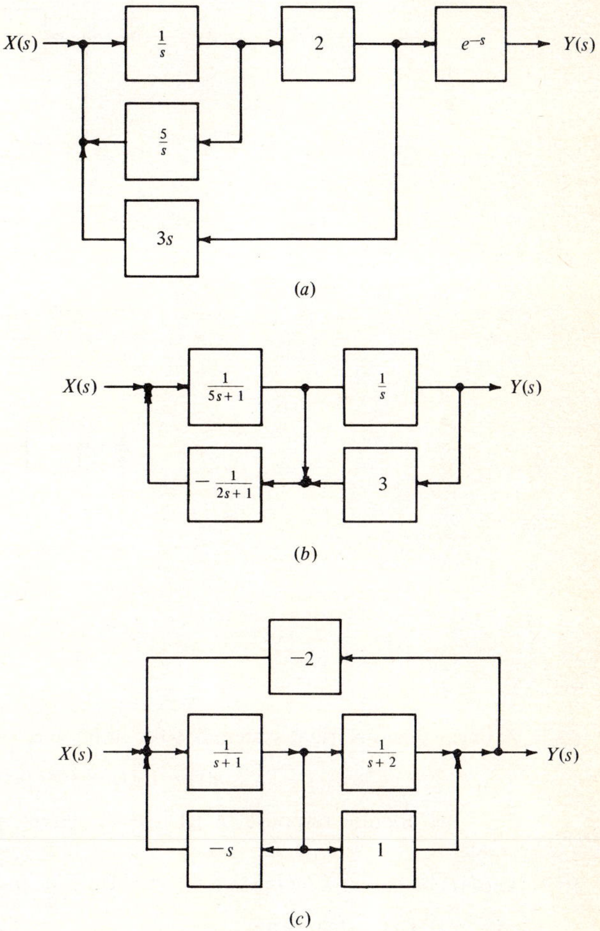

FIGURE 10P-5

10-6 Determine the transfer functions of the systems shown in Fig. 10P-6.

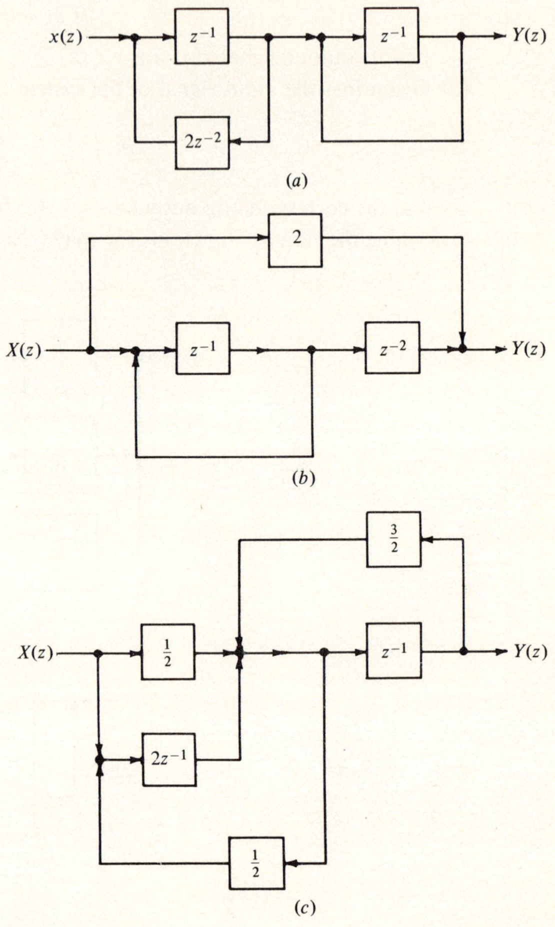

FIGURE 10P-6

10-7 A linear time-invariant system has the impulse response

$$h(t) = u_3(t) + 15e^{-t}u_{-1}(t)$$

Find the impulse response of its inverse system, given that the inverse system is stable.

10-8 Let $h(t)$ be the impulse response of a linear time-invariant system. It is known that:

(1) The system is not stable.

(2) The inverse of the system is unique.

(3) $h(t) = e^{5t}$, for $t \leq 0$.

(4) $H(s)$ has one finite zero. It is at $s = 3$.

(5) $H(s)$ has three finite poles. It is known that one of them is either at $s = +2$ *or* at $s = -2$; another of them is either at $s = +4$ *or* $s = -4$.

Determine $h(t)$ for $t \geq 0$.

10-9 Let $x(t)$ and $y(t)$ be the input signal and output signal, respectively, of the network shown in Fig. 10P-7.

(*a*) Draw the equivalent block diagram of the network as a series and/or parallel connection of the two subsystems.

(*b*) Suppose that the transfer function of the network $H(s) = Y(S)/X(S)$ is equal to 1. Find the transfer function and its region of absolute convergence of subsystem 1. What is the region of absolute convergence of the network?

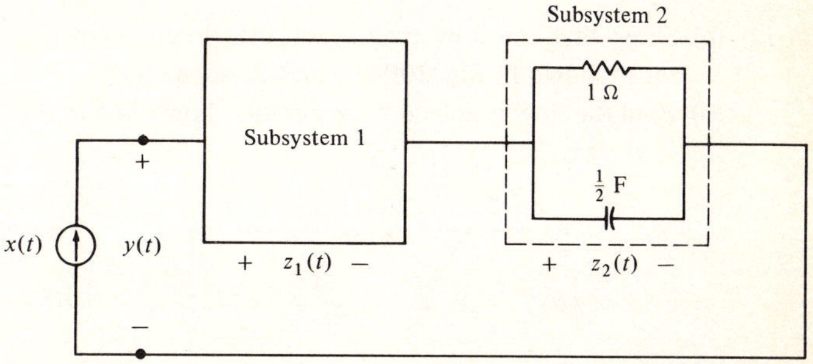

FIGURE 10P-7

10-10 (*a*) Find the output signal $y(t)$ of the causal linear time-invariant system shown in Fig. 10P-8, when its input signal is

$$x(t) = e^{-|t|}$$

(*b*) Write a differential equation relating $y(t)$ and $x(t)$.

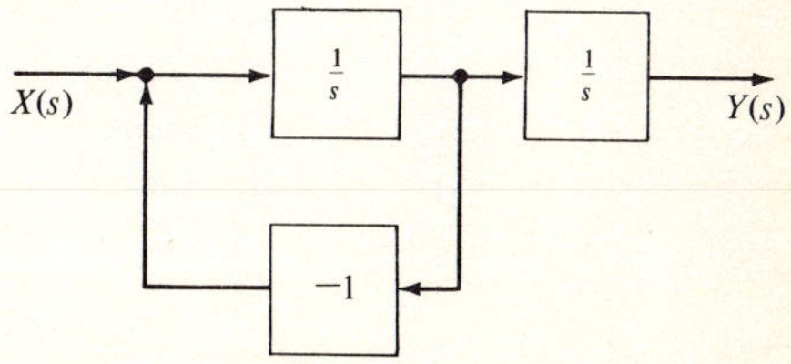

FIGURE 10P-8

10-11 Find the transfer function of the system shown in Fig. 10P-9.

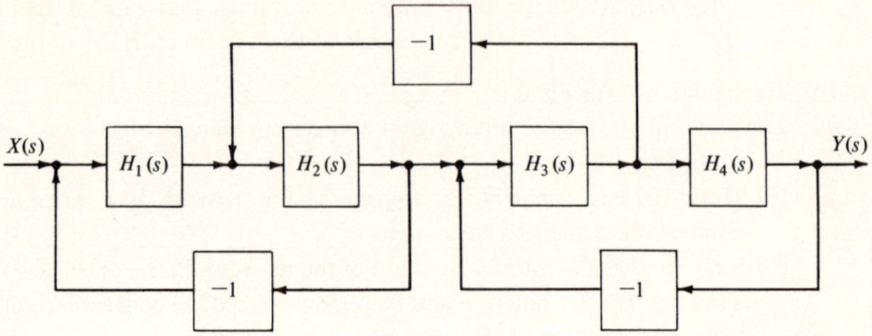

FIGURE 10P-9

10-12 (*a*) Show that the flow graphs for the circuits shown in Fig. 10P-10*a* and *b* are those shown in Fig. 10P-10*c* and *d*, respectively.

(*b*) Find the flow graphs for the circuits shown in Fig. 10P-11.

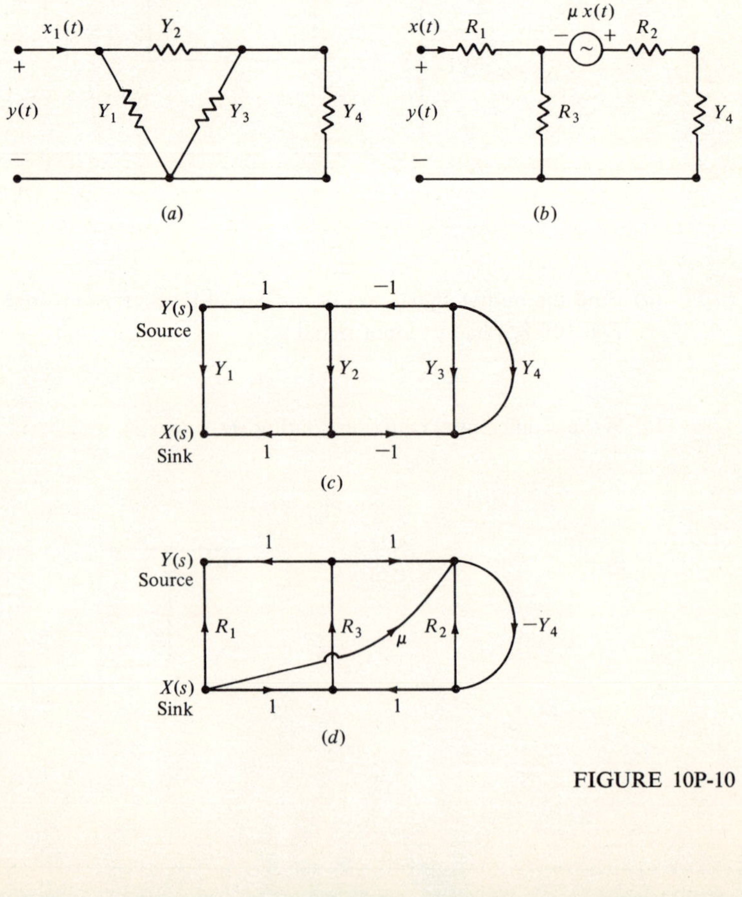

FIGURE 10P-10

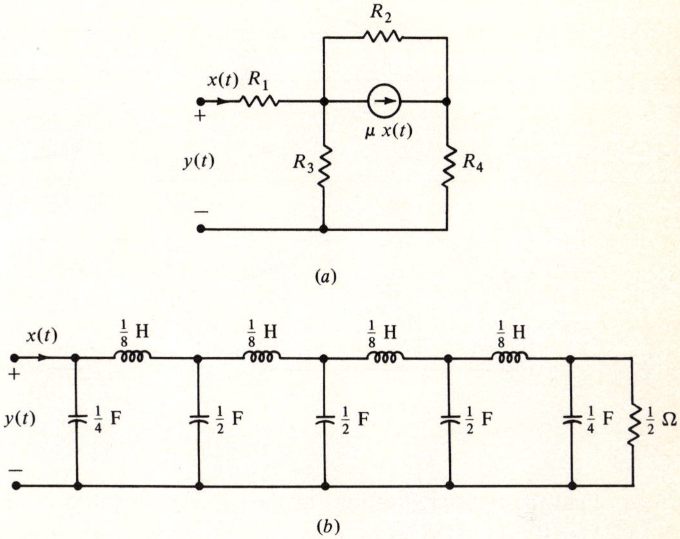

FIGURE 10P-11

10-13 In this problem, we consider flow graphs that have several source nodes and sink nodes. Two flow graphs are said to be equivalent if there is a one-to-one correspondence between their source nodes and sink nodes and if the transfer function between any pair of source node and sink node in one flow graph is equal to the transfer function between the corresponding pair of source node and sink node in the other flow graph. Show that the flow graphs in each part of Fig. 10P-12 are equivalent.

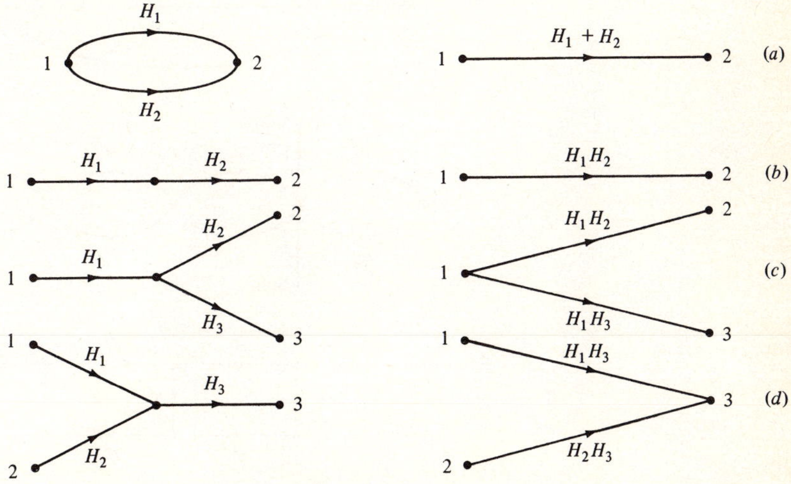

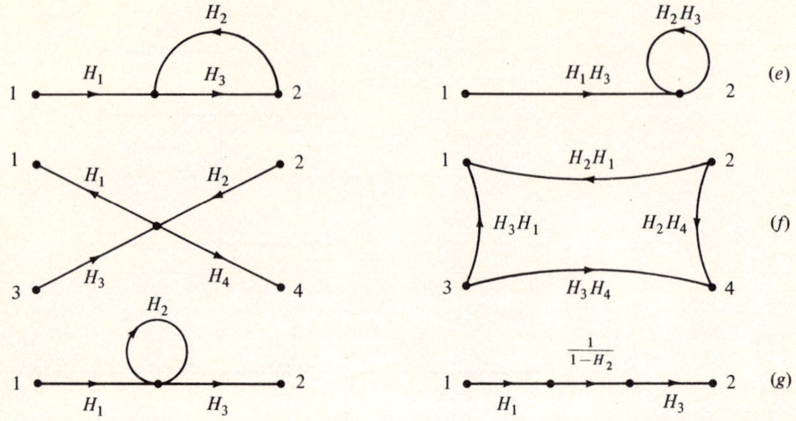

FIGURE 10P-12

10-14 Find the transfer functions of the systems described by the flow graphs shown in Fig. 10P-13. (Do so by inspection as much as possible. Use the results of Prob. 10-13.)

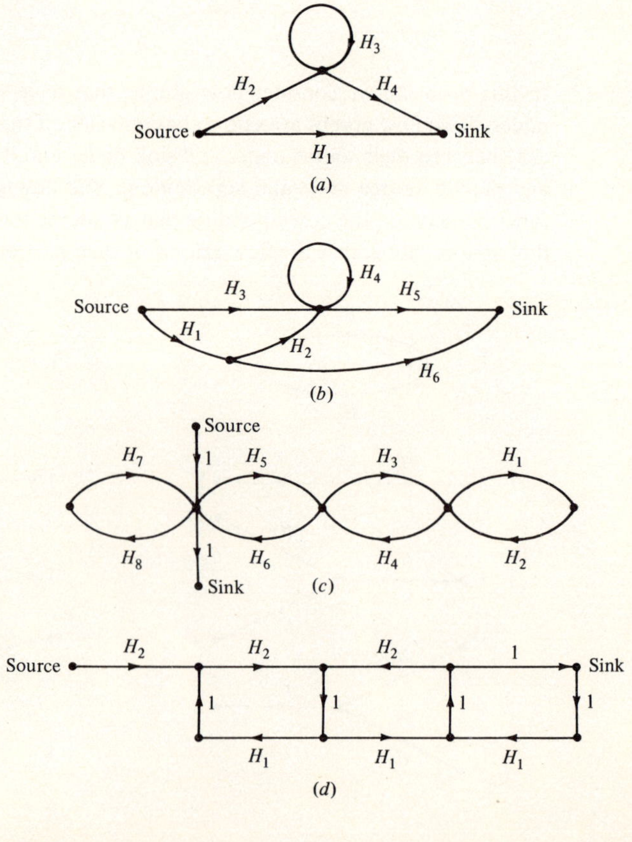

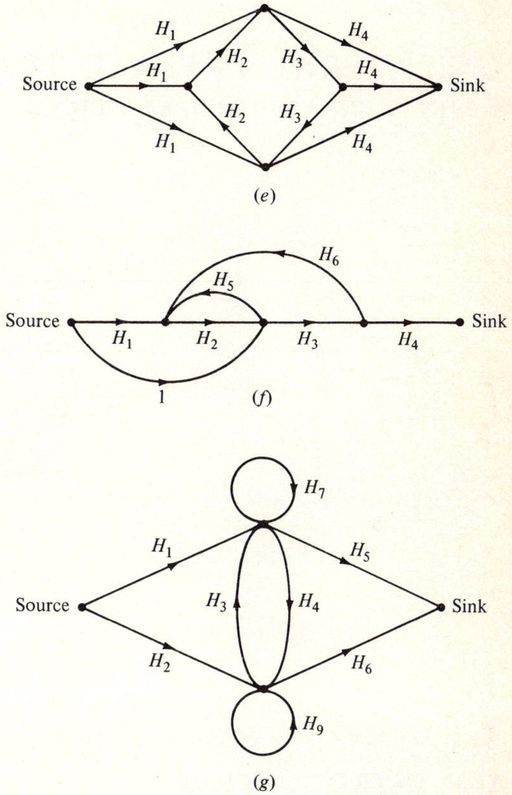

FIGURE 10P-13

10-15 Find the transfer function of the system described by the flow graph shown in Fig. 10P-14. (*Hint:* Note that because of the infinite structure of the flow graph, the transfer function of the system is equal to the transmission of the first loop.)

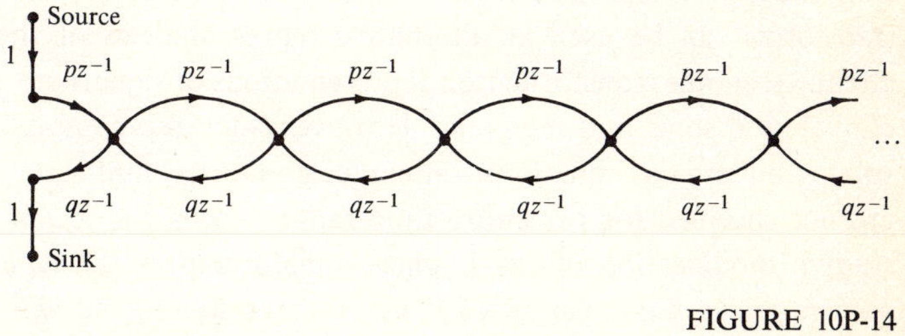

FIGURE 10P-14

11

UNILATERAL LAPLACE TRANSFORMATION AND z TRANSFORMATION

11-1 INTRODUCTION

In Chaps. 2 to 4, three different techniques for analyzing the behavior of linear time-invariant systems in the time domain were discussed. We recall that when signals are specified for $-\infty \leq t \leq \infty$ and when a system is initially at rest at $t = -\infty$, the convolution integral approach can be used. On the other hand, when signals are specified only for $t \geq t_0$ and when a system might not be at rest initially at $t = t_0$, the difference equation and differential equation approach or the state space approach can be used. In Chaps. 5 to 7 and 9, we found that for signals specified for the entire time range $-\infty \leq t \leq \infty$ their z transforms or their Fourier and bilateral Laplace transforms can be used as alternative representations of the signals. As we have already seen, the transformation representations of signals are quite useful in handling convolution sums and integrals. However, such representations are not suitable for solving differential equations, difference equations, and state equations when signals are not specified for the entire time range $-\infty \leq t \leq \infty$. In this chapter, we shall study a modification of the Laplace transformation, the *unilateral Laplace transformation*, and a modification of the z transformation, the *unilateral z transformation*, which are useful in solving differential equations, difference equations, and state equations.

11-2 UNILATERAL LAPLACE TRANSFORMATION

We define the *unilateral Laplace transform* of a time signal $x(t)$, denoted $X_I(s)$, to be

$$X_I(s) = \int_{0+}^{\infty} x(t)e^{-st}\,dt \tag{11-1}$$

Note that the lower limit of the integral in Eq. (11-1) is $0+$. Consequently, if $x(t)$ contains any singularity functions at $t = 0$, these singularity functions will *not* be included in the integral.

Conceptually, there is nothing new in the theory of unilateral Laplace transformation. For a signal $x(t)$ that is equal to zero for $t < 0+$ and contains no singularity functions at $t = 0$, its unilateral Laplace transform is the same as its bilateral Laplace transform. For a signal $x(t)$ that is not equal to zero for $t < 0+$, its unilateral Laplace transform is the same as the bilateral Laplace transform of a signal obtained by setting $x(t)$ to 0 for $t < 0+$.

Let $x(t)$ be a signal that has a finite number of maxima, minima, and discontinuities in any finite interval. The region in the s plane that contains all complex numbers s such that

$$\int_{0+}^{\infty} |x(t)e^{-st}|\,dt < \infty$$

is called the *region of absolute convergence* of the unilateral Laplace transform $X_I(s)$.† Note that the region of absolute convergence of the unilateral Laplace transform of a signal is always a right half plane. Thus, it follows directly from our discussion in Chap. 9 that the inverse transformation formula is

$$x(t) = \frac{1}{2\pi j}\int_{\sigma_0 - j\infty}^{\sigma_0 + j\infty} X_I(s)e^{st}\,ds \qquad t \geq 0+ \tag{11-2}$$

for some σ_0 in the region of absolute convergence of $X_I(s)$. It is not surprising that we can only recover $x(t)$ for $t \geq 0+$ from $X_I(s)$ since, according to Eq. (11-1), $X_I(s)$ is determined only by $x(t)$ for $t \geq 0+$. According to the definitions in Eqs. (11-1) and (11-2), the unilateral Laplace transform $X_I(s)$ of a time signal $x(t)$ is a function defined along a vertical line in the s plane. However, just as in the case of bilateral Laplace transformation, it is possible to extend the definitions so that $X_I(s)$ is defined as an analytic function of s in its region of absolute convergence.

† Recall that when we discussed the bilateral Laplace transformation of signals in Chap. 9 we talked about the region of absolute convergence of a signal and the region of absolute convergence of its bilateral Laplace transform interchangeably. Here, to save any possible confusion, we shall use only the term "the region of absolute convergence of the unilateral Laplace transform" of a signal.

For example, the unilateral Laplace transform of the signal

$$x(t) = e^{-\alpha t}u_{-1}(t)$$

is

$$X_I(s) = \frac{1}{s + \alpha}$$

with its region of absolute convergence being Re $(s) > -\alpha$. Note that the unilateral Laplace transform of the signal $e^{-\alpha t}$ is also $1/(s + \alpha)$ with its region of absolute convergence also being Re $(s) > -\alpha$. It is quite clear that a signal is not uniquely identified by its unilateral Laplace transform and its region of absolute convergence. Specifically, two time signals that are equal for $t \geq 0+$ will have the same unilateral Laplace transform although they might be different for $t < 0+$.

With the following exceptions, the properties of unilateral Laplace transforms are the same as those listed in Table 9-1 for bilateral Laplace transforms:

The unilateral Laplace transform of $dx(t)/dt$ is equal to

$$\int_{0+}^{\infty} \frac{dx(t)}{dt} e^{-st}\,dt = x(t)e^{-st}\Big|_{0+}^{\infty} + s\int_{0+}^{\infty} x(t)e^{-st}\,dt$$

Since $x(t)e^{-st} \to 0$ as $t \to \infty$ in the region of convergence of $X_I(s)$,† we obtain

$$\int_{0+}^{\infty} \frac{dx(t)}{dt} e^{-st}\,dt = -x(0+) + sX_I(s) \tag{11-3}$$

As an illustrative example, let

$$x(t) = e^{-\alpha t}u_{-1}(t)$$

Its unilateral Laplace transform is

$$X_I(s) = \frac{1}{s + \alpha}$$

Since $x(0+) = 1$, according to Eq. (11-3), the unilateral Laplace transform of $dx(t)/dt$ is

$$-1 + \frac{s}{s + \alpha} = \frac{-\alpha}{s + \alpha}$$

We can check the result by determining the unilateral Laplace transform of $dx(t)/dt$ directly. Since

$$\frac{dx(t)}{dt} = -\alpha e^{-\alpha t}u_{-1}(t) + u_0(t)$$

† Because, if this is not the case, the value of the integral $\int_{0+}^{\infty} |x(t)e^{-st}|\,dt$ will not be finite.

its unilateral Laplace transform is

$$\int_{0+}^{\infty} -\alpha e^{-\alpha t} e^{-st}\, dt = \frac{-\alpha}{s+\alpha}$$

As another example, the unilateral Laplace transform of the signal

$$x(t) = e^{-\alpha|t|}$$

is also $1/(s+\alpha)$. Since $x(0+) = 1$, the unilateral Laplace transform of $dx(t)/dt$ is

$$-1 + \frac{s}{s+\alpha} = -\frac{\alpha}{s+\alpha}$$

Again, we can check this result by determining the unilateral Laplace transform of

$$\frac{dx(t)}{dt} = \begin{cases} -\alpha e^{-\alpha t} & t > 0 \\ \alpha e^{\alpha t} & t < 0 \end{cases}$$

directly. Note that

$$\int_{0+}^{\infty} -\alpha e^{-\alpha t} e^{-st}\, dt = -\frac{\alpha}{s+\alpha}$$

The result in Eq. (11-3) can be extended immediately. According to Eq. (11-3), the unilateral Laplace transform of $d^2x(t)/dt^2$ is

$$-\frac{dx(0+)}{dt} + s[-x(0+) + sX_I(s)]$$

That is,

$$-\frac{dx(0+)}{dt} - sx(0+) + s^2 X_I(s)$$

As an example, for the signal $x(t) = e^{-\alpha t}u_{-1}(t)$, since $dx(0+)/dt = -\alpha$, the unilateral Laplace transform of $d^2x(t)/dt^2$ is

$$\alpha - s + s^2 \frac{1}{s+\alpha} = \frac{\alpha^2}{s+\alpha}$$

In general, the unilateral Laplace transform of $d^kx(t)/dt^k$ is equal to

$$-\frac{d^{k-1}x(0+)}{dt^{k-1}} - s\frac{d^{k-2}x(0+)}{dt^{k-2}} - \cdots - s^{k-1}x(0+) + s^k X_I(s)$$

We leave the simple derivation to the reader.

For $t_0 > 0$, the unilateral Laplace transform of the signal $x(t - t_0)$ is

$$\int_{0+}^{\infty} x(t - t_0)e^{-st}\,dt$$

Letting $\tau = t - t_0$, this integral can be written

$$\int_{-t_0+}^{\infty} x(\tau)e^{-s(\tau+t_0)}\,d\tau = e^{-st_0}\int_{-t_0+}^{\infty} x(\tau)e^{-s\tau}\,d\tau$$

Consequently, the unilateral Laplace transform of $x(t - t_0)$ is equal to $e^{-st_0}X_I(s)$ only if $x(t) = 0$ for $-t_0 \leq t \leq 0$. Similarly, the unilateral Laplace transform of $x(t + t_0)$ is equal to $e^{st_0}X_I(s)$ only if $x(t) = 0$ for $0 \leq t \leq t_0$. As an example, for

$$x(t) = e^{-\alpha t}u_{-1}(t)$$

and

$$X_I(s) = \frac{1}{s + \alpha}$$

the unilateral Laplace transform $x(t - t_0)$ is

$$\frac{e^{-st_0}}{s + \alpha}$$

because $x(t) = 0$ for $-t_0 \leq t \leq 0$. On the other hand, because $x(t) \neq 0$ for $0 \leq t \leq t_0$, the unilateral Laplace transform of $x(t + t_0)$ is not equal to

$$\frac{e^{st_0}}{s + \alpha}$$

In fact, since

$$x(t + t_0) = e^{-\alpha(t+t_0)}u_{-1}(t + t_0)$$

the unilateral Laplace transform of $x(t + t_0)$ is

$$\int_{0+}^{\infty} e^{-\alpha(t+t_0)}e^{-st}\,dt = \frac{e^{-\alpha t_0}}{s + \alpha}$$

As another example, let us determine the unilateral Laplace transform of the signal $x(t)$ shown in Fig. 11-1a. For the signal $w(t)$ shown in Fig. 11-1b, its unilateral Laplace transform is

$$W_I(s) = \int_{0}^{1} e^{-st}\,dt = \frac{1 - e^{-s}}{s}$$

Since

$$x(t) = w(t) + w(t - 3) + w(t - 6) + w(t - 9) + \cdots$$

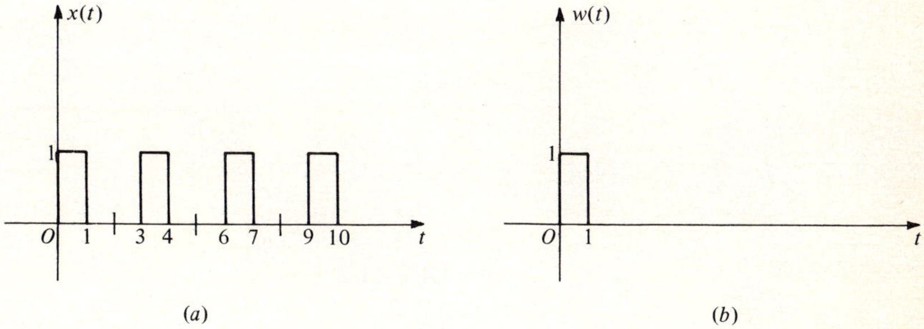

FIGURE 11-1

we obtain

$$X_I(s) = W_I(s)(1 + e^{-3s} + e^{-6s} + e^{-9s} + \cdots)$$
$$= \frac{1 - e^{-s}}{s(1 - e^{-3s})}$$

Note that the region of absolute convergence of $X_I(s)$ is Re $(s) > 0$.

The unilateral Laplace transform of the signal $\int_{-\infty}^{t} x(\tau)\, d\tau$ is equal to

$$\frac{1}{s} X_I(s) + \frac{1}{s} x^{-1}(0+)$$

where
$$x^{-1}(0+) = \int_{-\infty}^{0+} x(\tau)\, d\tau$$

To show this result, we note that

$$\int_{0+}^{\infty} e^{-st}\, dt \int_{-\infty}^{t} x(\tau)\, d\tau = \int_{0+}^{\infty} e^{-st}\, dt \int_{0+}^{t} x(\tau)\, d\tau + \int_{0+}^{\infty} e^{-st}\, dt \int_{-\infty}^{0+} x(\tau)\, d\tau$$
$$= \int_{0+}^{\infty} e^{-st}\, dt \int_{0+}^{t} x(\tau)\, d\tau + \frac{1}{s} x^{-1}(0+)$$

Interchanging the order of integration, we obtain

$$\int_{0+}^{\infty} e^{-st}\, dt \int_{-\infty}^{t} x(\tau)\, d\tau = \int_{0+}^{\infty} x(\tau)\, d\tau \int_{\tau}^{\infty} e^{-st}\, dt + \frac{1}{s} x^{-1}(0+)$$
$$= \frac{1}{s} \int_{0+}^{\infty} x(\tau) e^{-s\tau}\, d\tau + \frac{1}{s} x^{-1}(0+)$$
$$= \frac{1}{s} X_I(s) + \frac{1}{s} x^{-1}(0+)$$

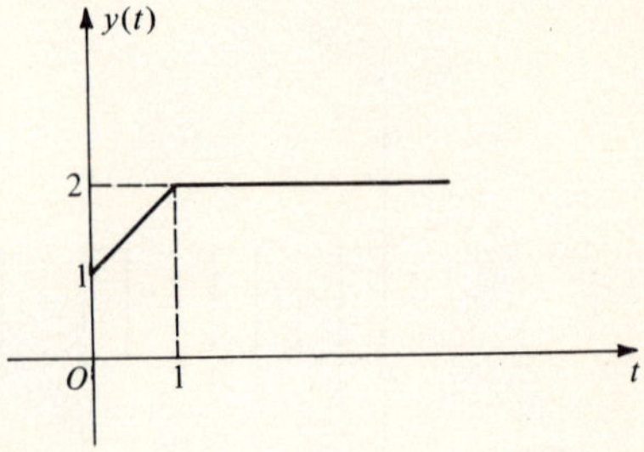

FIGURE 11-2

Suppose we are to determine the unilateral Laplace transform of the signal $y(t)$ shown in Fig. 11-2. A direct computation yields

$$Y_I(s) = \int_0^1 (1+t)e^{-st}\,dt + \int_1^\infty 2e^{-st}\,dt$$

$$= -\frac{1}{s^2}(1+s+st)e^{-st}\Big|_0^1 - \frac{2}{s}e^{-st}\Big|_1^\infty$$

$$= \frac{1}{s} + \frac{1-e^{-s}}{s^2}$$

To see an alternative way to compute $Y_I(s)$, we note that for the signal $x(t)$ shown in Fig. 11-3*a* the unilateral Laplace transform of the signal $\int_{-\infty}^t x(\tau)\,d\tau$, which is shown in Fig. 11-3*b*, is the same as that of $y(t)$. Since the unilateral Laplace transform of $x(t)$ is

$$X_I(s) = \frac{1-e^{-s}}{s}$$

the unilateral Laplace transform of $\int_{-\infty}^t x(\tau)\,d\tau$, which is equal to $Y_I(s)$, is

$$\frac{1}{s}\frac{1-e^{-s}}{s} + \frac{1}{s}\int_{-\infty}^{0+} x(\tau)\,d\tau = \frac{1-e^{-s}}{s^2} + \frac{1}{s}$$

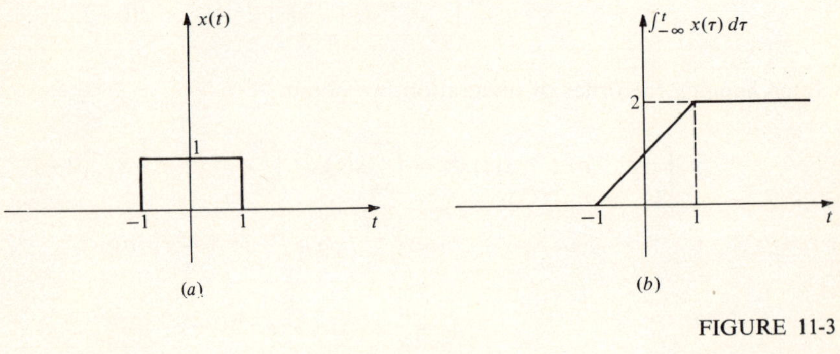

FIGURE 11-3

*11-3 AN ALTERNATIVE CONVENTION

We should point out that the definition of the unilateral Laplace transform of a signal in Eq. (11-1) is only one of many possible variations of the definition of the bilateral Laplace transform of a signal. The definition of Eq. (11-1) was adopted because in many problems of system analysis one is interested only in the portion of a signal $x(t)$ between $t = 0+$ and $t = \infty$. Indeed, it is quite possible to take a broader view and define a "truncated Laplace transform" of a time signal $x(t)$ to be

$$\int_{t_1}^{t_2} x(t)e^{-st}\,dt \tag{11-4}$$

for arbitrary t_1 and t_2. Instead of studying the transformation formula (11-4), which is not particularly useful anyway, let us consider a slightly different convention in the definition of the unilateral Laplace transform of a time signal by defining $X_I(s)$ to be

$$X_I(s) = \int_{0-}^{\infty} x(t)e^{-st}\,dt \tag{11-5}$$

Such a convention is indeed used in the literature. Because the lower limit of the integral in Eq. (11-5) is $0-$, we shall include singularity functions at $t = 0$ in our consideration.

To illustrate the difference between the two definitions in Eqs. (11-1) and (11-5), let us consider the unilateral Laplace transform of the signal

$$x(t) = u_0(t) + e^{-2t}u_{-1}(t)$$

According to the definition in Eq. (11-1)

$$X_I(s) = \frac{1}{s+2}$$

According to the definition in Eq. (11-5)

$$X_I(s) = 1 + \frac{1}{s+2}$$

Note that, because the region of absolute convergence of a singularity function is the entire s plane, the inclusion of singularity functions at $t = 0$ does not change the region of absolute convergence of the unilateral Laplace transform.

Also, according to the definition in Eq. (11-5), the unilateral Laplace transform of $dx(t)/dt$ is equal to

$$\begin{aligned}\int_{0-}^{\infty} \frac{dx(t)}{dt} e^{-st}\,dt &= x(t)e^{-st}\Big|_{0-}^{\infty} + s\int_{0-}^{\infty} x(t)e^{-st}\,dt \\ &= -x(0-) + sX_I(s)\end{aligned} \tag{11-6}$$

and, in general, the unilateral Laplace transform of $d^k x(t)/dt^k$ is equal to

$$-\frac{d^{k-1}x(0-)}{dt^{k-1}} - s\frac{d^{k-2}x(0-)}{dt^{k-2}} - \cdots - s^{k-1}x(0-) + s^k X_I(s)$$

As an example, let

$$x(t) = e^{-2t}u_{-1}(t)$$

According to Eq. (11-6), the unilateral Laplace transform of $dx(t)/dt$ is

$$\frac{s}{s+2}$$

Note that

$$\frac{dx(t)}{dt} = u_0(t) - 2e^{-2t}u_{-1}(t)$$

A direct computation also yields the unilateral Laplace transform:

$$1 - \frac{2}{s+2} = \frac{s}{s+2}$$

Although the definition in Eq. (11-5) enables us to include singularity functions at $t = 0$ in our consideration, we shall not follow this definition in our discussion, but rather that in Eq. (11-1). However, as will be seen in Sec. 11-6, depending on the circumstance, one convention might be more convenient than another.

11-4 THE INITIAL- AND FINAL-VALUE THEOREMS

Let us expand a time signal $x(t)$ as a Taylor series at $t = 0+$:

$$x(t) = x(0+) + \frac{dx(0+)}{dt}\frac{t}{1!} + \frac{d^2x(0+)}{dt^2}\frac{t^2}{2!} + \cdots$$
$$+ \frac{d^n x(0+)}{dt^n}\frac{t^n}{n!} + \frac{d^{n+1}x(0+)}{dt^{n+1}}\frac{t^{n+1}}{(n+1)!} + \cdots \qquad (11\text{-}7)$$

Computing the unilateral Laplace transforms of the two sides of Eq. (11-7), we obtain

$$X_I(s) = \frac{x(0+)}{s} + \frac{dx(0+)}{dt}\frac{1}{s^2} + \frac{d^2x(0+)}{dt^2}\frac{1}{s^3} + \cdots$$
$$+ \frac{d^n x(0+)}{dt^n}\frac{1}{s^{n+1}} + \frac{d^{n+1}x(0+)}{dt^{n+1}}\frac{1}{s^{n+2}} + \cdots \qquad (11\text{-}8)$$

Multiplying both sides of Eq. (11-8) by s and letting s approach ∞, we obtain the result

$$\lim_{s\to\infty} sX_I(s) = x(0+) \tag{11-9}$$

Equation (11-9) is known as the *initial-value theorem.* It states that the value of the function $sX_I(s)$ for large s is equal to the value of $x(t)$ at $0+$.

As an example, for

$$x(t) = e^{-\alpha t}u_{-1}(t)$$

we have

$$x(0+) = 1$$

Since

$$X_I(s) = \frac{1}{s+\alpha}$$

and

$$\lim_{s\to\infty} \frac{s}{s+\alpha} = 1$$

Eq. (11-7) is checked.

As another example, let

$$x(t) = t^2e^{-\alpha t}u_{-1}(t)$$

We have

$$X_I(s) = \frac{2}{(s+\alpha)^3}$$

and

$$x(0+) = \lim_{s\to\infty} \frac{2s}{(s+\alpha)^3} = 0$$

Again, Eq. (11-7) is checked.

A more general form of the initial-value theorem states that if

$$x(0+) = \frac{dx(0+)}{dt} = \frac{d^2x(0+)}{dt^2} = \cdots = \frac{d^{n-1}x(0+)}{dt^{n-1}} = 0$$

but

$$\frac{d^nx(t)}{dt^n} \neq 0$$

then

$$\lim_{s\to\infty} s^{n+1}X_I(s) = \frac{d^nx(0+)}{dt^n} \tag{11-10}$$

According to the assumption, Eq. (11-7) can be simplified as

$$x(t) = \frac{d^nx(0+)}{dt^n}\frac{t^n}{n!} + \frac{d^{n+1}x(0+)}{dt^{n+1}}\frac{t^{n+1}}{(n+1)!} + \cdots$$

Hence

$$X_I(s) = \frac{d^n x(0+)}{dt^n}\frac{1}{s^{n+1}} + \frac{d^{n+1}x(0+)}{dt^{n+1}}\frac{1}{s^{n+2}} + \cdots$$

That is,

$$s^{n+1}X_I(s) = \frac{d^n x(0+)}{dt^n} + \frac{1}{s}\frac{d^{n+1}x(0+)}{dt^{n+1}} + \cdots \tag{11-11}$$

Equation (11-10) follows immediately when we multiply both sides of Eq. (11-11) by s^{n+1} and let s approach ∞.

As an example, for

$$x(t) = t^2 e^{-\alpha t} u_{-1}(t)$$

and

$$X_I(s) = \frac{2}{(s+\alpha)^3}$$

we have

$$x(0+) = 0$$

$$\frac{dx(0+)}{dt} = 0$$

and

$$\frac{d^2x(0+)}{dt^2} = 2$$

Indeed, according to Eq. (11-10),

$$\lim_{s\to\infty} s^3 X_I(s) = \lim_{s\to\infty} \frac{2s^3}{(s+\alpha)^3} = 2$$

As another example, for

$$x(t) = tu_{-1}(t) - (t-T)u_{-1}(t-T)$$

and

$$X_I(s) = \frac{1-e^{-sT}}{s^2}$$

since

$$x(0+) = 0$$

and

$$\frac{dx(0+)}{dt} = 1$$

we have

$$\lim_{s\to\infty} s^2 X_I(s) = \lim_{s\to\infty} s^2 \frac{1-e^{-sT}}{s^2} = 1$$

Analogous to the initial-value theorem, we have the *final-value theorem:*

$$\lim_{t\to\infty} x(t) = \lim_{s\to 0} sX_I(s) \tag{11-12}$$

which relates the behavior of $X_I(s)$ at $s = 0$ to that of $x(t)$ at $t = \infty$. Let us consider the limit of the integral

$$\int_{0+}^{\infty} \frac{dx(t)}{dt} e^{-st}\, dt$$

as s approaches 0. We have

$$\begin{aligned}
\lim_{s\to 0} \int_{0+}^{\infty} \frac{dx(t)}{dt} e^{-st}\, dt &= \int_{0+}^{\infty} \frac{dx(t)}{dt}\, dt \\
&= x(t)\Big|_{0+}^{\infty} \\
&= \lim_{t\to\infty} x(t) - x(0+)
\end{aligned} \tag{11-13}$$

Also, according to Eq. (11-3),

$$\begin{aligned}
\lim_{s\to 0} \int_{0+}^{\infty} \frac{dx(t)}{dt} e^{-st}\, dt &= \lim_{s\to 0} [-x(0+) + sX_I(s)] \\
&= -x(0+) + \lim_{t\to 0} sX_I(s)
\end{aligned} \tag{11-14}$$

Combining Eqs. (11-13) and (11-14), we obtain

$$\lim_{t\to\infty} x(t) - x(0+) = -x(0+) + \lim_{s\to 0} sX_I(s)$$

Thus, Eq. (11-12) follows immediately. Note that Eq. (11-14) is valid only if the region of absolute convergence of $dx(t)/dt$ includes the $j\omega$ axis. Consequently, the final-value theorem can be applied only if $sX_I(s)$ does not contain a pole that is on or to the right of the $j\omega$ axis.

As an example, for

$$x(t) = e^{-\alpha t} u_{-1}(t)$$

and

$$X_I(s) = \frac{1}{s + \alpha}$$

we have

$$\lim_{t\to\infty} x(t) = \lim_{s\to 0} s \frac{1}{s + \alpha} = 0$$

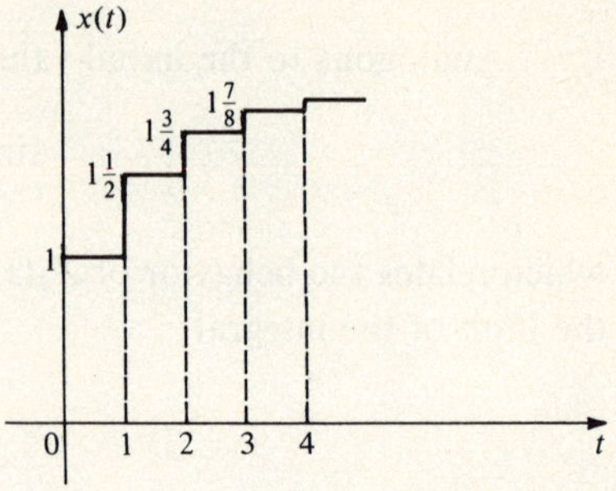

FIGURE 11-4

Similarly, for

$$x(t) = t^2 e^{-\alpha t} u_{-1}(t)$$

and
$$X_I(s) = \frac{2}{(s+\alpha)^3}$$

we have

$$\lim_{s\to\infty} x(t) = \lim_{s\to 0} \frac{2s}{(s+\alpha)^3} = 0$$

As another example, let us consider a system whose impulse response is

$$h(t) = e^{-3t} u_{-1}(t)$$

Suppose the input signal $x(t)$ is that shown in Fig. 11-4, and we want to know the value of the output signal $y(t)$ at $t = \infty$. Since

$$H_I(s) = \frac{1}{s+3}$$

and
$$X_I(s) = \frac{1}{s}\left(1 + \frac{1}{2}e^{-s} + \frac{1}{4}e^{-2s} + \frac{1}{8}e^{-3s} + \cdots\right)$$
$$= \frac{1}{s(1-0.5e^{-s})}$$

we obtain, according to the final-value theorem,

$$\begin{aligned}\lim_{t\to\infty} y(t) &= \lim_{s\to 0} sY_I(s)\\ &= \lim_{s\to 0} sX_I(s)H_I(s)\\ &= \lim_{s\to 0} \frac{1}{1-0.5e^{-s}}\frac{1}{s+3}\\ &= \tfrac{2}{3}\end{aligned}$$

Note that it is more tedious to determine $y(t)$ from $Y_I(s)$ and then compute the limit $\lim_{t\to\infty} y(t)$ directly.

11-5 UNILATERAL z TRANSFORMATION

Analogous to the unilateral Laplace transformation for continuous signals, we can define the unilateral z transformation for discrete signals. Let $x(n)$ be a discrete signal that is specified for $n \geq 0$. We define the *unilateral z transform* of $x(n)$, denoted $X_I(z)$, to be

$$\begin{aligned} X_I(z) &= x(0) + x(1)z^{-1} + x(2)z^{-2} + \cdots + x(k)z^{-k} + \cdots \\ &= \sum_{n=0}^{\infty} x(n)z^{-n} \end{aligned} \tag{11-15}$$

We also define the region of absolute convergence of $X_I(z)$ to be the area of the z plane containing all z such that the sum $\sum_{n=0}^{\infty} |x(n)z^{-n}|$ is finite. Clearly, the region of absolute convergence of $X_I(z)$ is always the exterior of a circle of finite radius.† For example, let

$$x(n) = \begin{cases} (\frac{1}{3})^n & n \geq 0 \\ (\frac{1}{2})^{-n} & n < 0 \end{cases} \tag{11-16}$$

The unilateral z transform of $x(n)$ is

$$\begin{aligned} X_I(z) &= 1 + \tfrac{1}{3}z^{-1} + (\tfrac{1}{3})^2 z^{-2} + \cdots + (\tfrac{1}{3})^k z^{-k} + \cdots \\ &= \frac{1}{1 - \frac{1}{3}z^{-1}} \end{aligned}$$

with its region of absolute convergence being $|z| > \frac{1}{3}$. Note that the signal

$$y(n) = \begin{cases} (\frac{1}{3})^n & n \geq 0 \\ 0 & n < 0 \end{cases}$$

will have the same unilateral z transform and the same region of absolute convergence.

All the properties of bilateral z transformation are carried over to unilateral z transformation with the exception that if $X_I(z)$ is the unilateral z transform of $x(n)$ then the unilateral z transform of $x(n-1)$ is

$$z^{-1}X_I(z) + x(-1)$$

and, in general, the unilateral z transform of $x(n-k)$, $k > 0$, is

$$z^{-k}X_I(z) + z^{-k+1}x(-1) + z^{-k+2}x(-2) + \cdots + z^{-1}x(-k+1) + x(-k)$$

Note that the unilateral z transform of $x(n-k)$ is equal to $z^{-k}X_I(z)$ only if $x(-1)$, $x(-2), \ldots, x(-k)$ are all equal to zero. Similarly, the unilateral z transform of $x(n+1)$ is

$$zX_I(z) - zx(0)$$

† The reader should recognize that this corresponds to the case of the region of absolute convergence of the unilateral Laplace transform of a signal being a right half plane.

and, in general, the unilateral z transform of $x(n+k)$, $k > 0$, is

$$z^k X_I(z) - z^k x(0) - z^{k-1}x(1) - \cdots - z^2 x(k-2) - zx(k-1)$$

Again, the unilateral z transform of $x(n+k)$ is equal to $z^k X_I(z)$ only if $x(0)$, $x(1)$, ..., $x(k-1)$ are all equal to zero. For example, for the signal $x(n)$ in Eq. (11-16), the unilateral z transform of $x(n-1)$ is

$$\tfrac{1}{2} + z^{-1} + (\tfrac{1}{3})z^{-2} + (\tfrac{1}{3})^2 z^{-3} + \cdots + (\tfrac{1}{3})^{k+1} z^{-k} + \cdots$$
$$= \frac{1}{2} + \frac{z^{-1}}{1 - \frac{1}{3}z^{-1}}$$

and the unilateral z transform of $x(n+1)$ is

$$\tfrac{1}{3} + (\tfrac{1}{3})^2 z^{-1} + (\tfrac{1}{3})^3 z^{-2} + \cdots + (\tfrac{1}{3})^{k+1} z^{-k} + \cdots$$
$$= \frac{\frac{1}{3}}{1 - \frac{1}{3}z^{-1}}$$

which, as the reader can readily check, is equal to $zX_I(z) - z$.

Setting z to ∞ in Eq. (11-15), we obtain

$$x(0) = \lim_{z \to \infty} X_I(z)$$

This result is known as the *initial-value theorem.* For example, let

$$x(n) = \begin{cases} 2^{n+1} & n \geq 0 \\ 0 & n < 0 \end{cases}$$

Then
$$X_I(z) = \frac{2}{1 - 2z^{-1}}$$

According to the initial-value theorem

$$x(0) = \lim_{z \to \infty} \frac{2}{1 - 2z^{-1}} = 2$$

The *final-value theorem* states that

$$x(\infty) = \lim_{z \to 1} (1 - z^{-1}) X_I(z) \tag{11-17}$$

provided that the region of absolute convergence of $(1 - z^{-1})X_I(z)$ includes the region $|z| \geq 1$. To prove the final-value theorem, let us consider the sum

$$\sum_{n=0}^{\infty} [x(n) - x(n-1)] z^{-n}$$

where $x(-1)$ is equal to 0. If we let z approach 1, then

$$\lim_{z\to 1} \sum_{n=0}^{\infty} [x(n) - x(n-1)]z^{-n} = x(\infty) \tag{11-18}$$

On the other hand,

$$\sum_{n=0}^{\infty} [x(n) - x(n-1)]z^{-n} = (1 - z^{-1})X_I(z)$$

Therefore, we have

$$\lim_{z\to 1} \sum_{n=0}^{\infty} [x(n) - x(n-1)]z^{-n} = \lim_{z\to 1} (1 - z^{-1})X_I(z) \tag{11-19}$$

Combining Eqs. (11-18) and (11-19), we obtain the result in Eq. (11-17). Note that the limit in Eq. (11-17) is meaningful only if $(1 - z^{-1})X(z)$ is defined at $z = 1$. In other words, the region of absolute convergence of $(1 - z^{-1})X(z)$ must include the circle $|z| = 1$. As an example, let

$$x(n) = \begin{cases} 1 & n \geq 0 \\ 0 & n < 0 \end{cases}$$

Then

$$X_I(z) = \frac{1}{1 - z^{-1}}$$

According to the final-value theorem

$$x(\infty) = \lim_{z\to 1} (1 - z^{-1}) \frac{1}{1 - z^{-1}} = 1$$

As another example, let

$$x(n) = \begin{cases} 1 & n \text{ is even} \\ 0 & \text{otherwise} \end{cases}$$

Then

$$X_I(z) = \frac{1}{1 - z^{-2}}$$

Although

$$\lim_{z\to 1} (1 - z^{-1}) \frac{1}{1 - z^{-2}} = \lim_{z\to 1} \frac{1}{1 + z^{-1}} = \frac{1}{2}$$

the value of $x(\infty)$ is clearly not $\frac{1}{2}$. Because the function $1/(1 + z^{-1})$ has a pole at $z = -1$, its region of absolute convergence does not include the unit circle. Consequently, the final-value theorem does not apply.

11-6 SOLUTION OF DIFFERENTIAL EQUATIONS

The method of unilateral Laplace transformation is most useful in solving linear differential equations with constant coefficients. Let us illustrate the solution procedure by an example. Consider the differential equation

$$\frac{d^2y(t)}{dt^2} + 4\frac{dy(t)}{dt} + 3y(t) = 4\frac{dx(t)}{dt} + 6x(t) \tag{11-20}$$

Suppose that

$$x(t) = e^{-2t}u_{-1}(t)$$

Also, we are given the boundary conditions $y(0+) = 0$ and $y'(0+) = 8$. To determine $y(t)$ for $t \geq 0$, we compute the unilateral Laplace transforms of the two sides of Eq. (11-20). We obtain

$$\begin{aligned} s^2 Y_I(s) - sy(0+) - y'(0+) + 4[sY_I(s) - y(0+)] + 3Y_I(s) \\ = 4[sX_I(s) - x(0+)] + 6X_I(s) \end{aligned} \tag{11-21}$$

Since

$$X_I(s) = \frac{1}{s+2}$$

Eq. (11-21) is simplified to

$$(s^2 + 4s + 3)Y_I(s) - 8 = \frac{4s}{s+2} + \frac{6}{s+2} - 4$$

or

$$Y_I(s) = \frac{8s+14}{(s^2+4s+3)(s+2)} = \frac{3}{s+1} - \frac{5}{s+3} + \frac{2}{s+2}$$

It follows that

$$y(t) = 3e^{-t} - 5e^{-3t} + 2e^{-2t} \qquad t \geq 0$$

Note that $3e^{-t} - 5e^{-3t}$ is the homogeneous solution of $y(t)$, where the coefficients 3 and -5 are determined by the boundary conditions at $t = 0+$, and $2e^{-2t}$ is the particular solution.

Suppose that we were given the boundary conditions at $t = 0-$.† We can determine $y(t)$ for $t \leq 0$ first, and then compute the boundary conditions at $t = 0+$ by matching the singularity functions at $t = 0$ on the two sides of Eq. (11-20). If we follow the convention in Sec. 11-3, we can also determine $y(t)$ for $t \geq 0$, using the boundary conditions at $t = 0-$ without computing the boundary conditions at $t = 0+$

† This example should be skipped if Sec. 11-3 was omitted in the discussion.

explicitly. Suppose that we were given that $y(0-) = 0$ and $y'(0-) = 2$ in the preceding example. Computing the unilateral Laplace transform of both sides of Eq. (11-20), we obtain

$$s^2 Y_I(s) - sy(0-) - y'(0-) + 4[sY_I(s) - y(0-)] + 3Y_I(s) = 4[sX_I(s) - x(0-)] + 6X_I(s)$$

or
$$(s^2 + 4s + 3)Y_I(s) - 2 = \frac{4s}{s+2} + \frac{6}{s+2}$$

It follows that

$$y(t) = 2e^{-t} - 4e^{-3t} + 2e^{-2t} \qquad t \geq 0$$

As another example, let us suppose that for the differential equation (11-20)

$$x(t) = \begin{cases} 1 & 0 < t < 1 \\ 0 & 1 < t < \infty \end{cases}$$

Since
$$X_I(s) = \frac{1 - e^{-s}}{s}$$

we obtain from Eq. (11-21)

$$(s^2 + 4s + 3)Y_I(s) - 8 = 4(1 - e^{-s}) + \frac{6(1 - e^{-s})}{s} - 4$$

or

$$Y_I(s) = \frac{8s + 6 - (4s + 6)e^{-s}}{s(s^2 + 4s + 3)} = \frac{2 - 2e^{-s}}{s} + \frac{1 + e^{-s}}{s+1} + \frac{-3 + e^{-s}}{s+3}$$

Thus,

$$y(t) = 2u_{-1}(t) - 2u_{-1}(t-1) + e^{-t}u_{-1}(t) + e^{-(t-1)}u_{-1}(t-1) - 3e^{-3t}u_{-1}(t) + e^{-3(t-1)}u_{-1}(t-1) \qquad t \geq 0$$

or
$$y(t) = \begin{cases} 2 + e^{-t} - 3e^{-3t} & 0 < t < 1 \\ (1 + e)e^{-t} - (3 - e^3)e^{-3t} & 1 < t < \infty \end{cases}$$

Note that if we use the method of solution presented in Chap. 2, we would determine the expression for $y(t)$ in the time region $0 < t < 1$ first, and then determine the expression for $y(t)$ in the time region $1 < t < \infty$, using the boundary conditions $y(1+)$ and $y'(1+)$ obtained by matching singularity functions in the two sides of Eq. (11-20) at $t = 1$. It is clear that the method of solution presented above is computationally much simpler.

These examples illustrate quite clearly a procedure for solving differential equations by the method of unilateral Laplace transformation. Instead of determining directly the output signal $y(t)$, we determine, from the given differential equation, first its unilateral Laplace transform, $Y_I(s)$. We can then determine $y(t)$ for $t \geq 0$ from $Y_I(s)$, using the inverse transformation formula. Note that the transformation technique reduces a *differential* equation relating two time signals to an *algebraic* equation relating their unilateral Laplace transforms. Moreover, in comparison with the method of solution presented in Chap. 2, the transformation approach not only determines the homogeneous solution and particular solution in one step, it also determines the multiplying constants in the homogeneous solution at the same time so that the given boundary conditions will be satisfied.

11-7 SOLUTION OF DIFFERENCE EQUATIONS

Similar to the case of applying the method of unilateral Laplace transformation to the solution of differential equations, the method of unilateral z transformation can be applied to solve difference equations. We present here some illustrative examples. As our first example, we recall the problem on the multiplication of rabbits studied in Sec. 2-6. Let $y(n)$ denote the number of pairs of rabbits there are at the beginning of the nth month. As was established in Sec. 2-6, we have

$$y(n) = y(n-1) + y(n-2) \tag{11-22}$$

with the boundary condition $y(0) = 0$ and $y(1) = 1$. Computing the unilateral Laplace transforms of the two sides of Eq. (11-22), we obtain

$$Y_I(z) = z^{-1} Y_I(z) + y(-1) + z^{-2} Y_I(z) + z^{-1} y(-1) + y(-2) \tag{11-23}$$

The values of $y(-1)$ and $y(-2)$ can be determined from Eq. (11-22) in a step-by-step manner: Since

$$y(1) = y(0) + y(-1)$$

we obtain

$$y(-1) = y(1) - y(0) = 1 - 0 = 1$$

Similarly,

$$y(-2) = y(0) - y(-1) = 0 - 1 = -1$$

Equation (11-23) is reduced to

$$Y_I(z) = z^{-1} Y_I(z) + 1 + z^{-2} Y_I(z) + z^{-1} - 1$$

or
$$Y_I(z) = \frac{z^{-1}}{1 - z^{-1} - z^{-2}}$$
$$= \frac{z^{-1}}{\left(1 - \frac{1+\sqrt{5}}{2} z^{-1}\right)\left(1 - \frac{1-\sqrt{5}}{2} z^{-1}\right)}$$
$$= \frac{1/\sqrt{5}}{1 - \frac{1+\sqrt{5}}{2} z^{-1}} - \frac{1/\sqrt{5}}{1 - \frac{1-\sqrt{5}}{2} z^{-1}}$$

It follows that
$$y(n) = \frac{1}{\sqrt{5}}\left(\frac{1+\sqrt{5}}{2}\right)^n - \frac{1}{\sqrt{5}}\left(\frac{1-\sqrt{5}}{2}\right)^n \qquad n \geq 0$$

Note that in the solution procedure shown above, we had to compute the values $y(-1)$ and $y(-2)$ from the given boundary conditions $y(0) = 0$ and $y(1) = 1$. There is an alternative way to take care of the boundary conditions: Multiplying both sides of Eq. (11-22) by z^{-n} and summing from $n = 2$ to $n = \infty$,† we obtain
$$Y_I(z) - y(1)z^{-1} - y(0) = z^{-1}[Y_I(z) - y(0)] + z^{-2}Y_I(z)$$
or
$$Y_I(z) = \frac{z^{-1}}{1 - z^{-1} - z^{-2}}$$

As another example, consider the difference equation
$$y(n) - 2y(n-1) = x(n-1) \tag{11-24}$$
Given that
$$x(n) = 4^n \qquad n \geq 0$$
we want to determine $y(n)$ for $n \geq 0$, if it is known that $y(0) = 1$. Multiplying both sides of Eq. (11-24) by z^{-n} and summing from $n = 1$ to $n = \infty$, we obtain
$$\sum_{n=1}^{\infty} y(n)z^{-n} - 2\sum_{n=1}^{\infty} y(n-1)z^{-n} = \sum_{n=1}^{\infty} 4^{n-1}z^{-n}$$
or
$$Y_I(z) = \frac{1}{1 - 2z^{-1}}\left(\frac{z^{-1}}{1 - 4z^{-1}} + 1\right) = \frac{\frac{1}{2}}{1 - 4z^{-1}} + \frac{\frac{1}{2}}{1 - 2z^{-1}}$$
It follows that
$$y(n) = \tfrac{1}{2}4^n + \tfrac{1}{2}2^n \qquad n \geq 0$$

† Note that computing the unilateral z transforms of both sides of Eq. (11-22) amounts to multiplying both sides of Eq. (11-22) by z^{-n} and summing from $n = 0$ to $n = \infty$.

We show another example in which the method of unilateral z transformation is applied to solve a *nonlinear* difference equation. Consider the problem of determining the number of ways to parenthesize the expression

$$w_1 + w_2 + \cdots + w_{n-1} + w_n$$

so that only two terms will be added at one time. For instance, the expression $w_1 + w_2 + w_3 + w_4$ can be parenthesized as

$$((w_1 + w_2) + (w_3 + w_4))$$
$$(w_1 + ((w_2 + w_3) + w_4))$$

and so on. Let $x(i)$ denote the number of ways of parenthesizing an expression with i terms. Consider the two subexpressions

$$w_1 + w_2 + \cdots + w_{n-r}$$
$$w_{n-r+1} + w_{n-r+2} + \cdots + w_n$$

There are $x(n-r)$ ways to parenthesize the first subexpression and $x(r)$ ways to parenthesize the second subexpression. It follows that there are $x(n-r)x(r)$ ways to parenthesize the overall expression in which the final parentheses added joins these two subexpressions. Letting r range from 1 to $n-1$, we obtain the difference equation

$$x(n) = x(n-1) + x(n-2)x(2) + \cdots + x(2)x(n-2) + x(n-1) \qquad (11\text{-}25)$$

Since the values $x(0)$ and $x(1)$ have no physical meaning, we can choose them in some arbitrary manner. If we let

$$x(0) = 0$$
$$x(1) = 1$$

then Eq. (11-25) can be written

$$x(n) = x(n)x(0) + x(n-1)x(1) + x(n-2)x(2) + \cdots + x(2)x(n-2) + x(1)x(n-1) + x(0)x(n) \qquad (11\text{-}26)$$

Multiplying both sides of Eq. (11-26) and summing from $n = 2$ to $n = \infty$, we obtain

$$\sum_{n=2}^{\infty} x(n)z^{-n} = \sum_{n=2}^{\infty} [x(n)x(0) + x(n-1)x(1) + \cdots + x(1)x(n-1) + x(0)x(n)]z^{-n}$$

$$X_I(z) - x(1)z^{-1} - x(0) = [X_I(z)]^2 - [x(0)]^2 - [x(1)x(0) + x(0)x(1)]z^{-1}$$

$$[X_I(z)]^2 - X_I(z) + z^{-1} = 0$$

$$X_I(z) = \frac{1 \pm \sqrt{1 - 4z^{-1}}}{2}$$

Although there are two solutions for $X_I(z)$, only the one corresponds to a sequence of nonnegative numbers $x(0)$, $x(1)$, $x(2)$, ... in the answer we are interested in. It turns out that we should choose

$$X_I(z) = \frac{1 - \sqrt{1 - 4z^{-1}}}{2}$$

It follows that

$$x(n) = \begin{cases} 0 & n = 0 \\ \dfrac{1}{n}\dbinom{2n-2}{n-1} & n > 0 \end{cases}$$

11-8 SOLUTION OF STATE EQUATIONS

Similar to the solution of difference and differential equations, the method of unilateral Laplace transformation and unilateral z transformation can also be applied to solve state equations. As an example, consider the state equations

$$\frac{d\lambda_1(t)}{dt} = 2\lambda_1(t) - \lambda_2(t) + x(t)$$

$$\frac{d\lambda_2(t)}{dt} = -4\lambda_1(t) + 5\lambda_2(t)$$

where $x(t)$ is equal to $\sin 2tu_{-1}(t)$. Also, it is given that $\lambda_1(0+) = 0$, $\lambda_2(0+) = 2$. From the state equations, we obtain

$$s\Lambda_{1I}(s) = 2\Lambda_{1I}(s) - \Lambda_{2I}(s) + \frac{2}{s^2+4}$$

$$s\Lambda_{2I}(s) - 2 = -4\Lambda_{1I}(s) + 5\Lambda_{2I}(s)$$

That is,

$$(s-2)\Lambda_{1I}(s) + \Lambda_{2I}(s) = \frac{2}{s^2+4}$$

$$4\Lambda_{1I}(s) + (s-5)\Lambda_{2I}(s) = 2$$

Solving these simultaneous equations for the unknown Λ_{1I} and Λ_{2I}, we obtain

$$\Lambda_{1I}(s) = \frac{-2s^2 + 2s - 18}{(s^2 - 7s + 6)(s^2 + 4)} = \frac{\frac{18}{25}}{s-1} - \frac{\frac{39}{100}}{s-6} + \frac{-\frac{33}{100}s - \frac{19}{50}}{s^2+4}$$

$$\Lambda_{2I}(s) = \frac{2s^3 - 4s^2 + 8s - 24}{(s^2 - 7s + 6)(s^2 + 4)} = \frac{\frac{18}{25}}{s-1} + \frac{\frac{39}{25}}{s-6} + \frac{-\frac{7}{25}s - \frac{2}{25}}{s^2+4}$$

Consequently, the inverse transformation formula yields

$$\lambda_1(t) = \tfrac{18}{25}e^{t} - \tfrac{39}{100}e^{6t} - \tfrac{33}{100}\cos 2t - \tfrac{19}{100}\sin 2t \qquad t \geq 0$$

$$\lambda_2(t) = \tfrac{18}{25}e^{t} + \tfrac{39}{25}e^{6t} - \tfrac{7}{25}\cos 2t - \tfrac{1}{25}\sin 2t \qquad t \geq 0$$

The reader should have no difficulty in seeing, from this example, a general procedure for solving a set of state equations by the method of unilateral Laplace transformation and unilateral z transformation. Again, we note that the transformation technique reduces a set of simultaneous differential or difference equations relating the state variables and the input signal to a set of algebraic equations relating their unilateral Laplace transforms or unilateral z transforms.

11-9 REMARKS AND REFERENCES

We have now completed our discussion of the various methods for analyzing the behavior of linear time-invariant systems. It is particularly interesting to see the relationship between a time domain approach and its corresponding frequency domain approach, as was pointed out in Sec. 11-1. To be able to handle a system analysis problem in different ways is more than just a mathematical convenience. Further insight will always be gained when one can look at a problem in several different ways.

Although our discussion of the application of the method of unilateral Laplace transformation and z transformation to the solution of differential equations, difference equations, and state equations is rather brief, we trust that the reader had no difficulty in understanding the basic ideas, which are actually quite simple and straightforward. In most cases, it is easier to obtain a description of a given system in terms of differential equations, difference equations, or state equations. However, these equations can be solved more directly when the Laplace transformation representation or z transformation representation is used.

As references, the reader may wish to consult those cited in Chaps. 5 and 9.

PROBLEMS

11-1 Determine the unilateral Laplace transform (or z transform) of each of the following time signals:

(*a*) $x(n) = \sin 2n$

(*b*) $x(n) = 2^{-|n|} \cos n$

(*c*) $x(n) = n^2 a^{-n}$

(*d*) $x(t) = \sinh at$

(*e*) $x(t) = \sin t \cosh t - \cos t \sinh t$

(*f*) $x(t) = \dfrac{e^{-at}}{t}$

(*g*) $x(t) = \dfrac{1}{1 + e^{t}}$

11-2 Determine the unilateral Laplace transforms (or *z* transforms) of the periodic signals shown in Fig. 11P-1.

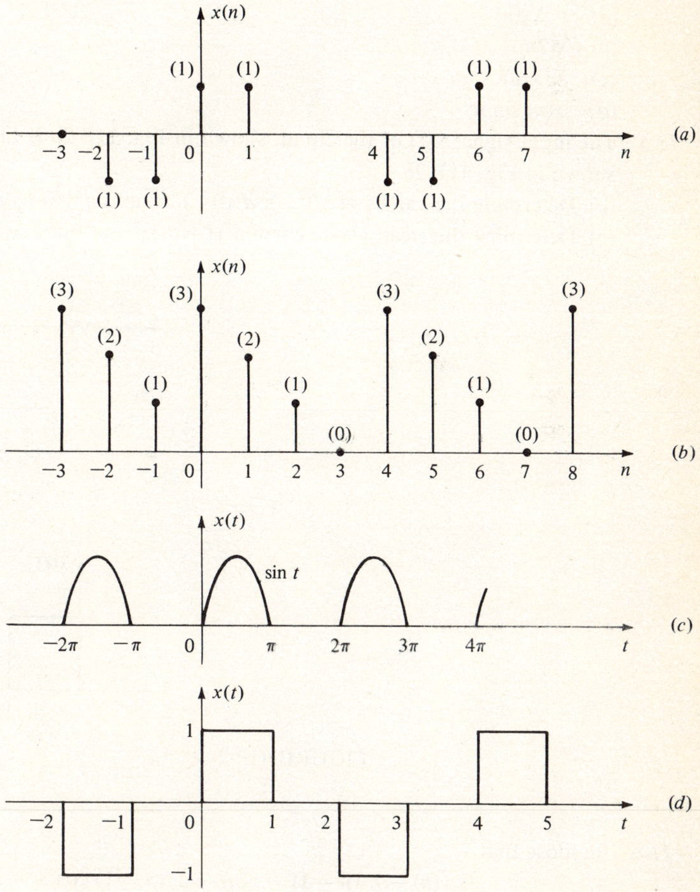

FIGURE 11P-1

11-3 Given that the unilateral Laplace transform of the signal $x(t)$ is $1/(s^2 + 4s + 8)$, find the unilateral Laplace transform of each of the following signals:

(*a*) $\dfrac{x(t)}{t}$

(*b*) $x(t) \sin t$

(*c*) $x(5t + 3)$

(*d*) $t \dfrac{d^2x(t)}{dt^2}$

11-4 Given that $x(n) = 0$ for $n = 0$ and its unilateral z transform is

$$\frac{e^2 - 1}{e} \frac{3}{(1 - ze)(1 - ze^{-1})}$$

find the unilateral z transform of each of the following signals:

(*a*) $n\, \Delta x(n)$

(*b*) $x(2n - 3)$

(*c*) $\nabla^2 x(n)$

(*d*) $x(n) \sin n$

11-5 The input signal $x(t)$ of the circuit shown in Fig. 11P-2*a* is the periodic square wave shown in Fig. 11P-2*b*.

(*a*) Determine the values of $y(0+)$, $dy(0+)/dt$, and $d^2y(0+)/dt$.

(*b*) Determine the steady-state current $y(+\infty)$.

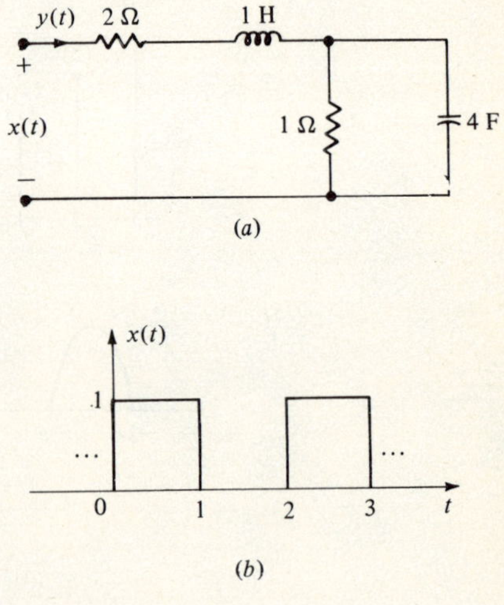

FIGURE 11P-2

11-6 Suppose that

$$y(n) + \tfrac{1}{2}y(n-1) = x(n) - 2x(n-1)\cos\omega + x(n-2)$$

with

$$y(-2) = y(-1) = 0$$

If $x(n) = \sin \omega n + \cos \omega n$ for $n \geq 0$, find $y(n)$ as $n \to \infty$.

11-7 Solve the differential (or difference) equation in each of the following problems, using the unilateral Laplace transform (or z transform) method.

(*a*) Problem 2-3, part (*a*).

(*b*) Problem 2-4.

(*c*) Problem 2-16.

(*d*) Problem 2-18.

11-8 Solve the set of state equations in each of the following problems, using the unilateral Laplace transform (or z transform) method.

(*a*) Problem 3-8, parts (*a*) and (*b*).

(*b*) Problem 3-14, part (*a*).

11-9 Let $x(n)$ be the number of nonoverlapping regions into which the interior of a convex n-gon is divided by its diagonals. Suppose that no three diagonals meet at one point.

(*a*) Show that

$$x(n) - x(n-1) = \frac{(n-1)(n-2)(n-3)}{6} + n - 2 \qquad n \geq 3$$

and $x(0) = x(1) = x(2) = 0$.

(*b*) Find the unilateral z transform of $x(n)$.

(*c*) Find the expression for $x(n)$ from its unilateral z transform.

11-10 Let $x(n)$ be the value of the $n \times n$ determinant of the form

$$x(n) = \begin{vmatrix} 2 & 1 & 0 & 0 & 0 & \cdots & 0 & 0 & 0 \\ 3 & 2 & 1 & 0 & 0 & \cdots & 0 & 0 & 0 \\ 0 & 3 & 2 & 1 & 0 & \cdots & 0 & 0 & 0 \\ 0 & 0 & 3 & 2 & 1 & \cdots & 0 & 0 & 0 \\ 0 & 0 & 0 & 3 & 2 & \cdots & 0 & 0 & 0 \\ \cdot & \cdot & \cdot & \cdot & \cdot & \cdot & \cdot & \cdot & \cdot \\ 0 & 0 & 0 & 0 & 0 & \cdots & 2 & 1 & 0 \\ 0 & 0 & 0 & 0 & 0 & \cdots & 3 & 2 & 1 \\ 0 & 0 & 0 & 0 & 0 & \cdots & 0 & 3 & 2 \end{vmatrix}$$

(*a*) Show that

$$x(n) = 2x(n-1) - 3x(n-2)$$

(*b*) What are the boundary conditions $x(1)$ and $x(2)$?

(*c*) Determine $x(n)$ for $n > 0$.

11-11 A sequence of binary digits is fed to a counter at the rate of 1 digit/s. The counter is designed to register 1 s in the input sequence. However, it is so slow that it is locked for exactly 7 s following each registration, during which time input digits are ignored. Let $x(n)$ be the number of binary sequences of length n at the end of which the counter is not locked. Find the difference equations which $x(n)$ satisfies, the boundary conditions, and the unilateral z transform of $x(n)$.

11-12 (*a*) Let $x(n)$ be the number of binary sequences of length n in which the pattern 100 occurs at the nth digit. (See Prob. 2-14.) Find the difference equation which $x(n)$ satisfies and the unilateral z transform $X_I(z)$ of $x(n)$.

(*b*) Let $y(n)$ be the number of binary sequences of length n in which the pattern 100 occurs for the first time at the nth digit, and $Y_I(s)$ be the unilateral Laplace transform of $y(n)$. Show that

$$x(n) = y(n) + y(n-3)x(3) + y(n-4)x(4) + \cdots + y(3)x(n-3) \qquad n \geq 6$$

and

$$Y_I(s) = 1 - \frac{1}{X_I(s)}$$

if we let $y(0) = y(1) = y(2) = 0$ and $x(0) = 1$, $x(1) = x(2) = 0$. Find the corresponding $Y_I(s)$ for the $X_I(z)$ in part (*a*).

(*c*) Let $w(n)$ be the number of binary sequences of length n in which the pattern 100 does not occur. Let $W_I(z)$ be its unilateral z transform. Find $W_I(z)$ and $w(n)$.

(*d*) Let $v(n)$ be the number of binary sequences of length n in which the pattern 100 occurs exactly once. Show that

$$v(n) = y(n) + y(n-1)w(1) + y(n-2)w(2) + \cdots + y(3)w(n-3) \qquad n \geq 3$$

and

$$V_I(z) = Y_I(z)[z^{-1}W_I(z) + 1]$$

Find $V_I(z)$.

11-13 Two players A and B gamble by tossing a coin which has probabilities p and q to turn up head and tail, respectively ($p + q = 1$). Initially, A has d dollars and B has $t - d$ dollars. At each toss, A wins a dollar if head shows and loses a dollar if tail shows. The game lasts as long as neither of the players is broke. Let $x(d, n)$ be the probability that A goes broke as a result of the nth toss.

(*a*) Show that

$$x(d, n) = px(d+1, n-1) + qx(d-1, n-1)$$

and determine the boundary conditions $x(0, n)$, $x(t, n)$, and $x(d, 0)$.

(*b*) Let the unilateral z transform $X_{Id}(z)$ of $x(d, n)$ for a fixed value of d be

$$X_{Id}(z) = \sum_{n=0}^{\infty} x(d, n)z^{-n}$$

Find the difference equation which $X_{Id}(z)$ satisfies and determine the boundary conditions $X_{I0}(z)$ and $X_{It}(z)$.

(*c*) Solve the difference equation found in part (*b*) for $X_{Id}(z)$.

11-14 A model train manufacturer sells $x(n)$ train sets during the nth month while it produces $y(n)$ sets. Let $w(n)$ denote the number of sets left in its warehouse at the end of the nth month. We have

$$w(n) = w(n-1) + y(n) - x(n) \qquad n > 0$$

Suppose that the production policy is such that

$$y(n+1) = Aw(n) + Bx(n)$$

where A and B are constants.

(*a*) Find the transfer functions $Y_I(z)/X_I(z)$ and $W_I(z)/X_I(z)$, given that $y(0)$ and $w(0)$ are equal to zero.

(*b*) Suppose that the sales $x(n)$ is equal to $u_{-1}(n)$. Find the values of A and B such that $w(n) = 0$ as $n \to \infty$.

(*c*) For the values of A and B found in part (*b*), find $y(n)$ and $w(n)$ corresponding to a sales $x(n) = 1000(1 - 3^{-n})u_{-1}(n)$.

11-15 In a study of the variation in the population size of inchworms in Westchester County, New York, in June, the birth and death rates of the inchworms were observed to be α per second. Let $p_n(t)$ denote the probability that there are n inchworms alive at the time instant t. $p_n(t)$ satisfies the equations

$$\frac{dp_0(t)}{dt} = -\alpha[p_0(t) - p_1(t)]$$

$$\frac{dp_n(t)}{dt} = \alpha[-2p_n(t) + p_{n-1}(t) + p_{n+1}(t)] \qquad n > 0$$

(*a*) Let

$$P(z, t) = \sum_{n=0}^{\infty} p_n(t) z^{-n}$$

be the unilateral z transform of $p_n(t)$. Find the differential equation which $P(z, t)$ satisfies if $p_1(0) = 1$ and $p_n(0) = 0$ for all $n \neq 1$.

(*b*) Find the unilateral Laplace transform of $P(z, t)$.

(*c*) Determine $p_n(t)$.

11-16 A telephone company has k telephone lines which can be used to carry k calls. If a new call is initiated when all the lines are busy, it joins a waiting line and waits until a line is freed. Let $p_n(t)$ denote the probability that at the time instant t there are n calls either being served or in the waiting line. The set of equations which describes the behavior of the system is

$$\frac{dp_0(t)}{dt} = -\alpha p_0(t) + \beta p_1(t)$$

$$\frac{dp_n(t)}{dt} = -(\alpha + n\beta)p_n(t) + \alpha p_{n-1}(t) + (n+1)\beta p_{n+1}(t) \qquad 1 \leq n < k$$

$$\frac{dp_n(t)}{dt} = -(\alpha + n\beta)p_n(t) + \alpha p_{n-1}(t) + k\beta p_{n+1}(t) \qquad n \geq k$$

(*a*) Let us consider the case where $\lim_{t\to\infty} p_n(t) = p_n$ exist for all n and $dp_n(t)/dt = 0$ as $t \to \infty$. Find the unilateral z transform and the expression of p_n in terms of p_0. For what values of α/β is $\sum_{n=1}^{\infty} p_n/p_0 < \infty$? What is the physical significance if $\sum_{n=1}^{\infty} p_n/p_0 = \infty$?

(*b*) Suppose that the number of telephone lines, k, is larger than the number of users in the system. (That is, for all practical cases, we can assume $k = \infty$.) Let $P(z, t)$ be the unilateral z transform of $p_n(t)$. Show that $P(z, t)$ satisfies the differential equation

$$\frac{\partial P(z, t)}{\partial t} = (1 - z^{-1})\left[-\alpha P(z, t) + \beta \frac{\partial P(z, t)}{\partial z^{-1}}\right]$$

Show that

$$P(z, t) = e^{-\alpha(1-z^{-1})(1-e^{-\beta t})/\beta}$$

if $\sum_{n=0}^{\infty} np_n(0) = 0$.

11-17 Let

$$\mathbf{A} = \begin{bmatrix} a_{11} & a_{12} \\ a_{21} & a_{22} \end{bmatrix}$$

be a 2×2 matrix and

$$\mathbf{A}^n = \begin{bmatrix} a_{11}(n) & a_{12}(n) \\ a_{21}(n) & a_{22}(n) \end{bmatrix}$$

be the nth power of the matrix **A**. Clearly, the elements of the matrix $\mathbf{A}^n$ satisfy the simultaneous difference equations.

$$a_{11}(n+1) = a_{11}(n)a_{11} + a_{12}(n)a_{21}$$

$$a_{12}(n+1) = a_{11}(n)a_{12} + a_{12}(n)a_{22}$$

$$a_{21}(n+1) = a_{21}(n)a_{11} + a_{22}(n)a_{21}$$

$$a_{22}(n+1) = a_{21}(n)a_{12} + a_{22}(n)a_{22} \qquad n \geq 1$$

Let $A_{11}(z)$, $A_{12}(z)$, $A_{21}(z)$, and $A_{22}(z)$ be the unilateral z transforms of the elements $a_{11}(n)$, $a_{12}(n)$, $a_{21}(n)$, and $a_{22}(z)$, respectively.

(*a*) Show that $A_{11}(z)$, $A_{12}(z)$, $A_{21}(z)$, and $A_{22}(z)$ are equal to the transfer functions of the flow graphs shown in Fig. 11P-3*a*, *b*, *c*, and *d*, respectively.

(*b*) Let **B** be an $m \times m$ matrix and $\mathbf{B}^n$ be the nth power of **B**. Let $B_{ij}(z)$ be the unilateral z transform of $b_{ij}(n)$, an element of $\mathbf{B}^n$. Show that $B_{ij}(z)$ is equal to the transfer function between node i and node j in the flow graph in Fig. 11P-3*c*.

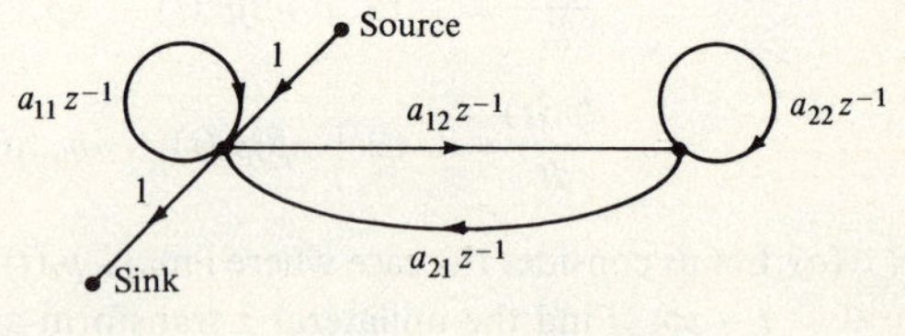

(*a*)

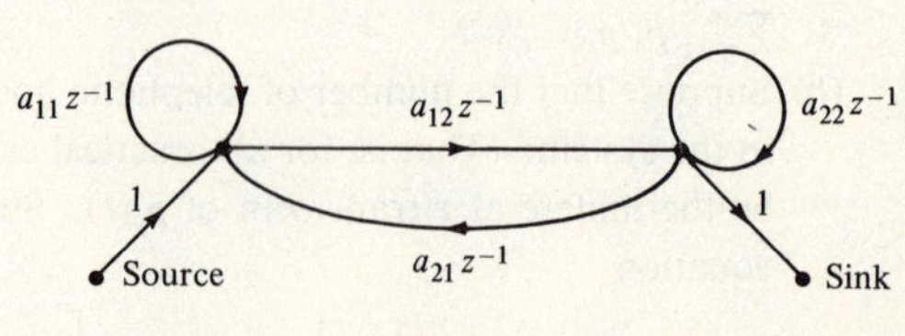

(*b*)

FIGURE 11P-3

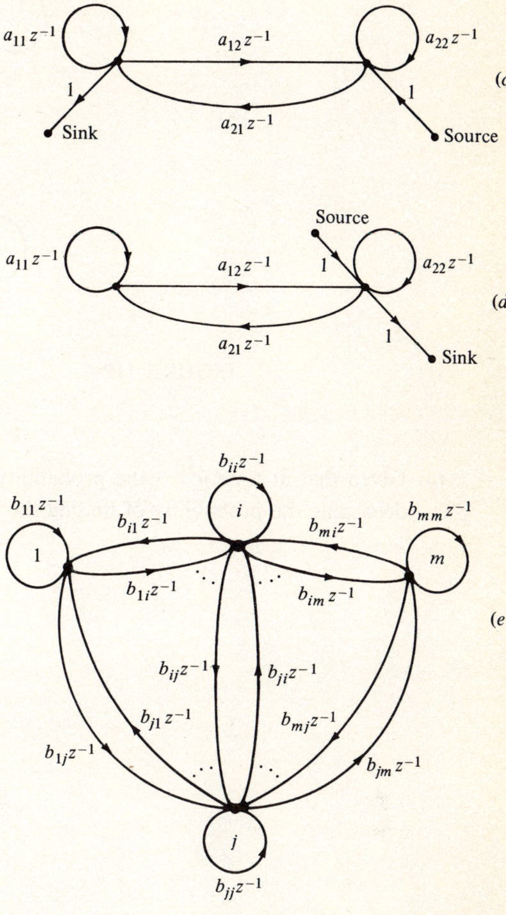

FIGURE 11P-3 (*continued*)

11-18 Consider a system whose behavior is described by a two-state Markov chain with transition matrix

$$P = \begin{bmatrix} 0.5 & 0.5 \\ 0.25 & 0.75 \end{bmatrix}$$

Its state transition diagram is shown in Fig. 11P-4*a*. Let $x_{12}(n)$ be the probability of finding the system in state 2 at time n, given that it was in state 1 at time 0.

(*a*) Find the unilateral z transform, $X_{12}(z)$, of $x_{12}(n)$. [$X_{12}(z)$ is equal to the transfer function of the system described by the flow graph in Fig. 11P-4*b*. See Prob. 11-17.]

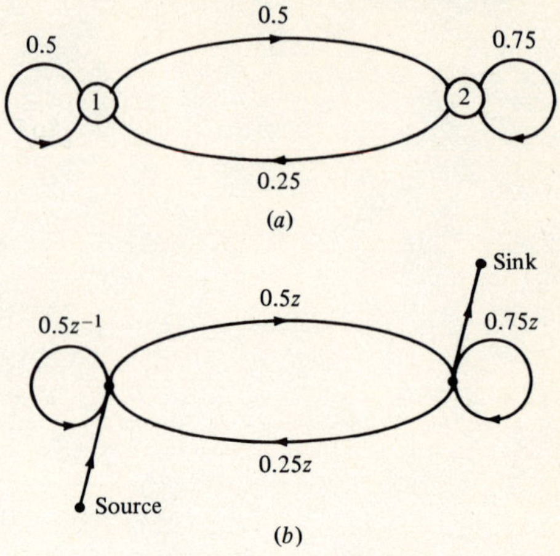

FIGURE 11P-4

(*b*) Given that at time $n = 3$ the probability of finding the system in state 2 is 0.1, determine the probability of finding the system in state 2 at time n.

INDEX